SURFACES, INTERFACES, AND THIN FILMS FOR MICROELECTRONICS

EUGENE A. IRENE

A JOHN WILEY & SONS, INC., PUBLICATION

Copyright © 2008 by John Wiley & Sons, Inc. All rights reserved

Published by John Wiley & Sons, Inc., Hoboken, New Jersey
Published simultaneously in Canada

No part of this publication may be reproduced, stored in a retrieval system, or transmitted in any form or by any means, electronic, mechanical, photocopying, recording, scanning, or otherwise, except as permitted under Section 107 or 108 of the 1976 United States Copyright Act, without either the prior written permission of the Publisher, or authorization through payment of the appropriate per-copy fee to the Copyright Clearance Center, Inc., 222 Rosewood Drive, Danvers, MA 01923, (978) 750-8400, fax (978) 750-4470, or on the web at www.copyright.com. Requests to the Publisher for permission should be addressed to the Permissions Department, John Wiley & Sons, Inc., 111 River Street, Hoboken, NJ 07030, (201) 748-6011, fax (201) 748-6008, or online at http://www.wiley.com/go/permission.

Limit of Liability/Disclaimer of Warranty: While the publisher and author have used their best efforts in preparing this book, they make no representations or warranties with respect to the accuracy or completeness of the contents of this book and specifically disclaim any implied warranties of merchantability or fitness for a particular purpose. No warranty may be created or extended by sales representatives or written sales materials. The advice and strategies contained herein may not be suitable for your situation. You should consult with a professional where appropriate. Neither the publisher nor author shall be liable for any loss of profit or any other commercial damages, including but not limited to special, incidental, consequential, or other damages.

For general information on our other products and services or for technical support, please contact our Customer Care Department within the United States at (800) 762-2974, outside the United States at (317) 572-3993 or fax (317) 572-4002.

Wiley also publishes it books in a variety of electronic formats. Some content that appears in print may not be available in electronic formats. For more information about Wiley products, visit our web site at www.wiley.com.

Library of Congress Cataloging-in-publication Data:

Irene, Eugene A.
　Surfaces, interfaces, and thin films for microelectronics / Eugene A. Irene.
　　　p. cm.
　Includes index.
　ISBN 978-0-470-17447-0 (cloth)
　1. Surface chemistry.　2.　Interfaces (Physical sciences)　3.　Thin films.
　4.　Microelectronics—Materials.　I.　Title.
　QD506.I74 2008
　621.381—dc22　　　　　　　　　　　　　　　　　　　　　　　　　　　2007024692

Printed in the United States of America

10　9　8　7　6　5　4　3　2　1

SURFACES, INTERFACES, AND THIN FILMS FOR MICROELECTRONICS

CONTENTS

PREFACE xi

PART I FUNDAMENTALS OF SURFACES AND INTERFACES 1

1 Introduction to Surfaces 3

    1.1 Introduction, 3
    1.2 Definition of a Surface, 5
    1.3 Preparing Surfaces, 8
    1.4 Simple Model for a Surface: Terrace–Ledge–Kink Model, 9
        1.4.1 Oxidation of Various Si Single-Crystal Orientations, 15
    1.5 Roughness, 18
    1.6 Summary of the Key Surface Concepts, 19
    References, 20
    Suggested Reading, 20

2 Structure of Surfaces 21

    2.1 Introduction, 21
    2.2 Reciprocal Space (RESP), 22
        2.2.1 Why Reciprocal Space?, 22
        2.2.2 Definition of RESP, 22
    2.3 The Ewald Construction, 26

2.4 Diffraction Techniques, 27
    2.4.1 Rotating Crystal Method, 28
    2.4.2 Powder Method, 28
    2.4.3 Laue Method, 29

2.5 Wave Vector Representation—$\mathbf{k}$ Space, 30

2.6 Diffraction from Surfaces, 31
    2.6.1 RESP and Ewald Construction for Surfaces, 31
    2.6.2 Low Energy Electron Diffraction, 34
    2.6.3 The LEED Pattern and Reconstruction, 38
    2.6.4 Indexing LEED Patterns and Surface Structure Nomenclature, 40
    2.6.5 LEED of Si(100), 45

2.7 Electron Microscopy, 45
    2.7.1 Transmission Electron Microscopy, 46
        2.7.1.1 Sample Preparation, 49
        2.7.1.2 TEM Results, 52
    2.7.2 Scanning Electron Microscopy, 58
        2.7.2.1 SEM Results, 63

References, 63
Suggested Reading, 63

3 Thermodynamics of Surfaces and Interfaces   65

3.1 Introduction, 65
3.2 Surface Energy, 65
3.3 Principles of Capillarity, 67
    3.3.1 Definitions, 67
    3.3.2 Curved Surfaces, 69
3.4 Surface Energy of Solids, 73
3.5 Interfaces and More Capillarity, 79
3.6 SiO_2–Si Interface Application, 83

References, 88
Suggested Reading, 88

4 Surface Roughness   89

4.1 Introduction, 89
4.2 Roughness Parameters, 90
    4.2.1 Height Parameters, 90
    4.2.2 Roughness Period or Wavelength, 92
    4.2.3 Shape of Rough Features, 93
    4.2.4 Statistical Descriptors of Roughness, 95

        4.2.5   Fractal Description of Roughness, 99
                4.2.5.1   Fractal Definitions, 99
                4.2.5.2   Extraction of the Fractal Dimension from
                          Experimental Data, 104
  4.3   Roughness Effects on the Properties of Materials, 106
        4.3.1   Effects of Roughness on Optical Properties, 106
        4.3.2   Roughness Effects on Electronic Properties, 108
        4.3.3   Roughness Effects on Other Physical and Chemical Properties, 112
                4.3.3.1   Roughness Effects on Contact Angle, 112
                4.3.3.2   Roughness Effects on Surface Thermodynamic
                          Properties, 113
                4.3.3.3   Roughness Effects on Chemical Reactivity, 115
  References, 121

5 **Surface Electronic States**   123

  5.1   Introduction, 123
  5.2   The Kronig–Penney Model, 124
        5.2.1   The KP Model for Infinite Solids, 124
        5.2.2   The KP Model Extended for Finite Solids, 131
  5.3   Other Models for Surface States, 135
        5.3.1   Extrinsic Surface States, 137
        5.3.2   Band Bending, 138
  5.4   Measurement of Surface Electronic States, 139
        5.4.1   Thermionic and Field Emission, 139
        5.4.2   Kelvin Probe, 146
        5.4.3   Photoemission, 148
                5.4.3.1   Ultraviolet Photoemission Spectroscopy (UPS) and
                          Inverse Photoelectron Spectroscopy (IPES), 151
                5.4.3.2   X-Ray Photoelectron Spectroscopy (XPS), 156
  References, 156
  Suggested Reading, 157

6 **Other Surface Probes**   159

  6.1   Surface Topology or Morphology: Scanning Probe Microscopy, 159
        6.1.1   Scanning Tunneling Microscopy, 161
                6.1.1.1   Electron Tunneling, 161
                6.1.1.2   Scanning Tunneling Microscopy Operation, 168
                6.1.1.3   Applications, 172
        6.1.2   Atomic Force Microscopy, 176
                6.1.2.1   Atomic Force Microscopy Operation, 177

6.2 Surface Composition: Auger Electron
Spectroscopy and Ion Scattering and Recoil Spectroscopy, 181
  6.2.1 Auger Electron Spectroscopy, 181
  6.2.2 Ion Scattering, 184
    6.2.2.1 Time-of-Flight Ion Scattering and Recoil Spectrometry, 185
    6.2.2.2 Ion Scattering Spectroscopy, 189
    6.2.2.3 Direct Recoil Spectroscopy, 190
    6.2.2.4 Mass Spectroscopy of Recoiled Ions, 190
    6.2.2.5 Mass Spectroscopy of Recoiled Ions Applications, 192
References, 197
Suggested Reading, 197

7 Charged Surfaces 199

7.1 Introduction, 199
7.2 Electrostatics and the Poisson Equation, 200
7.3 Two Simple Solutions to the Poisson Equation, 205
7.4 Metal–Oxide–Semiconductor Field-Effect Transistor and Fermi Level Pinning, 209
7.5 Metal–Oxide–Semiconductor Measurements for Interface Charge, 213
  7.5.1 Oxide Charges, 219
  7.5.2 Measurement of Charges in the $Si-SiO_2$ System, 224
References, 228
Suggested Reading, 228

8 Adsorption 229

8.1 Introduction, 229
8.2 Physisorption, 230
8.3 Chemisorption, 233
8.4 Vapor–Solid Equilibrium at Surfaces, 236
8.5 Adsorption Isotherms, 238
8.6 Surface Reaction Mechanisms, 243
8.7 Temperature-Programmed Desorption, 248
References, 255
Suggested Reading, 255

9 Ellipsometry and Optical Properties of Surfaces, Interfaces, and Films 257

9.1 Introduction, 257
9.2 What Is Ellipsometry? 258
9.3 What Does Ellipsometry Measure? 263
9.4 How Well Can Ellipsometry Measure? 264

9.5 Optical Models, 266
9.6 Manual and Automated Ellipsometry Techniques, 274
    9.6.1 Introduction, 274
    9.6.2 Isotropic Media, 276
    9.6.3 Linear Polarizer, 277
    9.6.4 Compensator, 279
    9.6.5 Detectors, 280
    9.6.6 Null Ellipsometry, 280
    9.6.7 Rotating Element Ellipsometry, 282
    9.6.8 Spectroscopic Ellipsometry, 284
    9.6.9 Ellipsometry Alignment and Calibration, 286
9.7 Ellipsometry Measurements, 287
    9.7.1 The Si Surface, 288
    9.7.2 Surface with Overlayer, 289
References, 293
Suggested Reading, 293

PART II MICROELECTRONICS APPLICATIONS 295

10 Films and Interfaces 297

10.1 Introduction, 297
10.2 Nucleation, 298
    10.2.1 Homogenous Nucleation, 299
        10.2.1.1 Mixing, 304
        10.2.1.2 Volmer–Weber Theory, 305
    10.2.2 Heterogeneous Nucleation, 308
    10.2.3 Nucleation Studies, 312
10.3 Film Formation, 321
    10.3.1 Film Growth, 324
    10.3.2 Film Deposition, 329
        10.3.2.1 Chemical Vapor Deposition, 329
        10.3.2.2 Physical Vapor Deposition, 333
References, 349
Suggested Reading, 349

11 Electronic Passivation of Semiconductor–Dielectric Film Interfaces 351

11.1 Introduction, 351
11.2 Interface Electronic States, 351
11.3 Electronic Passivation, 361

11.4 Semiconductor Passivation Studies, 363
    11.4.1 Si Thermal and Plasma Oxidation, 363
    11.4.2 Ge Passivation via Thermal and ECR Plasma Oxidation and SiO_2 Deposition, 372
    11.4.3 InP Passivation via Thermal and ECR Plasma Oxidation and SiO_2 Deposition, 382
    11.4.4 GaAs Passivation via Thermal and ECR Plasma Oxidation and SiO_2 Deposition, 392
11.5 High Static Dielectric Constant Gate Oxides, 403
    11.5.1 Barium Strontium Titanate, 405
    11.5.2 ZrO_2, HfO_2, and MgO as Potential High K Dielectrics, 419
References, 434
Suggested Reading, 435

12 The Si–SiO_2 Interface and Other MOSFET Interfaces   437

12.1 Introduction, 437
12.2 Nature of the Si–SiO_2 Interface, 438
12.3 Other Techniques for Interface Studies, 445
    12.3.1 Spectroscopic Immersion Ellipsometry (SIE), 445
    12.3.2 Tunneling Currents for Measurement of Refractive Index, Film Thickness, and Roughness, 453
        12.3.2.1 Tunneling Current Oscillation Measurement, 453
        12.3.2.2 Application to SiO_2: Refractive Index, Film Thickness, and Roughness, 457
    12.3.3 Interfacial and Film Stress, 478
12.4 Other Microelectronics Interfaces, 490
    12.4.1 Introduction, 490
    12.4.2 Electronic Characteristics of Junctions, 490
    12.4.3 Ideal Metal–Semiconductor Junctions, 492
    12.4.4 Ideal Semiconductor–Semiconductor PN Junctions, 495
    12.4.5 Nonideal Junctions, 496
References, 501
Suggested Reading, 502

INDEX   **503**

PREFACE

This book is first and foremost a text for classroom use or self-study aimed at providing an understanding of surfaces and interfaces that are important in the broad area of microelectronic materials. The subject matter selected for inclusion derives from a series of three courses offered at UNC that I have taught over 25 years on the materials science aspects of microelectronics. The first course in the series deals with the fundamentals of electronic materials. The content of that course is embodied in my previous text, *Electronic Materials Science*, published by Wiley, and this text is covered in its entirety in the first course. This first course and text introduces structure, diffraction, reciprocal and **k** spaces, defects, diffusion, phase equilibria, mechanical properties of materials, electronic structure, electronic properties, and some devices. This introduction to these subjects comprises the prerequisites to the two following courses and this text, which contains about two thirds of the material covered in the two following courses. These two courses include the fundamental aspects of surfaces and interfaces in one course and the physics and chemistry of microelectronics processing in the other.

    The second course is a surface science course where surface structure, both geometric and electronic structure, surface thermodynamics, morphology, and a sampling of surface science techniques provide the main topics. These subject areas are covered in Part I of the present text in Chapters 1 through 9. With the exception of vacuum technology, which is also covered in our UNC course, Chapters 1 to 9 are the main subject matter in the fundamentals of surface and interface science course at UNC. Vacuum technology is a crucial aspect of modern surface science and is discussed in the course but not included in this text. The reason is that there are several excellent texts available written by experts whose knowledge is extensive and current and that provide both the science and technology of vacuum science that cannot be presented in a chapter or part of a chapter. In fact, surface and interface science covers such a vast area of science and technology that I found it difficult to find a single book or even two that together contained all the aspects of electronic materials and microelectronics that I wanted to present. Hence, course notes were developed that were supplemented with parts of available textbooks. Finally, the course notes evolved into the present text. At the end of each chapter (except for Chapter 4) is a short list of my favorite texts in the areas covered in the chapter, and many of these texts were used in the courses over the years. The selection of the material for Part I of this book presented a challenge because of my attempt to present a reasonably complete picture of the field. Since I found this objective impossible to attain, the material selected is that which is covered in the courses with some omissions that particular instructors include as their areas of particular interest. For example, I also include ellipsometry as a sensitive surface, interface and thin-film technique, and include a full chapter on ellipsometry and optical properties while other instructors barely mention the topic.

The final course that is addressed by this text is the course dealing with microelectronics processes. At UNC this course has been co-taught by at least three instructors who lectured on various important topics such as crystal growth and silicon wafer fabrication and testing, thin-film growth and deposition, film measurements, substrate measurements, defects, ion beam analysis, electronic measurements, doping, and mechanical properties of films, and the topics sometimes varied year to year and instructor to instructor. Therefore, rather than trying to capture that elusive course, I have chosen to liberally add microelectronics applications (e.g., electronic measurement applications in Chapter 7, thin-film processing in Chapter 10, modern techniques in Chapters 6 and 12, and the like) and separate the text into Parts I and II. Part I captures the surface and interface course, and Part II has more applications on microelectronics materials studies that further illustrate the fundamentals presented in Part I.

A few words about applications are warranted. The UNC courses were originally aimed at microelectronics graduate students. Consequently, the applications used are predominantly from the field. I have found that I can use one device to discuss and illustrate almost all the surface and interface science that I present, and this device is the metal–oxide–semiconductor field–effect transistor, the MOSFET. In almost every chapter some aspect of the MOSFET is mentioned and applications discussed, and thus the entire book hinges on the MOSFET. I begrudgingly admit that there are other microelectronic devices, but none is so pervasive and none can be used better as a focus point for this book.

For this course material there are an enormous number of original studies to choose from that deal directly with various aspects of the fundamental issues addressed in this text. Obviously, I have included only a small subset of the available original research studies and reviews that I have used in the courses, but I have included the ones I used most often. I had passed out the articles for classroom discussion during lectures, excerpted problems and text for use in class notes, and/or assigned studies to be read and then discussed in class. Many of the articles that I have used repeatedly in the courses are listed in the References at the end of each chapter, and used figures and the authors units from the original studies (consequently, several different units are used for the same variables and constants throughout the book and especially in Part II). I have tried to include some experimental techniques and experimental details to give students a better feel for experiments and the experimental results. With the use of Internet search capability at universities, I am hopeful that students will be encouraged to access the original studies and examine the original research and results while they read the text. Many of the included studies are from my own research over the years since in using these I was able to include insights often missed in the details when reading articles, and I was able to extract the important points for this book without insulting or misrepresenting the authors. For these research studies I owe a great debt of gratitude to all my graduate students and postdoctorals that have been in my research group since 1982 when I came to UNC and before that to my colleagues at the IBM Research Center at Yorktown Heights where I learned about microelectronics. Most of these people are included in the citations to specific studies and figures. Many of the drawings and figures were developed in my research group for various talks, papers, and theses and usually referenced, but some origins are unknown.

The courses and this text are aimed at first- and second-year graduate students, and it is assumed that the students have had introductory courses in thermodynamics and quantum mechanics that are typical of chemistry and physics undergraduate curricula as well as the material in my introductory text mentioned above.

PART 1

FUNDAMENTALS OF SURFACES AND INTERFACES

1

INTRODUCTION TO SURFACES

1.1 INTRODUCTION

A surface is intuitively defined as the boundary of a condensed phase, liquid, or solid. In this text the main concern is with solid surfaces. Typically, a solid surface displays many properties that are different from the bulk material. Surface structure is often different, and compared to the bulk material usually leads to a density and stoichiometry difference at the surface. With both structural and chemical differences, it is straightforward to conclude that the potential at the surface that binds the constituent atoms is different from that in the bulk, that is, the bonding at the surface is different from the bulk. The direct consequence is a different electronic structure and the subsequent altered properties derived therefrom. Among the most important surface properties that are different from the bulk material properties are chemical reactivity resulting from different thermodynamic and kinetic properties and different electronic properties resulting from different electronic energy band structure. Therefore, in the study of surfaces it is important to first understand the structure of surfaces. Structure can be understood from different perspectives. First is the geometric structure that is essentially the arrangement of atoms or molecules. Many properties are dependent upon the specifics of this arrangement since different surface sites result from different arrangements. The arrangement of atoms and/or molecules and the nature of the atoms and molecules at a surface yield a potential specific to the surface. This potential leads to different surface electronic structure and different electronic states than those that exist in the bulk. Both perspectives, sites and states, yield a measure of

---

Surfaces, Interfaces, and Thin Films for Microelectronics. By Eugene A. Irene
Copyright © 2008 John Wiley & Sons, Inc.

understanding about surface chemical and physical properties. In this text different perspectives will be considered as well as techniques that yield information about surface sites and states. In addition, relevant examples mostly from the arena of microelectronics and the author's research experiences are included mainly in Part II to illustrate various surface and interface principles. Chapter 1 will present several of the overarching ideas that will be covered in more depth throughout the text with the intent in Chapter 1 being the introduction of ideas that help to define surfaces and interfaces and some tools for evaluating those differences and examples from the research literature.

With different geometrical and electronic structure at surfaces and the resulting alteration of thermodynamic and kinetic properties, it is useful to question why one often encounters the chemical reactivity of solids being described using only bulk thermodynamic properties. It is clear that a reaction among solids first involves the contact of the solids, which is via their respective surfaces. Furthermore, it is obvious that the contact of surfaces and the prediction of surface reactivity will involve the use of surface properties rather than bulk properties. One explanation for the use of bulk properties rather than surface properties is that often the relevant surface properties are unknown. Thus, one uses bulk properties to achieve a first approximation. Surface properties usually require special techniques, and many of the most powerful techniques have been developed in the past 30 years. Therefore, much of the older literature on surface science and interfaces does not benefit from the modern advanced techniques, and with a relatively short history of measurement many surface properties are unknown. Often these surface-sensitive techniques involve the use of high (below about 10^{-4} atm) or ultra-high (below about 10^{-8} atm) vacuum, so as to preserve a pristine surface for examination. Thus, knowledge of vacuum techniques is a part of the study of surfaces. For the most part this text will introduce surface characterization techniques that are useful for various surface measurements, along with the major surface topic with which the technique(s) is associated. For example, surface structure techniques will be introduced in Chapter 2 that deals with geometrical surface structure and electronic structural techniques are introduced in Chapter 5 that deals with electronic structure. Chapter 6 discusses several techniques that are also useful for surface analysis. Chapter 9 discusses ellipsometry, which is useful for films and interfaces. There is no attempt to be exhaustive relative to surface and interfaces science techniques but rather to introduce techniques that are both popular and that are used to obtain results used in this book. It should be understood that there are many surface, interface, and film techniques that are not covered in this book.

Surfaces are typically more reactive than the same bulk material. For example, it is difficult to maintain a pristine metal surface for study in air, and as was mentioned above ultra-high vacuum is necessary to prevent or retard surface oxidation or nitridation and enable surface properties to be assessed. For this reason it is rare to actually deal with pristine surfaces in applications. Usually, there is at least a thin oxide on metals, a reaction layer on nonmetals, and almost always a film of contamination on all surfaces exposed to laboratory ambient. Thus, real solid materials comprise a surface covered with some kind of a film. Underneath the surface film there is an interface between film and substrate surface. This interface can be abrupt, extending to fractions of a nanometer, or it can be graded and extend several nanometers or more. The interface can represent a pure, a mixed, or a chemically reacted phase or a combination; and, because it is typically thin, its physical and chemical properties can be quite different from the underlying material, its surface, or the overlayer film. Therefore, a realistic study of surfaces would be incomplete without a study of interfaces and thin films. This means a study of film formation on surfaces that commences with nucleation and the formation of an interface and continues with film growth. There are many methods for film formation on surfaces and several prominent methods will be covered in Chapter 10 as well as techniques to observe film formation from nucleation onward. However, before real surfaces are discussed, that is, surfaces with films and interfaces, the fundamentals of surface structure, thermodynamics, and reactivity and morphology as well as techniques to characterize surfaces are covered in Chapters 2 through 9. Chapter 10 deals with film formation from nucleation onward toward coalescence and film growth, and Chapter 11 considers more about the most important interface in microelectronics, the $Si-SiO_2$ interface. Finally Chapter 12 discusses important tools that can access information about thin films.

1.2 DEFINITION OF A SURFACE

Throughout this text the definition for a surface will evolve and broaden as different aspects of surfaces are explored, and each new aspect will contribute to the breadth of an elusive simple and concise definition. It is useful to start the evolutionary process of understanding surfaces with some first-order ideas upon which a more complete picture can be gradually built. We commence with the notion of surface energy, that is, the net energy it takes to create a surface and is characteristic of all surfaces. For this purpose we consider Fig. 1.1a, which is a side view of an idealized monatomic crystal where a two-dimensional representation of the atoms is seen as filled circles. In the bulk of the crystal well below the surface, the adjacent nearest-neighbor bonds are indicated as lines. The circled atom (dotted circle) in the bulk is seen to have four bonds to its nearest neighbor. To compare this bulk atom with a surface atom, we perform the thought experiment to remove the bulk atom by breaking its four bonds (indicated by hatch marks), keeping track of the energy required, and reinsert this atom in a surface position. This process is illustrated by the path indicated in Fig. 1.1a. To free the bulk atom, four units of bond energy (4 BE) are required. Then when the extracted atom is reinserted into a surface position, the atom remakes three bonds thereby returning 3 units of BE (3 BE). Thus the process costs a net energy expenditure of 1 BE to move a bulk atom to the surface, and this energy is called the surface energy and is typically given in units of energy per area of surface created. For such a simple process it can be concluded that it always costs energy from some source to create a surface. Also, for the creation of a surface, the most likely path will be that path that minimizes the required energy. In our simple example, the path chosen requires the minimum energy of 1 BE, but for more complex process more paths might be accessible. Since it always requires energy to create a surface, the formation of a surface is not spontaneous. BEs of

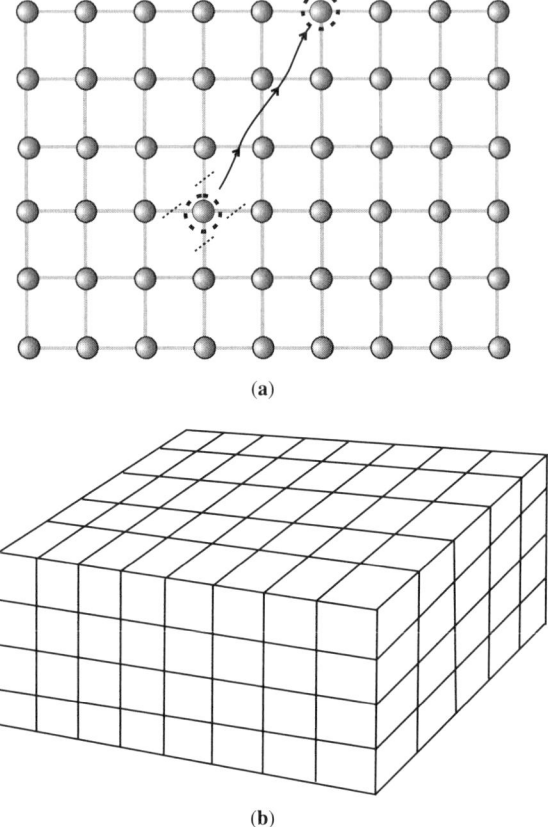

FIGURE 1.1 Idealized surfaces: (**a**) side view of a surface emphasizing chemical bonding and (**b**) a surface represented by atoms as cubes with six nearest neighbors and six bonds.

1 eV and higher are commonplace for surface energies, and compared with the available room temperature energy of only about 0.025 eV, the substantial amount of additional energy required to create a surface reduces the spontaneity of the surface creation process. Furthermore, for the most probable processes to produce a surface, the amount of surface or the surface area produced is minimized so as to minimize the total energy required. Another way to say all this is that the ratio of the surface area produced (S) relative to the volume of solid from which the surface is produced (V) is minimized in the more probable pathways: S/V is minimized. This notion of the minimization of the ratio S/V will permeate many ideas and mechanisms in the following chapters, and it is worthwhile to understand its basis early in the study of surfaces. It should be realized that this simple model for surface energy depicted in Fig. 1.1a ignores all but nearest-neighbor interactions and the energy changes associated with the structural difference for surfaces, and while it does provide justification for the notion that it costs energy to produce a surface, this simple model does not accurately quantify the energy required.

1.2 DEFINITION OF A SURFACE

There are other aspects of the simple two-dimensional picture of surface formation portrayed in Fig. 1.1a. It is seen that the environment for an atom in the bulk of this two-dimensional crystal is different from the environment of the surface atom. Specifically, there are four bonds for the bulk atom and three for the surface atom considering only nearest neighbors. Therefore, we conclude that the binding potential (V) is different for a surface atom compared to a bulk atom. Thus, later (Chapter 5) when we consider the electronic structure of surfaces, we will take this difference into consideration. We will learn that a different potential will lead to different allowed and disallowed electronic states and, consequently, a different electronic structure for surface than for the bulk. Specifically, surface atoms are less tightly bound than are bulk atoms, and thus V for a surface is less than that for the bulk material. If we consider next nearest neighbors and next next neighbors, we can also conclude that the atomic potential changes as the surface is approached and not simply abruptly at the surface. Thus, there is a graded interface in terms of the potential, extending from the bulk uniform potential for crystalline materials deep within the material to a reduced value as the surface is neared, to zero potential away from the surface in vacuum. We can similarly expect some properties to also change from the bulk to surface in some graded fashion.

Figure 1.1b shows a three-dimensional version of a solid where each cube is considered to be an atom in a monatomic solid with a nearest-neighbor bond at each face of the cube, that is, six coordinate. This model is useful to identify different sites at the surface. In this simplistic nearest-neighbor representation we can see that a bulk atom has six nearest-neighbor bonds while a surface atom can have a different number of bonds but always less than six. For example, a surface atom away from the edges of the solid in the Fig. 1.1 has five nearest-neighbor bonds while one at the edge has either four bonds or, if at a corner, only three bonds. Also, if next-nearest-neighbor bonding is considered, then the surface is even more complicated since the amount of bonding would depend on the specific location on a finite surface. We will return to the implications of this notion of specific surface sites later.

In considering the structural implication of the reduced interatomic potential near a surface, it is also expected that the structure or arrangement of atoms for a monatomic solid or the building blocks for a molecular solid will be different near the surface. After all there is no bonding on one side of the surface. In fact for most solids the structure at the surface will be different than the bulk. The surface symmetry is usually similar to that in the bulk, but the surface lattice is often larger since there is some relaxation in the arrangement due to the reduced binding potential. Overlayers or films can affect the surface or interface structure by providing chemical bonding but usually different chemical bonding on the originally free side of the surface as compared to the bulk material. Surfaces that have a different structure than the bulk are called reconstructed surfaces, and the process is called reconstruction. This will be discussed further in Chapter 2.

In summary then we expect that a surface will form as the boundary of a solid in such a way as to minimize energy and, therefore, its area, and it will have structure and properties consistent with the smaller interatomic potential, and the surface will typically have different properties than the bulk.

1.3 PREPARING SURFACES

Although Fig. 1.1a depicts a simple process of creating a surface, namely removing atoms one by one to the surface, it is understood that this process is fictitious and that a real surface is likely to be created in other ways. For example, a straightforward way to create a surface is by cleavage of the solid producing a fresh surface composed of atoms that were bulk atoms prior to the cleavage. Once having performed the cleavage, the new surface atoms are not in the same equilibrium state as they were in the bulk since the bonding of these formerly bulk atoms is altered. As such, the bonding at the newly formed surface will adjust to the new potential, and consequently a rearrangement will occur. As was mentioned above this rearrangement is called reconstruction. A part of the large free energy expenditure necessary for the creation of the surface (the energy to cleave the solid) will be compensated by the rearrangement of the surface atoms. If the surface were created in a reactive environment such as the cleavage of a metal to produce a metal surface in air, then it is likely that the surface will react with the ambient and produce a surface film (e.g., an oxide or nitride) sometimes with a lower surface energy than that for the pristine fresh surface and sometimes with a higher surface energy, with the extra energy derived from the chemical reaction. Even where a chemical reaction does not take place, a surface will seek to lower its surface energy by being covered with a lower surface energy substance if such a substance is available. For example, a freshly prepared surface of SiO_2 will often be covered with adsorbed water obtained from the laboratory ambient. As discussed in Chapter 3, the freshly produced pristine solid surface has a large surface energy often exceeding 10^3 dynes/cm (or erg/cm^2) for many solids including SiO_2. Water has a surface energy more than $10\times$ smaller, and water is a constituent of ambient air. Thus, it should be expected that water in the laboratory air will cover the freshly prepared SiO_2 surface since this process lowers the surface energy of the solid by a substantial factor and lowers the system energy (the solid and ambient). Indeed, this is precisely what is observed, that is, most solid surfaces exposed to ambient become covered with water, albeit a thin layer of water. If a solid is immersed in liquid water, the phenomenon of wetting can occur where the surface energy of the solid is lowered due to uniform coverage with water. Of course, there are solids that have a very low surface energy such as Teflon for which the coverage with water actually raises the surface energy. In this case the process is not favored, and, therefore, the Teflon surface is not wet by water. The topic of wetting will also be discussed in Chapter 3. At this point it is useful to only become acquainted with the expected behavior of surfaces.

Thus a surface can be prepared by cleavage of a solid. If the process takes place in ambient, then it is likely that the final surface will be both structurally and chemically different than the pristine cleaved surface. In order to avert any chemical reaction or even wetting of the pristine surface, ultra high vacuum (UHV) that is devoid of reactant or other species must be used to prepare the cleaved surface.

Besides cleavage a surface can be prepared by depositing one substance (the same substance as the bulk or different) onto another. As was discussed above, this can occur inadvertently by simply exposing a surface to an ambient that contains

1.4 SIMPLE MODEL FOR A SURFACE: TERRACE–LEDGE–KINK MODEL

atoms or molecules that can form a new surface upon the original pristine surface either by chemical reaction or simply by physically adhering to the surface. On the other hand a pristine surface can be prepared in UHV and then a desired substance(s) to be deposited can be controllably introduced into the system. In such a way, the controlled preparation of a surface that is different from the starting or substrate surface can be accomplished. While a few ways to form new surfaces have been described, there are more ways to prepare new surfaces on substrates or surface films, and some important film formation methods will be discussed in later chapters.

As discussed above the chemical bonding or binding potential for an atom will depend on exactly where an atom is located. For simplicity we consider a monatomic material that has an isotropic binding potential in the bulk. Thus, for all the interior atoms that have at least nearest and a few levels of next nearest neighbors, the binding potential in the bulk will be the same for all interior atoms. However, near the surface even for the simple monatomic solid, the interior bonding uniformity changes near the surface, and at the surface there are several distinctly different surface sites. Figure 1.2 summarizes the different surface sites in a so-called terrace–ledge–kink (TLK) model, which provides a useful albeit simplistic visualization of the major diversity in surface bonding.

First we notice that Fig. 1.2, like Fig. 1.1b, depicts the atoms in the monatomic solid to be bonded to nearest neighbors at the face of each cube. Next nearest neighbor and other bonding levels are ignored. Then, referring to Fig. 1.2, the easiest to

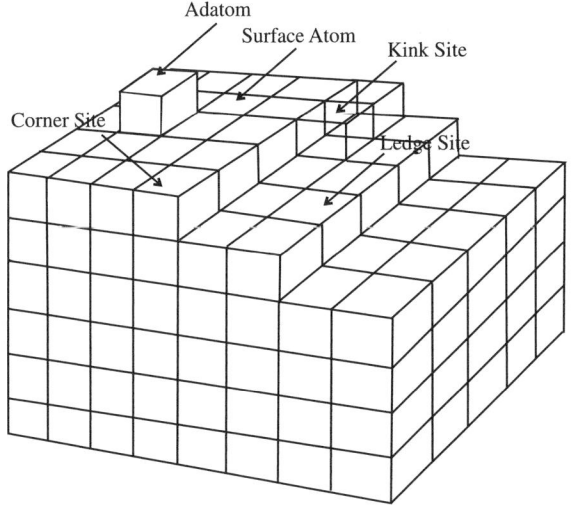

FIGURE 1.2 Surface sites on an idealized surface.

visualize surface site, called the surface atom site, is seen to be located on a flat terrace and bound to five nearest neighbors. Because this site is at the surface, it has one free face available for bonding. Now in three dimensions this site is the same surface, atom site discussed earlier using Fig. 1.1a and 1.1b. The surface site has the greatest bonding for a surface atom, five bonds. At the other extreme is the adatom site that has one bond. The adatom site is essentially a site on top of the surface with the weakest bonding to the solid. This site could represent the last site occupied prior to desorption and/or the first site occupied in adsorption, and for these reasons the adatom site is prominent. In Fig. 1.2 there are two ledges (or risers) that delineate the three flat terraces (or steps). An atom in a ledge site has four bonds. The ledges can also be discontinuous where at the end of the ledge there is a three-bond site called a kink site. A lone atom at a ledge would occupy a two-bond site. At the corner of a ledge is another three-bond corner site. However, if the solid surface is large, then the number of corner sites will be insignificant. The kink site also takes on special significance because it represents bonding midway between bulk bonding (six bonds) and total freedom (zero bonds).

Real surfaces can have many diverse sites available, and it is understood that the different sites possess different amounts of surface energy. Figure 1.1b can be considered to be the (100) surface for a cubic structure. From a study of crystallography, the low index planes are those with the greatest atomic density and thus the closest bonding. Using the simple model above for surface energy, we can deduce that the planes with the most and closest bonding, these low index planes, represent planes with the lowest surface energy. Considering that nature will choose processes and results that involve the smallest expenditure of energy, these low index planes will be the ones most often observed. The surfaces represented by low Miller index planes are called singular surfaces.

While Fig. 1.2 shows a surface with two ledges running in one direction yielding three terraces and single atom steps, it should be clear that in principle the terraces can run in multiple directions and can have different size risers. Figure 1.3a shows three (100) terraces with steps also within the <100> family. For comparison Fig. 1.3b shows the same terraces but now running in another direction. In this case all the terraces and the steps have the same orientation, but this is not necessarily always the case. In contrast to singular or low Miller index surfaces, a crystal can, in principle, be cut at any angle relative to the (100) to obtain a higher Miller index plane and surface. The angle θ between any two planes with the Miller indexes $h_i k_i l_i$ can be calculated for a cubic system using the following formula:

$$\cos \theta = \frac{h_1 h_2 + k_1 k_2 + l_1 l_2}{\sqrt{(h_1^2 + k_1^2 + l_1^2)(h_2^2 + k_2^2 + l_2^2)}} \tag{1.1}$$

For example, the angle between the (100) and the (311) is given as

$$\cos \theta = \frac{3}{\sqrt{11}} = 0.90453$$

1.4 SIMPLE MODEL FOR A SURFACE: TERRACE–LEDGE–KINK MODEL

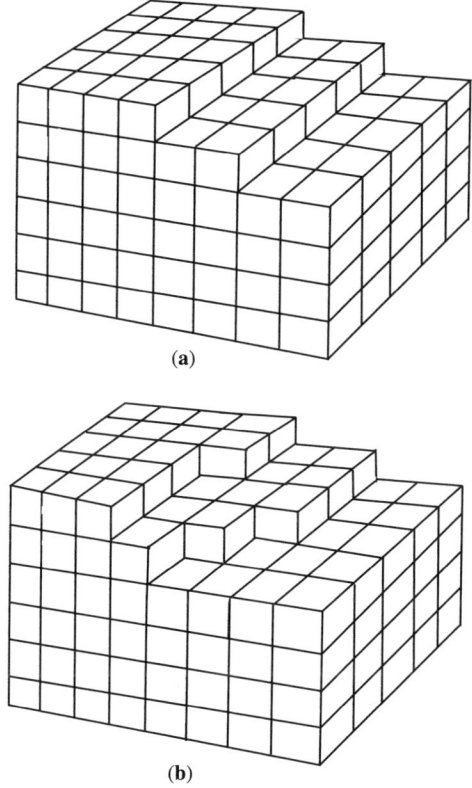

FIGURE 1.3 Idealized surfaces with ledges (**a**) in one direction and (**b**) in two directions.

which yields $\theta = 25.2°$ as indicated in Fig. 1.4a. As shown in Fig. 1.4a, the (311) (crosshatched) with the intercept of $\frac{1}{3}$ on the a axis ($w/3$) is a plane at 25.2° from the (100) and about 29.5° from the (111) (shaded). The (311) is called a vicinal plane to the (100) since the (100) is the closest singular surface or low index plane. Similarly, the (511) plane would have an intercept of $\frac{1}{5}$ on the a axis ($w/5$ is not shown in the figure) and is positioned in between the (100) and (311) at an angle of about 16° from the (100). The (511) is also vicinal to the (100). While the (311) and (511) angles can be cut relative to the low index planes, typically these high Miller index planes exhibit higher surface energy and are thus not as stable as the lower index planes. In many cases these planes, if permitted to relax, will reconstruct to form a series of steps and risers consisting of lower Miller index components. For example, it has been found using low energy electron diffraction (LEED) (to be covered in the next chapter) that the (311) and (511) planes on a Si surface will form (100) terraces or steps with (111) risers (1). That this occurs is a result of the tendency to lower the surface energy by forming high atom density and low surface energy planes.

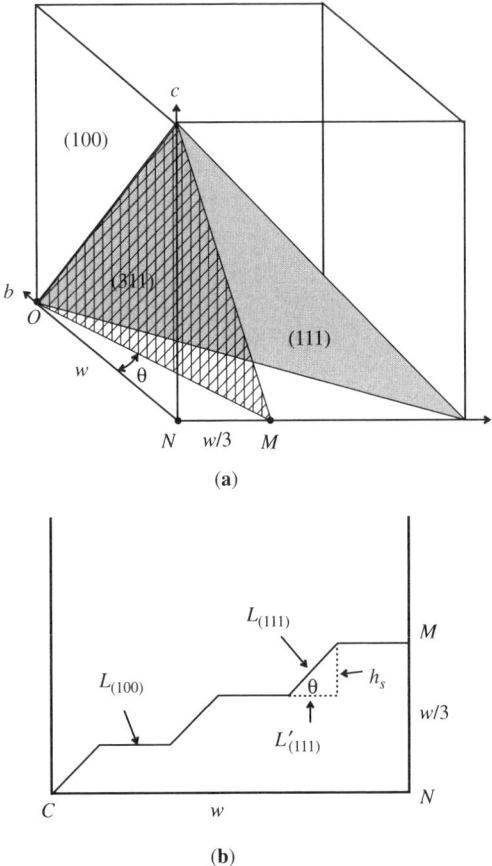

FIGURE 1.4 (a) (100), (311), and (511) planes for a cubic solid and (b) the side view of steps and risers for the (311) plane.

Because so much surface chemistry and physics can be related to the atom densities on a surface, it is useful to now consider how the atom density can be obtained, both for singular and vicinal surfaces. For singular surfaces, the number of atoms in an area of surface are counted considering sharing by adjacent areas. For example, for Si the lattice parameter is $a_0 = 0.5431$ nm and the structure is diamond cubic (DC) as shown in Fig. 1.5. The area of the (100) surface of a unit cell is a_0^2 or 0.2950 nm². The (100) surface of the DC unit cell has four corner atoms shared by four adjacent areas and a central atom for a total of two atoms for the (100) area of the DC unit cell. This yields $2/2.9496 \times 10^{-15}$ cm² or the atom density of 6.78×10^{14} cm^{-2} as shown in the bottom row of Table 1.1. The numbers for the (110) and (111) are similarly obtained and left as a useful exercise. For this problem one must consider that the three-dimensional DC unit cell has eight atom positions. One atom position arising from the corner positions each shared with eight adjacent unit cells, plus the six

1.4 SIMPLE MODEL FOR A SURFACE: TERRACE–LEDGE–KINK MODEL

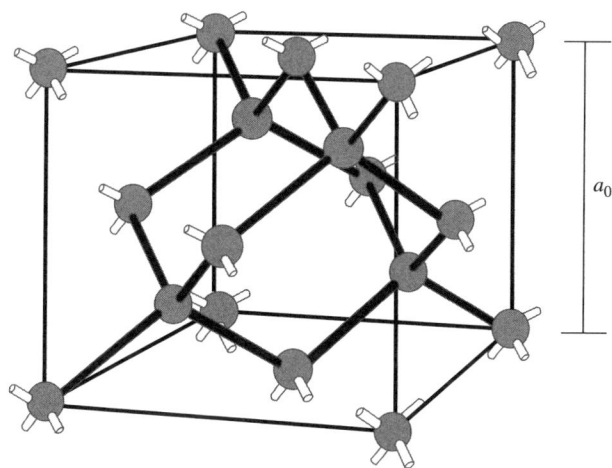

FIGURE 1.5 Si diamond cubic (DC) structure showing sp^3 hybridized Si atoms.

face-centered atom positions each shared with an adjacent unit cell, and four interior atom positions at $(\frac{1}{4}\frac{1}{4}\frac{1}{4})$, $(\frac{3}{4}\frac{3}{4}\frac{3}{4})$, $(\frac{1}{4}\frac{3}{4}\frac{3}{4})$, and $(\frac{1}{4}\frac{3}{4}\frac{3}{4})$. Locating the interior atom positions on planes is sometimes a matter of best judgment. However, the vicinal (311) and (511) surfaces consist of both (100) and (111) planes, and as such the counting of atoms is more involved. Table 1.1 shows the atom densities for three low index planes and the (311) and (511) planes.

For the calculation of the atom density on the vicinal (311) surface of Si, we refer to Fig. 1.4b, which shows a side view of the (311) shown in Fig. 1.4a. In Fig. 1.4b the (100) flat terraces or steps are joined to the (111) risers to yield the intercept of $w/3$, where w is the lattice parameter. The problem reduces to calculating the relative amount of (100) and (111) areas or the ratio $A_{(100)}/A_{(111)}$. This ratio can be written in terms of the number of steps or risers, N_s, which is assumed equal for a large number of steps and risers, as

$$\frac{A_{(100)}}{A_{(111)}} = \frac{N_s L_{(100)} \cdot 1}{N_s L_{(111)} \cdot 1} \tag{1.2}$$

TABLE 1.1 Si Atom Area Density for Selected Planes in Single-Crystal Si

Si Surface Orientation	Atom Density (10^{14} atoms/cm^2)
(110)	9.59
(111)	7.83
(311)	7.21
(511)	7.05
(100)	6.78

where L is the length of the step or riser as shown in Fig. 1.4b. It is assumed that the widths of the steps and risers are the same so the value 1 is assumed and $L \cdot 1$ is the area of each step or riser, which is then multiplied by the total number of steps or risers N_s. Thus, from Equation 1.2 the area ratio is essentially the ratio of the lengths for each component, (100) and (111) that form the (311).

It is noticed in Fig. 1.4b that

$$N_s h_s = \frac{w}{3} \tag{1.3}$$

where h_s is the riser height. It is also noticed that

$$N_s(L_{(100)} + L'_{(111)}) = w \tag{1.4}$$

where $L'_{(111)}$ is the length of $L_{(111)}$ projected onto the (100) plane. Equating these two expressions for w (Equations 1.3 and 1.4) and considering that $\theta = 45°$ (see Fig. 1.4a) and is the angle between L_{111} and L'_{111} and thus $L'_{111} = h_s$, we obtain

$$2h_s = L_{(100)} \tag{1.5}$$

Also $L_{111} = \sqrt{2}h_s$. Substituting these results in Equation 1.2, the area ratio is calculated:

$$\frac{A_{(100)}}{A_{(111)}} = \frac{2h_s}{\sqrt{2}h_s} = \sqrt{2} \tag{1.6}$$

Now the total number of Si atoms on the (311) can be calculated using the following relationship:

$$N_{Si} = \frac{A_{(100)}}{A_{tot}} N_{(100)} + \frac{A_{(111)}}{A_{tot}} N_{(111)} \tag{1.7}$$

where $N_{(100)}$ is the number of Si's per area on the (100) from Table 1.1 and $N_{(111)}$ is the corresponding number of Si's per area on the (111). The total area $A_{tot} = A_{(100)} + A_{(111)}$ and the equation above for N_{Si} can be transformed and readily evaluated in the form:

$$N_{Si} = \frac{1}{1 + A_{(111)}/A_{(100)}} N_{(100)} + \frac{1}{1 + A_{(100)}/A_{(111)}} N_{(111)} \tag{1.8}$$

where the area ratios in the denominator are given by Equation 1.6. This yields the value in Table 1.1 for (311) of 7.21×10^{14} cm^2. The same calculation can be repeated for the (511) remembering to use $w/5$ for the intercept rather than $w/3$ as for the (311) and the angles appropriate to the (511). In the following section an example

1.4 SIMPLE MODEL FOR A SURFACE: TERRACE–LEDGE–KINK MODEL

of the effect of crystal orientation and the area density of Si atoms on Si surface reactivity will be discussed.

This example for Si illustrates a method to evaluate atom densities on various reconstructed surfaces. However, the nature of a specific reconstruction depends on an intimate knowledge of specific surface energies and the effects of experimental conditions and ambient. Consequently, reconstructions are usually experimentally determined.

1.4.1 Oxidation of Various Si Single-Crystal Orientations

The oxidation of single-crystal Si to form a film of SiO_2 on the Si surface is a crucial step in the construction of metal–oxide–semiconductor field-effect transistors (MOSFETs), which comprise the majority of electronic devices presently manufactured. Much more will be said about MOSFETs, films, and SiO_2 in later chapters. At this point it is useful to know that the MOSFET family of devices are constructed on the surface of Si and therefore depends on both the chemical interactions at surfaces and on the formation of interfaces. In addition the devices operate by altering the Si surface potential and depend on the electronic states that exist on the Si surface and the Si–SiO_2 interface. Thus the MOSFET serves as a nearly perfect vehicle from which to understand the implications of a large body of surface and interface science and therefore will be used throughout this text.

In the 1960s in connection with the emerging microelectronics industry based on MOSFETs, the Si single-crystal substrate orientation dependence for the oxidation of single crystalline Si to form SiO_2 was reported (2) and showed that for high pressure steam oxidation the Si <111> surface was the fastest oxidizing surface with the <110> next fastest, followed by the <100> Si surface for the major Si orientations. According to the accepted Si oxidation model (linear-parabolic (LP) model discussed in Chapter 10), the Si surface orientation dependence derives from the reaction of oxidant with Si atoms at the exposed Si surface. This would mean that the oxidation kinetics should scale with the area density of Si atoms on the surface. However, as seen in Table 1.1, the (110) Si plane has the greatest area density of Si atoms for the planes listed and not the (111), which was found to be the faster oxidizing plane. With this knowledge about area densities of Si atoms and with the above experimental observations, Ligenza (2) constructed a more convoluted model that relied not simply on the area density of Si atoms but rather on the number of available bonds at the Si surface. In the LP model the orientation dependence of the Si oxidation rate should be incorporated in the linear rate constant k_i, which derives from the reaction of Si with oxidant at the Si surface (the details of the LP model will be presented in Chapter 10). Thus in the earliest stages of Si oxidation, before transport of oxidant through the growing SiO_2 film becomes rate limiting (the slowest step and therefore the one observed in a series process), the substrate surface ought to dominate the surface reaction and yield the following order for the rates of oxidation R for the major Si orientations:

$$R_{110} > R_{111} > R_{100} \tag{1.9}$$

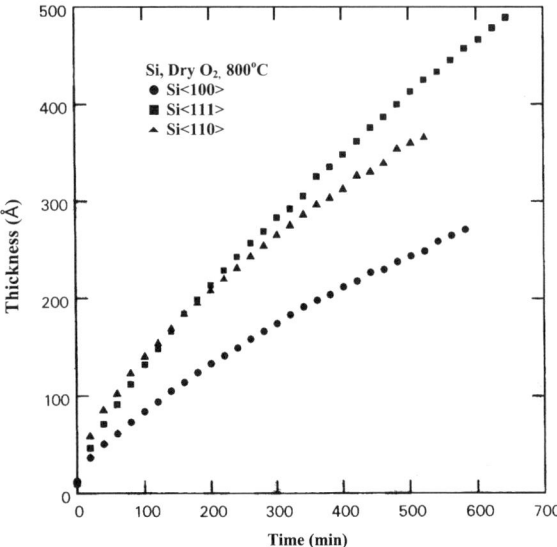

FIGURE 1.6 Oxidation of (100), (110), and (111) oriented single-crystal silicon in pure O_2 in terms of the SiO_2 film thickness versus oxidation time. [Adapted from Irene et al. (4), Figure 1.]

In the 1970s Si oxidation experiments were done (3) using *in situ* real-time ellipsometry (ellipsometry is an optical technique with great surface and thin-film sensitivity and will be covered in Chapter 9) with some of relevant Si oxidation data shown in Fig. 1.6, which was obtained in dry flowing O_2. It is comforting to note that initially the Si <110> oxidizes fastest, but there is a crossover in the order of the Si<110> and <111> orientations for thicker films (near 15 nm), yielding agreement with the early studies that were performed on thicker films (hundreds of nanometers). An explanation for this crossover in the oxidation rates is briefly discussed below and was based on the interface reaction and the effects of stress on Si oxidation (4). The initial oxidation rate should and does scale with the area density of Si atoms as predicted by the LP model. Beyond the crossover, however, another physical mechanism is required to dominate. The model invoked is based on the observation that Young's modulus E for Si varies with orientation as

$$E_{111} > E_{110} > E_{100} \tag{1.10}$$

and is also based on the observation that an intrinsic stress develops during thermal oxidation. The observed tensile stresses in the Si surface stretch the Si–Si bonds in the surface and thereby increase the Si surface reactivity resulting in an accelerated surface reaction rate. The relative magnitude for the observed stress should have the same order as for E above, considering a constant strain resulting from the

1.4 SIMPLE MODEL FOR A SURFACE: TERRACE–LEDGE–KINK MODEL

molar volume change for Si oxidation. The actual force is obtained from the product of the stress and film thickness; hence the force grows with the oxide film growth. Thus, the dominance of this stress-modified kinetics should occur when the forces rise as the oxide grows. In the early stage of oxidation where the SiO_2 film is thin, the kinetics are dominated by the area density of Si atoms; but as both the oxide and the surface force grows, the dominance switches to that of stress dominance and hence the crossover in the oxidation rates. In short, this stress model seems to predict the qualitative aspects of the crossover observations and is even relatively quantitative with the $Si<111>$ and $<110>$ orientations. In Chapter 12 intrinsic SiO_2 film stress is discussed, and experimental results indicate that the intrinsic compressive stress that exists at the oxidation temperature is lowest for the Si(111). This lower compressive stress may also enable more rapid diffusion of oxidant and therefore an increased oxidation rate for thicker films where diffusion is important (see Chapter 10 for a further discussion of the Si oxidation model).

In an effort to confirm this model, the Massoud et al. (3) *in situ* ellipsometry data has been extended down to 600°C oxidation temperature where the stresses are larger and to include the $Si<311>$ and $<511>$ orientations (5). Previously, the crossover has been seen in the oxidation data from 1100°C down to 750°C. However, for lower oxidation temperatures, the films were too thin to exhibit the crossover. The $<311>$ and $<511>$ orientations were chosen by the researchers because ostensibly they would have higher Si atom densities if the planes are flat. However, the assumption of flatness for these planes is erroneous as was discussed above because these planes are actually vicinal planes of the low index major planes, the $Si<111>$ and $<100>$ planes. We now know that these vicinal planes are stepped with $<111>$ risers

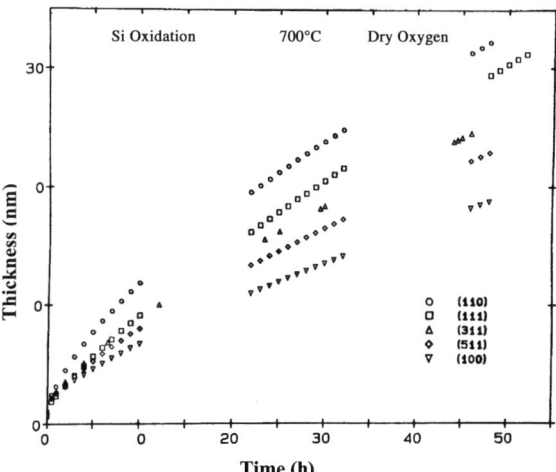

FIGURE 1.7 Oxidation of (100), (110), (111), (311), and (511) oriented single-crystal silicon in pure O_2 in terms of the SiO_2 film thickness versus oxidation time. [Adapted from Lewis and Irene (5), Figure 1.]

and <100> terraces, and therefore the actual area density of Si atoms is between <111> and <100> as given in Table 1.1. As shown in Fig. 1.7 the experimentally determined oxidation rate order is

$$R_{110} > R_{111} > R_{311} > R_{511} > R_{100}$$

which confirms that the area densities for these vicinal planes scales with the initial oxidation rate. This order parallels the area density of Si atoms without resorting to convoluted arguments.

These experimental studies from the 1960s through the 1980s provide convenient examples about the importance of surface structure on practical issues such as film growth for microelectronics. In addition, there is the lesson that many other material science factors need to be considered when considering surfaces and interfaces such as mechanical issues (stresses), electronic issues, and even specific experimental conditions.

1.5 ROUGHNESS

Throughout the text and in particular in Chapter 4, the subjects of surface roughness and the implications of rough surfaces will be considered. In Chapter 4 and throughout the text it will be shown that many properties are changed for rough surfaces, and in some cases it is understood why certain surface properties are altered by roughness. However, at this point only a broad view is desired with the details reserved for later chapters.

First, roughness is an intuitive idea, and we can all imagine a rough surface as compared with an atomically smooth flat surface. Returning to the TLK model, one can imagine all kinds of assemblies of atoms yielding a disordered array of hills and valleys on the surface. Included in this profile are tall and short features as well as wide and sharp features. Essentially, that is roughness. Figure 1.8 shows

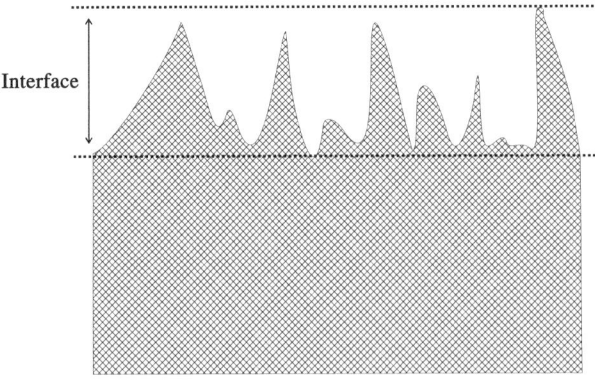

FIGURE 1.8 Side view of a rough surface with an interfacial roughness layer delineated.

a single profile of roughness where the tops of the assemblies are connected to form the boundary at the rough surface. Of course, this profile is for one direction on the surface, and a different profile could be obtained in a different direction on the surface. With potentially very different profiles in different directions, it seems impossible to adequately describe a particular surface in terms of its roughness. Without such a description, namely a metric, it would be virtually impossible to determine how properties change with roughness. However, it might still be possible to compare a rough surface property with that from a smooth surface, but finer distinctions require a reliable roughness metric.

Furthermore, it is easy to imagine that relative to a smooth surface, a rough surface will have more atoms in the projected area than a smooth surface. By projected area, we mean the area of an encircled region to be taken as if it were smooth. The real area will consider all the surface features and nooks and crannies. It will be important to be able to consider both the projected area to compare with the area of a smooth surface and also to know the real area so as to determine surface-area-dependent property changes.

There are many methods in use that attempt to describe rough surfaces. One particularly elegant concept and method is fractals. A fractal is a fractional or noninteger dimension. For surfaces the fractal can vary between the integer values 2 and 3 with 2 being a Euclidian flat surface and 3 being a solid. A noninteger dimension between these integer limits defines a rough surface in terms of its ability to fill volume. Later fractals will be discussed to describe roughness and other methods will also be discussed. There have been many methods developed to measure surface roughness, and these will also be discussed in the context of rough surfaces and surface properties. Most notable has been the evolution since the mid-1980s of scanning probe microscopies (SPM) that can achieve atomic resolution in some cases. Also roughness can be described as an interface between a smooth flat surface and vacuum, as is shown by the dashed lines in Fig. 1.8. This interface region can be thought of as a mixture of substrate and vacuum as constituents of the interface. The extent of roughness is a function of the amounts of these two constituents. This idea has been useful for constructing optical models of rough surfaces and interfaces, and this idea will also be discussed further in the following chapters.

1.6 SUMMARY OF THE KEY SURFACE CONCEPTS

This chapter enumerated and briefly introduced several key ideas about surfaces that will be developed in individual chapters and within other chapters and further illustrated with results from research studies in the field of microelectronics with many from the author's experiences.

The first idea relates to the intuitive definition of a surface as a boundary. At and near the boundary the chemical bonding and hence the binding potential is different from the bulk and graded. This fact gives rise to different arrangements of atoms at a surface and different electronic band structures that lead to different surface sites and different surface states. Also as a consequence of different bonding, the thermodynamic properties at surfaces are different from the bulk properties. Consequently, the reactivity of surfaces is different from the bulk. It always costs energy to create a surface, and thus the processes that minimize the surface area to volume are favored processes. Surface roughness needs to be included for real surfaces since roughness affects the sites and states at the surface.

Now we proceed to develop the key ideas and provide more examples.

REFERENCES

1. Ueda K, Inoue M, Surf Sci 1985;161:L578.
2. Ligenza JR. J Phys Chem 1961;65:2001.
3. Massoud HZ, Plummer JD, Irene EA. J Electrochem Soc 1985;132:1745;1985;132:2685; 1985;132:2693.
4. Irene EA, Massoud HZ, Tierney E. J Electrochem Soc 1986;133:1253.
5. Lewis EA, Irene EA. J Electrochem Soc 1987;134:2332.

SUGGESTED READING

J. B. Hudson, 1992. *Surface Science an Introduction*, Butterworth Heinemann. Many of the ideas in this chapter are also covered in the Hudson book.

E. A. Irene, 2005. *Electronic Materials Science*, Wiley Interscience. A suitable materials science textbook at the correct level for the material in this book. This text is used at UNC for the prerequisite course to the surface science course.

M. Prutton, 1994. *Introduction to Surface Physics*, Oxford Science Publications. A general text that applies to several chapters in this book.

2

STRUCTURE OF SURFACES

2.1 INTRODUCTION

As mentioned in Chapter 1 the structure of surfaces is often different from the structure of the underlying bulk material. Consequently, those properties that depend on structure will be different at the surface than in the bulk. For example, the energy band structure of crystalline materials depends on the binding potential and its periodicity, both of which depend on structure. Therefore, knowledge of surface structure is fundamental to a study of surfaces and interfaces, just as knowledge of structure is fundamental to all of materials science. The surface structure of crystalline materials is usually obtained using diffraction techniques that are rendered surface sensitive. For example, the usual x-ray diffraction techniques using high energy x-rays have large penetration depths in materials typically exceeding 100 nm. Consequently, the information derived from the use of this kind of radiation is most representative of the bulk material. However, by choosing lower energies and/or grazing angles, surface sensitivity can be improved. For the study of the surface structure of noncrystalline materials, no single technique can provide as complete information as that derived from diffraction on crystalline materials. However, there are surface-sensitive techniques that can yield atom positions, composition, and some structural information, and several of these techniques will be discussed in Chapters 5, 6, 9, 11, and 12. In this chapter the discussion focuses on diffraction from surfaces, and to broach the subject of surface structure from this perspective, we commence with a review of diffraction in reciprocal space that

---

Surfaces, Interfaces, and Thin Films for Microelectronics. By Eugene A. Irene
Copyright © 2008 John Wiley & Sons, Inc.

provides the proper coordinates for diffraction. Then diffraction techniques that can yield surface structure such as low energy electron diffraction (LEED), transmission electron microscopy, and scanning electron microscopy are presented as well as the nomenclature associated with surface structures. Finally, in Chapter 6 and elsewhere in the text other important surface structural analysis techniques are covered comprising scanning probe microscopies (SPM) such as scanning tunneling microscopy (STM), atomic force microscopy (AFM), and a less used but powerful structural tool ion scattering spectroscopy (ISS).

2.2 RECIPROCAL SPACE (RESP)

2.2.1 Why Reciprocal Space?

Starting with Bragg's law, which is the fundamental relationship uniting planar indices (hkl), planar spacings (d_{hkl}), and the measurable diffraction angles (θ_{hkl}). When rearranged appropriately, they reveal the natural coordinates in which to interpret a diffraction experiment. Bragg's law can be written as

$$\sin(\theta_{hkl}) = n\lambda \left(\frac{1}{2d_{hkl}} \right) \tag{2.1}$$

where λ is the radiation wavelength used for diffraction and n is an integer indicating the order for diffraction. This form shows that the experimental diffraction angles, θ_{hkl}, are a function of the reciprocal of the interplanar spacing (d spacing), $1/d_{hkl}$. Also the d spacings are reciprocally related to the Miller indices, which for a cubic system are given as

$$\frac{1}{d_{hkl}} = \frac{(h^2 + k^2 + l^2)^{1/2}}{a_0} \tag{2.2}$$

where a_0 is the cubic lattice parameter. Also the superposition of waves depends on the reciprocal size of the unit cell. Thus, reciprocal distances are related to the essential physics of diffraction, and a reciprocal distance coordinate system provides the simplest system. Therefore, reciprocal space (RESP) is defined and used to represent diffraction.

2.2.2 Definition of RESP

A reciprocal lattice, REL, is defined by reciprocal lattice vectors $\mathbf{a}^*$, $\mathbf{b}^*$, and $\mathbf{c}^*$ using the following formulas:

$$\mathbf{a}^* = \frac{\mathbf{b} \times \mathbf{c}}{\mathbf{a} \cdot (\mathbf{b} \times \mathbf{c})} \quad \mathbf{b}^* = \frac{\mathbf{c} \times \mathbf{a}}{\mathbf{b} \cdot (\mathbf{c} \times \mathbf{a})} \quad \mathbf{c}^* = \frac{\mathbf{a} \times \mathbf{b}}{\mathbf{c} \cdot (\mathbf{a} \times \mathbf{b})} \tag{2.3}$$

where $\mathbf{a}$, $\mathbf{b}$, and $\mathbf{c}$ are real space lattice vectors. Figure 2.1a shows that the magnitude of the vector $\mathbf{b} \times \mathbf{c}$ is the area of the plane (shaded) defined by the real space vectors $\mathbf{b}$ and $\mathbf{c}$, and the direction is normal to the $\mathbf{b} \times \mathbf{c}$ defined plane; $\mathbf{a} \cdot (\mathbf{b} \times \mathbf{c})$ is the volume of the solid defined by $\mathbf{a}$, $\mathbf{b}$, and $\mathbf{c}$. Thus, the area of the base of a three-dimensional

2.2 RECIPROCAL SPACE (RESP)

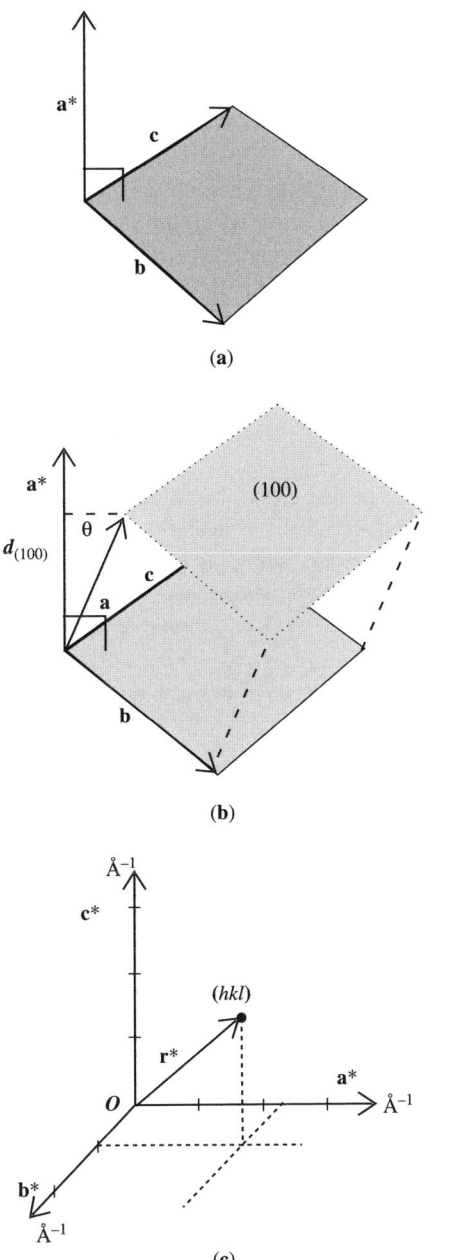

FIGURE 2.1 Reciprocal space (RESP) representations: (**a**) the definition of **a***; (**b**) **a*** in a nonorthogonal system; (**c**) the reciprocal lattice vector **r*** in RESP. **a***, **b***, and **c*** RESP vectors and **a**, **b**, and **c** are real space lattice vectors.

figure divided into the volume yields the height. This height is perpendicular to the basal area and has a magnitude of $1/a$, and for a cubic system we obtain

$$\mathbf{a}^* \perp \mathbf{b} \text{ and } \mathbf{c} \quad \mathbf{b}^* \perp \mathbf{c} \text{ and } \mathbf{a} \quad \mathbf{c}^* \perp \mathbf{a} \text{ and } \mathbf{b} \tag{2.4}$$

The solid figure defined by $\mathbf{a}$, $\mathbf{b}$, and $\mathbf{c}$ has a basal area $\mathbf{b} \times \mathbf{c}$ and volume $\mathbf{a} \cdot (\mathbf{b} \times \mathbf{c})$ can be thought of as a unit cell bounded on top and bottom by planes with indices (hkl) and with areas $\mathbf{b} \times \mathbf{c}$. The distance between the planes is the interplanar spacing, d_{hkl}, which is the height of the three-dimensional figure and is the perpendicular distance from the bottom to top plane. Now as was determined above, this height has the direction of $\mathbf{a}^*$ as is obtained from the following:

$$|\mathbf{a}^*| = \frac{1}{\mathbf{a} \cos \theta} = \frac{1}{d_{100}} \tag{2.5}$$

For orthogonal vectors $\mathbf{a}$, $\mathbf{b}$, and $\mathbf{c}$, the height also has the magnitude $|\mathbf{a}|$ or $|1/\mathbf{a}^*|$. This geometry is clarified further by considering a triclinic cell in Fig. 2.1b where the top plane is the (100) plane, and d_{100} or $|1/\mathbf{a}^*|$ is shown at an angle θ to $\mathbf{a}$. The volume of this triclinic cell is the area of the base, $\mathbf{b} \times \mathbf{c}$, dotted by the height, which is d_{100} or $\mathbf{a} \cos \theta$ or $|1/\mathbf{a}^*|$. Some other relationships are

$$|\mathbf{a}^*| = \frac{1}{d_{100}} \quad \mathbf{a}^* \cdot \mathbf{a} = 1 \quad \text{but } \mathbf{a}^* \perp \mathbf{b} \text{ and } \mathbf{c} \text{ thus } \mathbf{a}^* \cdot \mathbf{b} = 0 \quad \mathbf{a}^* \cdot \mathbf{c} = 0 \tag{2.6}$$

since $\cos 90° = 0$,

$$|\mathbf{c}^*| = \frac{1}{d_{001}} \quad |\mathbf{b}^*| = \frac{1}{d_{010}} \tag{2.7}$$

It is useful to define a reciprocal lattice vector, $\mathbf{r}^*$, as follows:

$$\mathbf{r}^*_{(hkl)} = h\mathbf{a}^* + k\mathbf{b}^* + l\mathbf{c}^* \tag{2.8}$$

where $\mathbf{r}^*$ is drawn from the origin of RESP to the hkl point, which is a plane in real space, (hkl) as is shown in Fig. 2.1c. With the help of Fig. 2.2 it can be shown that

$$\mathbf{r}^* \perp (hkl) \quad \text{and} \quad |\mathbf{r}^*| = \frac{1}{d_{hkl}} \tag{2.9}$$

From Fig. 2.2 the plane ABC is defined in real space with indices (hkl) and the following relationships obtain among the defining vectors:

$$\mathbf{OA} + \mathbf{AB} = \mathbf{OB} \quad \text{and} \quad \mathbf{OA} + \mathbf{AC} = \mathbf{OC} \tag{2.10}$$

2.2 RECIPROCAL SPACE (RESP)

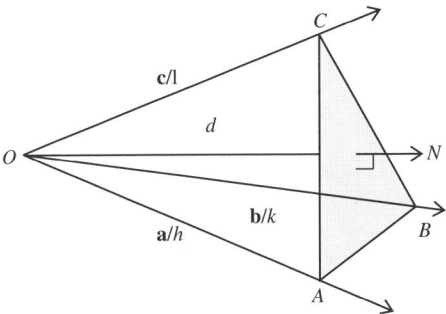

FIGURE 2.2 Normal N to plane ABC with fractional intercepts.

If $\mathbf{r}^*$ is $\perp$ to plane ABC, then $\mathbf{r}^*$ is also $\perp$ to all the lines in the plane such as AB, AC, and so on. We can express the various vectors in terms of fractional intercepts as

$$\mathbf{AB} = \mathbf{OB} - \mathbf{OA} = \frac{\mathbf{b}}{k} - \frac{\mathbf{a}}{h} \quad \text{and} \quad \mathbf{AC} = \mathbf{OC} - \mathbf{OA} = \frac{\mathbf{c}}{l} - \frac{\mathbf{a}}{h} \quad (2.11)$$

Then, we can form the scalar (dot) products of $\mathbf{r}^*$ with lines in the plane in order to test the orthogonality:

$$\mathbf{r}^* \cdot \mathbf{AB} = h\mathbf{a}^* + k\mathbf{b}^* + l\mathbf{c}^* \cdot \left(\frac{\mathbf{b}}{k} - \frac{\mathbf{a}}{h}\right) = -1 + 1 + 0 = 0 \quad (2.12)$$

$$\mathbf{r}^* \cdot \mathbf{AC} = h\mathbf{a}^* + k\mathbf{b}^* + l\mathbf{c}^* \cdot \left(\frac{\mathbf{c}}{l} - \frac{\mathbf{a}}{h}\right) = -1 + 1 + 0 = 0 \quad (2.13)$$

Thus, $\mathbf{r}^*$ is $\perp$ to two lines in the ABC plane; hence $\mathbf{r}^*$ must be $\perp$ to the plane ABC or (hkl).

With d_{hkl} as the distance from the origin O to the plane ABC, we observe that d is the projection of $\mathbf{a}/h$ on the ABC plane normal $\mathbf{N}$, and we have the following relationships:

$$d_{hkl} = \left(\frac{\mathbf{a}}{h}\right)\mathbf{N} = \left(\frac{\mathbf{a}}{h}\right)\left(\frac{\mathbf{r}^*}{|\mathbf{r}^*|}\right) = \frac{1}{r^*} \quad (2.14)$$

From these equalities we can see clearly the importance of $\mathbf{r}^*$ as the vector from the origin of reciprocal space to an hkl point with magnitude $1/d$, and hence related to the interplanar spacings. Furthermore, the hkl point is a plane in real space. Figure 2.3 displays a partial map of RESP where the space is seen to be an array of points that are planes identified with Miller indices in real space separated by reciprocal space distances but with the same symmetry as real space.

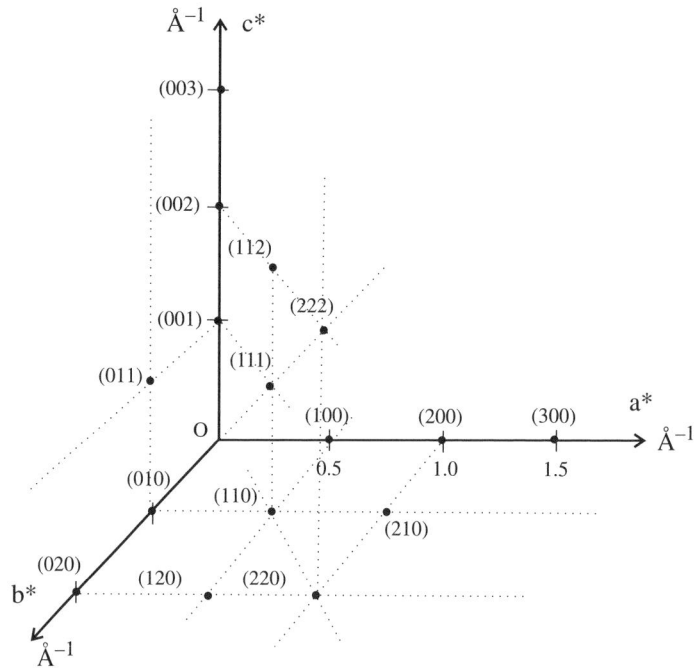

FIGURE 2.3 A map of reciprocal space (RESP) showing selected planes as points in RESP. Points in RESP correspond to planes in real space with cubic lattice and $a_0 = 0.2$ nm.

2.3 THE EWALD CONSTRUCTION

The Ewald construction summarizes diffraction by considering the geometrical consequences of electromagnetic radiation (emr) with magnitude and direction $\mathbf{S}_0/\lambda$, interacting with a fixed REL. Geometrically, this means that for certain directions the emr wavelength conforms to distances between lattice points so as to yield constructive interference. We proceed with the Ewald construction using Fig. 2.4, which shows an array of REL points in two dimensions that are planes in real space. Consider any REL point as the origin for the construction and label this point O. Now consider a wavelength λ and construct a circle (in two dimensions) with radius $1/\lambda$, that touches O. With the center of the circle labeled C, label the line CO, as the vector $\mathbf{S}_0/\lambda$ pointing from C to O, that is, originating at C and terminating on O. Now from O draw a vector $\mathbf{OP}$ to any other point on the REL that lies exactly on the circle. Notice that $\mathbf{OP}$ to point P_{hkl} is an REL vector, $\mathbf{r}^*_{hkl}$. The diffracted beam $\mathbf{S}/\lambda$ completes the construction when drawn from C to P. The fact that this construction can be made with the severe restriction that an $\mathbf{r}^*$ exists on the circle implies that λ is appropriate to the REL to effect diffraction from certain REL points at the appropriate direction(s). The points P_{hkl} and directions 2θ are identified in the construction once the vector $\mathbf{S}_0/\lambda$ is identified.

2.4 DIFFRACTION TECHNIQUES

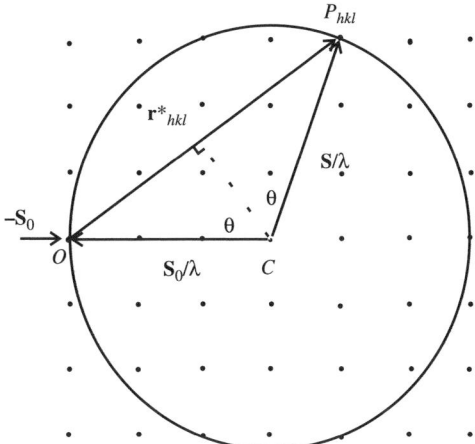

FIGURE 2.4 Ewald construction (circle) in two-dimensional RESP.

From this construction there are several useful relationships:

$$\mathbf{OP} = \mathbf{r}^*_{hkl} = \frac{\mathbf{S} - \mathbf{S}_0}{\lambda} \tag{2.15}$$

$$\sin(\theta) = \frac{1/2|\mathbf{r}^*|}{(1/\lambda)} \tag{2.16}$$

but

$$|\mathbf{r}^*_{hkl}| = \frac{1}{d_{hkl}}$$

Hence we conclude with Bragg's law:

$$\lambda = 2d \sin(\theta) \tag{2.17}$$

Thus, those points that lie on the Ewald circle (sphere in three dimensions called an Ewald sphere) result in constructive interference for the wavelength λ and direction $\mathbf{S}_0$ chosen, and diffraction occurs for those points in RESP (planes in real space).

2.4 DIFFRACTION TECHNIQUES

The great power of the RESP representation with the Ewald construction is the simplicity with which all the experimental diffraction techniques are represented. The experimental methods are directly distinguished by using a RESP representation and considering the different ways REL points, namely the planes in real space, are brought onto the surface of the Ewald sphere, that is, brought into diffracting conditions. Among the parameters that affect the Ewald sphere are the wavelength, λ, which determines the radius as $1/\lambda$, and the direction of the incident radiation or more generally the orientation of the Ewald sphere relative to the REL.

2.4.1 Rotating Crystal Method

This method is used for determining entire crystal structures. It uses monochromatic x-rays and requires a single crystal of material. As seen in Fig. 2.5, the crystal, as represented by its REL, is being rotated about the [0 0 1], which is equivalent to rotation of REL about the $\mathbf{c}^*$ axis. All crystal planes having indices $(h\ k\ l)$ are represented by points on a plane called the $l = 1$ layer in REL normal to $\mathbf{c}^*$. When the REL rotates, this plane cuts the sphere in a small circle, and, when the points touch the sphere, at a particular orientation(s), they give rise to diffraction spots with diffraction vectors $\mathbf{S}/\lambda$ on the surface of a cone. It is typical to rotate around several axes to produce a map of REL and deduce symmetry and structure from the symmetry relationships among the spots and relative intensities.

2.4.2 Powder Method

This method, shown in Fig. 2.6, also uses monochromatic x-rays, but in this method the sample is finely ground. Imagine that one starts with a single crystal, and as this is ground to a finer and finer powder, the crystals get smaller and smaller. In a pinch of

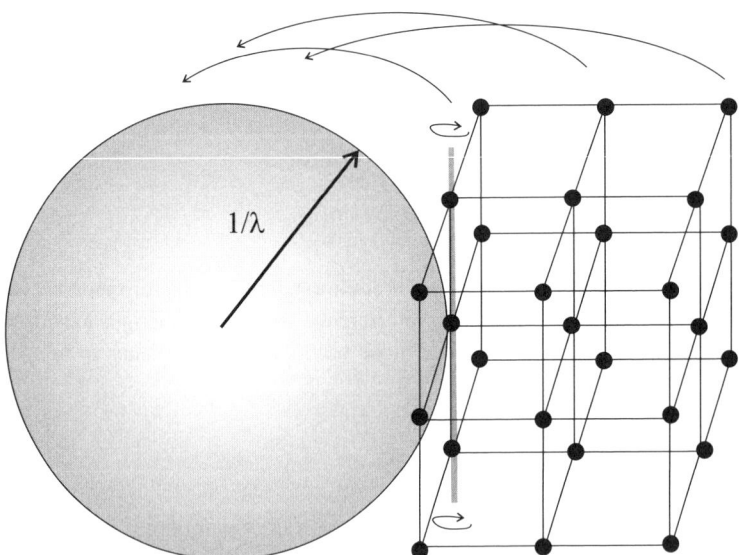

FIGURE 2.5 Rotating crystal method in which a materials reciprocal lattice (REL) is rotated so that REL points intersect the Ewald sphere (shaded) of radius $1/\lambda$.

2.4 DIFFRACTION TECHNIQUES

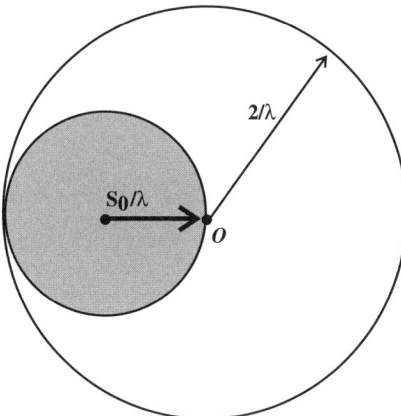

FIGURE 2.6 Power method where an Ewald sphere (shaded) of radius $1/\lambda$ is rotated about O to produce diffraction from various orientations of crystals in the powder.

fine powder one has all (or a large number) crystal orientations. Hence all these orientations of the REL of the material can be simultaneously painted by the x-ray beam of several millimeters in diameter without the need for rotation. This experimental situation can be thought of as having the REL fixed and $\mathbf{S}_0/\lambda$ rotating about the origin O. In this way a new sphere of radius $2/\lambda$ is generated, called the limiting sphere, which contains all the possible spheres of reflection and thus defines all possible $h\,k\,l$'s for λ that will diffract. This method produces all the cones at once. If a film strip is placed to intercept the cones and record the diffracted radiation, arcs, called Debye arcs, are seen on the film. From the positions of the arcs, hkl indices are obtained as well as d spacings from which the identity of the powder can be deduced by comparison with available libraries of powder patterns.

2.4.3 Laue Method

In the Laue method "white" radiation (many λ's) is used with single-crystal samples. Each of the many λ's defines a sphere with the family of spheres having radii from $1/\lambda$ for the K edge for Ag, assuming that a photographic emulsion is used to record the diffraction to, $1/\lambda$ for the shortest wavelength in the x-ray spectrum (the short wavelength limit, SWL), and this situation is depicted in Fig. 2.7. The wavelengths select planes from the REL with which to diffract for a given orientation of the REL at O. From the diffraction spot pattern, the orientation of the REL is deduced. This method is typically used to obtain the crystal orientation. For example, a Si crystal boule is pulled from the melt yielding a large cylindrically shaped Si single crystal. It is required to slice the boule into precisely oriented Si wafers for microelectronics processing. To do this a first cut is made, and then the Laue pattern from the flat boule surface is obtained to determine the precise sawing angle for all the slices. Also the Laue method is useful in surface science where an investigator usually wants to know the crystallographic orientation of the surface under study.

30   STRUCTURE OF SURFACES

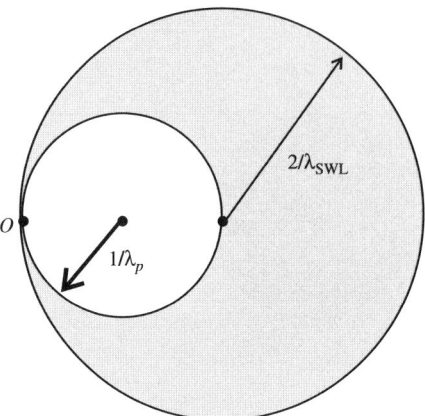

FIGURE 2.7 Laue method where radiation with wavelength from the shortest limit λ_{SWL} to the adsorption edge of the emulsion λ_p define those Ewald spheres in the shaded region that yield diffraction.

2.5 WAVE VECTOR REPRESENTATION—k SPACE

The wave vector or **k** space representation of reciprocal space is particularly important to represent electron energy band structures. The RESP development above sets the stage for this representation. The **k** space is obtained by simply expanding RESP by the factor 2π. To generate this space define a new vector, **k**, called the wave vector:

$$\mathbf{k} = \frac{2\pi}{\lambda} \mathbf{S} \,(\text{or } \mathbf{S}_0) \tag{2.18}$$

and thus it follows for an orthogonal system:

$$\mathbf{a}^* = \frac{2\pi}{a}, \quad \mathbf{b}^* = \frac{2\pi}{b} \quad \mathbf{c}^* = \frac{2\pi}{c} \tag{2.19}$$

In RESP $\mathbf{r}^* = h\mathbf{a}^* + k\mathbf{b}^* + l\mathbf{c}^*$, but now with the RELs expanded by 2π, the new REL vector is labeled **G**, the diffraction vector in **k** space. The situation analogous to the Ewald diffraction sphere is shown in **k** space in Fig. 2.8 where:

$$\mathbf{k}' - \mathbf{k} = \mathbf{G} \tag{2.20}$$

and

$$\Delta \mathbf{k} = \mathbf{G} \tag{2.21}$$

Squaring both sides yields

$$\mathbf{k}'^2 = \mathbf{k}^2 + 2\mathbf{k}\mathbf{G} + \mathbf{G}^2 \tag{2.22}$$

2.6 DIFFRACTION FROM SURFACES

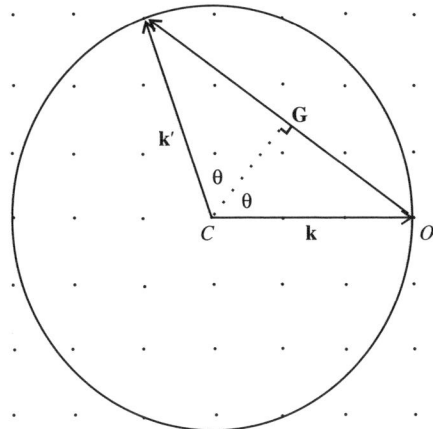

FIGURE 2.8 **k** space representation of RESP where **G**, **k**, and **k'** are analogous to **r***, S_0/λ, and S/λ, respectively.

and the condition for diffraction is when $2\mathbf{kG} + \mathbf{G}^2 = 0$, which means that $\mathbf{k} = \mathbf{k}'$ and that **G** ends at a point of the REL. The solution of this quadratic is $\mathbf{k} = \pm \mathbf{G}/2$. Thus, the wave vector that bisects a REL vector **G** will be diffracted with $\mathbf{a}^* = (2\pi/a)\mathbf{x}$, where **x** is a unit vector. Then for a linear array of regularly spaced lattice points in the REL, the diffraction condition is given as

$$\mathbf{G} = n\mathbf{a}^* = \left(\frac{2\pi}{a}\right)n\mathbf{x} \qquad (2.23)$$

$$\mathbf{k} = \pm\frac{n\pi}{a}, \quad n = 1, 2, \ldots \qquad (2.24)$$

Electrons propagating in a crystal lattice often have deBroglie wavelengths commensurate with the subnanometer and nanometer crystal dimensions. Thus, the electron waves are diffracted by the lattice under certain conditions of length in reciprocal space and electron energy, which gives the λ that would define **G** in **k** space. When the electron with a certain energy diffracts, it is as if it were prevented from propagating. So there are energy regions in which electrons can propagate, called *energy bands* and energy regions where electrons cannot propagate, indeed the electrons are diffracted out of the crystal in these regions and they are called *band gaps*. The bandgap regions of electron energy are said to have no allowed energy states to and from which electrons can flow.

2.6 DIFFRACTION FROM SURFACES

2.6.1 RESP and Ewald Construction for Surfaces

Figure 2.9a shows a real-space three-dimensional (3D) array of atoms (filled circles) in a crystal where the **c** direction is noted, and the atoms on a rectangular planes are

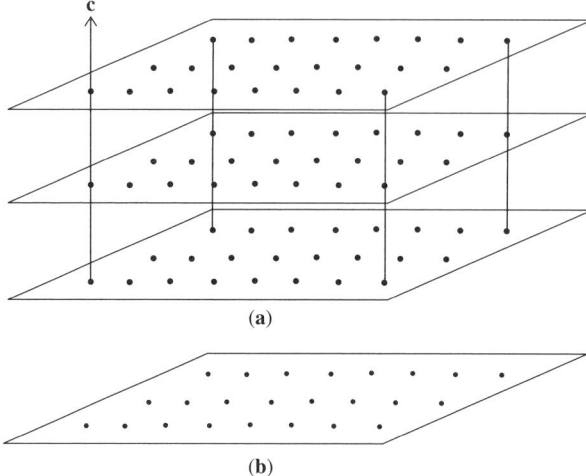

FIGURE 2.9 (a) Three planes in real space with the **c** direction indicated and (b) shows a surface formed as a result of stretching and cleaving in the **c** direction.

indicated. We could imagine stretching the crystal in the **c** direction. Ultimately, the crystal will cleave with the creation of a fresh surface. It is useful to follow this production of a new surface in RESP. One way to think about the cleaving process is to first cut the crystal in Fig. 2.9a somewhere perpendicular to **c**, and then extend **c** toward infinity. In essence one part of the cut crystal, say the top part, is removed to infinity. That will leave a freshly exposed surface for the bottom part as is shown in Fig. 2.9b. To transfer this process into RESP, we recall Equation 2.5 where here **c** is given in terms of **c*** as

$$\mathbf{c}^* = \frac{1}{\mathbf{c}\cos\theta} \tag{2.25}$$

Thus, as **c** increases toward infinity in our cutting and extension process in real space described above, **c*** decreases toward zero. The reciprocal lattice points indicating the planes in Fig. 2.9a that in real space were separated by **c** will in RESP collapse to zero separation of **c***. The result is that an array of RESP points will collapse along the **c*** direction and pile up forming lines or rods as is shown in Fig. 2.10. Remember that a reciprocal lattice is also a three-dimensional array of points, but in RESP the points are real space planes. For simplicity in Fig. 2.10 one plane of points in RESP is depicted. The rods are called diffraction rods and can now be used along with an Ewald construction to formulate diffraction from this surface in RESP shown in Fig. 2.10. Figure 2.11a shows the diffraction rods in RESP piercing the newly formed surface. To define the diffraction conditions, we construct an Ewald sphere at O as is shown. To accomplish this, we first choose a radiation wavelength for

2.6 DIFFRACTION FROM SURFACES 33

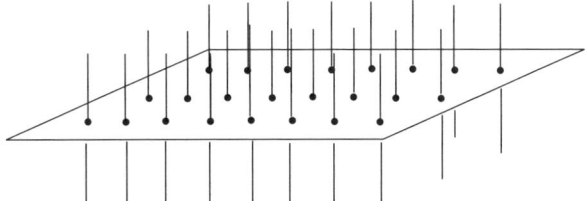

FIGURE 2.10 RESP lattice points with lattice rods formed as a result of cleaving a real space lattice in the **c** direction.

diffraction and then impinge this radiation at O with a specific direction in reciprocal space. In RESP the radiation would be a vector of magnitude $1/\lambda$ and in **k** space with a specific direction $\mathbf{k} = 2\pi/\lambda$. As was done above assigning a direction to the radiation and naming the endpoint of the vector **k** as is shown in Fig. 2.11b, an Ewald sphere is constructed at point of origin O. Now we observe that some of the diffraction rods emanating from the planes pierce the sphere. At the point at which they pierce the sphere, the Bragg condition for diffraction is met and the plane also pierced by the rod diffracts. So once again changing the position of the sphere relative to RESP or its size will bring other planes into diffracting conditions. Thus, we have formulated diffraction from surfaces similar to bulk diffraction. The next task is to obtain diffraction angles and intensities.

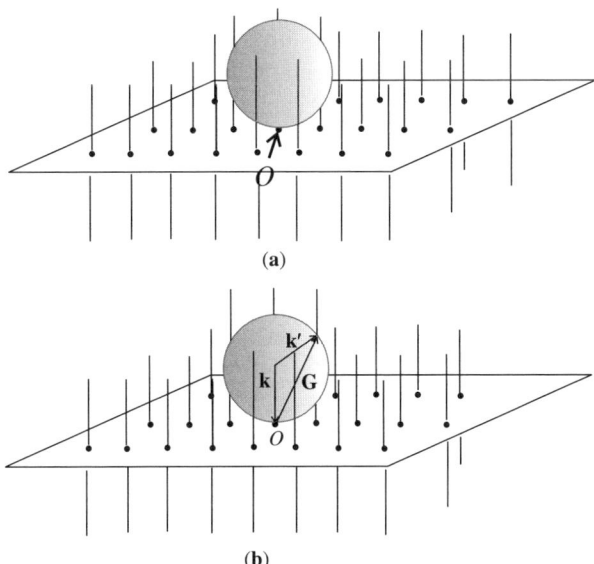

FIGURE 2.11 (a) RESP with lattice points and diffraction rods with an Ewald sphere at O and (b) in **k** space.

2.6.2 Low Energy Electron Diffraction

Diffraction from a bulk crystal depends on the phase difference of coherently scattered radiation of appropriate wavelength from different atomic positions. Figure 2.12 show a typical bulk diffraction scheme where the path difference of the coherently scattered radiation from the two planes shown, *ABC*, is set to be an integral number of wavelengths and results in the scattered waves 1' and 2' being in phase at the detector. This geometry and conditions yield Bragg's law as given in Equation 2.1 and hence ensure that diffraction occurs. In this geometry the incident angle θ was set equal to the exiting angle. However, this geometry, which is typically used to describe bulk diffraction from a crystal, does not work for diffraction from a surface. Figure 2.13a shows the same geometry as for Fig. 2.12 with incident and diffracted beams at the same angle, but with only surface atoms shown. With this geometry diffraction from the surface atoms introduces no phase shift and hence no diffraction. However, as shown in Fig. 2.13b, if the incident beam is near normal incidence, then diffracted beams can be obtained at certain angles where the path differences (indicated in the figure) are integral units of the incident wavelength. This is the preferred geometry for surface diffraction since it can yield diffraction information from surface atoms.

The next issue to address for surface diffraction is that of the specific kind of electromagnetic radiation to use. For bulk diffraction x-rays are often preferred because they can be produced with relatively inexpensive sources, they can be readily monochromatized, and thus, be produced with both accurate and varied wavelengths. However, x-rays penetrate deeply into most materials and thereby yield diffraction information averaged over the bulk of the material. For surface diffraction not only is radiation with appropriate wavelength required, but also the penetration into the material must be minimal. Otherwise, diffraction from the bulk that has many more rows of atoms could possibly obscure the desired surface diffraction from the topmost one or two rows. A suitable radiation source for this task are electrons. To illustrate the suitability of electrons for surface diffraction, we first calculate the wavelength appropriate for diffraction from the kinetic energy of the electrons to determine

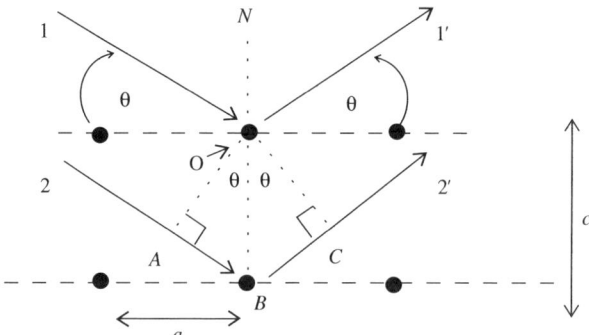

FIGURE 2.12 Diffraction geometry from multiple planes.

2.6 DIFFRACTION FROM SURFACES

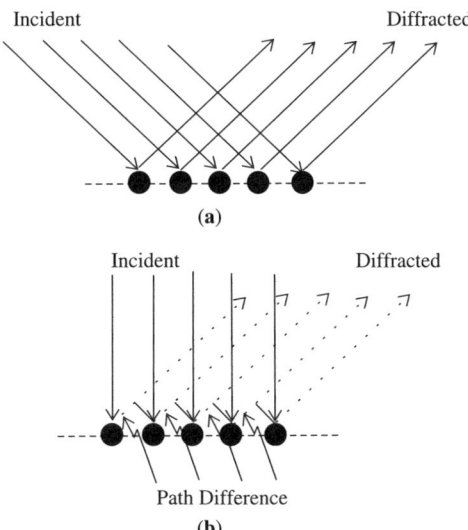

FIGURE 2.13 (a) Nonnormal incidence for diffraction from surfaces and (b) normal incidence with path differences.

the range of usable electron energies. Using the deBroglie relationship for electrons where the electron mass is $m_e = 9.11 \times 10^{-31}$ kg and the wavelength λ is given as

$$\lambda = \frac{h}{\mathbf{p}} = \frac{h}{m_e \mathbf{v}} \quad (2.26)$$

where $\mathbf{p}$ is the momentum and $\mathbf{v}$ the electron velocity. The electron kinetic energy corresponding to an electron wavelength can be obtained using the velocity from the formula above as follows:

$$E_{KE} = eV = \frac{m\mathbf{v}^2}{2} \quad \mathbf{v} = \sqrt{\frac{2\,eV}{m}} \quad (2.27)$$

then using the values for $\mathbf{v}$ from Equation 2.27 in 2.26 we obtain for the electron wavelength (in angstroms) the following:

$$\lambda = \frac{h}{m_e \sqrt{\frac{2\,eV}{m_e}}} = \sqrt{\frac{150}{V}} \quad (2.28)$$

with the electron energy in volts, V. From this formula electrons with a kinetic energy from tens of an electron volt upwards yield a wavelength conformable to unit cell dimensions and hence suitable for diffraction. For example, 50 eV yields $\lambda \cong 0.2$ nm and 1×10^6 eV yields $\lambda \cong 0.001$ nm.

Finally, we need to consider the surface sensitivity for the electron-derived radiation used for surface diffraction. While we do no consider the details here, x-rays typically can penetrate more than hundreds of nanometers into a solid, with the precise penetration depending on many factors such as the energy of the x-rays used and the x-ray mass absorption coefficient of the material, which is a strong function of the wavelength (λ^3), of the x-ray photon, and of the atomic number (Z^3) of the material as well as reciprocally on the materials density. For example, 30-keV x-rays that are typically used for x-ray diffraction penetrate on the order of 10^5 nm in a crystal lattice. Thus, except for grazing incidence x-ray techniques, x-rays will penetrate well below the surface and will be sampling at least many layers beneath the surface atomic layer, and, consequently, the usual x-ray techniques are unsuitable for surface diffraction. In certain energy regimes electrons are well suited for surface diffraction. To evaluate the appropriate energy regime for electron diffraction, we need to consider that diffraction depends on the coherent scattering from the material. Thus, we need to know how far a scattered electron will travel in a solid before its energy hence wavelength and coherency can change. This distance is called the *escape depth* (and sometimes is called the *mean free path* for electrons) for an electron, and it is the distance that electrons can travel without losing energy. Figure 2.14 is a plot of the electron escape depth versus electron energy. A minimum in the escape depth of about 0.5 nm near 50 to 100 eV is seen. Thus, electron energies in this range are suitable for obtaining diffraction information from the near surface region, and low electron energies typically around 100 eV are used. Hence, the diffraction using this radiation is called low energy electron diffraction, or LEED. Figure 2.15 shows that incident radiation can penetrate to varying

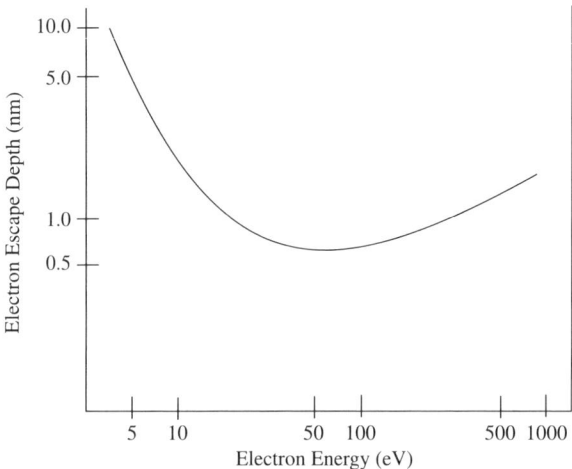

FIGURE 2.14 Election mean free path or escape depth versus incident electron energy.

2.6 DIFFRACTION FROM SURFACES

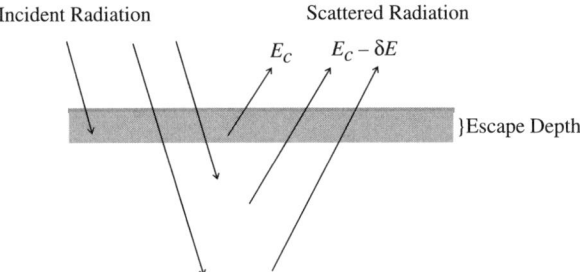

FIGURE 2.15 Incident radiation scattered from within the escape depth with no energy change E_c and outside the escape depth with energy loss $E_c - \delta E$.

distances, at which point electron scattering occurs. The scattering leads to all kinds of processes, some of which produce coherent radiation (without change in energy) and others produce incoherent radiation. Throughout our discussion of surface science we will see that many of the processes can lead to useful surface and bulk material information (see especially Chapters 5 and 6). However, for the immediate purpose of diffraction that depends on phase differences in the coherently scattered radiation, only the coherently scattered radiation is useful. From Fig. 2.15 we see that this coherent component, E_c, comes only from near surface regions within the escape depth as shown in Fig. 2.14 while modified radiation, $E_c - \delta E$, comes from the deeper regions within the solid.

Now we have seen that there is a preferred geometry and a suitable source of radiation that yields high sensitivity for diffraction from surfaces, namely low energy electrons. The typical geometry for LEED is shown in Fig. 2.16. An incident

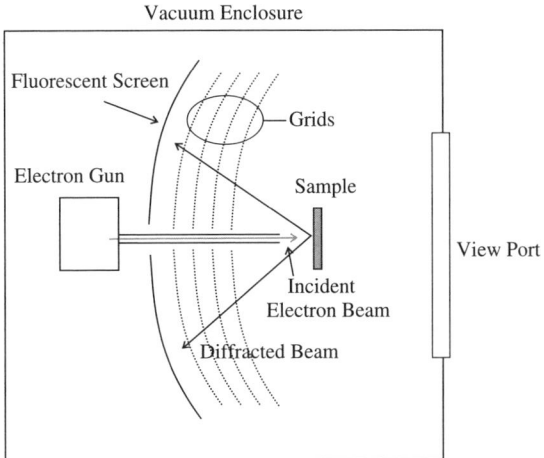

FIGURE 2.16 LEED geometry.

monochromatic electron beam is formed in an electron gun and impinged upon a sample surface that is contained within a vacuum enclosure. The diffracted beams of coherent radiation pass through retarding grids with potentials adjusted to remove or at least significantly reduce incoherent radiation. The coherently scattered radiation from the sample that passes the first few retarding grids is then accelerated by the remaining grids to impinge upon and excite the fluorescent screen to form a diffraction pattern image. This pattern which reveals the surface periodicity and overall surface symmetry may be viewed and photographed from the view port. The entire LEED apparatus is housed within a vacuum system both to enable the electron beam to reach the sample before being scattered by the atmosphere and to enable a sample surface to be viewed with minimum contamination. The precise number and kinds of grids can vary with the specific instrumental design. The fluorescent screen can be replaced with electron detecting equipment that can record both the pattern and specific diffraction spot intensities that can be used to determine atomic positions.

2.6.3 The LEED Pattern and Reconstruction

First, the relationship between a real space arrangement of atoms on a surface and the reciprocal space representations is considered, keeping in mind that the RESP representation will represent the diffraction pattern actually acquired using LEED. Figure 2.17a shows a row of RESP points that are planes in real space with the associated diffraction rods that have a spacing $1/d_1$ in RESP or d_1 in real space. Once a λ is chosen for diffraction, an Ewald circle (in two-dimensional) is defined and shown in

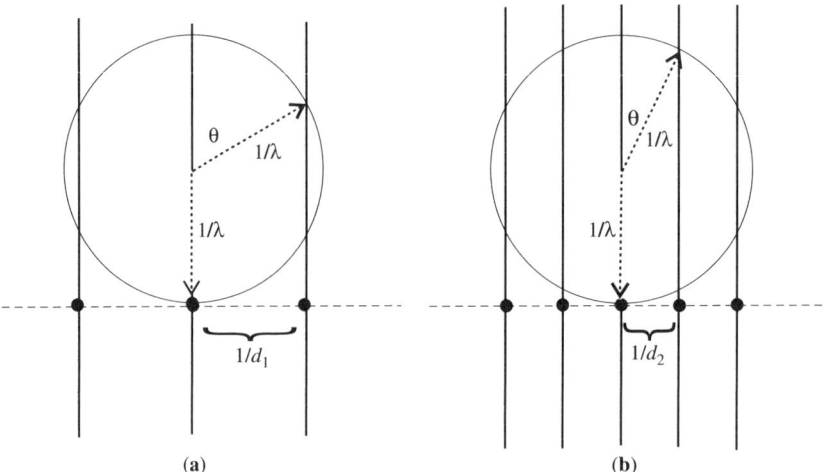

FIGURE 2.17 (a) Diffraction rods in RESP piercing Ewald circle with interplanar spacing d_1 in real space and (b) with real space spacing increased to d_2.

2.6 DIFFRACTION FROM SURFACES

Fig. 2.17, a diffraction rod pierces the Ewald circle shown and the Bragg angle for diffraction is readily calculated from Fig. 2.17 as

$$\sin\theta = \frac{1/d_1}{1/\lambda} = \frac{\lambda}{d_1} \tag{2.29}$$

The result will be a diffraction spot on the LEED screen that has the appropriate angle. Now suppose that the surface reconstructs so that the d spacing in real space is doubled or in RESP $1/d$ is halved, as shown in Fig. 2.17b. With the same λ the Ewald circle is the same, but the new Bragg angle is smaller, and since $d_2 = 2d_1$ is given as

$$\sin\theta = \frac{\lambda}{2d_1} \tag{2.30}$$

Thus, with a reconstruction that expands the surface structure in one direction, the corresponding diffraction spots after the reconstruction will be inside the ones before reconstruction. Also in Fig. 2.17b the next rods also intersect the Ewald circle, and these have spacing $1/d_1$ and thus yield the original spot for the unreconstructed surface as was shown in Fig. 2.17a. Thus one would observe simultaneously two sets of diffraction spots, one between the other.

It is useful to consider this result in real space. Figure 2.18a shows an array of lattice points or atoms in real space with separation a. In the LEED geometry with radiation normal to the linear array or normal to this surface, the detector is at an angle θ. The path difference between adjacent scattered rays is shown as δ. For diffraction to occur, δ must be equivalent to an integral number of λ's. For first-order diffraction ($n = 1$) we can express the diffraction condition as follows:

$$\sin\theta = \frac{\delta}{a} = \frac{\lambda}{a} \tag{2.31}$$

Now we consider three cases for the ratio λ/a as is shown in real space in Fig. 2.18b. For case A, the starting structure, each atom is separated as shown in real space and the ratio of λ/a is given as $1/3$ so as to yield a diffraction angle $\theta = 19.5°$. If this starting surface reconstructs to case B where the surface lattice is expanded and separation is doubled (a goes from 3 to 6) so that $\lambda/a = 1/6$, the diffraction angle changes to $\theta = 9.6°$, or if the starting surface reconstructs to case C where the spacing is halved so that $\lambda/a = 1/1.5$, which yields $\theta = 41.8°$. Thus, as the spacings in real space increase the diffraction angle decreases and vice versa leading to the diffraction patterns shown in Fig. 2.18c that shows the LEED or RESP representations for the situations in Fig. 2.18b. This reasoning can be extended to two-dimensional as is shown in Fig. 2.19. On the left are relative atom positions (real space) and on the right a corresponding LEED pattern (RESP). The surface in the top left, the starting surface, reorders to that in the bottom left. Consequently, the RESP representation or LEED pattern changes from the top right to that at the bottom right of Fig. 2.19. Figure 2.20 displays the changes that occur in diffraction

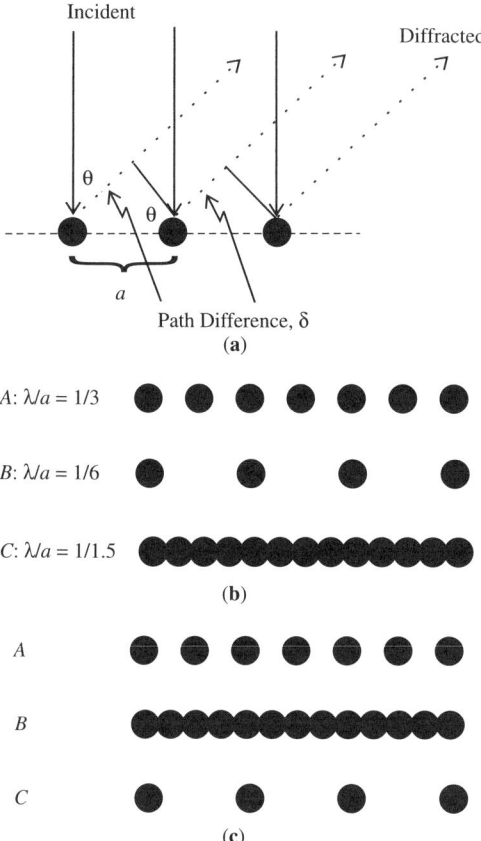

FIGURE 2.18 (a) LEED geometry with spacing a; (b) real space spacings $a = 3, 6, 1.5$, respectively; and (c) the LEED or RESP representation of the situation in (b).

patterns due to surface ordering that is different from the bulk. This reordering can occur due to reconstruction as discussed above or due to an overlayer forming on the starting surface. The starting structure is shown at the top left in real space and the real space changes are seen as the gray circles on the starting structure in the left-hand frames with the corresponding RESP or LEED structures to the right. Note that diffraction is from both the top surface structure and the underlayer (starting surface). This is due to the finite escape depth for coherently scattered electrons. On the far right is the nomenclature that is used to describe the surface structure and is discussed in the following section.

2.6.4 Indexing LEED Patterns and Surface Structure Nomenclature

The identity of diffraction spots is straightforward and follows from Figs. 2.11 and 2.17, which show the diffraction rods piercing an Ewald sphere and having

2.6 DIFFRACTION FROM SURFACES

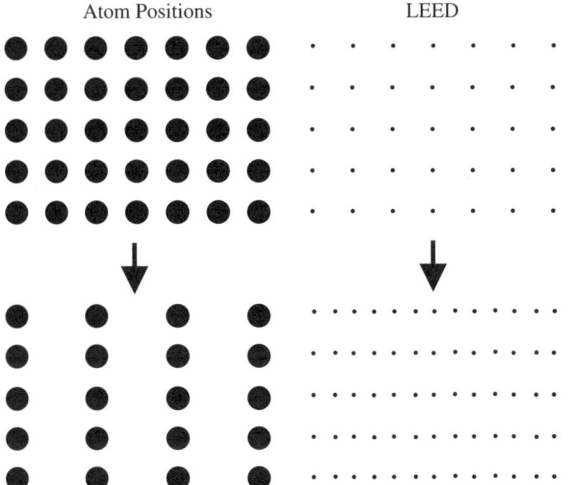

FIGURE 2.19 Reconstruction on left with corresponding LEED pattern changes on the right.

separations of $1/d$. Each rod that pierces the Ewald sphere gives rise to the observed diffraction spots and in turn the LEED pattern. The symmetry of the array of diffraction spots directly reflects the symmetry of the surface, and distances on the diffraction pattern are reciprocal distances in the surface. Thus, Fig. 2.21 shows a diffraction pattern with a cubic lattice with appropriate h,k indices for the two-dimensional surfaces. If the pattern were rectangular, it would have the same indexing. For substrates that are reconstructed or that have an overlayer with a different structure, a nomenclature that describes the surface structure is needed. The most common nomenclature will be described, and this system is derived by comparison of the surface mesh with the bulk structure. The form for the notation is as follows:

$$M \; (hkl) - (n \times m) - C$$

where M is the chemical symbol for the substrate surface and C for the overlayer, (hkl) is the substrate surface plane under examination, and $(n \times m)$ are the multipliers for the new surface periodicity, which is n and m times the original surface or substrate periodicity in the **a** and **b** directions, respectively. Sometimes another letter, either p or c, is inserted before the $(n \times m)$ to indicate primitive or centered. Often p is omitted even when it applies. If the new surface unit cell or lattice is rotated relative to that of the substrate plane, an angle is included to indicate the amount of rotation. Figure 2.22 shows the 5 possible surface lattices or meshes as compared to the 14 Bravais lattices for three-dimensional structures. Only these 5 are needed and others can be expressed in terms of these 5 lattices.

42 STRUCTURE OF SURFACES

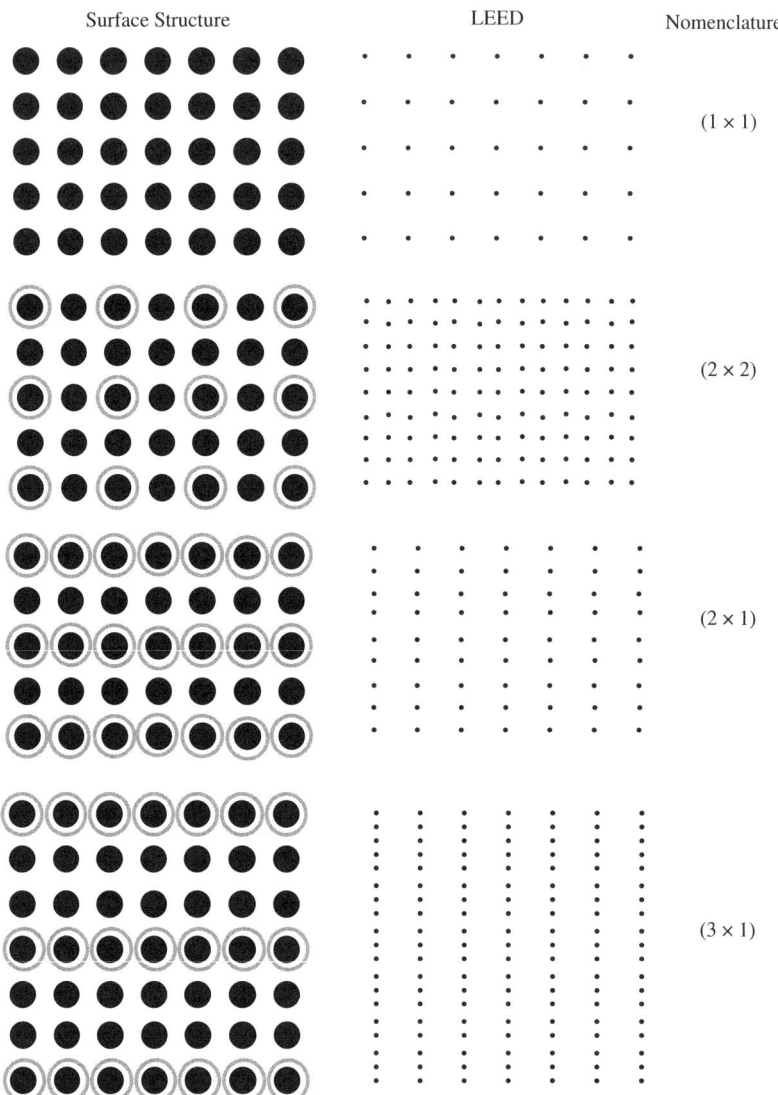

FIGURE 2.20 Various surface structures (reconstruction or overlayers) on the left with LEED patterns and nomenclature for the surface structure.

Applying this nomenclature to the structures and LEED patterns in Fig. 2.19, the top right pattern is a (1×1), which simply reflects the square surface geometry of the atoms in the top left atom position image. This is the initial surface plane that upon reconstruction yields the bottom left atom positions and the bottom right LEED pattern. The LEED pattern indicates that $\mathbf{a}^*$ is $1/2\mathbf{b}^*$. In real space the lattice parameter $\mathbf{a}$ would be twice the length of $\mathbf{b}$ and would therefore yield a (2×1) surface mesh or lattice.

2.6 DIFFRACTION FROM SURFACES

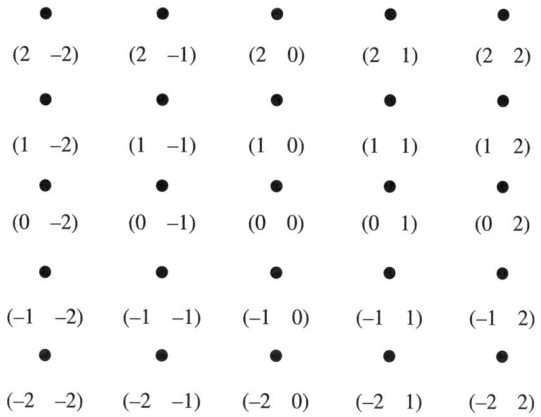

FIGURE 2.21 Indexed LEED pattern.

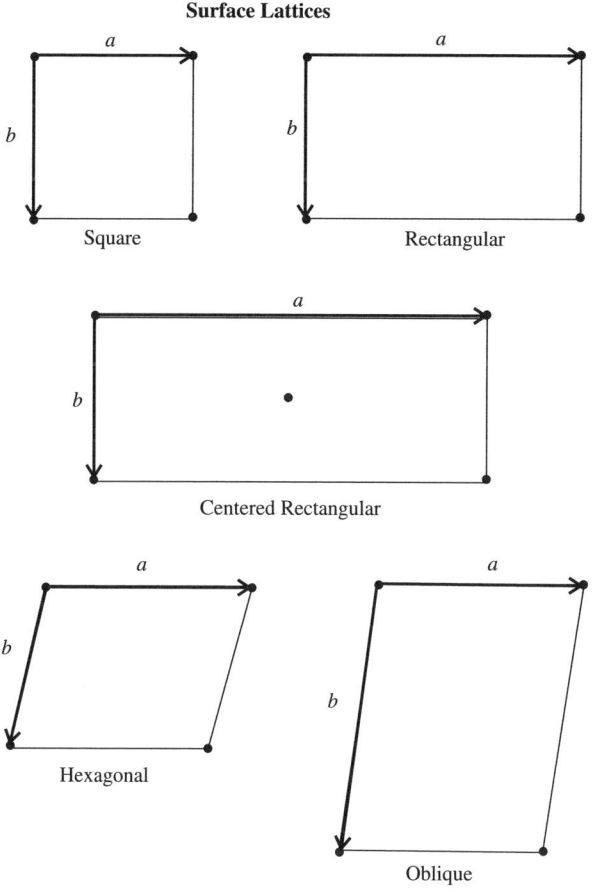

FIGURE 2.22 Five fundamental lattices for surface structures.

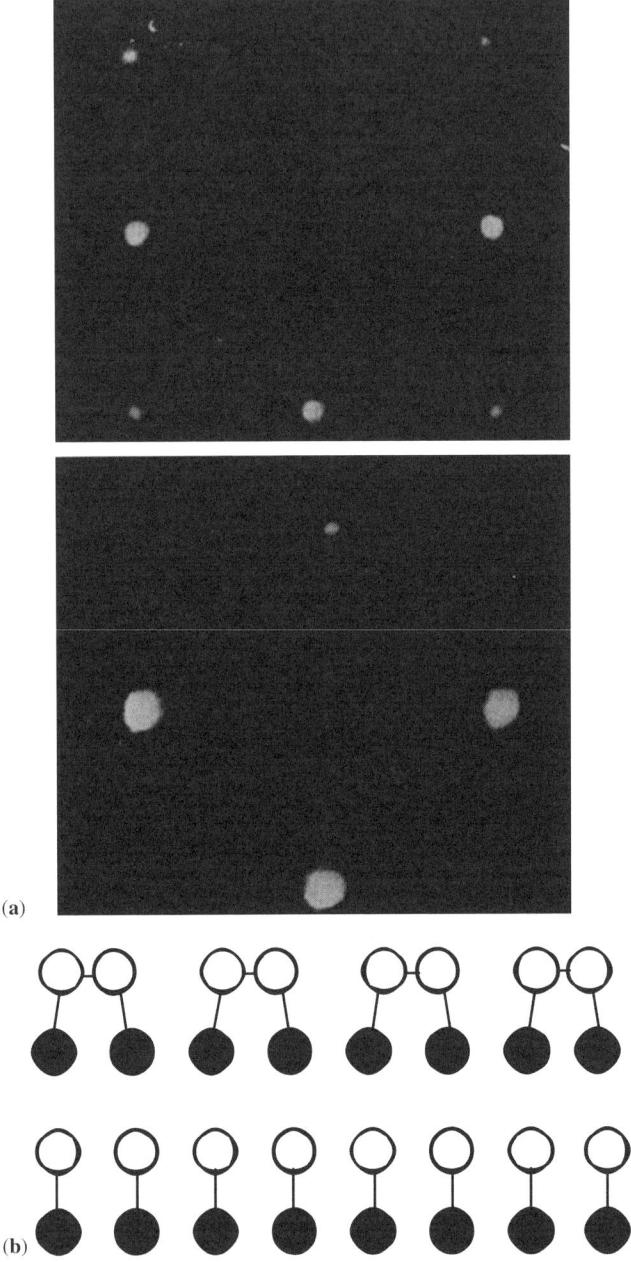

FIGURE 2.23 (a) Si LEED patterns: Top is clean Si(100) that has been reconstructed yielding a 2×1 pattern and the bottom is the result of H on the cleaned surface yielding a 1×1 surface structure; (b) side view of surface structure with the top plate showing dimer formation corresponding to the 2×1 LEED pattern and bottom showing the bulk or 1×1 structure. [(a) was adapted from White and Woodruff (1), Figure 1.]

Diffraction from a pristine unreconstructed surface yields a LEED pattern with the spots associated with its surface structure, which is very close to the bulk structure as shown by Fig. 2.19 top right. Upon reconstruction the LEED pattern has the original spots and also new spots about $1/2$ way between $\mathbf{a}^*$ direction spots, so-called $1/2$ order spots. At this point it is also possible to introduce a reactant that will be adsorbed either physically or chemically on the surface. A further discussion of these chemical and/or physical adsorption and/or reaction processes is reserved for later chapters. This overlayer, however formed, could result in a periodicity different from the underlying reconstructed surface and different from the original surface periodicity. From the viewpoint of diffraction this is readily handled using the above nomenclature and the method developed above. Figure 2.20 shows several surface structures in real space that have resulted from overlayer formation (larger gray circles) along with the corresponding LEED patterns, and the nomenclature associated with the surface structure.

2.6.5 LEED of Si(100)

Si(100) is the Si orientation of choice for modern Si-based MOSFET technology. As will be discussed in Chapters 5 and 7 the reason for the predominance of this Si orientation is that Si(100) displays the lowest number of intrinsic interface electronic states after the Si surface is oxidized. The interface electronic states deleteriously alter the performance and reliability of MOSFET devices. Therefore, the nature of this orientation has received considerable attention.

Figure 2.23a shows LEED patterns corresponding to a clean Si(100) surface (top) that shows the 2×1 reconstruction due to dimerization of the Si top row of atoms. This is shown schematically in the side view of Fig. 2.23b. The Si(100) 1×1 is found when H atoms adsorb to the top surface, as shown in Fig. 2.22a bottom panel, and this is reflective of the bulk structure shown in Fig. 2.23b bottom panel. The dimer formation reduces the surface energy for the bare Si(100) surface, and this surface energy lowering apparently takes place with H atoms on the Si surface. Researchers ((1)) indicated that the LEED patterns could be reproduced back and forth by desorbing and readsorbing the H. It should be noted that LEED can thus be used to determine the cleanliness of a surface.

2.7 ELECTRON MICROSCOPY

Electron microscopy techniques are used for a wide variety of analytical tasks in materials and surface science. For the present purposes surface and interface applications are germane; and after a brief introduction to transmission and scanning electron microscopies, these techniques are discussed with respect to surface and interface applications.

Throughout this text a variety of surface and interface analytical techniques will be discussed. Several important techniques use electrons impinging upon a sample. Figure 2.24 displays some of the processes that can occur when electrons impinge

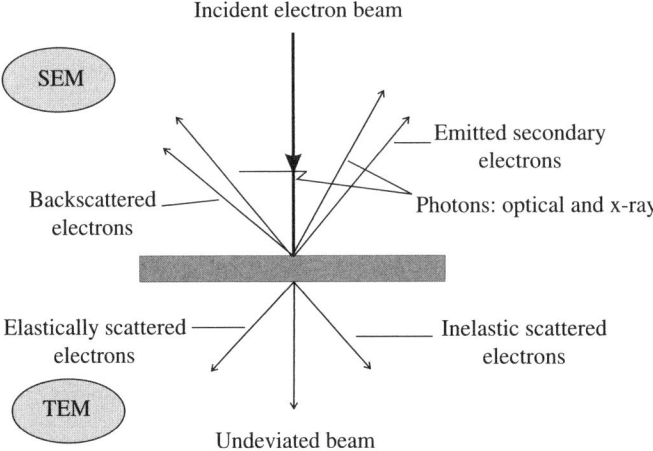

FIGURE 2.24 Incident electron beam upon a sample with numerous scattered electrons.

on a solid sample. The incident electron beam can be transmitted without any interaction (undeviated) and transmitted with changes in direction (coherent elastic scattering) and/or energy (inelastic scattering). Recall that coherent elastic scattered radiation leads to diffraction and it is this radiation that is discussed below in the section on transmission electron microscopy (TEM) and the transmitted radiation is labeled as TEM in Fig. 2.24. The incident beam can also yield emitted secondary electrons from the sample as well as backscattered electrons. These electrons are used to form images of the surface from which the electrons arise and together yield the technique called scanning electron microscopy (SEM) and discussed in a separate section below. Both modern TEM and SEM instruments include other techniques that yield a variety of useful information, and some of these will be discussed when appropriate in this text.

2.7.1 Transmission Electron Microscopy

Transmission electron microscopy employs the elastically scattered electrons that transmit through the sample. Thus, the diffraction that occurs is mainly from the bulk of a sample, and it is not obvious at this juncture how this technique yields relevant surface and/or interface information. However, that important point will be made clear in the following section when the subject of sample preparation and TEM results are discussed. In this section we concentrate on the TEM technique itself. Figure 2.25a shows a schematic of a conventional microscope used for TEM, and it depicts the diffraction mode for TEM. Typically, the electron microscope column that contains the electron source, sample, and associated electron optics is contained within a vacuum enclosure primarily to enable electrons to be produced and transmitted through the sample efficiently and also to prevent the sample from reacting. Starting from the top of the electron microscope column is a source of

2.7 ELECTRON MICROSCOPY

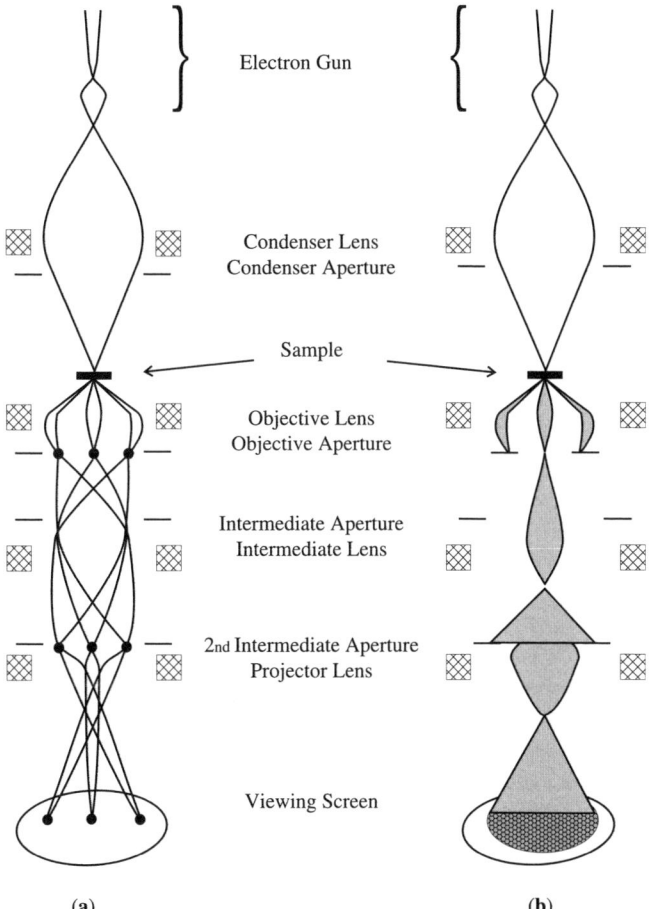

FIGURE 2.25 Transmission electron microscopy columns with major components, electron beams (lines), and diffraction spots (filled circles) indicated with both (**a**) diffraction and (**b**) imaging modes.

electrons that is typically a heated metal (often W) filament that can emit electrons when the thermal energy exceeds the binding energy of electrons. This process is called thermionic emission and will be discussed further in Chapter 5. Suffice for the present that some metals can efficiently emit electrons when heated. These emitted electrons are accelerated to the desired velocity and hence kinetic energy using a series of grids that both accelerate the electrons and monochromatize the energetic electron beam. The electron emission hardware and associated grids are labeled as the electron gun in Fig. 2.25. Typically, electrons with energies of the order of 50 keV and higher are used for TEM and 100 keV is typical and can penetrate samples that are hundreds of nanometers thick. Microscopes up to 10^6 eV are commercially available, and the higher energies are used to penetrate thicker samples where information from the bulk sample is desired. The high resolution obtainable with TEM images is primarily due to the short wavelength achievable with high energy electrons. Using Equation 2.28 and an electron energy of 100 keV, an approximate electron wavelength $\lambda = 0.004$ nm is calculated that in principle can yield excellent diffraction results.

Figure 2.25a shows that after an energetic electron beam is formed it is collimated and focused on the sample surface using electromagnetic lenses. An aperture is used to eliminate any unwanted radiation. The coherently scattered electron waves emanating from the sample can yield diffraction at certain angles, the so-called Bragg angles. Three diffraction beams are seen after the sample. Through a series of other lenses and apertures a number of the diffraction beams are brought to the fluorescent screen at the base of the microscope column where the diffraction pattern can be viewed and photographed. By magnifying one of the diffraction beams coming from the sample, an image can be formed. This is seen in Fig. 2.25b where an objective aperture selects the central beam passing through the sample to be magnified. The lenses below the objective aperture are adjusted to magnify this beam. Electron intensity variations found in the magnified beam image arise from the variability of electron transmission in the sample and provide image contrast. This kind of image is called a bright-field image. Dark-field images can be obtained by moving the objective aperture to allow one of the diffracted beams to pass and become magnified. The dark-field image can also be obtained by tilting the incident beam. Another very useful technique is called *selected area diffraction* (SAD), and this is accomplished by the intermediate aperture that selects a small area of the sample from which to obtain a diffraction pattern. It will be seen below how SAD can be used to identify the material near a surface or interface.

Thus, as shown above, not only can TEM provide diffraction patterns for crystalline material but also images that have electron intensity contrast that are similar in appearance (but arising from very different physics) to optical images. These different modes, diffraction and imaging, are readily obtained by controlling the electromagnetic lenses and apertures. Considering that atoms are tenths of nanometers apart, atomic resolution should be readily achievable with TEM due mainly to the short wavelengths used. However, atomic resolution can only be achieved by the decrease of aberrations that reduce the theoretical resolution that is based only on the wavelength of the electrons. The three main aberrations are:

1. Chromatic aberration due to an image being formed from different energy electrons. The different energy waves have different focal lengths and leads to some blurriness. The monochromatic nature of the incident electrons largely determines the magnitude of this aberration.
2. Spherical aberration occurs where there is distortion in projection of a spherical wave front onto a flat screen. This aberration can be reduced in TEM through the use of apertures that remove the most distorted parts of the image that are near the edge of an image.

3. Astigmatism is the most difficult aberration to deal with and arises from rays coming from different paths in the sample that have different focal lengths. This aberration can be reduced in modern electron microscopes with the use of electromagnets appropriately placed near the beam and appropriately powered.

2.7.1.1 Sample Preparation Sample preparation for TEM examination is often more time consuming than the actual TEM examination itself. Both the art and science associated with sample preparation are variable according to the sample being investigated, and there is a considerable wide-ranging literature on the preparation of all kinds of samples for TEM. The intention here is not to be exhaustive but rather to illustrate the key steps in the preparation of solid nonbiological samples that have thus far been typical for surface and interface studies in the field of electronic materials. The three major steps for these kinds of samples include cutting a larger sample to size to fit on the electron microscope sample stage and thinning the sample for thin-film and substrate studies, while for surface and interface studies there is the additional step of the preparation of cross-sectioned samples.

Modern electron microscopes can accommodate samples that are about 3 mm in diameter. Thus, the first step in preparation is to cut a sample to the dimension and shape accommodated by the specific microscope. For example, Si samples are cut with an ultrasonic cutter in circular 3-mm-diameter pieces that can fit in the microscope sample holder and is usually held in place with a spring-type retaining ring. Metallurgical cutting wheel saws are often used and even more simply solid samples can be cleaved or broken to approximate size using diamond scribes. Free-standing already sufficiently thin samples can be carried on metal (usually Cu) grids especially made to fit in a given microscope. This is often used for polymer film samples and biological samples that have been microtomed.

In the discussion above it was mentioned that electrons can penetrate hundreds of nanometers at usual TEM electron energies. Most samples are initially much thicker; for example, and Si wafers could be millimeters thick (depending on the wafer diameter). Therefore, most samples need to be thinned for TEM examination. Considering that the greater the path length in the sample the more electrons will lose coherency, and thus the fewer will be available for diffraction and imaging, and more electrons will contribute to the background noise arising from multiple scattering events, thinning is a crucial step in TEM preparation. There are many methods used to thin samples and only a few of the commonly used procedures are discussed.

One simple thinning method that does not involve elaborate equipment is chemical etching. The use of this method presupposes that a suitable chemical etchant is available for the desired sample. Si is again used as an example. The chemical etchant frequently used for thinning Si is called "white etch" or "acid etch" in the microelectronics industry and is a mixture of HNO_3 and HF and often in 3 : 1 proportions where the starting HNO_3 is 100% and the HF is 48% by weight (typical stock chemicals). However, the proportions for Si etching can vary and the etching can be made faster with more HF. The procedure for thinning TEM samples is to mask most of the

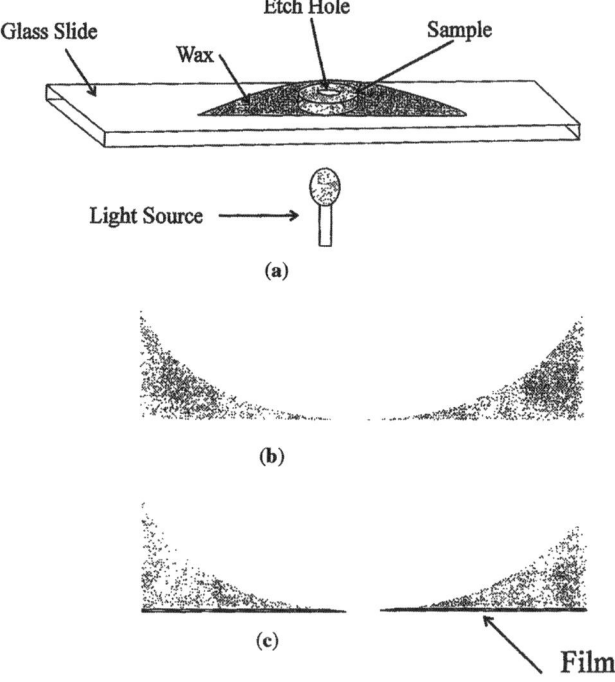

FIGURE 2.26 (a) Sketch of sample affixed to slide with wax for TEM thinning; (b) sample etched through; and (c) etching a sample with a surface film.

cut 3-mm piece of Si and then unmask the central portion of the sample for exposure to the etchant. This procedure can be visualized using Fig. 2.26a, which shows a side view of a Si sample that is held in place on a glass slide using wax. Wax is used as a mask since it is resistant to the etchant. A sample is prepared for thinning by first placing a glass slide on a hot plate and melting a drop of wax on the exposed side of the slide. Then the Si sample (already cut to proper size) is placed in the molten drop, and the slide with sample is removed from the hot plate for the wax with sample to dry. The Si sample is placed with the side to be viewed in TEM facing down on the glass slide. If, for example, one desires to observe the polished side of the Si, then the polished side is placed face down on the slide. Also for observing a film deposited on the polished side of Si, the film side must be facing the glass slide. After the wax dries, the sample is placed on the stage of a simple inspection optical microscope with an intense light directed to the underside of the slide from the bottom as is shown in Fig. 2.26. Looking through the optical microscope (a low power stereo inspection microscope is ideal), a small circular hole is scratched in the top of the wax with a sharp tool exposing a small circular area in the center of the backside of the sample. This now exposed sample area is covered with a drop of etchant. In the case of Si with white etch, bubbling will be observed as the etchant does its work. The spent drop is removed after the etching slows and replaced

2.7 ELECTRON MICROSCOPY

with a fresh drop. Now as the front surface of the sample is neared by the etching and the Si is progressively thinned, red light can pass through the sample from the source underneath the Si sample. This is an indication that the sample is nearly thin enough and the etching should be slowed. When the very center of the etched hole permits the passage of white light, the procedure is finished. The wax is melted and the sample removed and cleaned repeatedly using solvents, and the sample is ready for TEM examination. An ideal finished sample profile is shown in Fig. 2.26b. Notice that there is a central hole completely etched through the sample. Adjacent to the hole is the very surface of the sample that is thicker away from the hole. The sample a is wedge shaped sample with varying thickness. The thinnest part is nearest the sample surface (the surface that was originally nearest the slide surface) or if the sample was a film-covered Si sample, then this thin region may be all or mostly film as is shown in Fig. 2.26c. Moving to thicker regions more substrate is included. Thus in one thinned sample several important areas can be observed. It should be noted that at first this method of thinning appears simplistic and not capable of repeated success. However, using a suitable etchant and with a little practice, it is virtually foolproof.

As an alternative to chemical etching, the back side of the sample can first be mechanically abraded to quickly thin the sample, and then the sample can be further thinned using a common TEM preparation technique called ion milling. The mechanical abrasion can cause damage to the sample, so while it is a fast method it is best not to thin too much using this technique. Ion milling is slow but less aggressive. An ion milling tool is contained within a vacuum chamber that contains an ion gun and a sample holder. An ion gun has a source of electrons and accelerating grids that produce and accelerate electrons that collide with Ar atoms bled into the chamber. The collisions remove electrons from the Ar atoms forming Ar^+ ions that can be accelerated by negatively charged grids and shaped into an energetic beam of ions. This beam is focused on the back of the sample to be thinned. The impinging energetic Ar^+ ions simply erode the sample producing a tapered hole similar to the chemical etch procedure above. With adjustable angle of impingement and rotating sample a uniform taper with a central hole can be produced. Often commercial equipment is equipped with a light source focused on the sample and a photocell on the opposite side of the sample. Thus, the first vestige of a hole in the sample can be detected and the process automatically stopped.

Now we return to the crucial question of how TEM, a technique that uses high energy electrons, can be used to obtain surface information. In the discussions above, the thinning techniques leave a wedge-shaped thinned area that is mostly surface and interface, and this does yield unambiguous information specific to a surface, interface, and/or thin film. Another more powerful preparation technique called *cross-sectional TEM*, or XTEM, provides even better and less ambiguous surface and interface information albeit requiring more sample preparation time. The general procedure for XTEM sample preparation can be understood with the help of Fig. 2.27. The first step is to cut the sample perpendicular to the surface or interface that is to be observed, and this is shown in Fig. 2.27a. The slices are made with a metallurgical saw capable of 'cuts less the 1 mm. The slices are then

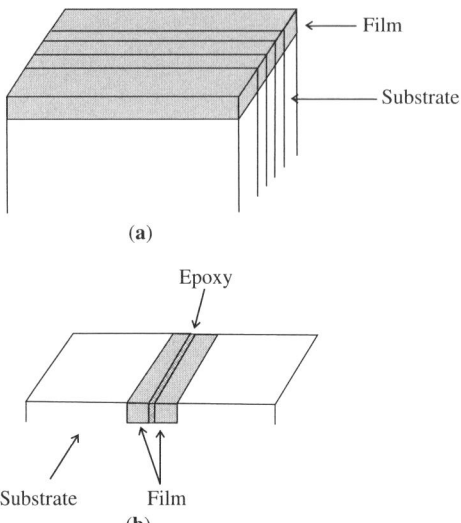

FIGURE 2.27 XTEM sample preparation: (**a**) samples are cut normal to film covered surface and (**b**) glues face to face for thinning.

literally glued together with epoxy with the desired surfaces glued face to face as is shown in Fig. 2.27b. Then this glued assembly is cut to fit the microscope and thinned by chemical or ion milling as was described above. The result is that a side view of the sample is obtained, and in the vicinity of the surface or interface specific information can be obtained from that region of the sample and compared with areas away from the surface or interface. High magnification TEM images can be obtained from the very surface or interface regions for comparison with adjacent regions, and SAD diffraction patterns can be obtained from the various regions also for comparison. XTEM has evolved as a routine technique for interface and thin-film studies. Not only is information obtained about the surfaces and interfaces, but film thicknesses can be estimated and the distribution of atoms as well. Some examples are given below.

2.7.1.2 TEM Results As was discussed above, two different but complimentary kinds of information can be obtained from TEM: images and diffraction. Figure 2.28 shows an example of each kind of information from a sample thinned by the chemical etch method described above. This sample consisted of a single-crystal Si wafer that was oxidized to form a SiO_2 layer of around 100 nm. Then using a gaseous mixture of SiH_4 in Ar, a polycrystalline Si film was deposited upon the SiO_2 covered Si by the method of chemical vapor deposition (CVD). Essentially, CVD uses a tube that contains the substrate and enables the injection of a reactive gas mixture that flows past a heated section containing the substrate to be coated. A combination of gas-phase reactions and gas-to-solid transformations cause a film to form on the substrate.' This method will be discussed further in

2.7 ELECTRON MICROSCOPY

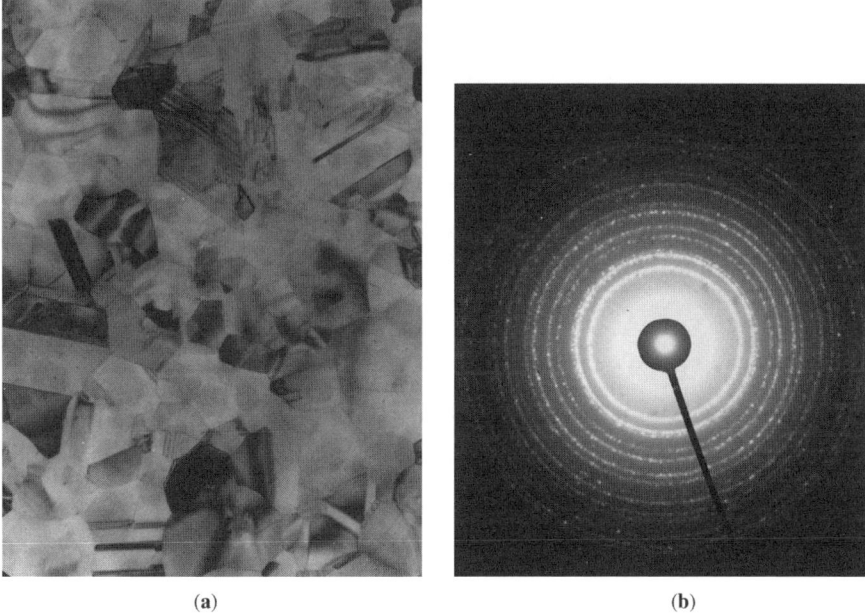

FIGURE 2.28 Polycrystalline Si film: (**a**) bright field (20,000) and (**b**) diffraction.

Chapter 10. The CVD Si film coated wafer was cut to proper sample size and thinned from the Si side leaving a free-standing polycrystalline Si film for TEM examination. Figure 2.28a shows a bright-field TEM image of the film magnified 20,000 times, where the grain size and morphology are revealed, and Fig. 2.28b shows the diffraction pattern for the polycrystalline film sample that will yield the identity of the film. Polycrystalline Si films that are heavily doped are used in microelectronics to define contacts to the gate, source, and drain regions of MOSFET devices. Often the polycrystalline contacts will be exposed to high temperature and oxidizing gases and thus will become oxidized. The oxidation process of polycrystalline Si was studied by successively oxidizing the exposed surfaces of free standing polycrystalline Si films such as shown in Fig. 2.28a ((2)). The results of successive oxidation are shown in bright-field TEM images in Fig. 2.29. Grain boundaries are seen in these images and the darker sharper central region of each grain is the unoxidized Si. As oxidation proceeded, the Si grew smaller while the SiO_2 regions surrounding the Si grew larger. From this study it was concluded that oxidation takes place both on the free Si grain surfaces and also at the grain boundaries, yielding a thickness gradient in the formed SiO_2, yielding thicker SiO_2 in the midgrain regions, and yielding thin oxide at the grain boundaries where lateral oxidation takes place. Also it is seen that after some oxidation time spherical shaped grains are formed (the darker features) imbedded in an amorphous SiO_2 matrix as is shown at higher magnification in Fig. 2.29c. This indicates that the oxidation process at the interface was not dependent on the grain crystallography and therefore the oxidation process is likely diffusion controlled.

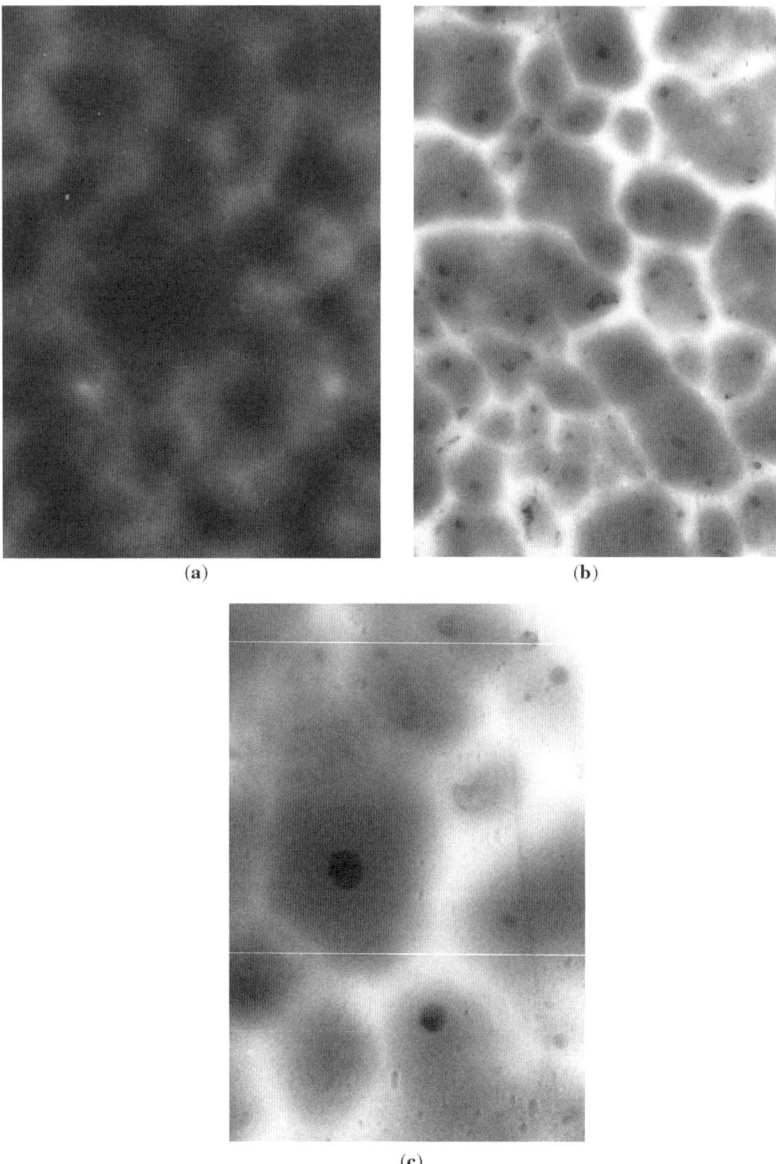

FIGURE 2.29 Progressively oxidized polycrystalline Si films: (**a**) 20,000×, (**b**) 25,000×, and (**c**) 50,000×. [Adapted from Irene et al. (2), Figure 8.]

Figure 2.30 displays TEM results for a study of chemically etched porous Si. Porous Si can be obtained by electrochemically etching Si. The porous Si surface was then mildly oxidized. The images shown in Fig. 2.30 show a free-standing region of the porous Si layer. Figure 2.30a shows a free-standing region defined by a circular aperture. The pores can be seen from close scrutiny of this image.

2.7 ELECTRON MICROSCOPY

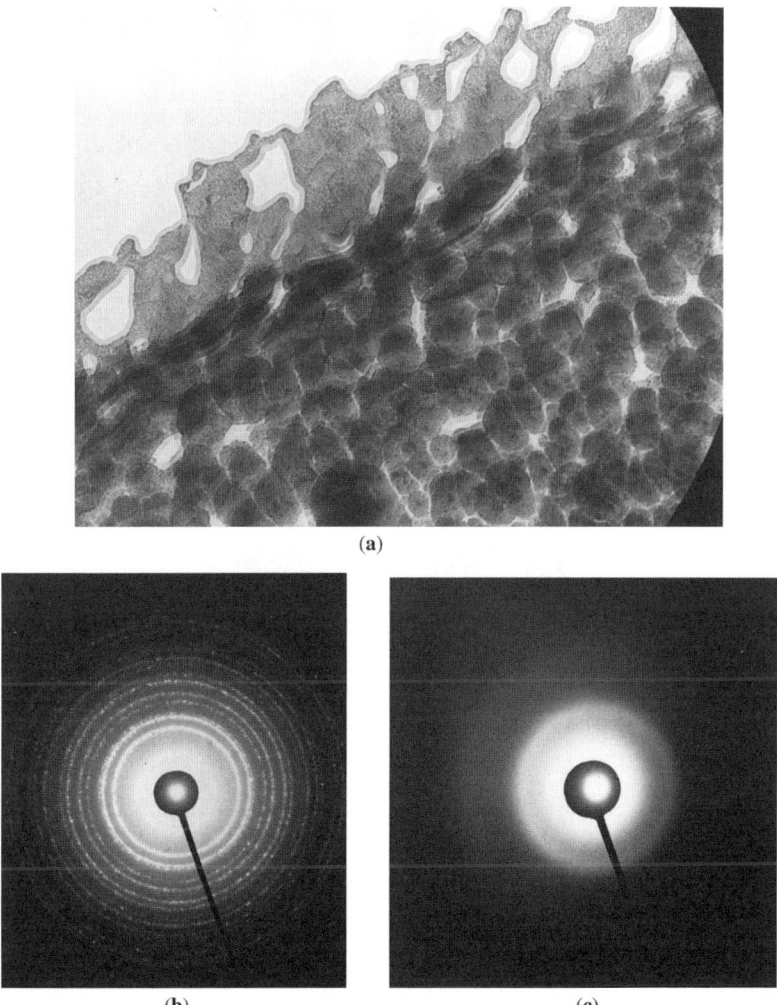

FIGURE 2.30 Porous Si partly oxidized: (**a**) bright–field (50,000×), (**b**) SAD in crystalline area, and (**c**) SAD in amorphous area.

Figure 2.30b shows a selective area diffraction (SAD) defined by an aperture over a predominantly crystalline region near the bottom right portion of the sample shown in Fig. 2.30a. This portion of the sample shows that the material is polycrystalline Si. Figure 2.30c shows an SAD pattern from a region of Fig. 2.30a at the top left of the sample that shows an amorphous diffraction pattern indicative of oxidized Si, but interestingly with the pores intact. Thus, moving to different regions of a sample can yield both morphological and diffraction information.

Figure 2.31 shows CVD-prepared Si-rich SiO_2 films that have been prepared by chemically etching the Si substrate from the backside ((3)). Figure 2.30a shows a bright-field image that yields no real information about the nature of the material.

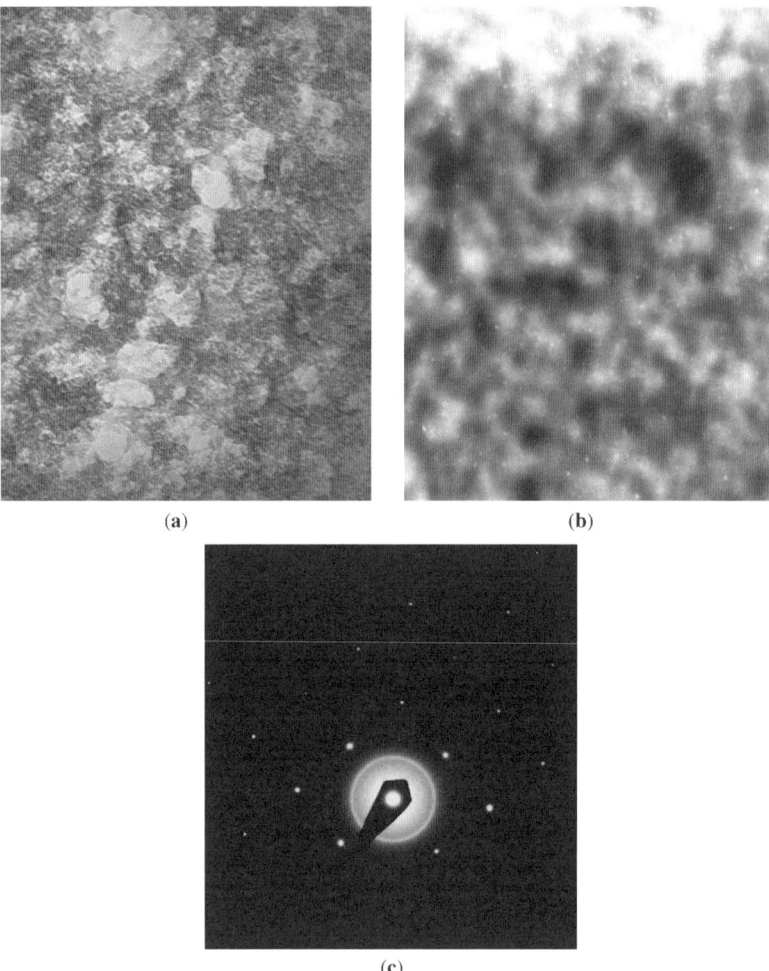

FIGURE 2.31 Si-rich SiO_2 film on Si(100) by CVD and annealed: (a) bright field at 100,000×, (b) dark field at 100,000×, and (c) diffraction. [Adapted from Irene et al. (3), Figures 1 and 2.]

However, Fig. 2.31b shows a dark-field image that upon close examination reveals tiny bright specks of diffracting material. In Fig. 2.31c the diffraction pattern that includes some overhanging single-crystal Si(100) substrate reveals the substrate diffraction and a superimposed polycrystalline ring pattern that is identified as Si. These diffraction rings correspond to the Si crystals imbedded in the SiO_2 matrix shown in Fig. 2.31b. The combination of dark field and SAD yielded important information about the complex multiphase surface film.

The TEM diffraction patterns can be indexed in several ways. One important way when indexing diffraction patterns from the surface structure or from a film on a substrate is to simultaneously obtain the diffraction pattern from the known substrate and the unknown film. This allows comparison under exactly the same instrument conditions. Ideally, one can get both the known substrate pattern and unknown surface or film pattern in the same photographs as was done in Fig. 2.31c. Alternatively, one can obtain photographs of diffraction patterns from known materials and compare them with photographs of unknown material and if possible under the same instrument conditions. Sometimes this can be accomplished by moving the sample to different areas using only the TEM translation stage.

Several XTEM results are now presented to show how both interface and surface information can be obtained. Figure 2.32 shows an XTEM sample thinned by ion milling ((4)). The sample is a $YB_2Cu_3O_7$ (YBCO) the so-called 123 superconductor that was prepared by reactive sputter deposition in an O_2-containing ambient on a single-crystal MgO substrate. Sputter deposition uses a collimated ion beam of heavy ions such as Ar^+ or Kr^+ that are made to impinge on a target that has atoms desired in the film. Atoms from the target are ejected and permitted to condense on an appropriately placed substrate. More on this technique will be discussed in Chapter 10. An MgO substrate was chosen as a relatively inert substrate for YBCO. From the transmitted electron contrast shown in the bright-field image of Fig. 2.32a the interface between MgO and YBCO is clearly seen. Interface and surface roughness can be assessed as well as the columnar large grained morphology of the YBCO film. SAD of the film and of the substrate at the circled region in Fig. 2.32b shows diffraction from both the MgO substrate (shown alone in Fig. 2.32c) and in between those large spots diffraction from the columnar YBCO grains is also seen. No other diffraction is seen attesting to the inertness of the substrate. Figure 2.33a shows a bright-field XTEM image from the ZrO_2 film–MgO interface ((5)). In this study the film thickness from ellipsometry (covered in Chapter 9) of 24.2 nm and from TEM of 25 nm are compared. The interface is

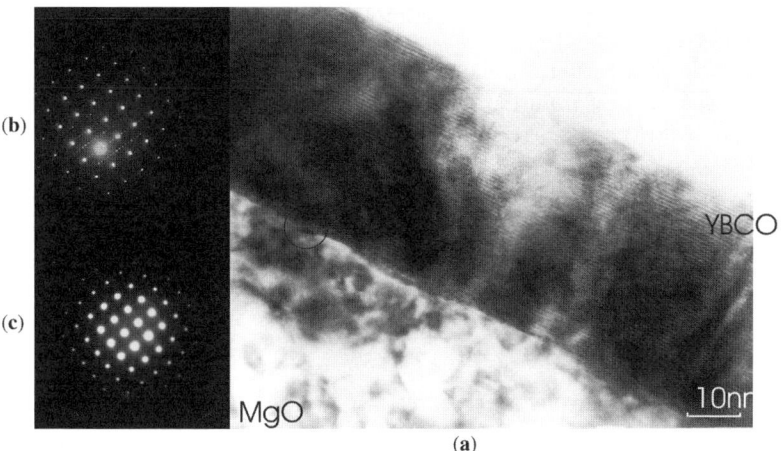

FIGURE 2.32 YBCO (**a**) XTEM, (**b**) SAD in circled area, and (**c**) SAD of MgO substrate. [Adapted from Gao et al. (4), Figure 1.]

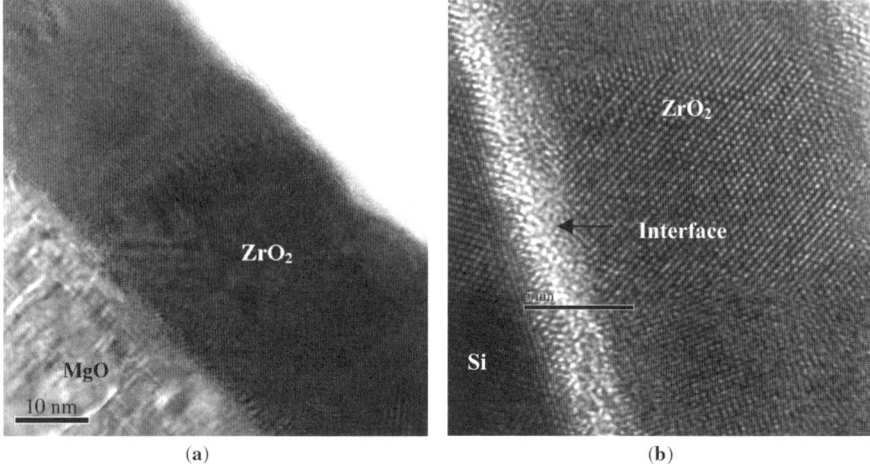

FIGURE 2.33 XTEM images of reactively sputtered ZrO_2 on (**a**) single–crystal MgO and (**b**) single–crystal Si surfaces. [Adapted from Lopez et al. (5), Figure 6.]

sharp and film morphology and roughness are seen. Figure 2.33b shows the same ZrO_2 film but on Si rather than MgO. In this case an amorphous interface formed. In other interface studies the composition of the interface was determined using XTEM ((6)). Figures 2.34a and 2.34b compare sputter-deposited barium strontium titanate ($Ba_{0.5}Sr_{0.5}TiO_3$, BST) films on different initial thickness SiO_2 films all on Si single-crystal substrates. It is seen in this study that after deposition in an O_2-containing ambient the underlying oxide films have the same thickness. This occurs due to subcutaneous Si oxidation resulting from the diffusion of O_2 to the interfacial region during deposition. Another important technique associated with TEM and useful in surface and interface studies is energy-filtered TEM (EFTEM) where electrons cause the emission of characteristic x-rays from the region to be analyzed. The emitted x-rays have an energy characteristic of the element in the sample emitting and, therefore, can be used to indicate the presence of the elements in a film, for example. The contrast from the different elements is seen in Fig. 2.34c and then colors (not shown) can be assigned to one emission or another, and extra contrast can be obtained as seen in Fig. 2.34d.

In the studies excerpted above a wide variety of surface and interface information can be obtained using various manifestations of TEM and sample preparation. Thus, TEM has emerged as a versatile mainstay surface and interface characterization technique.

2.7.2 Scanning Electron Microscopy

Scanning electron microscopy (SEM) uses a similar electron source with filament and anode, but the microscope column, also contained in a vacuum envelope, is simpler and only imaging of the surface occurs, that is, there is no diffraction information

2.7 ELECTRON MICROSCOPY 59

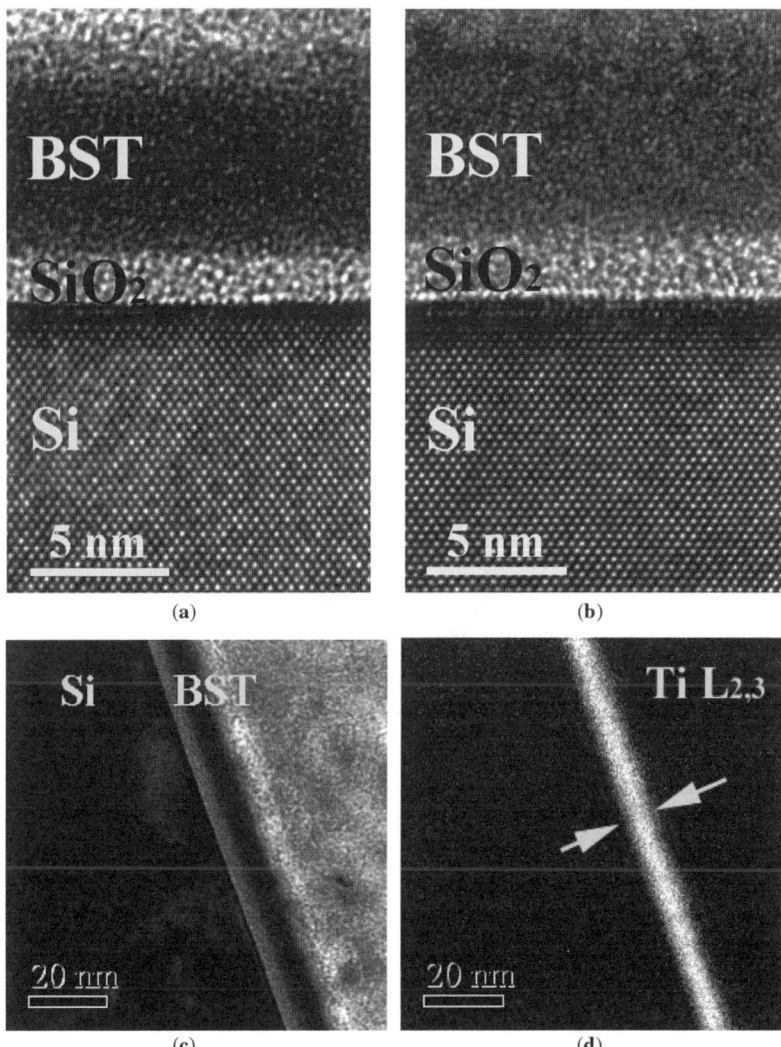

FIGURE 2.34 <110> XTEM of Si(100)/SiO$_2$/(Ba,Sr)TiO$_3$ (BST) correponding to (**a**) 1.0 nm and (**b**) 3.5 nm initial SiO$_2$ layer thicknesses and EFTEM images: (**c**) zero-loss image and (**d**) elemental map showing Ti distribution. [Adapted from Suvorova et al. (6), Figures 2 and 3.]

from a conventional SEM. Although combined scanning and transmission electron microscopes (STEM) are available and can perform many important functions, they are not discussed in this text. Figure 2.35 shows a simplified SEM schematic, the incident electron beam is collimated, and then focused onto the sample surface in a small spot. Then blanking and scanning coils are used to raster the spot across the sample. The signal from each spot on the surface is imaged at the display, keeping the same raster position as the spot on the sample. In this way an image is

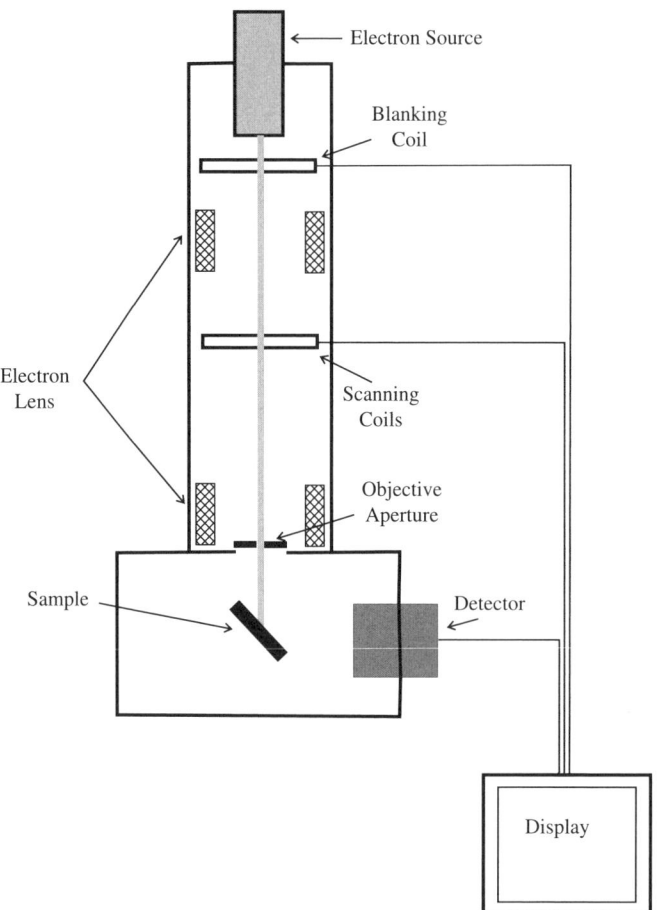

FIGURE 2.35 SEM sketch showing major components.

formed from the interaction of the incident electron beam at each position on the sample surface. Magnification is achieved by rastering over a smaller area and displaying the smaller raster area on the entire screen. Incident beam size and the interaction area at the surface determine the resolution. To form an image two interactions of the beam with the surface are usually used: secondary electron emission or backscattered electrons. Secondary electrons come from the sample surface as a result of collisions with the incident electron beam. The contrast in the image arising from secondary electrons is due to the yield of secondary electrons from each point probed on the sample surface. A particular point on the surface may contain atoms that yield more secondary electrons from an incident electron dose than an adjacent spot, and thus this spot will have a brighter image than the adjacent spot. Therefore, the particular kind of atoms is important. Backscattered electrons depend upon the scattering

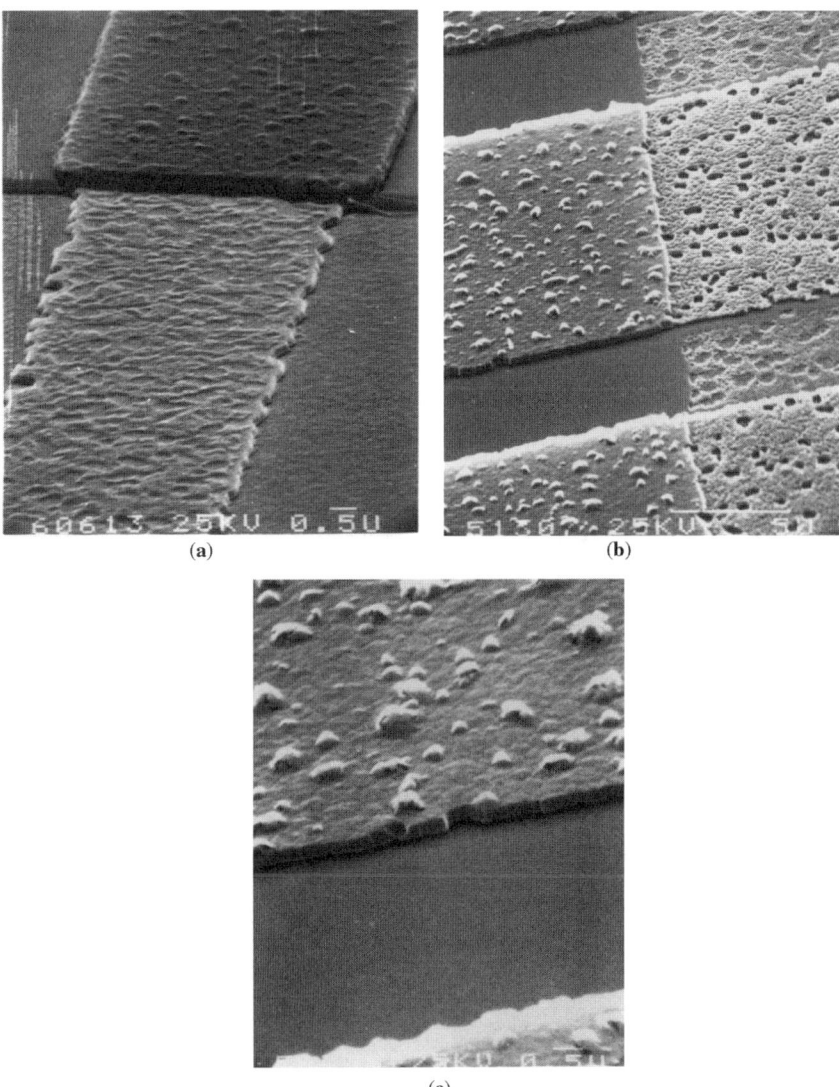

FIGURE 2.36 SEM of doped polycrystalline Si after plasma etching: (**a**) PSG stripped, O_2 anneal; (**b**) PSG in place, O_2 anneal; (**c**) same as (**b**). [Adapted from Irene et al. (7), Figures 4 and 5.]

efficiency of the incident electrons into the detector. Since these modes of forming an image depend on different physics, each may provide different contrast, and it is sometimes useful to compare the images from the two modes. The ideas presented for SEM image formation assume that the sample can conduct electronic charge. If that is not the case, then the sample surface will charge from the beam and the

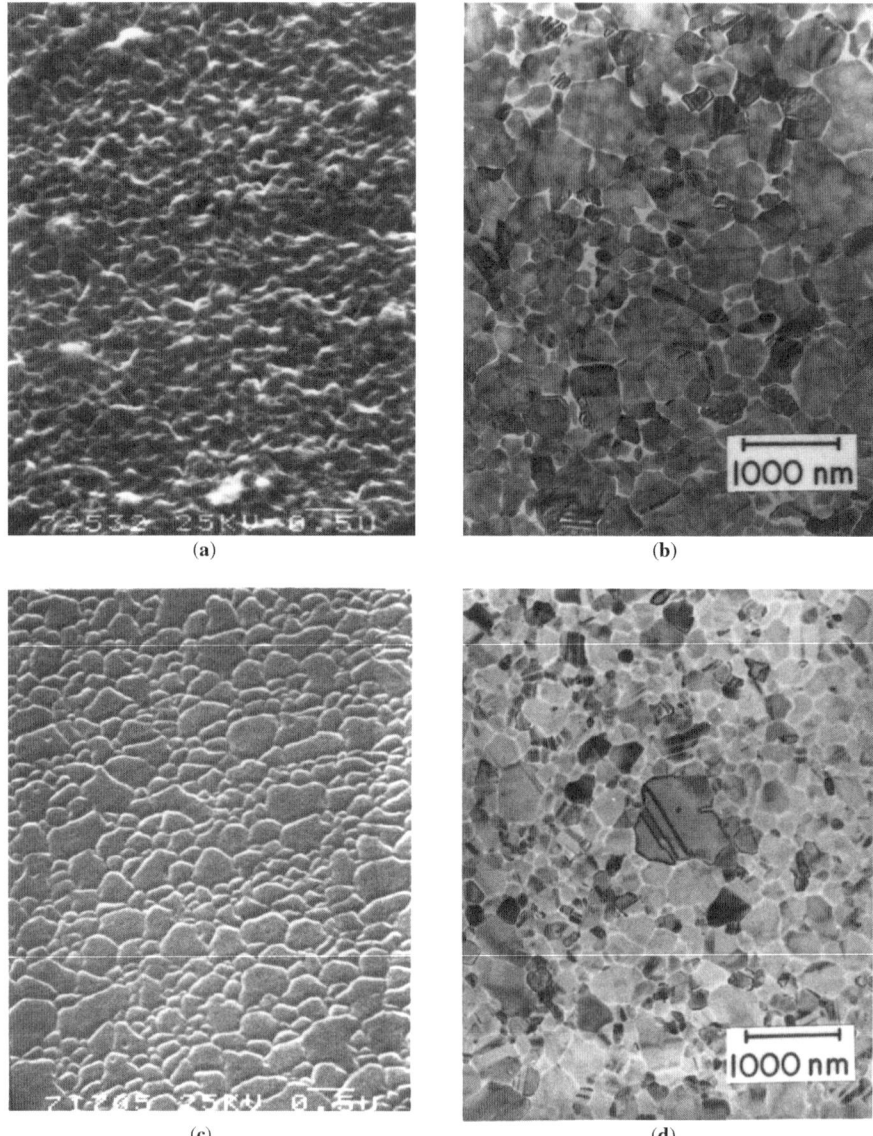

FIGURE 2.37 SEM of poly-Si, $POCl_3$ PSG stripped, O_2 anneal, BHF strip, plasma etch; (b) TEM of (a); (c) SEM of (a) but N_2 anneal; and (d) TEM of (c). [Adapted from Irene et al. (7), Figure 2.]

charging will distort the incident beam and the subsequent image. Modern instruments are equipped with techniques that limit surface charging such as surface clips that can discharge the surface and electron flood guns that render the surface scanned conductive.

2.7.2.1 SEM Results Scanning electron microscopy is most useful to assess surface structures, surface roughness, and surface morphology. Figure 2.36 shows the result of plasma etching polycrystalline Si lines that have been heavily doped with P to render it conductive and thus useful as conducting lines in microelectronics circuits ((7)). Figure 2.36a shows a line that has been doped using phosphosilicate glass (PSG) deposited on the surface. The PSG was removed and then the sample was annealed in O_2, the oxide removed, and then plasma etched. The top part in the figure has been masked. The etching caused roughening and jagged line edges. In Fig. 2.36b the PSG was left in place during the O_2 anneal producing heavier P doping. After plasma etch the unmasked portion of the line (to the right) is much rougher than in Fig. 2.36a. A closeup of Fig. 2.36b for the masked region is seen as Fig. 2.36c where the nature of the roughness is better observed at an angle to observe the surface topography.

Figure 2.37 shows a comparison of SEM at left [(a) and (c)] and TEM [(b) and (d)] on the same samples of polycrystalline Si that has been PSG doped, PSG stripped, annealed, and plasma etched. Figures 2.37a and 2.37b were annealed in O_2 as were the samples in Fig. 2.36, but the sample in Figs. 2.37c and 2.37d were annealed in N_2. Profound differences are observed in the surface topography as is seen in the SEM images in Figs. 2.37a and 2.37c. The grain structure is relatively unaffected as is seen in the TEM images in Figs. 2.37b and 2.37d.

REFERENCES

1. White SJ, Woodruff DP. Surf Sci 1977;63:254.
2. Irene EA, Dong DW, Tierney E. J Electrochem Soc 1980;128:705.
3. Irene EA, Chou NJ, Dong DW, Tierney E. J Electrochem Soc 1980;127:2518.
4. Gao Y, Mueller AH, Irene EA, Auciello O, Krauss AR, Schultz JA. J Appl Phys 1999;86:6979.
5. Lopez CM, Suvorova AA, Saunders M, Irene EA. J Appl Phys 2005;98:033506.
6. Suvorova NA, Lopez CM, Irene EA, Suvorova AA, Saunders M. J Appl Phys 2004;95:2672.
7. Irene EA, Tierney E, Blum JM, Aliotta CF, Lamberti AC, Ginsberg BJ. J Electrochem Soc 1981;128.

SUGGESTED READING

L.C. Feldman and J.W. Mayer, 1986. *Fundamentals of Surface and Thin Film Analysis,* North Holland. An excellent book on surface science techniques where both theory, hardware, and results are presented in a readable manner. This book was often used and referred to in the surface science courses at UNC.

3

THERMODYNAMICS OF SURFACES AND INTERFACES

3.1 INTRODUCTION

As pointed out in Chapter 1, a surface is the boundary of a condensed phase. As such, the bonding and local environment of atoms at a surface are different from those in the bulk material. Thermodynamic properties of materials are a strong function of the chemical bonding. Consequently, it is anticipated that one will find different thermodynamic properties at surfaces than in the bulk. Furthermore, since surfaces provide the first portion of a condensed phase that is exposed to reactants, surface reactions can be different than predicted from thermodynamic and kinetics information derived from bulk materials. The intent of this chapter is to clarify thermodynamic definitions that apply to surfaces and then to apply these definitions to solid surfaces and interfaces.

3.2 SURFACE ENERGY

In Chapter 1 an atomistic approach to surface energy was taken where the differences in bonding between surface and bulk atoms yielded an extra energy term attributed to the surface. Now a different approach is taken where we consider the creation of a surface with a large number of atoms present, a thermodynamic approach. We can imagine a three-dimensional solid with infinite extension in all directions. This solid is represented by only bulk thermodynamic properties. To create a surface from this infinite solid, we need to cleave the material and thereby terminate its

---

Surfaces, Interfaces, and Thin Films for Microelectronics. By Eugene A. Irene
Copyright © 2008 John Wiley & Sons, Inc.

infinite extension in one direction. In this way two surfaces are created and the energy needed to accomplish this task per area of surface created is called surface energy. The surface energy γ is the energy required to create a surface, and it can be expressed in terms of the Gibbs free energy G and area A, as

$$\gamma = \left(\frac{\partial G}{\partial A}\right)_{T,P,n} \tag{3.1}$$

It is useful to consider how this energy is expended to create a surface. For a liquid the thermodynamic process is straightforward to envisage. If a container of water is tipped as shown in Fig. 3.1a, the area of the liquid increases from A_1 to A_2 where it is clear that $A_2 > A_1$. Thus, this simple process creates new surface area at the expense of the energy required to tip the container. This energy divided by the new area produced is γ. The liquid molecules rapidly adjust to the new morphology such that water from below the surface rises to fill in the new surface.

The surface tension or surface stress refers to the work or energy needed to stretch a surface. In the process described above to create a surface from a liquid it is as if the surface were stretched simultaneous with the creation of area, and thus the surface energy and surface tension are the same. Consequently, the terms *surface energy* and *surface tension* are often used interchangeably for liquids. However, for solids the process of creating a new surface is not as straightforward, and typically the surface energy is not equal to the surface tension or stress. Consider Fig. 3.1b in which an isotropic solid is stretched along the x and y coordinates. If a small force per unit area is applied (a stress below the elastic limit for the material), the material

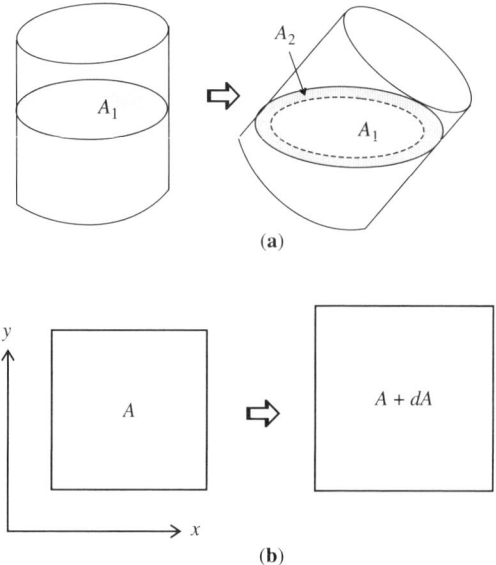

FIGURE 3.1 Creation of new surface area: **(a)** in a liquid and **(b)** in a solid.

3.3 PRINCIPLES OF CAPILLARITY

will elastically deform and increase its surface area. However, after the small force is released the material will elastically return to its original shape leaving no new surface area. During the deformation, the atoms or molecules in the material are not in equilibrium positions. Bonds are stretched and the density will decrease as the material is stretched. If on the other hand a stress exceeding the elastic limit is used, the material will permanently deform, and, if stretched as in Fig. 3.1b, the surface area will be permanently increased. Like the elastic case the chemical bonds will distend and atoms or molecules from other parts of the solid are usually not able to flow into the stretched region and take up the vacated regions as was the case for liquids. Thus, the density will change if the surface and/or volume is changed without adding new atoms or molecules. The process of forming a new surface for solids is clarified by considering the creation of a new surface as a two-step process. In the first step a solid can be cut so as to reveal a new surface with the expenditure of γ and with the same structure and atom/molecule density as the bulk. Then this new surface readjusts (negatively or positively), $d\gamma$, to its new bonding environment that is essentially the result of different interatomic bonding potentials than in the bulk. For a solid the surface tension τ_s is the sum of the surface energy γ that produced a new surface (or that the original surface was stretched) plus a rearrangement term (for either a positive or negative change in area) as the new surface relaxes and can be written as

$$\tau_s = \gamma + A\left(\frac{d\gamma}{dA}\right) \qquad (3.2)$$

For a liquid or even a solid at high temperature where the atoms or molecules can rearrange as the surface is produced, the change in γ with A is zero and the surface energy and surface tension are equal. It should also be noted that if the solid were not isotropic, then the values of the surface tension would be different in different directions, and these quantities would be represented by tensors.

3.3 PRINCIPLES OF CAPILLARITY

Strictly speaking, capillarity refers to the behavior of liquids in capillaries or at the liquid–solid interface where the liquid can readjust to an equilibrium shape. Small capillaries filled with liquid are dominated by the interface interactions since the surface area for the interaction is large. Thus, the emphasis of capillarity is on the thermodynamics of liquids in contact with solid surfaces, and it is found that many of the basic principles of the thermodynamics of solid surfaces and interfaces are revealed from a consideration of capillarity.

3.3.1 Definitions

Since liquids provide a simpler vehicle for understanding surfaces, it is useful to commence a study of the principles of the thermodynamics of surfaces with liquids, and then progress to liquids in contact with solids, and finally solids and

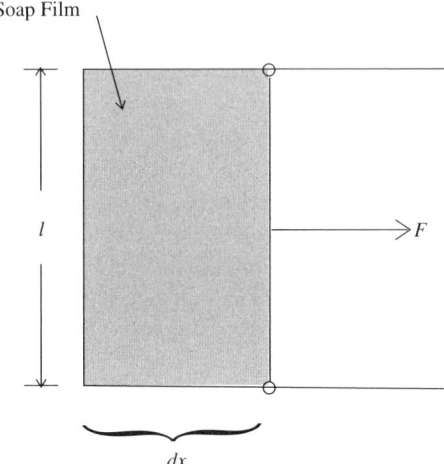

FIGURE 3.2 Wire frame to create surface area in soapy solution. Area produced by the application of the force F is ldx.

solid surfaces. For liquids the surface energy γ or surface tension (since these quantities are equal) has units of energy per area (J/m², ergs/cm², dynes/cm are commonly found units for $\dot{\gamma}$).

As discussed in Chapter 1, energy is needed to create a new surface, hence γ is positive. A classical way to produce a surface is to create a soap film on a wire frame as is shown in Fig. 3.2 where a force F is applied to the movable wire through a distance dx from which we can describe the process of creating the soap film in terms of the work w as

$$dw = F\,dx = \gamma l\,dx = \gamma\,dA \tag{3.3}$$

This implies that γ is the change in work with the area produced. This is the same as the definition for γ given above, and we can obtain the specific relationship by considering the following thermodynamic development.

Starting from definitions of the Gibbs free energy G, the enthalpy H, and the internal energy E with S the entropy, T the temperature, q the heat, w the work (negative for work done by the system), and P and V pressure and volume, and no change in the number of moles n ($dn = 0$), we can obtain an expression for the free energy starting from the definitions as follows:

$$G = H - TS \qquad H = E + PV \qquad E = q + w \qquad dw = -P\,dV \tag{3.4}$$

Then the derivative of G is expressed as

$$\begin{aligned} dG &= dq - P\,dV + P\,dV + V\,dP - T\,dS - S\,dT \\ &= dq + V\,dP - T\,dS - S\,dT \end{aligned} \tag{3.5}$$

3.3 PRINCIPLES OF CAPILLARITY

Now using the expression for the change in S, $dS = dq/T$, Equation 3.5 becomes

$$dG = V\, dP - S\, dT \tag{3.6}$$

If we also include the effect on dG of a possible change in the number of moles, we obtain

$$dG = V\, dP - S\, dT + \sum_i \mu_i\, dn_i \tag{3.7}$$

where μ_i is the chemical potential for the ith component, and the expression derives from the definition of the chemical potential for the ith species given by Equation 3.8:

$$\mu_i = \left(\frac{\partial G}{\partial n_i}\right)_{T,P} \tag{3.8}$$

Now if we include all other possible work terms such as electric work and work against gravity, we simply add these terms to Equation 3.7, the formula for dG and obtain

$$dG = V\, dP - S\, dT + \sum_i \mu_i\, dn_i + \sum(\text{all other work}) \tag{3.9}$$

For our present purpose we desire to add only the work required to produce the soap film shown in Fig. 3.2, or $dw = \gamma\, dA$, and thus obtain

$$dG = V\, dP - S\, dT + \sum_i \mu_i\, dn_i + \gamma\, dA \tag{3.10}$$

Now for the case where T, P, and n are constant, we obtain the desired result as was stated in Equation 3.1:

$$\left(\frac{\partial G}{\partial A}\right)_{T,P,n} = \gamma \tag{3.1}$$

3.3.2 Curved Surfaces

Most real solid surfaces are not perfectly flat. The issues associated with how to characterize nonflat or rough surfaces is covered in Chapter 4. Here we consider the effect of roughness on thermodynamic properties. We commence with a nearly perfect spherical surface that can be obtained from a small soap bubble (gravity distorts the shape of the bubble, thus a small bubble is a more perfect sphere). A soap bubble is formed by blowing air into a soap film allowing the film to inflate and close upon itself so that a pressure difference, ΔP, exists between the inside and outside of

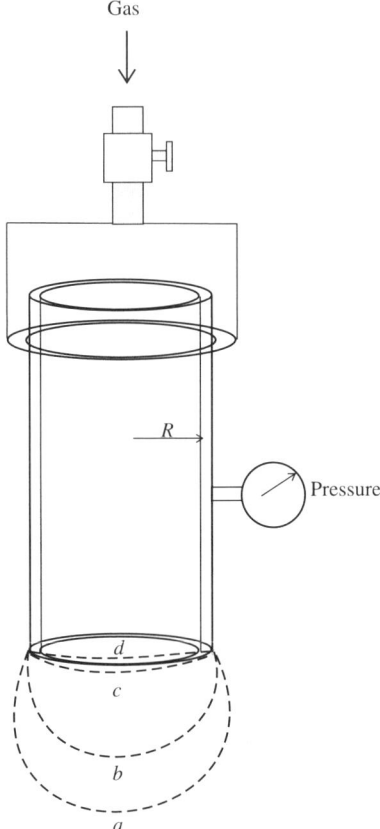

FIGURE 3.3 Capillary tube of radius R with gas inlet at the top and bubble forming at the bottom with various radii (a–d) as a function of gas pressure.

the bubble with a higher pressure inside to keep the bubble from collapsing. This higher pressure inside overcomes the surface tension of the film that would reduce the area. Consider the bubble in Fig. 3.3 expanding from an initial radius r to $r + dr$. For the initial bubble the surface energy is given as

$$G = \gamma 4\pi r^2 \tag{3.11}$$

The change in the free energy with dr is given as

$$dG = \gamma 8\pi r \, dr \tag{3.12}$$

If r changes, then ΔP must change ($\Delta P = P_{in} - P_{out}$). For example, to make a bigger bubble with more area, we need to blow more air inside the bubble. Now there must

3.3 PRINCIPLES OF CAPILLARITY

be an equivalence between the change in the surface energy (work) from the change in r or $8\pi r\gamma\, dr$ and the work done in the change in the pressure that changes the size of the bubble or $\Delta P\, 4\pi r^2\, dr$. At equilibrium this expression must be equal to Equation 3.12, and this equivalence is written as

$$\Delta P\, 4\pi r^2\, dr = 8\pi r\gamma\, dr \tag{3.13}$$

The result is the so-called Young–Laplace equation:

$$\Delta P = \frac{2\gamma}{r} \tag{3.14}$$

Equation 3.14 teaches that a larger ΔP is needed to establish a given bubble radius for higher materials. For a general nonspherical curved surface the formula can be expressed with two axes defining the curvature as

$$\Delta P = \gamma\left(\frac{1}{r_1} + \frac{1}{r_2}\right) \tag{3.15}$$

The implications of Equation 3.14 (or 3.15) as well as a method to measure γ can be explored with the help of a thought experiment depicted in Fig. 3.3. A capillary tube of radius R is connected through a valve to a source of an inert gas such as Ar. The other end of the capillary tube is covered with a soap film. The pressure in the capillary can be measured with the pressure gauge. With the outside pressure of 1 atm and the inside pressure measured with the gauge, ΔP can be obtained from the formula above, $\Delta P = P_{in} - P_{out}$. Figure 3.3 shows four different radii (r) for the spherically curved soap film. Case a has $r \gg R$, case b shows $r > R$, case c shows $r = R$, and case d shows the soap film tight across the mouth of the capillary. The radius of this last surface is the largest of all $r \ggg R$. Using the Young–Laplace equation (Equation 3.14), we can estimate relatively what the pressure gauge should read for P_{in}. To maintain an outwardly curved surface, the inside pressure P_{in} must always be greater than the outside atmospheric pressure P_{out}. For case a for large r, ΔP is smaller than cases b or c where r decreases from a to c. The minimum radius occurs at $r = R$ for case c, which yields the largest ΔP and the smallest for case d, where r is largest. This experiment suggests a method to obtain the surface energy from a measurement of R and P_{in}. The radius of the bubble, r, will be equal to that of the capillary when P_{in} is a maximum. Thus, the bubble size is changed with P_{in} recorded. For the maximum value of P_{in}, $r = R$ and γ is obtained from the following formula:

$$\gamma = \frac{R\, \Delta P_{max}}{2} \tag{3.16}$$

Another way to measure γ is the method of capillary rise. Figure 3.4a shows a thin glass tube of small radius or capillary immersed in water. Water wets the glass. That means that water will cover or adhere to the surface of glass. The result will

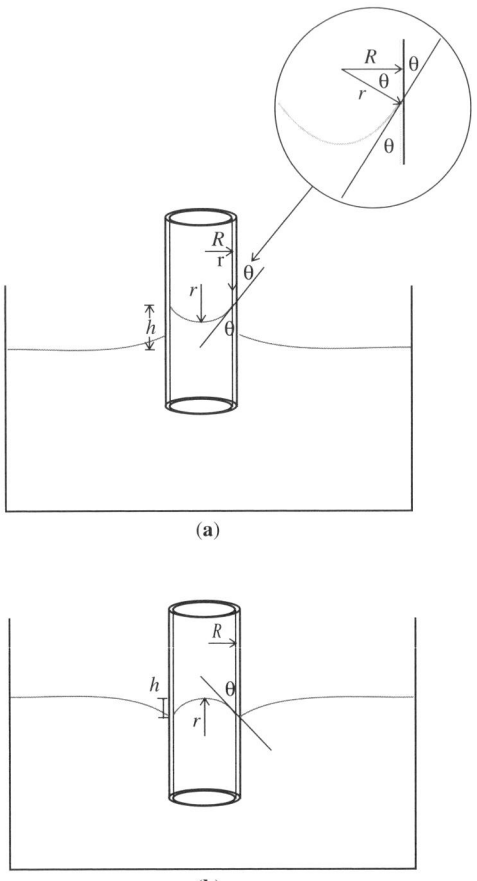

FIGURE 3.4 Capillary with radius R immersed in liquid: (**a**) where the interaction is hydrophillic, forming concave upward meniscus (inset shows closeup of contact angle θ and (**b**) where the interaction is hydrophobic, forming concave downward meniscus.

be that a concave upward boundary of the water–air interface will form as shown in Fig. 3.4 as the water adheres to the glass at the capillary inner walls. This interface is called a meniscus. The radius of the capillary is R and the radius of the meniscus is r. The meniscus makes an angle θ with the inner tube wall. Because of the adhesion of the water and glass, the water will rise a distance h in the capillary tube. Using the Young–Laplace formula above (Equation 3.14) for a symmetrical (spherical) interface, ΔP is the pressure difference across the interface and γ is the surface tension/energy for the water. From an observation of the shape of the meniscus, the ambient pressure is greater than the pressure of the liquid in the capillary:

$$P_{\text{ambient}} > P_{\text{liquid}} \tag{3.17}$$

3.4 SURFACE ENERGY OF SOLIDS

Furthermore, the height of the water in the capillary provides a measure of the pressure difference so that we can write for water with a density ρ and the acceleration of gravity g:

$$\Delta P = \rho g h = \frac{2\gamma}{r} \qquad (3.18)$$

Now the angle θ that the meniscus makes with the capillary can be expressed in terms of the radii as

$$\cos\theta = \frac{R}{r} \qquad (3.19)$$

Then the pressure difference can be written in terms of known or measured quantities:

$$\Delta P = \rho g h = \frac{2\gamma}{r} = \frac{2\gamma \cos\theta}{R} \qquad (3.20)$$

So from the experiment in Fig. 3.4a, a measurement of h for a known liquid (where ρ is known) yields ΔP, and with a known capillary radius R and measured contact angle θ, the surface energy of the liquid γ is calculated. Of course, with different liquids different contact angles are expected. When θ exceeds $90°$, $\cos\theta$ becomes negative, and the shape of the meniscus would be concave down and is called a negative meniscus, and h would be below the water level. In this case the liquid does not wet the glass. This case is shown in Fig. 3.4b. The case where water wets the glass surface and yields the positive meniscus as shown in Fig. 3.4a yields a so-called hydrophilic solid surface while the surface where water does not wet is called hydrophobic (such as a Teflon surface). We will return to more capillarity later after a brief discussion of the surface energy of solids.

3.4 SURFACE ENERGY OF SOLIDS

In the previous sections of this chapter definitions and measurements mainly of γ for liquids was considered. Now we return to the more complicated case of solids and consider first a model for surface energy for the simple solid shown in Fig. 1.1b. We consider that an atom in a solid has n number of bonds per area and each with bond energy Ω. If the solid is cut to produce two fresh surfaces, the surface energy can be written as

$$\gamma = \frac{n\Omega}{2} \qquad (3.21)$$

Now if we make the calculation specific and consider the cubic solid in Fig. 1.1b where each atom is approximated by a cube with six nearest neighbors and considering only the nearest neighbor bonding n has the value of 6/(sum of area of the sides of the cubes), which is $6/6a^2$, where a is size of the square that contains the bond, and we obtain for the surface energy:

$$\gamma = \frac{n\Omega}{2} = \frac{\Omega}{2a^2} \tag{3.22}$$

Furthermore, the bond energy Ω can be related to the latent heat of sublimation, ΔH_s, in that as atoms leave the solid to the vapor, bonds are broken. The use of ΔH_s for Ω partially accounts for any energy changes that occur upon reorganization of a newly formed solid surface because that would contribute to ΔH_s. For the number of bonds needed to be broken to remove an atom, which is 6 for our simple solid in Fig. 1.1b, and considering that for every bond broken two atoms are affected, the bond energy Ω can be expressed as

$$\Omega = \frac{2\Delta H_s}{6} \tag{3.23}$$

Now γ can be written in terms of ΔH_s as

$$\gamma = \frac{\Omega}{2a^2} = \frac{\Delta H_s}{6a^2} \tag{3.24}$$

It is useful to compare this calculation for a flat or singular surface with that from a stepped or vicinal surface such as that shown in Fig. 3.5. For this case we need to consider the contribution to n the number of bonds broken from both the steps and risers. For the steps there is one bond per a^2 as for the singular surface case above. The risers collectively yield a misorientation that is quantified in Fig. 3.5 by the angle ξ. Figure 3.5 is similar to Fig. 1.2 but with higher risers for illustrative purposes. The contribution from a riser is proportional to h and hence to ξ. The height h for each riser can be expressed as

$$\tan \xi = \frac{h}{a} \tag{3.25}$$

The number of bonds from a riser is then calculated from the product of a and h that yields the number per area or ah/a^2, which is h/a or $\tan \xi$. The total number of bonds, n, is then given by the sum of the contributions from the steps and risers:

$$n = \frac{1 + \tan \xi}{a^2/\cos \xi} = \frac{\cos \xi + \sin \xi}{a^2} \tag{3.26}$$

3.4 SURFACE ENERGY OF SOLIDS

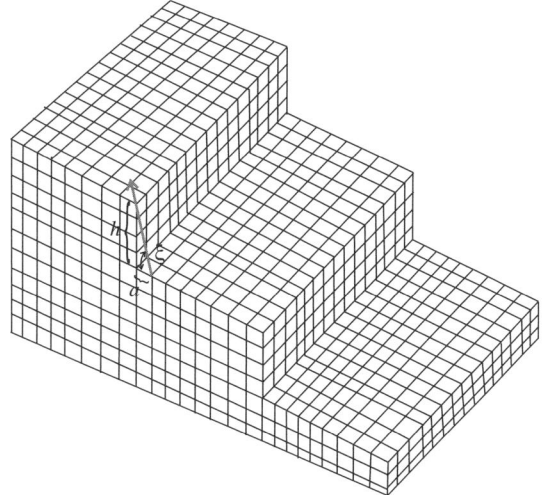

FIGURE 3.5 Stepped surface structure with ξ defined to account for area contributions from steps.

Notice the area for the vicinal surface as compared to the singular surface is increased by the projection angle ξ, hence a^2 is divided by $\cos \xi$ in Equation 3.26. Then as was done above with the general expression for γ, the number of bonds n, and that $\Omega = \Delta H_s/3$ (Equation 3.23), we can obtain γ for this vicinal surface as

$$\gamma = \frac{\cos \xi + \sin \xi}{a^2} \frac{\Delta H_s}{6} \tag{3.27}$$

Now note that the difference between γ for the singular and vicinal surfaces is the factor $\cos \xi + \sin \xi$ term in the numerator for the vicinal surface. This factor is always >1 for any angle $\xi > 0°$ and $<90°$. Hence the vicinal surface has a higher surface energy than its singular surface counterpart. When $\xi = 0$, the surface is singular.

The factor $\cos \xi + \sin \xi$ indicates that the surface energy varies with orientation. This can be understood by referring again to the formula above for γ for vicinal surfaces. Figure 3.6 shows a plot of this function. Cusps are seen at $\xi = 0°, 90°, 180°,$ and $270°$ indicating smaller γ values for the major low index planes and larger γ values for the planes in between. Of course, Fig. 3.6 and formula 3.27 are based on the simplistic model for solids considering nothing other than bond energy and nearest-neighbor interactions. While this behavior may be closely followed for some materials, it by no means represents the majority of materials, which would show many more cusps. Also as one extends even this simple model to other directions more cusps will occur. Nevertheless, the simple model yields a level of understanding that can then be applied to more complex materials.

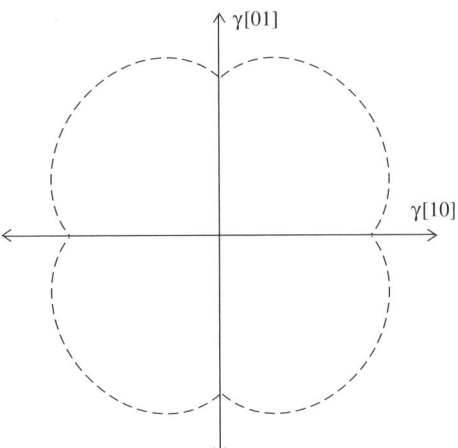

FIGURE 3.6 The γ plot for a cubic material.

Another approach to obtain values for γ of covalent solids is that of Harkin (1) in which everything but bond energy is neglected. Covalent solids best exemplify this approximation since the bond energy is large and therefore dominant and where reconstruction if any is small. Figure 3.7a shows the side view of the bonding at the (111) of diamond. Notice that to form a new surface by cleavage at this plane, bonds oriented normal to the (111) plane need to be broken. To calculate γ, we first determine the number of bonds per area and multiply this by half of the bond energy since two surfaces are formed via cleavage. Figure 3.7b shows a view of the (111) with equal shaded and cross-hatched triangle areas. The central shaded triangle area has one broken bond associated with the bond oriented normal to it and centered above this shaded triangle. However, that atom with the broken bond is also associated with one-third of each of the cross-hatched triangles for a total area of two of the triangles associated with the one broken bond. Figure 3.7b also shows the lattice parameter for the diamond unit cell of $a_0 = 0.3560$ nm and other cell dimensions readily obtained from the lattice parameter and the unit cell symmetry. In parentheses in Fig. 3.7b are values for the lattice parameter or other cell dimensions for Si, which will be discussed below. For diamond the area of the shaded triangles is calculated as

$$A = \frac{bh}{2} = \frac{(0.2517)^2 \sin 60°}{2} = 0.02743 \text{ nm}^2 \qquad (3.28)$$

For two triangles the total area is 0.05487 nm². Now from Fig. 3.7b there is one broken bond for two shaded triangular areas and yields the number of bonds $n = 1.82 \times 10^{15}$ cm^{-2} bonds converting area from nanometers squared to centimeters squared. For diamond the approximate bond energy Ω is 90 kcal/mol and converted

3.4 SURFACE ENERGY OF SOLIDS

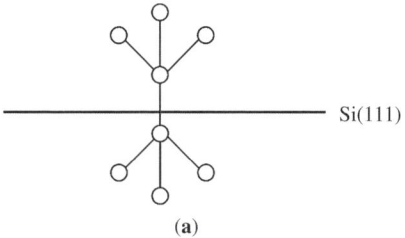

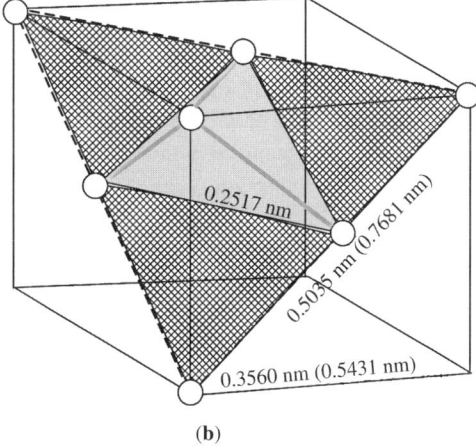

FIGURE 3.7 (a) Side view of bonding at a Si(111) surface and (b) the area associated with the bond in (a).

to ergs (4.18×10^{10} ergs/kcal) yields about 3.76 ergs/mol. Now with $\gamma = n\Omega/2$ (from Equation 3.21) we obtain

$$\gamma = \frac{(3.76 \times 10^{12} \text{ ergs/mol})(1.82 \times 10^{15} \text{ bonds broken/cm}^2)}{2N_A}$$

$$= 568 \text{ ergs/cm}^2 \tag{3.29}$$

where N_A is Avagadro's number (6.023×10^{23} atoms/mol). Since no experimental values for the surface energy for diamond are available, this calculated value cannot be checked. However, for Si(111) some values are available, and Si is also a covalently bonded solid with the diamond cubic structure, and thus the calculation above can be readily performed by merely using the numbers in parentheses in Fig. 3.7. Thus, for Si(111) that has a bond energy $\Omega = 42.2$ kcal/mol or 1.77×10^{12} erg/mol and an atom density, hence a broken bond number $n = 7.83 \times 10^{14}$, γ is given as

$$\gamma = \frac{(1.77 \times 10^{12} \text{ ergs/mol})(7.83 \times 10^{14} \text{ bonds broken/cm}^2)}{2N_A}$$

$$= 1148 \text{ ergs/cm}^2 \tag{3.30}$$

The experimental value (2) for cleaving a Si crystal to reveal the Si(111) is $\gamma[Si(111) = 1230\, ergs/cm^2]$, which is reasonable agreement considering the assumptions that were made. For materials that are not as covalent as diamond or Si, the problem is much more complicated and involves contributions for several levels of neighboring atoms. Nevertheless, these approximations isolate one of the essential ingredients that comprise the surface energy for solids, namely the bonding at the surface.

Besides simply cleaving a solid and measuring the energy required to create the new surfaces, another interesting method has been reported using metal wires (3). As shown in Fig. 3.8a several wires of a metal with the same radii (r) have different weights attached. The wires are heated to the point where the atoms have considerable mobility. This is typically about 90% of the melting point for metals. This enables the metal atoms to relax as new surface is created by the stretching of the wires, much like the earlier experiment where a beaker of water is tipped to create a new surface. Now there are two opposing forces acting on the wires. One is the surface energy driven force to minimize the surface area that will cause the wires to shrink in area and is given as $\gamma\, dA$. The second is the force exerted by the weights of mass m to lengthen the wires by dl in gravity (g) and create new surface, and this is given as $mg\, dl$. A plot of Δl versus the weight applied as shown in Fig. 3.8b will enable the location of the null point of $\Delta l = 0$ where

$$mg\, dl = \gamma\, dA \qquad (3.31)$$

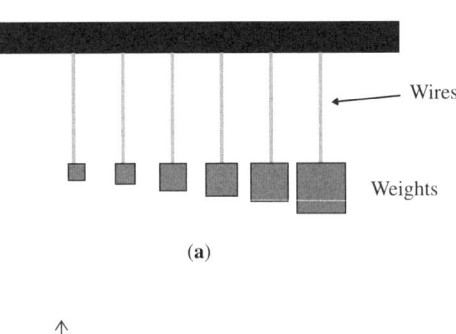

(a)

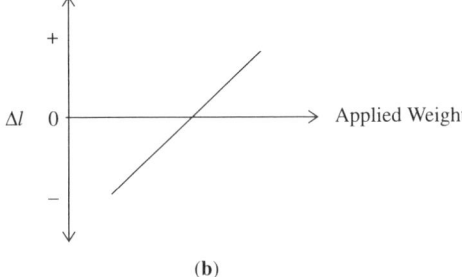

(b)

FIGURE 3.8 Method to obtain γ for metal wires: (a) sketch of experiment and (b) plot of the wire elongation Δl versus the applied weight.

3.5 INTERFACES AND MORE CAPILLARITY

which is where the forces are equal. Then with r as the radius of the wires this equation can be solved for γ using expressions for the area A and the volume V for the wires as follows:

$$A = 2\pi r l \quad \text{and} \quad dA = 2\pi(r\, dl + l\, dr) \tag{3.32}$$

also

$$V = \pi r^2 l \quad \text{and} \quad dV = 0 = 2\pi l r\, dr + \pi r^2\, dl \tag{3.33}$$

This yields an expression for dr:

$$dr = -\frac{r}{2l} dl \tag{3.34}$$

This can be substituted for dA to yield

$$dA = \pi r\, dl \tag{3.35}$$

Then the force equivalence Equation 3.31 becomes

$$mg\, dl = \gamma \pi r\, dl \tag{3.36}$$

From this the surface energy is obtained from known or measurable quantities (m and r at $\Delta l = 0$):

$$\gamma = \frac{mg}{\pi r} \tag{3.37}$$

3.5 INTERFACES AND MORE CAPILLARITY

Now with γ values for liquids (L) and solids (S) we consider the liquid–solid–vapor (L–S–V, L–S, and S–S) interfaces. Starting from Equation 3.10:

$$dG = V\, dP - S\, dT + \sum_i \mu_i\, dn_i + \gamma\, dA \tag{3.10}$$

which for T, n, and A constant yields

$$dG = V\, dP \tag{3.38}$$

where V is the volume for a mole of substance, or the molar volume. Using Equation 3.38 we consider the solid–vapor (S–V) interface where a vapor is in equilibrium

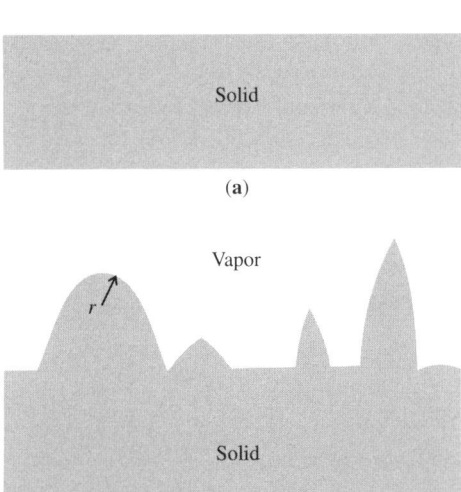

FIGURE 3.9 (a) Flat interface where r is infinite and (b) a rough interface where the radius of the features is variable and finite.

with its solid as is shown in Fig. 3.9a. For a vapor that can be considered to be an ideal gas so that the molar volume is given by the ideal gas law $V = RT/P$, we can integrate Equation 3.38 and obtain ΔG_V for the vapor from some initial pressure, P_0, to a final pressure, P as

$$\Delta G_V = \int_{P_0}^{P} \frac{RT}{P} dP = RT \ln \frac{P}{P_0} \tag{3.39}$$

For a solid, the molar volume for the solid V_S is nearly unchanged with a change in P at least relative to a vapor and thus can be approximated as a constant to yield:

$$\Delta G_S = V_S \Delta P \tag{3.40}$$

However, now we have an expression for ΔP from the Young–Laplace equation: $\Delta P = 2\gamma/r$ and inserting this into Equation 3.40 and considering S–V equilibrium, (viz. $\Delta G_S = \Delta G_V$), we obtain a very important equation, namely the Kelvin equation:

$$\frac{2V_S \gamma}{r} = RT \ln \frac{P}{P_0} \tag{3.41}$$

The Kelvin equation teaches that the vapor in equilibrium with the solid is a function of the radius of curvature for the S–V interface. For example, a flat solid surface such

3.5 INTERFACES AND MORE CAPILLARITY

as shown in Fig. 3.9a, which has a flat interface, $r = \infty$ and therefore $P = P_0$, the equilibrium vapor pressure. However, at a surface roughness feature such as shown in Fig. 3.9b that is not flat, $r < \infty$ and thus $P > P_0$, the vapor pressure at the rough feature is increased above the flat surface value. Consider having a surface with roughness features distributed along with flat regions in between as seen in Fig. 3.9b. If this surface at some temperature T is in equilibrium with its vapor at all points on the surface, the vapor pressure at all points is not the same. The Kelvin equation indicates that the vapor pressure above the sharpest features, those with the smallest r's, is the highest. Thus, heating such a surface will eventually lead to smoothing of the surface where the sharpest features vaporize at the expense of the less sharp features. In the next chapter on roughness we will return to this point and show that this smoothening effect is also the manifestation of reducing the surface-to-volume ratio discussed in Chapter 1. Another important conclusion that derives from the Kelvin equation is that the free energy for the rough features is greater than smooth regions, and consequently the chemical reactivity is also greater as is the melting point lower at the rough features.

In the discussion above two-phase equilibrium was considered. Now we turn attention to three-phase equilibrium among three unspecified condensed phases, α, β, and δ as shown in Fig. 3.10a. The drawn tangent vectors show the magnitude and direction for the interfacial energies (units of force/area), γ_{ij}, that exists between any two phases. At equilibrium the sum of the vectors must equal 0 as

$$\sum_{i,j} \gamma_{ij} = 0 = \gamma_{\alpha\beta} + \gamma_{\alpha\delta} + \gamma_{\beta\delta} \tag{3.42}$$

Figure 3.10b shows the application of the three-phase equilibrium relationship to the S–L–V equilibrium. For this situation the equilibrium yields the following equation:

$$\gamma_{LV} + \gamma_{SV} + \gamma_{SL} = 0 \tag{3.43}$$

Also, from Fig. 3.10 the following important relationship is obtained:

$$\gamma_{SV} = \gamma_{SL} + \gamma_{LV} \cos\theta \tag{3.44}$$

and written in terms of the contact angle θ that is measurable, it is called the Young equation:

$$\cos\theta = \frac{\gamma_{SV} - \gamma_{SL}}{\gamma_{LV}} \tag{3.45}$$

It is observed that for $\theta < 90°$ $\cos\theta$ is positive (+) and $\gamma_{SV} > \gamma_{SL}$, and the liquid wets the solid and is called a hydrophobic interaction. For systems where $\theta > 90°$

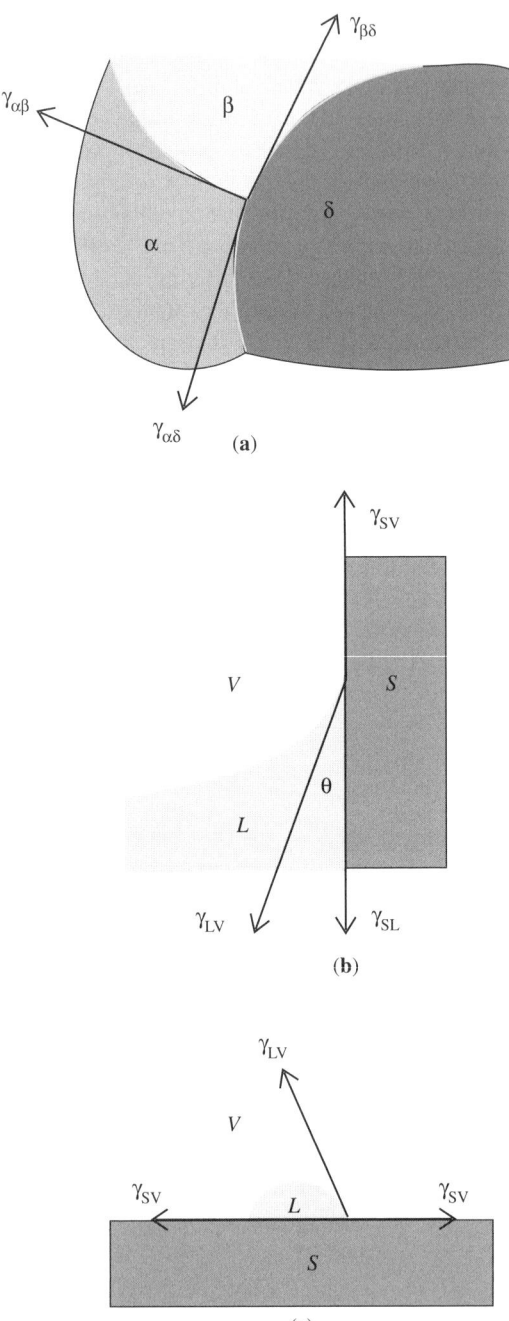

FIGURE 3.10 Three phase equilibria: (**a**) three condensed phases, (**b**) S–L–V equilibrium, and (**c**) also S–L–V equilibrium but with L a drop of liquid.

3.6 SiO$_2$–Si INTERFACE APPLICATION

$\cos\theta$ is negative (−) and $\gamma_{SV} < \gamma_{SL}$, and the liquid does not wet the surface and this interaction is called a hydrophobic interaction.

It is interesting to apply these ideas to the water–Si interaction, and the water–Teflon interaction. The surface energy for water is about 72 ergs/cm^2 while that for Si(111), as mentioned above, is $\gamma \cong 1200$ ergs/cm^2. Thus, if water covered the Si surface, the newly formed surface would have a surface energy closer to that for water than for Si, since the exposed surface would be water. This is a favored thermodynamic interaction that results in a lower surface energy, and consequently water would be expected to wet Si and many other high surface energy solids. On the other hand with γ for Teflon less than 20 ergs/cm^2, water would not be expected to wet Teflon and it does not.

Figure 3.10c shows the same situation as Fig. 3.10b but for a drop of liquid on a surface. The solid surface is exposed to vapor. It is likely that at least a monolayer of vapor would absorb on the solid surface. Also it is possible that a monolayer (or more) of atoms or molecules from the liquid would spread on the surface of the solid. This coverage of the solid surface would alter the surface energy for the solid so that $\gamma_S \neq \gamma_{SV}$. This fact can be expressed in terms of another variable called the film pressure, π, that is defined as

$$\pi = \gamma_S - \gamma_{SV} \tag{3.46}$$

Then from the Young equation above, we can write

$$\gamma_{LV} \cos\theta = \gamma_{SV} - \gamma_{SL} = \gamma_S - \gamma_{SL} - \pi \tag{3.47}$$

This formula explicitly shows the difference in the surface energy for the solid from that of a solid covered with another moiety.

The ideas developed are useful for estimating relative γ values, For evaluating interactions among phases at interfaces, for exploring nucleation, and for understanding surface roughness phenomena and a specific application will be discussed below and other applications in future chapters.

3.6 SiO$_2$–Si INTERFACE APPLICATION

As was mentioned in Chapter 1 the interface formed by the oxidation of Si, the SiO$_2$–Si interface, is perhaps the most technologically important interface in the area of electronic materials and microelectronics (see Chapters 11 and 12 for more specifics on this interface). Also the metal–oxide–semiconductor field-effect transistor (MOSFET) is the main microelectronics device in computers, and the SiO$_2$–Si interface dominates in the operation of the device. More will be discussed about this device and the Si–SiO$_2$ interface throughout this book. For the present it is sufficient to understand the multiple roles of SiO$_2$ to act as a dielectric to support an electric field to operate the MOSFET device and to provide electronic passivation of the Si surface. Electronic passivation refers to the reduction of electronic states at the surface, so-called surface states, to sufficiently low levels to enable controlled electron or hole conduction at the Si surface. Controlled conduction is not possible at a bare Si surface due to the large number of surface electronic states, and these states comprise the subject matter for Chapters 5 and 11. However, upon oxidation of the Si surface with the formation of the SiO_2–Si interface, the number of surface states (now interface states) is reduced by about 5 orders of magnitude, and this is sufficient to render the interface electronically passivated. Consequently, the formation and nature of the SiO_2–Si interface is at the heart of MOSFET science and technology.

Throughout the fabrication of Si-based MOSFETs, SiO_2 is grown on the Si surface and removed from selected areas so as to define the various device regions. SiO_2 is usually removed by wet etching in an aqueous HF solution. Contact angle measurements as discussed above in Section 3.4 have shown that profound changes occur at the SiO_2–Si interface during the HF etching process, and these changes can be understood using the thermodynamics principles discussed throughout this chapter. To observe the changes in the contact angle in one study of the etching of SiO_2 films on Si, a special *in situ* cell and associated hardware were constructed (4). Figure 3.11 shows a sketch of the apparatus that consists of a low power (20 to 100×) stereo inspection microscope and a transparent cell to hold the sample and liquid etchant. The cell is mounted on a *xz* translation stage ($TS_{x,z}$) on a leveling platform (LP) that enables the positioning of the sample for observation. The sample cell was illuminated (using light L and mirror M) from the side opposite the microscope, and photographs of the sample could be obtained through one eyepiece of the stereo microscope. Figure 3.12 shows a side view of the sample cell. In this cell a sample that is typically a SiO_2-film-coated Si wafer segment is suspended in the etchant solution film-side down by the notched fused silica sample holder posts. Etching

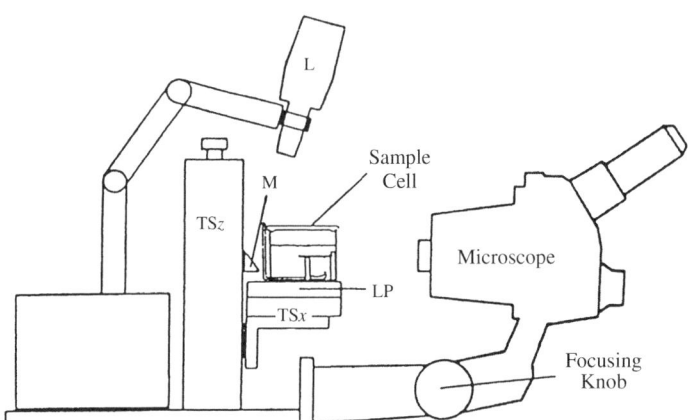

FIGURE 3.11 Microscope and sample cell apparatus for *in situ* contact angle observations. The cell is moved for viewing using the two translation stages: TS*x* and TS*z*.

3.6 SiO$_2$–Si INTERFACE APPLICATION

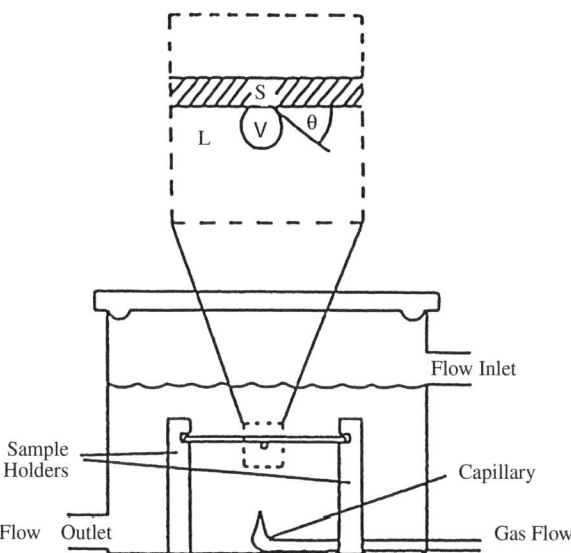

FIGURE 3.12 Side view of the cell in Fig. 3.11. The inset shows a magnified view of the sessile bubble (V) in the liquid L and attached to the sample S. [Adapted from Gould and Irene (4), Figure 2.]

solution is flowed through the cell from the top inlet to the bottom outlet. The flowing of fresh etch solution eliminates possible depletion effects that may arise with the use of dilute etchant. Beneath the inverted sample is a capillary through which a gas (typically N$_2$) can flow. This capillary is used to form a bubble that rises through the etch solution and clings to the film side of the sample. This clinging bubble is called a sessile bubble. Notice that in Fig. 3.10c a drop of liquid was put on the surface of a solid all within laboratory air ambient. This situation results in the three phases, solid–liquid–vapor, in equilibrium and the interfacial vectors, γ_{SL}, γ_{SV}, and γ_{LV}, were resolved accordingly yielding the contact angle defined by the Young Equation 3.45. Returning to Fig. 3.12 with the sessile bubble, the closeup shown in the inset indicates that the same three phases are present, S, L, and V, and thus the same three-phase equilibrium is established as with the sessile drop. However, the sense of the angle is changed but not the definition as the angle between the liquid and solid as is shown in the inset. This cell design using the inverted sample and sessile bubble enables maximum coverage of the surface etched by the etching solution, and the microscope enables continual monitoring of the contact angle. In this study it was estimated that the contact angle could be measured to within $\pm 3°$.

Etching experiments were performed with the apparatus described above with a solution of water and concentrated HF solution (49% by weight) of 500:1 by volume. This dilution provided slow etching of SiO$_2$. The starting SiO$_2$ film thickness was 23 nm as measured by ellipsometry (discussed in Chapter 9). Initially with the sample in deionized water a contact angle of 39° was measured. Remembering that

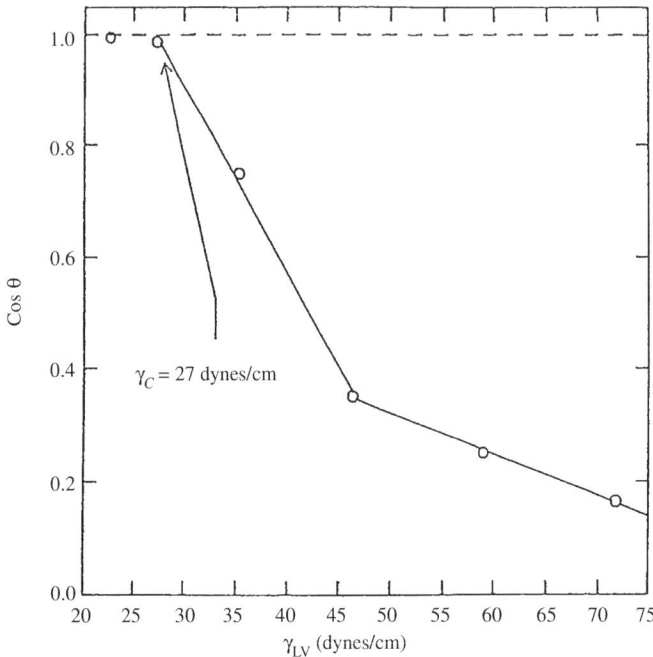

FIGURE 3.13 Zisman plot of the cosine of the contact angle cos θ for the etching of SiO_2 on Si in HF–Methanol–water solutions of various surface tension γ. [Adapted from Gould and Irene (4), Figure 5.]

TABLE 3.1 Surface Terminations for Low Energy Films Adsorbed on Metal Surfaces from Zisman

Surface Termination	γ_C erg/cm^2 @ 20°C
Fluorocarbon Surfaces	
—CF_3	6
—CF_2H	15
—CF_3 and —CF_2—	17
—CF_2—	18
—CH_2—CF_3	20
—CF_2—CFH—	22
—CF_2—CH_2—	25
—CFH—CH_2—	28
Hydrocarbon Surfaces	
—CH_3 (crystal)	22
—CH_3 (monolayer)	24
—CH_2—	31
—CH_2— and —CH—	33
—CH— (phenyl ring edge)	35

Source: Adapted from Gould and Irene (4), Table 1.

3.6 SiO$_2$–Si INTERFACE APPLICATION

a contact angle of 90° indicates no wetting (hydrophobic interaction) and 0° is for complete wetting (hydrophilic interaction), 39° is indicative of intermediate wetting behavior. Upon the introduction of the etching solution the contact angle changed rapidly from the initial value of 39° to 8°. The value of 8° remained constant until the SiO$_2$ was removed and then changed abruptly to 78°, and this value persists so long as etch solution is maintained in the cell.

The initial intermediate wetting condition of the SiO$_2$ film surface with a contact angle of 39° cannot be explained based solely on a SiO$_2$ water interaction. Inorganic solids have γ values greater than 500 ergs/cm^2 and as mentioned above Si is 1200 ergs/cm^2. Most oxides have higher γ values than their parent metal, and it is expected that SiO$_2$ has a higher γ than Si. Also as mentioned above γ for water is 72 ergs/cm^2. Therefore, a strongly hydrophilic surface interaction is expected for the interaction of water with SiO$_2$ resulting in a small contact angle. Rather an intermediate wetting interaction is observed with a contact angle value of 39°. However, a strong hydrophilic interaction occurs after brief contact with etchant, causing a change from 39° to an 8° contact angle. This indicates that the top surface of SiO$_2$ is different and perhaps contaminated, but, after a few layers are removed and a fresh SiO$_2$ surface is exposed, the strong anticipated interaction is realized. This strong hydrophilic interaction persists as is expected until the SiO$_2$ is removed. The ellipsometric results indicate that as the last vestige of oxide is removed an optically different film forms. When this occurs, the contact angle abruptly changes from 8° to 78° indicating a strongly hydrophobic interaction. This interaction would not be expected for bare Si. An angle slightly above 8° would be anticipated since γ for Si is expected to be less than that for SiO$_2$. The strongly hydrophobic interaction can be attributed to the new optically distinct film that forms, and its nature can be obtained from further analysis of the contact angle results.

The analysis followed in this study was due to Zisman (5), who has shown that there is a relationship between the cosine of the contact angle and the liquid surface tension γ_{LV} for low energy surfaces wet by liquids that have a variety of surface tensions. This relationship leads to a critical surface tension γ_C for the low energy surface that is obtained by extrapolating a plot of $\cos\theta$ versus γ_{LV} to the value $\cos\theta = 1$, where θ is the contact angle. Recall that when $\cos\theta = 1$, $\theta = 0°$, which is total wetting. To use the Zisman analysis several solutions of methanol ($\gamma = 23$ ergs/cm^2) and water ($\gamma = 72$ ergs/cm^2) were made with γ varying from the pure solvent values, and these solutions were each used individually for contact angle measurements. Figure 3.13 shows this Zisman plot for SiO$_2$-coated Si samples with 6 solutions and where the SiO$_2$ has been removed using HF. The abrubt change in slope in the middle of the curve is attributed by Zisman to hydrogen bonding between the solutions and the surface. The data extrapolated to $\cos\theta = 1$ yields $\gamma_C = 27$ dynes/cm or ergs/cm^2. Table 3.1 shows a variety of fluorocarbon and hydrocarbon surfaces with γ_C values taken from Zisman. The experimental value for γ_C is not as small as a fluorine-terminated surface but indicative of a partially H-terminated surface. Later experiments have determined that the HF-etched SiO$_2$ on Si surface is largely H terminated in agreement with the contact angle assessment. The fluorine from the HF hydrolyzes leaving mostly H at the Si surface.

REFERENCES

1. Harkin, X. J Chem Phys 1942;10:268.
2. Gilman JJ. J Appl Phys 1960;32:2208.
3. Butner FH, Funk ER, Udin H. J Phys Chem 1952;56:657.
4. Gould G, Irene EA. J Electrochem Soc 1988;135:1535.
5. Zisman WA. In: Fowkes FM, editor. Contact angle: Wettability and adhesion. Volume 43, Advances in chemistry series. Washington, DC: ACS; 1987; Chapter 1.

SUGGESTED READING

A. W. Adamson and A. P. Gast, 1997. *Physical Chemistry of Surfaces*, Wiley Interscience. This is an update of a classical text in surface science. This book is especially good for surface thermodynamics and was often used and referred to in the surface science courses at UNC.

4

SURFACE ROUGHNESS

4.1 INTRODUCTION

The concept of roughness is intuitive in that it is easy to imagine a rough surface as compared to a smooth surface. If someone claims to have roughened a surface or to have a very rough surface, everyone can readily form a mental picture of the roughness as being a surface with convolutions of various shapes and sizes. However, it is likely that each person's mental image of the roughness will differ in the details. Therefore, to achieve the next level of understanding of the concept of roughness, a quantitative description of roughness is required that is far less intuitive. Seemingly straightforward questions such as which factor quantitatively defines roughness or what exactly do you mean by stating that one surface is rougher than another often lead to imprecise and confusing answers. When comparing rough surfaces, it is not at all clear which surface features can be used to claim that one surface is rougher than another. For example, is it the size of convolutions or asperities that distinguish roughness, or is it the spatial frequency, or distribution of features, or the shape of the features, whether sharp or rounded, rather than size that are more important? It is clear that to address these questions, an unambiguous description of roughness is required. Even a precise measurement of surface topography is not enough. It was seen it Chapter 2 that TEM and SEM techniques can measure surface features, and the scanning probe microscopies (SPMs) to be discussed in Chapter 6 can produce dramatic and accurate images of the surface topography and texture even at the atomic scale. Yet while all of these powerful techniques yield accurate and

---

Surfaces, Interfaces, and Thin Films for Microelectronics. By Eugene A. Irene
Copyright © 2008 John Wiley & Sons, Inc.

impressive pictures of rough surfaces, none reveal the essence of roughness, although they provide the information to do so. In addition for the determination of the topography of rough surfaces through the use of modern analytical techniques, what is crucially required is the one or perhaps several parameters that when considered together provide an accurate, complete, and understandable description of roughness. Since it is only through the use of an accurate description of roughness can the effects of roughness on properties be determined and compared, this chapter commences with a definition of several commonly used roughness parameters. Each parameter enables the quantification of an aspect of roughness. This discussion is followed by a discussion of the various properties that are affected by roughness, and then some literature examples are presented.

4.2 ROUGHNESS PARAMETERS

In the scientific and engineering literature there are literally hundreds of possibly useful parameters that enable some level of quantification of roughness or surface texture. In the following paragraphs a selected subset of the available parameters is defined and discussed. This subset has been chosen on the basis of both widespread use in scientific publications and general scientific merit. Also, as was mentioned above, there are many aspects of roughness that need to be quantified. The selection of parameters addresses the most important aspects of roughness such as size, shape, density, and statistics of rough features. The last roughness descriptor covered is the fractal representation. In principle, the fractal representation can represent roughness as a single parameter. Thus, using the fractal representation it should be possible in principle to unambiguously compare the roughness-related properties of surfaces.

4.2.1 Height Parameters

One obvious parameter is the size of the roughness features. One aspect of the size is the height and the other is the width. Typically, the height is quantified either by the average height as is given by the parameter called the roughness average, R_a, or by a more statistical average called the root-mean-square roughness, or rms roughness, R_q. With the use of Fig. 4.1 that shows an edge view or profile of a surface in one direction, R_a is expressed as an integral or approximately in digitized form as a sum:

$$R_a = \frac{1}{L} \int_0^L y(x)\, dx$$
$$= \frac{1}{N} \sum_{i=1}^{N} |y_i| \qquad (4.1)$$

The average is obtained from the integral of the analytical expression for the surface profile over some distance L in a direction in the profile. Often, the analytical expression is unavailable, and, even if available for one direction, it is likely to be

4.2 ROUGHNESS PARAMETERS

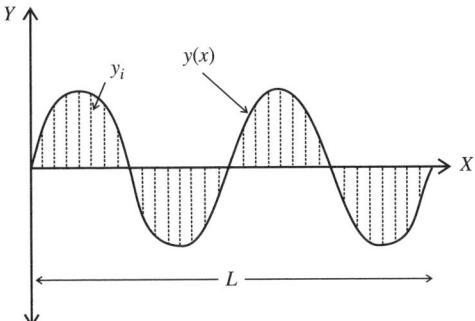

FIGURE 4.1 Periodic surface profile described by y(x).

different in other directions and hence not representative. Alternatively, the heights, y_i, of roughness features relative to the mean line are collected without regard to the direction from the mean line (positive or negative deviation), and these are averaged over some distance L on the surface. Since as these expressions for R_a are written, there is one profile from 0 to L to yield an average, and as mentioned there is no reason to believe that all of the profiles on a surface are equivalent in different directions, it is often useful to obtain the digitized R_a over the entire surface by sampling y_i in 2D over the surface and then performing the average calculation rather than merely using the 1D value for a single profile.

The rms roughness is obtained either from the analytical or digitized expressions as follows:

$$R_q = \left[\frac{1}{L}\int_0^L y^2(x)\,dx\right]^{1/2}$$
$$= \left(\frac{1}{N}\sum_{i=1}^N y_i^2\right)^{1/2} \tag{4.2}$$

where R_q represents the standard deviation of the roughness profile heights in the 0 to L interval. The statistical nature of R_q and the fact that it is proportional to R_a has made R_q one of the more popular parameters used to quantify roughness. In fact, R_q is so popular that it is often used as the only measurement of roughness. This is a very dangerous situation because neither R_a nor R_q reveal the complete essence of roughness. Figure 4.2a illustrates the major issue that arises with the sole use of R_q (or R_a) as a measure of roughness. Two very different roughness profiles are shown (solid and dotted lines). One is pointed while the other is smooth and sinusoidal. Yet both have exactly the same rms value. Thus, from R_q there is no spatial information or information about shape. Another less obvious problem with R_q and R_a is shown in Fig. 4.2b. This figure shows a randomly rough profile with a large step (halfway down the profile L at $L/2$). If the rms is calculated for the left portion of the profile prior to the step, an rms value of around 12 is obtained, but an rms of

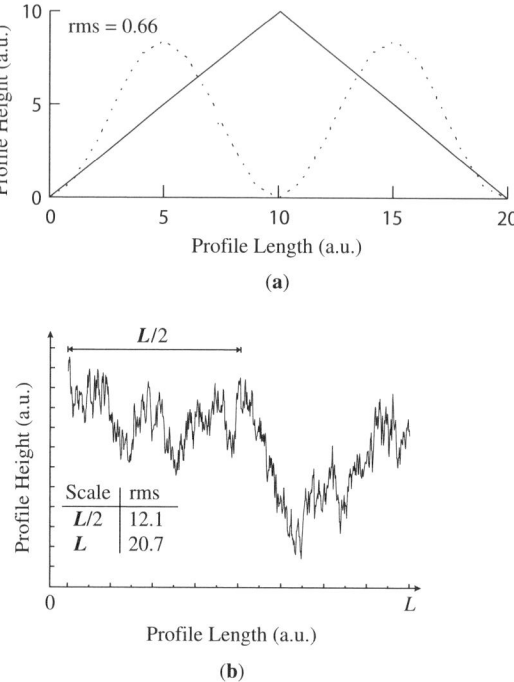

FIGURE 4.2 Roughnes profiles: (**a**) compares two different profiles that have the same rms roughness and (**b**) a profile that has very different rms for different L values.

nearly double that is obtained from the entire profile. Therefore, the sampled interval affects the value. Consequently, it would be more useful to know beforehand which kinds of features are of interest, the smaller features represented well with short L's or the larger features that typically bias the rms upward.

4.2.2 Roughness Period or Wavelength

One method to estimate the spatial extent of roughness is to average the wavelengths. As shown in Fig. 4.3 each length of a variation S_{m1}, S_{m2}, S_{m2}, and so on is tabulated as it crosses the mean plane. The lengths are averaged and this average is expressed mathematically as

$$S_m = \frac{1}{n}\sum_{i=1}^{n} S_{mi} \qquad (4.3)$$

where S_{mi} is the length of a particular ith wavelength on the surface, and n such wavelengths are averaged. This parameter can also be biased by a large period feature, and, as illustrated in the figure, there is no accounting for small feature variations that do not cross the mean plane. As will be discussed below and was briefly discussed in

4.2 ROUGHNESS PARAMETERS

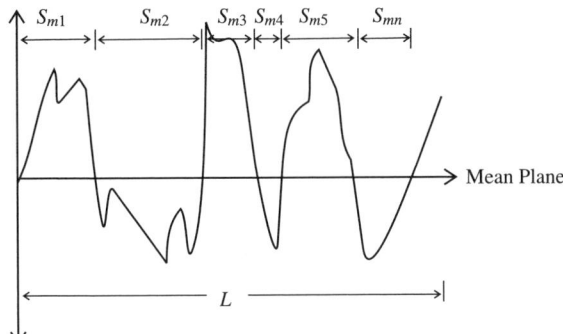

FIGURE 4.3 Roughness profile with different wavelengths S_{mi} indicated.

Chapter 3 using the Kelvin equation, small roughness features may be the most important features relative to chemical reactivity, but these are the very features ignored by the wavelength-averaging method.

4.2.3 Shape of Rough Features

The two main roughness parameters that are used to give information about the shape of roughness features are skewness and kurtosis. Before defining these parameters it is useful to consider the distribution of roughness heights, sometimes called the amplitude distribution function (ADF). The ADF gives the probability that a surface profile has a certain height z at some position in the x,y plane, or for a 1D profile at some position x in the profile. Figure 4.4 shows that the ADF is simply a histogram of heights across the profile, and it relates how much of the roughness profile lies at a particular height. The ADF is shown to be perfectly symmetrical in Fig. 4.4 but it need not be. Skewness, R_{sk}, measures the asymmetry of the ADF and can be calculated from the following formula:

$$R_{sk} = \left(\frac{1}{R_q^3}\right) \frac{1}{N} \sum_{i=1}^{N} y_i^3 \tag{4.4}$$

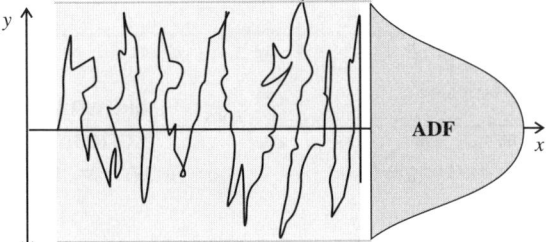

FIGURE 4.4 Complex surface profile with the distribution of size features at the right.

where R_q is as before the rms for the profile. Surfaces with positive skewness have high peaks or spikes above the average while those with negative skewness have more valleys below the mean line, and random surfaces have near zero skewness with the obvious changes in the symmetry of the ADF. Another often used shape parameter related to the asymmetry of the ADF is the kurtosis, R_{ku}, and it is given as

$$R_{ku} = \left(\frac{1}{R_q^4}\right)\frac{1}{N}\sum_{i=1}^{N} y_i^4 \tag{4.5}$$

Essentially, kurtosis measures the number of features above and below the mean line. A large kurtosis indicates many high features while a lower one indicates smaller features whose heights are closer to the mean line. A perfectly Gaussian ADF would yield a kurtosis of 3, and higher values indicate many heights away from the mean, far above and below the mean line. Figure 4.5 shows two surface profiles with the height distrubition histogram or ADF and rms, R_{sk}, and R_{ku} values. It is

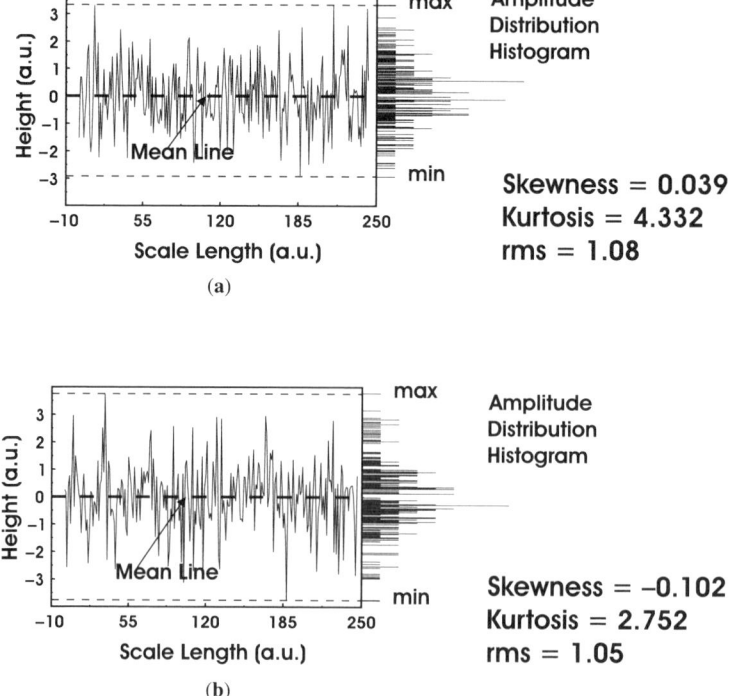

FIGURE 4.5 Two profiles (**a**) and (**b**) that have nearly the same rms but each with different skewness and kurtosis. At the right of each profile is an amplitude distribution histogram.

4.2 ROUGHNESS PARAMETERS

seen that both profiles have nearly identical rms values. The top profile has positive (+) skewness or more of its histogram above the 0 or mean line, while the bottom has negative (−) skewness with more of its profile below the mean line. The top profile has a higher kurtosis, indicating a higher proportion of high peaks than the bottom curve. Thus, while there is significant differences in the shape of the profiles, the rms was unable to make any real distinction, and visually the profiles are too complicated to glean the differences obtained from the parameters.

While the height and shape parameters defined above yield particular perspectives about roughness, what is lacking is a way to summarize the various parameters. Statistical parameters greatly help with this task.

4.2.4 Statistical Descriptors of Roughness

A surface profile is essentially height versus distance from 0 to L as was shown in Fig. 4.1 to 4.3. A Fourier transform (FT) of a profile will yield the frequency spectrum, and the square of the FT is the power spectrum or power spectral density (PSD) parameter. The PSD can be expressed as

$$\text{PSD}(F) = \lim_{L \to \infty} \left(\frac{1}{L}\right) \left| \int_0^L y(x) \cdot e^{-2i\pi Fx} dx \right|^2$$

$$\text{PSD}(k) = \frac{1}{N\Delta} \left| \sum_{j=1}^{N} y_j \cdot e^{-2i\pi kj/N} \right|^2$$

(4.6)

where Δ is the data point spacing of N points and $L = N\Delta$, F is a set of spatial frequencies, and for the digitized form is k/L where k is an integer from 1 to $N/2$. A PSD for a roughness profile is shown in Fig. 4.6, and it consists of a histogram of frequencies with 0 frequency at the left origin indicated as DC, and higher frequency components to the right. This spectrum shows the relative amounts of high (to the right) and low frequency features (to the left). The relative heights of the features indicate the relative amounts of that particular spectral frequency. Thus, the PSD that gives the frequency distribution along with R_q that gives the magnitude together yield a good representation of the kind of roughness with which one is dealing. The drawback is that the PSD is an image, albeit a revealing one, and therefore difficult to use in calculations and/or direct analytical comparisons.

As an example Fig. 4.7a shows the PSD for a sinusoidal surface profile (1). The PSD displays essentially one dominant peak indicative of the single-frequency roughness. However, closer examination of Fig. 4.7a reveals several smaller peaks indicating imperfect sinusoidal roughness. Figure 4.7b shows a random appearing roughness profile, and Fig. 4.7c shows the PSD corresponding to the profile. The PSD reveals peaks corresponding to the dominant roughness frequencies. Thus, the PSD yields the intimate details relative to the frequency of the roughness undulations.

Power Spectrum

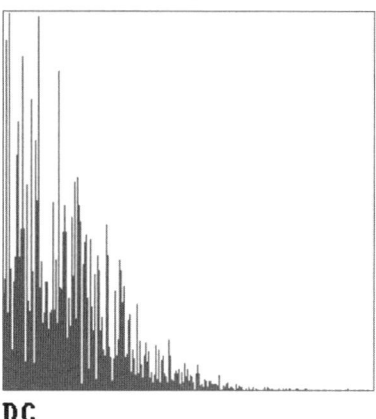

FIGURE 4.6 Power spectral density plot that is a histogram of frequencies found in a surface roughness profile with the amplitude proportional to the occurrence of that spectral frequency. DC indicates zero frequency with higher frequency features to the right.

Another revealing statistical roughness parameter is called the autocovariance, $B(\tau)$, and it can be obtained from the surface profile and is given as

$$B(\tau) = \frac{1}{L} \int_{-\infty}^{\infty} y(x)y(x+\tau)\,dx \qquad (4.7)$$

where τ is the lag or offset between profiles. In essence $B(\tau)$ measures the extent to which a surface profile repeats. This extent is called *self-similarity* and we will discuss it further later. If, for example, a profile has a dominant feature that repeats, the function shows a maximum when τ approaches that repeat value. If there are several repeating features, then there will be several corresponding features in $B(\tau)$. If on the other hand the profile is random with no repeating features, then $B(\tau)$ will drop rapidly to zero. The correlation length is that distance where the autocovariance, $B(\tau)$, first crosses zero. The autocorrelation function (ACF) is simply a normalized version of $B(\tau)$ and is given as

$$\text{ACF} = \frac{B(\tau)}{R_q^2} = \frac{\left(\frac{1}{L}\right)\int_0^L y(x)y(x+\tau)\,dx}{\int_0^L y^2(x)\,dx} \qquad (4.8)$$

Figure 4.8 displays at the top two different surface profiles, *a* and *b*. The power spectrum, or PSD, for each profile shows a different distribution of roughness frequencies with profile *b* having more low frequency features than profile *a*. The autocovariance shows a larger correlation length for spectrum *b*.

4.2 ROUGHNESS PARAMETERS

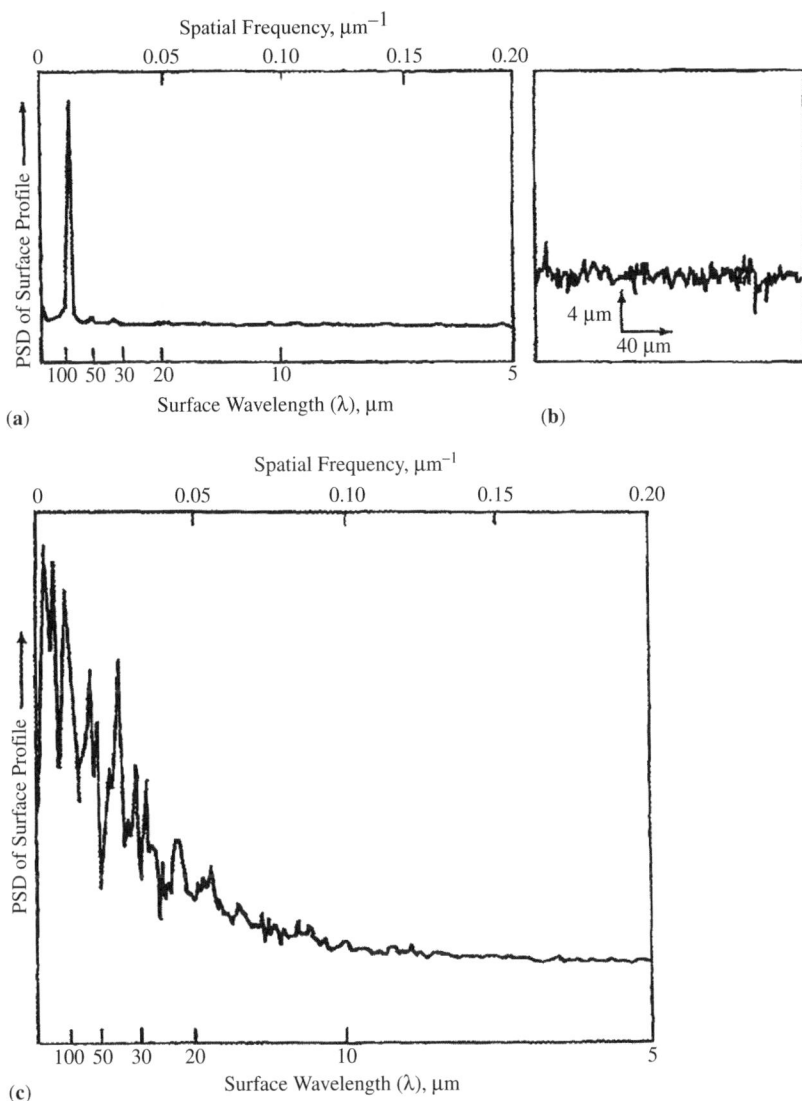

FIGURE 4.7 Power spectral density (PSD) for various surfaces: (**a**) a nearly perfect sinusoidal surface with (**b**) a closeup of the PSD showing some variation and (**c**) the PSD for a rough surface. [Adapted from Song and Vorburger (1), Figure 4.]

After considering all of the above parameters that collectively provide a description of roughness and that are frequently encountered in the literature, the question of which one is best arises. The simple answer is that a single parameter is insufficient, and several parameters are required to glean a more complete picture of roughness. The selection of parameters also is dependent on the kind of information

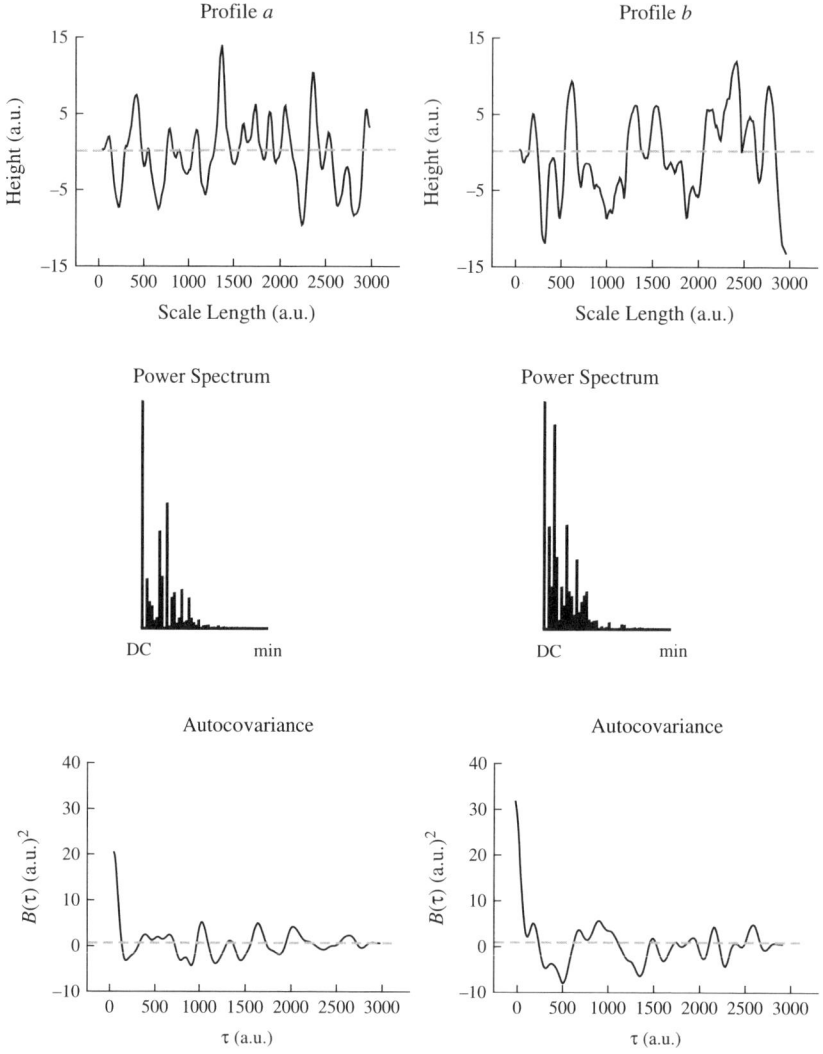

FIGURE 4.8 Two different profiles are shown as profiles *a* and *b*. Below each profile is the PSD and below that is the autocovariance $B(\tau)$ versus the offset τ. The correlation lengths are obtained where $B(\tau)$ crosses 0.

desired. For example, if one knows *a priori* that the magnitude of the roughness is important, then one would select a parameter that yields size such as R_a or R_q. Also, several of the important parameters such as PSD and ACF are in themselves images, and although they convey considerable information about the roughness, they are difficult to use in attempting to correlate properties with different levels of roughness. In principle, this dilemma of which roughness parameters are sufficient to accurately describe roughness can only be resolved if there existed a quantifiable

4.2 ROUGHNESS PARAMETERS

parameter that uniquely and thoroughly describes roughness. Potentially such a parameter, the fractal dimension for a rough surface, can accomplish this task and will be discussed below.

4.2.5 Fractal Description of Roughness

4.2.5.1 Fractal Definitions The mathematical concept of fractals and the fractal dimension, D_F, was first introduced by Mandelbrot (2) to describe certain types of irregular objects that Euclidean geometry (with integer dimensions 0, 1, 2, and 3 for points, lines, planes, and cubes, respectively) does not adequately describe due to the irregularity of the objects. The fractal dimension being a noninteger dimension is a measure of the irregularities of rough or "fractal" objects by examining the space-filling ability of the objects. The magnitude of D_F is a reflection of the shape complexity of the object. The fractal dimension is an intrinsic property of an object in the same way that a Euclidian dimension is an intrinsic property of a Euclidean object. For example, the Euclidean dimension 2 is a property of a perfect Euclidean surface. The fractal dimension of a real surface would have a noninteger value greater than 2 but less than 3. Therefore, using the fractal dimension enables the comparison of the complexity (one important aspect of roughness) of different rough objects with a simple, single-value parameter.

The fractal dimension (D_F) characterizes the deviation of a profile or a surface from its corresponding Euclidean dimension (D_E). D_E is the number of coordinates required to locate the position of a point contained by an object, and it must be an integral value. As illustrated in Fig. 4.9, D_E of a point, a line, a plane, and a box are 0, 1, 2, and 3, respectively, whereas D_F values being nonintegers lie in the D_E gaps.

All fractal objects exhibit dilation symmetry, which means that the fractal object looks the same at any magnification or scale, and the fractal dimension quantifies this dilation symmetry; hence, it is scale independent, and the following scaling law obtains:

$$N = R^{D_F} \tag{4.9}$$

where N is some property of the system, R is a variable, and D_F is the dimension of the system. Rearranging Equation 4.9 and solving for D_F yields the following:

$$D_F = \frac{\ln N}{\ln R} \tag{4.10}$$

An example of the use of these formulas and of dilation symmetry can be obtained by considering the construction of a fractal object, namely a triadic Koch curve as is shown in Fig. 4.10c. This curve is created by taking a straight line as shown in Fig. 4.10a, replacing it with a segmented structure (four segments) shown in Fig. 4.10b, and repeating this operation at each segment over and over. In this

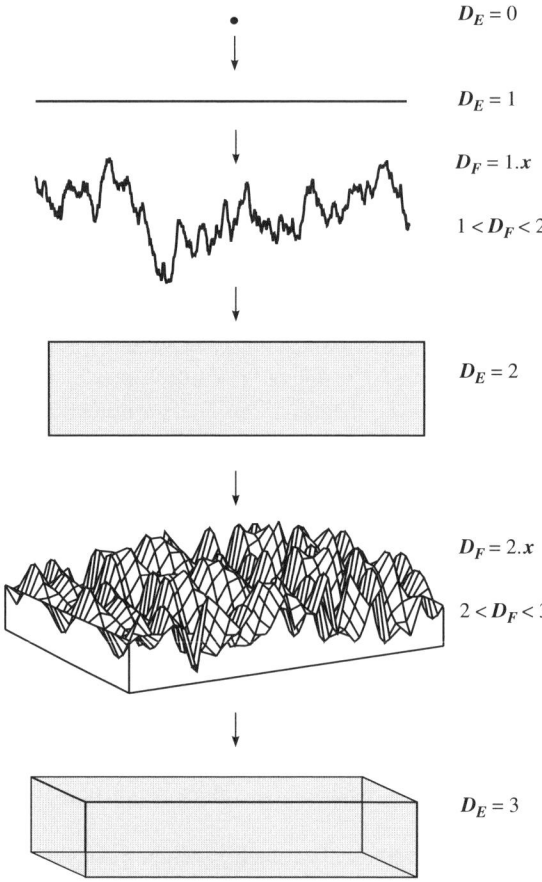

FIGURE 4.9 From the top is the Euclidian point, line, surface, and 3D solid with dimension $D_E = 0, 1, 2$, and 3, respectively. In between the line and surface, and surface and 3D object are the rough surface profile and rough surface, respectively, with D_F values in between the Euclidian values.

example, every straight line is of length λ and is replaced with 4 new lines of length $\lambda/3$, so with $N = 4$ and $R = 3$, $D_F = 1.262$ for the triadic Koch curve. The result, Fig. 4.10c, when viewed at higher magnification, looks like many of the Fig. 4.10b structures end to end and at various angles.

A fractal object can be self-similar in that it remains invariant under isotropic (uniform) scale transformation. For example, when part of a self-similar object is enlarged, it can overlap exactly with the original object. At every scale the object has the same features as did the Koch curve discussed above. However, many objects, such as Brownian motion, that exist in nature are random, and they may be self-similar but in a statistical sense. There is a broader class of fractal objects called self-affine fractals in which the object is only self-similar under anisotropic

4.2 ROUGHNESS PARAMETERS

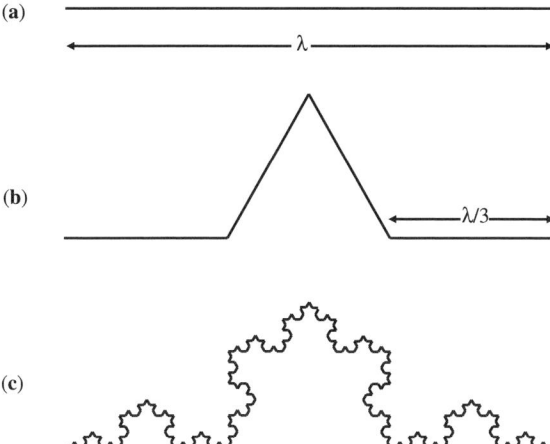

FIGURE 4.10 Steps to produce a fractal object, the Koch curve. A line in (**a**) is segmented to make the object in (**b**). This procedure is repeated on each segment of (**b**) to produce (**c**).

transformations. This means that part of an object can become identical with the original objects by unequal stretching. Furthermore objects may be self-similar or self-affine over limited scale ranges. These latter fractals are sometimes referred to as physical fractals since for many real systems true self-similarity cannot be adequately checked, yet an object seems to be self-similar or self-affine under some scales that can be checked. Examples of physical fractals are irregular curves and rough surfaces where under reasonably attainable magnifications the irregularity or roughness seems the same. This notion will permit the use of the fractal designation to describe real objects, keeping in mind that while it may be useful it is inexact.

It should be understood that the fractal dimension gives no information about the vertical magnitude of roughness, but rather the fractal dimension is related to the spatial complexity of roughness. That this is the case can be shown using a mathematical model that can be used to generate fractal profiles. The model chosen to generate fractal profiles is the Weierstrass–Mandelbrot model, and it is given as

$$f(x) = \sum_{n=-\infty}^{\infty} b^{-nH}[1 - \cos(b^n x)] \qquad (4.11)$$

where b is a constant, $b > 1$, and $H \in (0, 1)$. D_F is related to H as $H = 2 - D_F$. Different D_F values (1.1 to 1.8) are used in the model to produce the different profiles shown in Fig. 4.11 in which it illustrates that as D_F increases the profile complexity increases, so D_F is a measure of the complexity of the profile.

Although for a given profile there is a unique fractal dimension due to the scale independence of this parameter, different profiles can have the same fractal

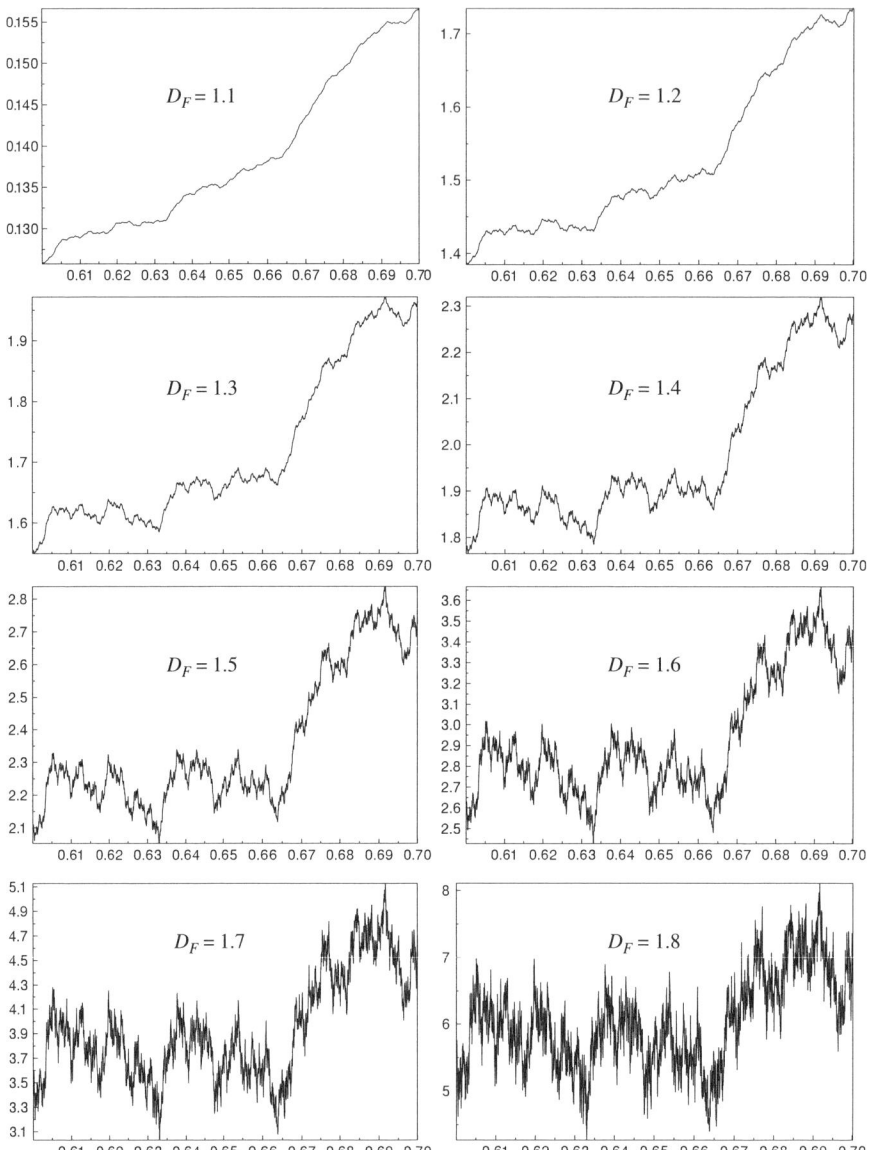

FIGURE 4.11 Various fractal profiles generated using the Weierstrauss–Mandelbrot model Equation 4.11.

dimension. To illustrate this point, Fig. 4.12 shows two Weierstrauss–Mandelbrot curves that were generated with different b values that change the scale of the object, namely 2.0 and 2.1, but they both have the same $D_F = 1.3$. Therefore, the fractal dimension can be used for comparison of the spatial complexity even for profiles with different sized features (different scales).

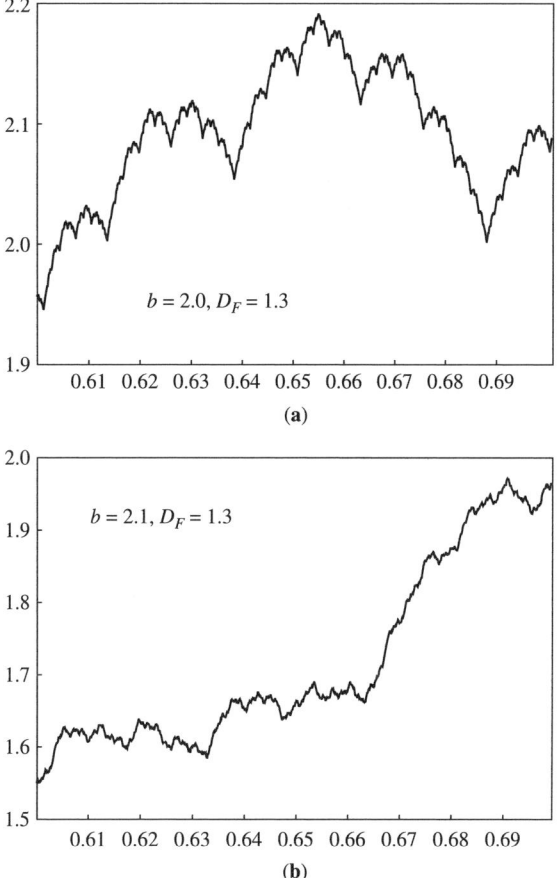

FIGURE 4.12 Weierstrauss–Mandelbrot curves (Equation 4.11) with the same D_F but with different appearance: (a) and (b) have different b values of 2.0 and 2.1, respectively.

Figure 4.13 shows computer-simulated rough profiles surfaces and the corresponding D_F's. From Fig. 4.13 and Fig. 4.11, it is clear that as the surface/profile becomes more irregular, D_F increases and more volume/area is filled by the surface/profile. D_F for the surface increases toward 3, and toward 2 for a profile.

In summary, D_F provides a unique and complete description of the complexity of the surface topography. However, D_F does not give any information about the size of roughness features. Thus, D_F along with R_q provides a reasonably complete description of roughness that has the advantage of being expressed by two numbers. For example, the variation of some determinable parameters P can be measured separately as a function of D_F and R_q to identify which roughness feature is dominant for P.

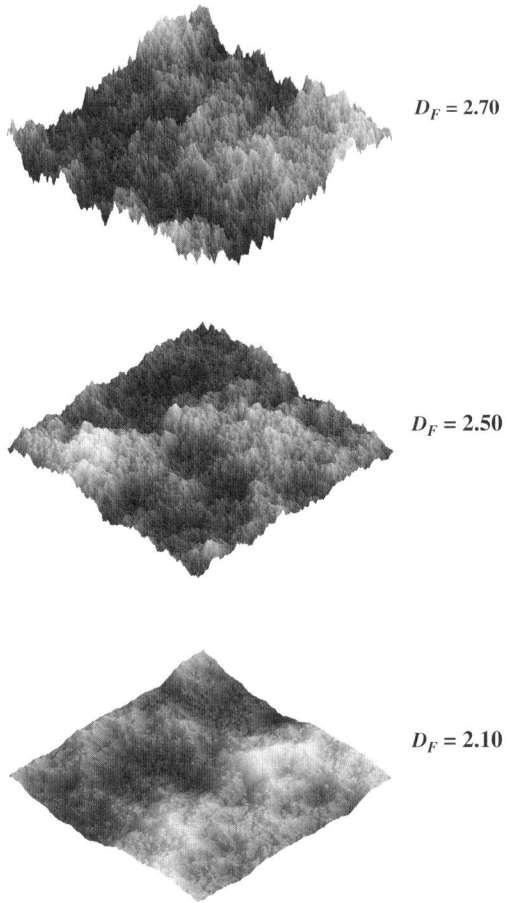

FIGURE 4.13 Computer-simulated rough surfaces with different D_F values.

4.2.5.2 Extraction of the Fractal Dimension from Experimental Data Many algorithms exist for extracting the fractal dimension from the roughness profile data obtained from a given object. The most common method is the box-counting method. This method is perfectly acceptable when applied to ideal or nearly ideal fractal objects. Data from ideal fractal objects have no resolution limits. However, when the surface topography of a real object is measured with a real instrument, a finite set of data is obtained with a finite resolution imposed by the instrument. In many instances it has been shown that real surfaces with roughness can be approximated as fractal objects over a scale range. Now since a real surface cannot be measured with infinite resolution over all scales, the roughness data is imperfect. With imperfect data different mathematical algorithms can yield different values for the fractal dimension for the same rough surface. Thus, the method used to obtain D_F can be important.

4.2 ROUGHNESS PARAMETERS

The box-counting method is the most commonly discussed algorithm for extracting fractal dimensions, simply because it is the easiest to understand. When applied to a rough profile, this method superimposes a uniform two-dimensional array of squares on the profile such that the squares completely encompass the profile as is illustrated in Fig. 4.14a. The number of squares intersected with the profile is recorded. A rougher profile would intersect more squares than the smoother profile. This process is continued with the size of the squares, ε, slowly decreasing, hence increasing the number of intersections. The number of squares intersected by the profile, $N(\varepsilon)$, is recorded as a function of ε. Then, the log of the number of intersected squares is plotted against the log of the size of the squares, and the slope of this log–log plot gives the fractal dimension of the profile as is illustrated in Fig. 4.14b. Notice that

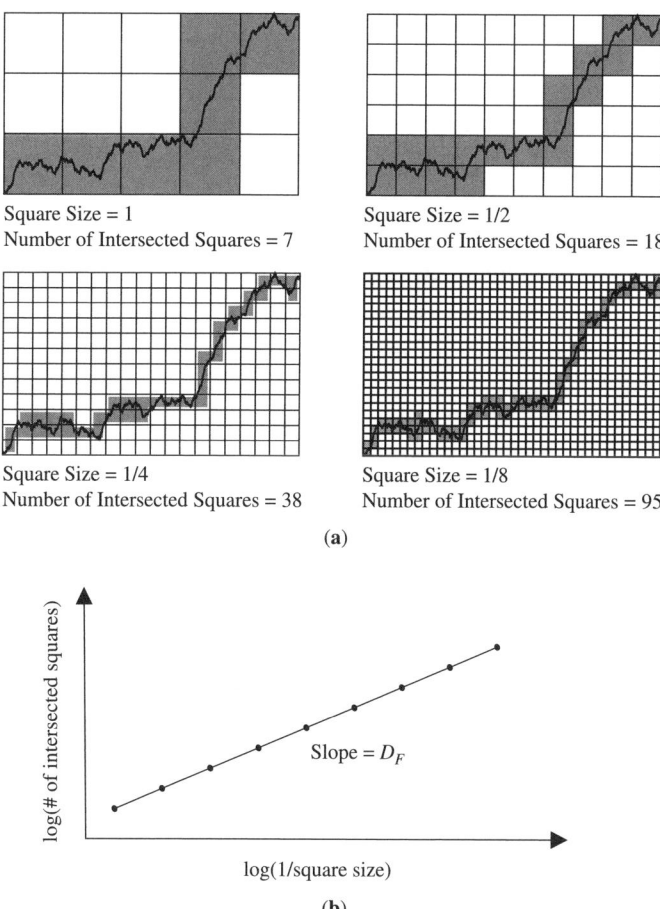

FIGURE 4.14 Illustration of the box-counting method for obtaining D_F: (**a**) shows the grids that are used to overlay the profile and (**b**) shows a plot of the intersections versus the size of the boxes.

this is another application of the formula for D_F previously discussed (Equation 4.10). When this method is applied to rough surfaces, a uniform three-dimensional array of boxes is used instead, and the procedures are similar to that applied for a rough profile. Other methods are not covered here, but while the box-counting method is easy to follow, it is not among the most accurate methods for extracting D_F from the most reliable surface topographic instruments such as atomic force microscopy (AFM), which has become the most widely used and accurate surface profile technique (AFM and other surface probes will be covered in Chapter 6). A variation method developed by Spanos and co-workers (3,4) for use with AFM has been used successfully to extract D_F values from high quality AFM surface profile data.

4.3 ROUGHNESS EFFECTS ON THE PROPERTIES OF MATERIALS

It should be realized that a discussion of the effects of surface roughness on the properties of materials covers a broad range of investigations, and many of the applications have significant technological relevance. The intent here is to first show some applications relevant to the broad field of microelectronics materials.

4.3.1 Effects of Roughness on Optical Properties

Optical properties are important to elucidate the electronic energy band structure of solids and even to determine the quality of a material's surface and interface and for films, and the thickness is often accurately determined using optical techniques (see Chapter 9 on ellipsometry). For the transmission or reflection of light from a material, Beer's law is often used to characterize the ratio of the measured light intensity I to the incident intensity I_0. For the transmission of light through a material the transmissivity T is defined in terms of the ratio of the transmitted light, I_T, and I_0, and Beer's law can be written as follows:

$$T = \frac{I_T}{I_0} = e^{-\alpha x} \tag{4.12}$$

where α is called the absorption coefficient and x is the distance traveled in the material or the thickness of the material; α characterizes the amount of light absorbed per length for a given material and is therefore characteristic of the material. The coefficient α is related to the more fundamental property of a material, the absorption constant k as follows:

$$\alpha = \frac{4\pi k}{\lambda_0} \tag{4.13}$$

where λ_0 is the vacuum wavelength for the light used. The absorption constant k (not to be confused with the wave vector $\mathbf{k}$) along with the refractive index n, which is the

4.3 ROUGHNESS EFFECTS ON THE PROPERTIES OF MATERIALS

ratio of the speed of light in vacuum, c_v, to the speed of light in the material, c_m as

$$n = \frac{c_v}{c_m} \tag{4.14}$$

When taken together n and k define the optical properties for a material through the complex dielectric function, ε given as

$$\varepsilon = \varepsilon_1 + i\varepsilon_2$$
$$\varepsilon_1 = n^2 - k^2 \text{ and } \varepsilon_2 = 2nk \tag{4.15}$$

For the reflection of light from a surface, the reflectivity R at normal incidence is given as a ratio of the reflected intensity I_R to I_0, and also as a function of n and k as

$$R = \frac{I_R}{I_0} = \frac{(n-1)^2 + k^2}{(n+1)^2 + k^2} \tag{4.16}$$

With these relationships we can explore the effect of roughness on the determination of optical properties. Figure 4.15a shows a surface without roughness, a

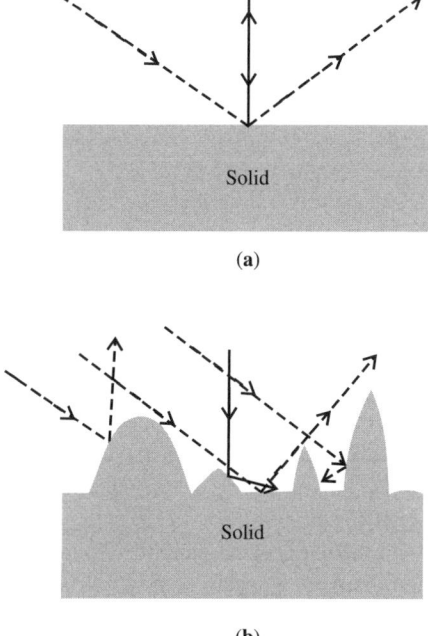

FIGURE 4.15 Reflection of light at normal and oblique incidence from (**a**) a smooth surface and (**b**) a rough surface.

so-called specular surface, and light rays at normal (solid line) and oblique (dashed lines) incidence. Light intensities measured in transmission and reflection will enable a determination of T and R to be made from which α (from T and the sample thickness), k (calculated from α) and then n (from k and R) can be determined leading to the complete description, ε. However, if the material has a rough surface, as shown in Fig. 4.15b, the situation is not as simple. Some of the light transmitted through or reflected from this nonspecular surface will be lost due to scattering in unintended directions. Thus, both I_T and I_R and hence T and R will be smaller than is representative of the bulk material. The direct consequence is that the determination of α will yield a larger value than for the smooth surface, and k will also be larger. If the light scattered away from the detectors were unaccounted for, the result would be an apparently (not actually) larger k for the material. If the surface roughness features are much smaller or much larger than the wavelength of the probing light, the effect of scattering would be small and a nearly accurate k can be determined.

4.3.2 Roughness Effects on Electronic Properties

There are many modern electronic devices that operate based on the conduction of electrons or holes at a surface or interface. One of these devices, shown in Fig. 4.16a, is a metal–oxide–semiconductor field-effect transistor, or MOSFET, as mentioned in previous chapters and will be discussed further in future chapters. Specifically, this figure shows an N-channel MOSFET that is constructed on P-type Si. A P-channel MOSFET is constructed on N-type Si. The MOSFET device is presently the most important device in microelectronics and is at the heart of both memory and logic devices, and the computer chips incorporate large densities of these kinds of devices.

The MOSFET device is usually a three-terminal device as shown with the terminals named source (S), drain (D) and gate (G). A back side contact is sometimes used but not necessary for device operation. The MOSFET is most often used as a switch, where the source-to-drain current (I_{SD}) is switched on and off using the potential at the gate (V_G) to effect the switching. The channel region for the N-channel MOSFET in Fig. 4.16a separates the N-type source and drain, and this is the region where carriers flow from source to drain to turn on the device. Notice that the N-type doping at the source S and drain D is labeled N^+ where the superscript + indicates heavy doping. Initially, the N-type source and drain are separated by a P-type region. Thus, if one injects a majority carrier, an electron, from the source into the channel for the purpose of effecting conduction, the electron will eventually recombine with the majority holes in the P-type channel region before the electron ever reaches the drain. If a negative potential is applied to the gate, $-V_G$, the majority carriers that are holes in the Si substrate will be attracted to the channel. This condition is called accumulation and refers to majority carriers that will accumulate in the channel. Accumulation prevents source–drain current, and thus maintains the off state for the MOSFET. Now as the V_G is made more positive,

4.3 ROUGHNESS EFFECTS ON THE PROPERTIES OF MATERIALS

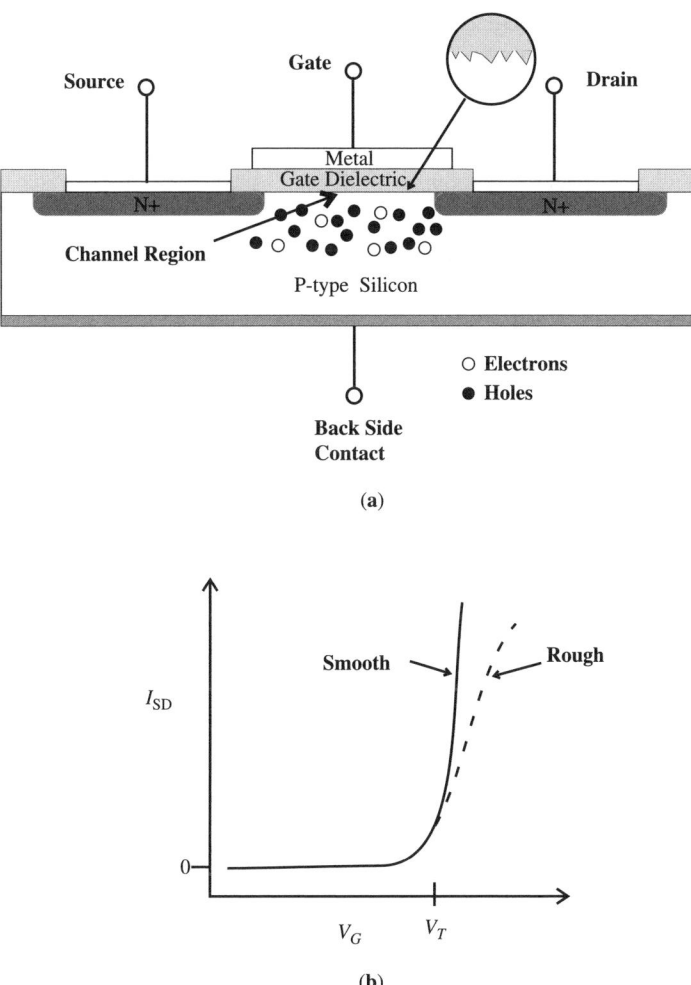

FIGURE 4.16 (a) N-channel MOSFET device with rough interface (inset) and (b) the operating characteristic of the MOSFET in terms of source-to-drain current I_{SD} versus gate voltage V_G with a smooth and rough interface. The turn-on or threshold voltage is V_T.

the majority carrier holes in the substrate will be repelled from the channel. When the number of holes in the channel becomes less than the equilibrium number, the channel is said to be depleted. The condition in the channel goes from accumulation with $-V_G$ to depletion with a more positive V_G. As V_G gets even more positive, the minority carrier electrons in the P-type Si in the channel region will outnumber the holes that are being depleted. The resulting condition is called inversion because the majority carrier type in the channel (and only in the channel) has been changed from holes to electrons, namely the carrier type has been inverted by virtue of changing V_G.

The important aspect of inversion is that now electrons from the source injected into the inverted channel can traverse the now N-type channel to the drain and thus turn on the device. The ideal I-vs.-V characteristic is shown by the solid line (labeled smooth) in Fig. 4.16b, and it is seen that the source–drain current, I_{SD}, is zero as V_G increases but then rises steeply when V_G is sufficiently positive in this N-channel device to effect inversion and turn on of the device. A V_G value at which I_{SD} is sufficiently high to indicate unambiguous turn-on is called the threshold voltage and labeled as V_T. The on and off states are all that are necessary to perform computer memory operations based on Boolean 1 and 0.

It is observed that the MOSFET operates via the conduction of carriers at the interfacial region between the Si substrate and the gate dielectric (SiO_2), the so-called channel region. In the discussion above it was assumed that this region has a smooth interface. However, if the interface were rough, as is indicated in the inset for Fig. 4.16a, the carriers traversing the channel from source to drain would experience scattering that would tend to randomize the carrier trajectories. In this case of an N-channel device, many of the electrons at the rough interface would need to travel a greater distance due to the scattering at the asperities to get to the drain and be counted as I_{SD}. Consequently, I_{SD} would drop as is shown by the dashed line (labeled rough) in Fig. 4.16b. This effect is attributed to a reduction of the carrier mobility. Carrier mobility μ is defined as the carrier velocity, $\mathbf{v}$ per electric field $\mathbf{E}$ applied:

$$\mu = \frac{\mathbf{v}}{\mathbf{E}} \tag{4.17}$$

Thus, if the distance traveled is increased for an electron, then the velocity in the intended direction would decrease since it would take longer to reach the desired position, and μ would also decrease.

At rough features or asperities there is also a local electric field enhancement effect. Figure 4.17a shows the bunching of electric field lines in the projected area above a single rough feature. Recall that electrostatics conventions require that the electric field line originate at a positive charge and terminate at negative charge and that the field lines must be perpendicular to the conducting surfaces. The electric field enhancement depends on the height-to-width ratio (the aspect ratio) of the asperity. Figure 4.18a shows the results of calculations (5) with this ratio, $b/a = 1$ (solid line) and 10 (dashed line). These calculations show that near the top of the asperity (small z) the field enhancements are about 3 and 10, respectively, and can drop rapidly for larger z. This could result in electric device failure because an applied field on a uniform smooth surface could be greatly magnified near an asperity, causing unexpected large currents and possible device failure. However, if the period of the roughness is small, that is, the features are closely spaced, then the electric field amplification effect is reduced. The electric field line distribution for adjacent close asperities is shown in Fig. 4.17b. It is seen that despite the occurrence of roughness the electric field is nearly uniform. Figure 4.18b shows sample calculations for adjacent asperities with center-to-center separation of $2d$.

4.3 ROUGHNESS EFFECTS ON THE PROPERTIES OF MATERIALS 111

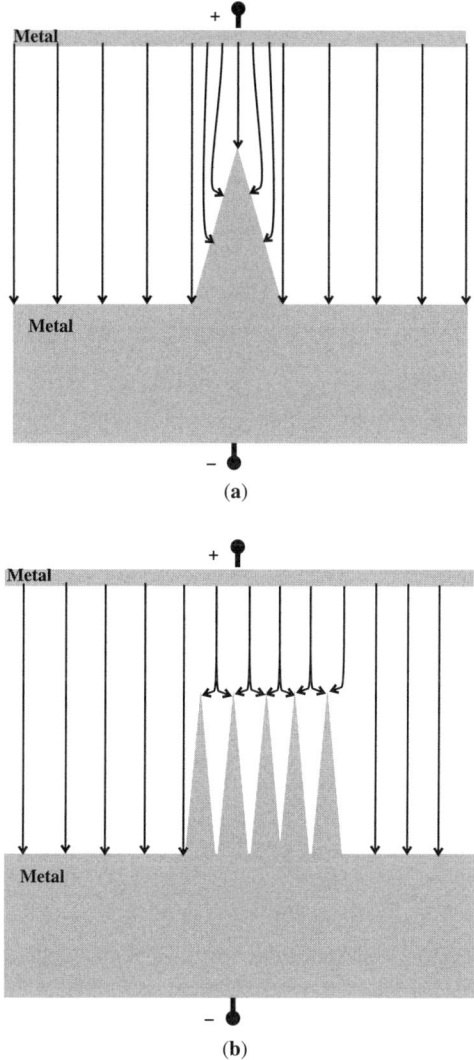

FIGURE 4.17 Electric field lines for smooth and rough features: (**a**) shows the case for a single asperity and (**b**) shows the case for closely spaced asperities.

The asperities have an aspect ration of 5 (in between the cases in Fig. 4.18a with enhancements of 3 and 10), but for a separation of 0.3 display, almost no electric field enhancement. When these asperities are separated to $d = 6$, an electric field enhancement of about 7 is calculated. These results clearly indicate that the frequency of the asperities as well as the aspect ratios are important in determining electric field enhancements, which could be important for a uniformly rough surface with small and close asperities.

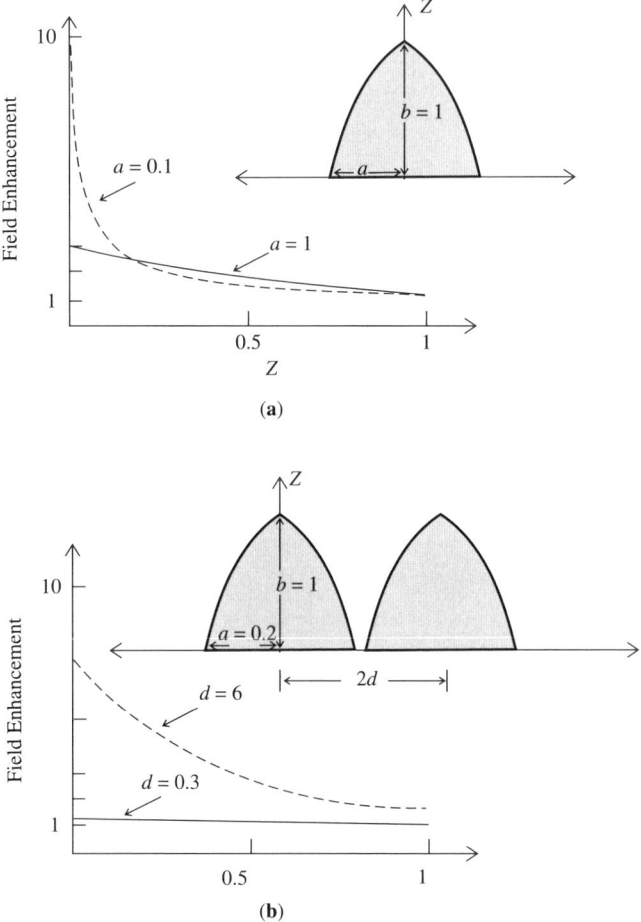

FIGURE 4.18 Electric field enhancement effects for (**a**) a single asperity with different aspect ratios b/a and (**b**) adjacent closely spaced asperities with different spacing d. [Adapted from Lewis (5), Figures 1, 2, 3, and 4.]

4.3.3 Roughness Effects on Other Physical and Chemical Properties

4.3.3.1 Roughness Effects on Contact Angle As discussed in Chapter 3, the contact angle θ yields information about the interactions occurring at an interface. The formula for $\cos\theta$ in terms of the γ's for the three interfaces, as was given in Equation 3.47 and corrected for the film pressure π, is written as follows:

$$\cos\theta = \frac{\gamma_{SV} - \gamma_{SL}}{\gamma_{LV}} = \frac{\gamma_S - \gamma_{SL} - \pi}{\gamma_{LV}} \qquad (4.18)$$

It is clear that for a rough surface the area of the surface is larger than the projected area, which is the area for a smooth surface. Thus, if one measures γ_{Sj} where j

4.3 ROUGHNESS EFFECTS ON THE PROPERTIES OF MATERIALS

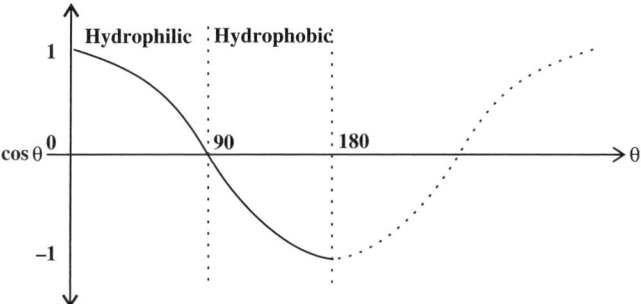

FIGURE 4.19 Cosine function with regions where the contact angle as cos θ is indicated as hydrophobic and hydrophillic.

is for liquid or vapor phases, the value obtained will be larger than for the projected area for the surface. With this in mind as well as the formula above for cos θ, we first consider a hydrophilic rough surface where $0° < θ < 90°$. For this rough surface of some material, $γ_S$ will appear larger than for a smooth surface of the same material. A larger $γ_S$ in the formula above will yield a larger cos θ. Using Fig. 4.19, which displays a cosine function, it is seen that a larger cos θ for $0° < θ < 90°$, yields a smaller θ. Hence a rough hydrophilic surface appears more hydrophilic than a smooth one of the same material. For hydrophobic surface where $90° < θ < 180°$ and formula 4.18 above, we see that $γ_{SL}$ dominates because for a hydrophobic surface $γ_S < γ_L$. As a consequence of roughness $γ_{SL}$ will appear larger and in Equation 4.18 $γ_{SL}$ is negative so cos θ will increase in the negative direction toward -1. This will yield a larger θ. Hence a rough hydrophobic surface will appear to be more hydrophobic than a smooth one of the same material. In both the hydrophilic and hydrophobic cases, roughness increases both the hydrophilic and hydrophobic interactions.

4.3.3.2 Roughness Effects on Surface Thermodynamic Properties Directly from the Kelvin equation as discussed in Chapter 3, the vapor pressure of small drops or surface features can be calculated. Starting from the Kelvin equation (Equation 3.41) written as:

$$\ln \frac{P}{P_0} = \frac{2V_m γ}{RTr} \tag{4.19}$$

the vapor pressure for small droplets of water can be calculated. With the molar volume of water $V_m = 18 \, \text{cm}^3/\text{mol}$, $γ_{H_2O} = 72 \, \text{ergs/cm}^2$, $R = 8.314 \times 10^7 \, \text{ergs/mol-K}$ and $T = 298 \, \text{K}$:

$$\ln \frac{P}{P_0} = \frac{1.05 \times 10^{-7}}{r \, (\text{in cm})} \tag{4.20}$$

TABLE 4.1 Calculation of Vapor Pressure Ratios for Droplets of Water of Various Radii r

r (Å)	P/P_0
10 (10^{-7} cm)	2.86
100 (10^{-6} cm)	1.12
1000 (10^{-5} cm)	1.06
10000 (10^{-4} cm)	1.001

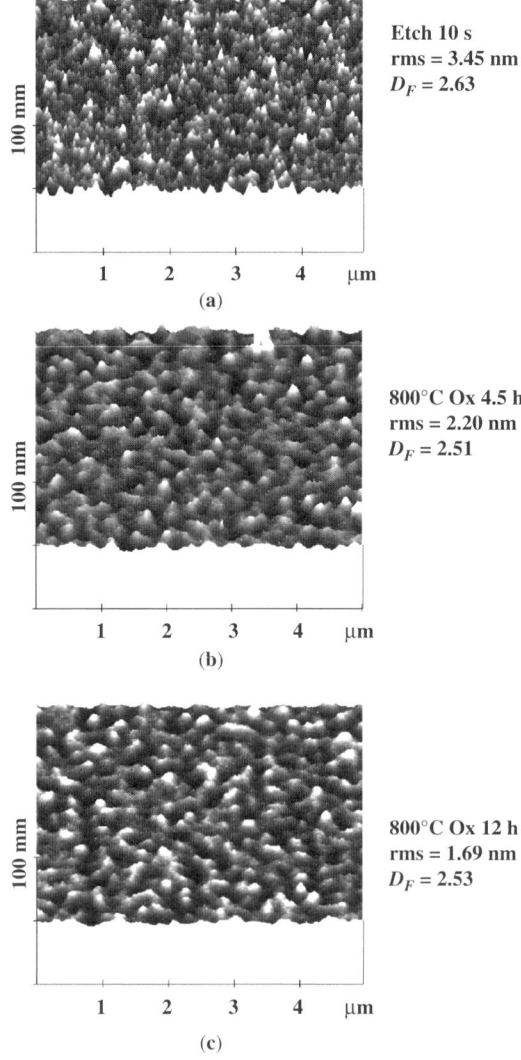

FIGURE 4.20 Si surfaces roughened (**a**) by etching in HF–HNO$_3$ and then (**b**) oxidized in pure O$_2$ and then (**c**) further oxidized in pure O$_2$. Roughness was measured using atomic force microscopy and rms and D_F values obtained. All samples were given a second HF dip prior to measurement. [Adapted from Liu et al. (6), Figure 4.]

4.3 ROUGHNESS EFFECTS ON THE PROPERTIES OF MATERIALS

The values in Table 4.1 indicate that very small particles, smaller than 100 Å in radius, can exhibit vapor pressures considerably higher than for flat surfaces, while the effect is negligible for larger particles. The conclusion is that small droplets and indeed small features are less stable than larger features due to their higher vapor pressures. Later we will reintroduce these ideas when the subject of nucleation is broached in Chapter 10. Nuclei can be clusters of a few atoms where considerable instability is observed.

4.3.3.3 Roughness Effects on Chemical Reactivity Recall that the Kelvin equation, Equation 4.19, was derived in Chapter 3 based on the condition of

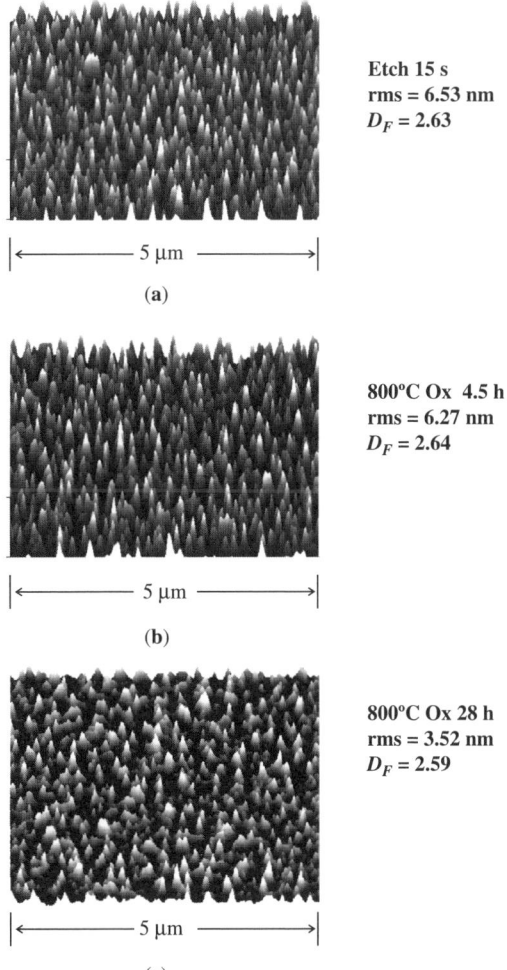

FIGURE 4.21 Si surfaces roughened (**a**) by etching in HF–HNO$_3$ and then (**b**) oxidized in pure O$_2$ and then (**c**) further oxidized in pure O$_2$. Roughness was measured using atomic force microscopy and rms and D_F values obtained. All samples were given a second HF dip prior to measurement. [Adapted from Liu et al. (6), Figure 9.]

equilibrium between solid and vapor phases and that the Kelvin equation expresses the local Gibbs free energy at surface asperities. The increase in vapor pressure is an indication of not only a change in surface energy but also of an increase in chemical potential and possibly local chemical reactivity. This relationship between surface feature curvature and reactivity could lead to a prediction that not only would a surface smoothen due to the higher vapor pressure for sharp small features but also smoothen through enhanced chemical reactivity of these same small curved features with suitable reactants. Figure 4.20 shows a Si single-crystal sample that has been roughened by exposure to a chemical etchant (a mixture of HF and HNO_3) for 10 s in Fig. 4.20a (6). The surface rms and D_F values are noted. Then this sample was oxidized in 100% O_2 for 4.5 h, and then 12 h with the results shown in Figs. 4.20b and 4.20c, respectively. Figure 4.21 shows similar results but starting with a rougher surface after 15 s chemical etch and longer oxidation time. The oxidized samples were given a few seconds HF dip prior to atomic force microscopy (AFM is discussed in Chapter 6) examination. This short dip removed the SiO_2 and improved image quality but did not affect the Si surface roughness as was determined in separate experiments. It is seen that both the rms and D_F values decrease after oxidation at the Si surface. During the oxidation Si is consumed via conversion to SiO_2. While this is occurring, the sizes of the surface features are decreasing with the rms decreasing by nearly half of its original value. In addition the decrease in D_F indicates that small features are reacting at the expense of larger ones, thereby reducing the complexity of the surface. Both factors decreasing, rms and D_F, lead to smoothing of the surface. In addition, the changing morphology of the surface features that are measured by D_F are also visually observable with fewer sharp asperities seen in Figs. 4.20c and 4.21c as compared with Figs. 4.20a and 4.21a, respectively. When an electric field is applied so as to increase electron flow to the Si surface (+ on Si), it is known to increase the oxidation rate for the oxidation of Si that is exposed to an oxygen plasma (7,8). When starting with rough Si surfaces, the electric field also enhances the smoothing effect, as shown in Fig. 4.22, with similar

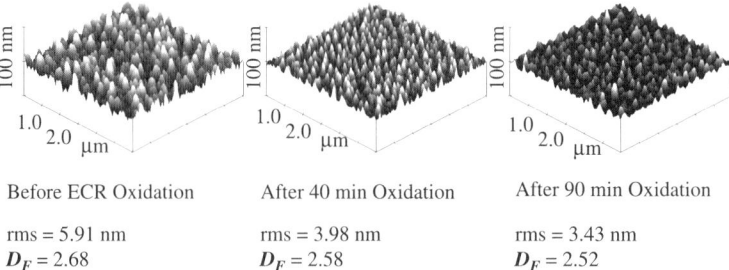

Before ECR Oxidation   After 40 min Oxidation   After 90 min Oxidation

rms = 5.91 nm   rms = 3.98 nm   rms = 3.43 nm
$D_F = 2.68$ $D_F = 2.58$ $D_F = 2.52$

FIGURE 4.22 AFM images of (**a**) HF–HNO_3 roughened Si sample, (**b**) after oxygen plasma oxidation at 200°C and bias of +60 V for 40 min, and (**c**) after 90 min plasma oxidation. All samples were given a second HF dip prior to AFM measurement. [Adapted from Zhao et al. (8), Figure 10.]

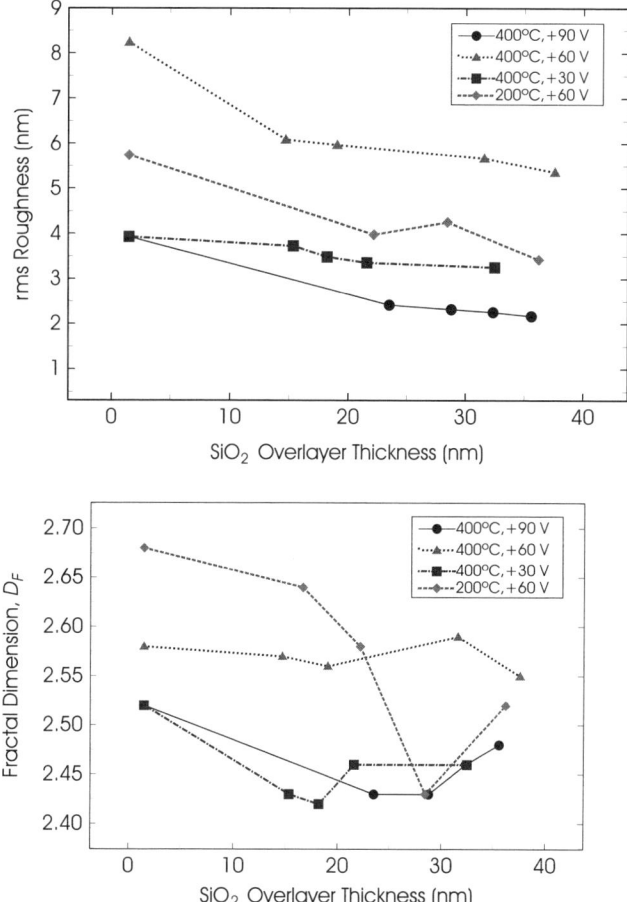

FIGURE 4.23 The rms and D_F versus SiO_2 film thickness of plasma-oxidized initially roughened Si surfaces. The samples were plasma oxidized under different conditions. All samples were given a second HF dip prior to AFM measurement. [Adapted from Zhao et al. (8), Figure 5.]

smoothing taking place with less oxidation time. Figure 4.23 shows the results of the changes in rms and D_F, summarized for a variety of plasma oxidation conditions on initially roughened Si surfaces. It is seen from this collection of results that the greater roughness changes due to electric field are observed at the lower plasma oxidation temperatures, while at higher temperatures the oxidation rate enhancement due to temperature dominates the smoothing process. Thus, there is a combined effect of enhanced chemical reactivity and enhanced electric field at the sharper asperities.

The discussion above about the smoothing that takes place due to chemical reaction at the Si surface dealt exclusively with the smoothing of previously roughened surfaces. The situation is more complicated in that initially smooth surfaces actually

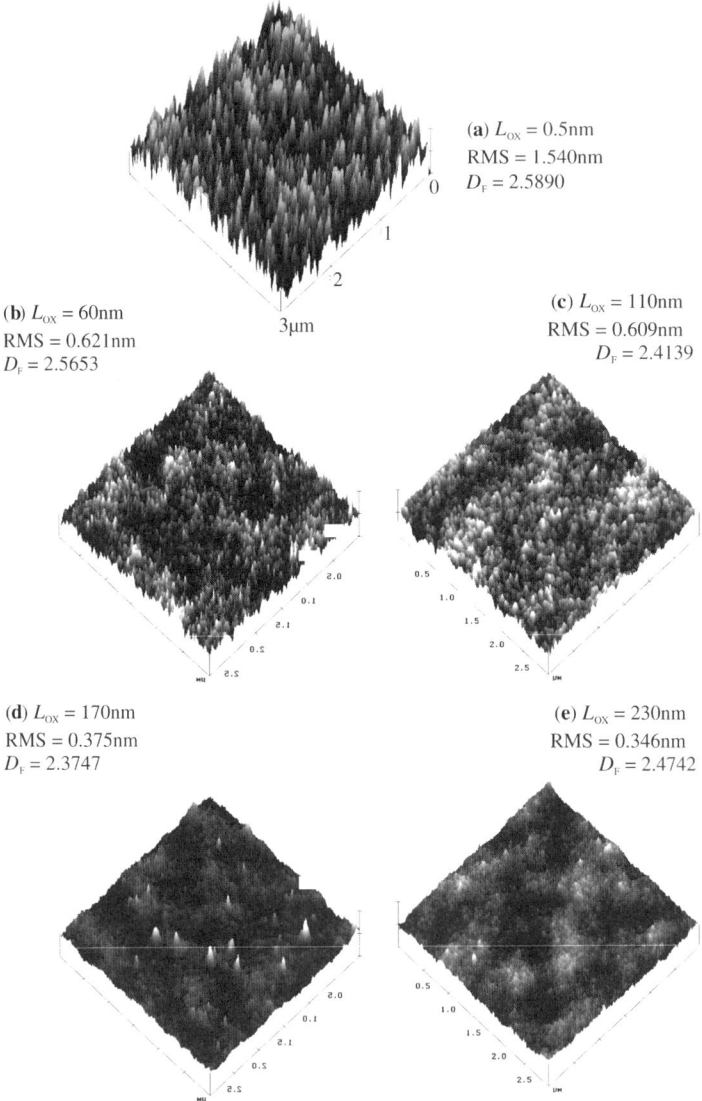

FIGURE 4.24 AFM images of (**a**) initially roughened Si that was progressively thermally oxidized at 1000°C in O_2 (**b**)–(**e**) to various SiO_2 thicknesses (L_{ox}). All samples were given a second HF dip prior to AFM measurement. [Adapted from Lai and Irene (9) Figure 3.]

roughen, while as discussed above, initially rough surfaces smoothen due to chemical reaction (9,10). Figure 4.24 shows a set of results for a Si sample that has been initially roughened (Fig. 4.24a) and then progressively oxidized at 1000°C in 100% O_2 (Figs. 4.24b to 4.24e). Overall there is a decrease in both rms and D_F as expected from the Kelvin equation (Equations 3.41 and 4.19) as was discussed

4.3 ROUGHNESS EFFECTS ON THE PROPERTIES OF MATERIALS 119

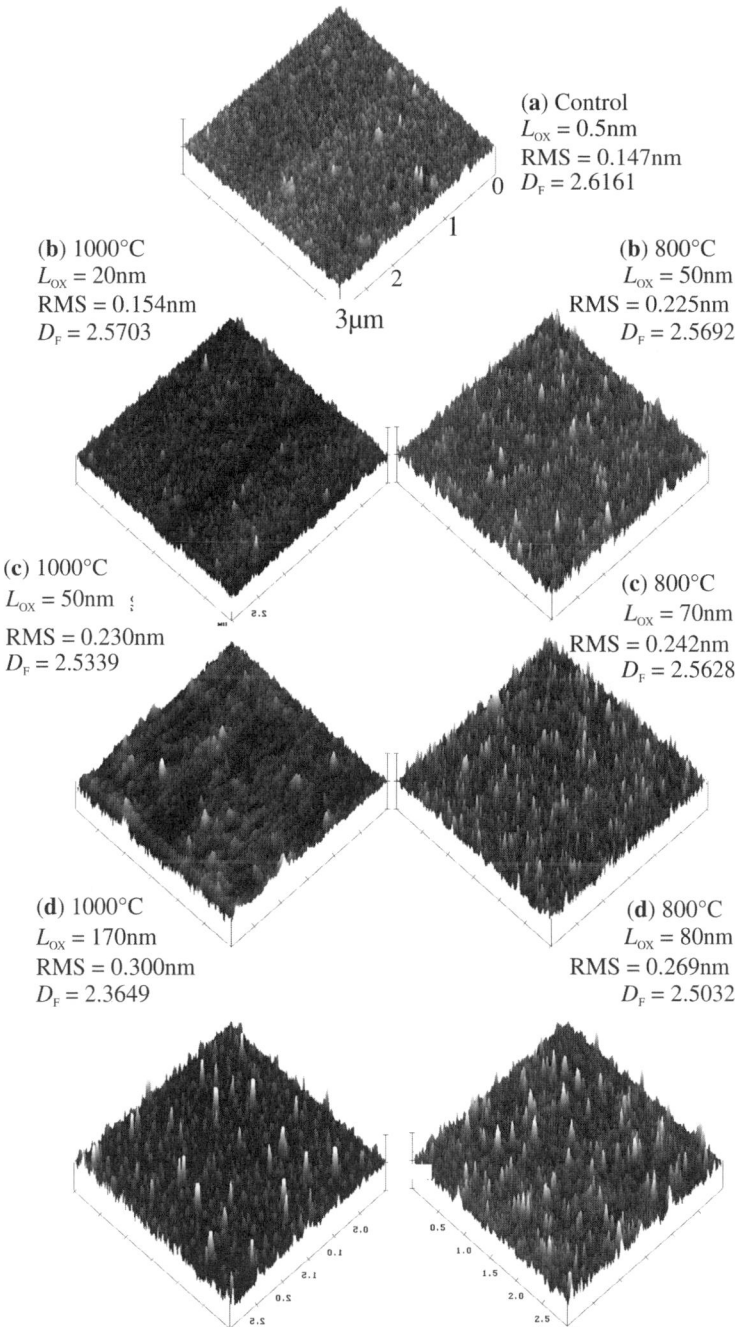

FIGURE 4.25 AFM images of (**a**) initially smooth Si that was progressively thermally oxidized in O_2 at different temperatures. The left side at 1000°C and the right at 800°C to various SiO_2 thicknesses (L_{ox}). All samples were given a second HF dip prior to AFM measurement. [Adapted from Lai and Irene (9), Figure 5.]

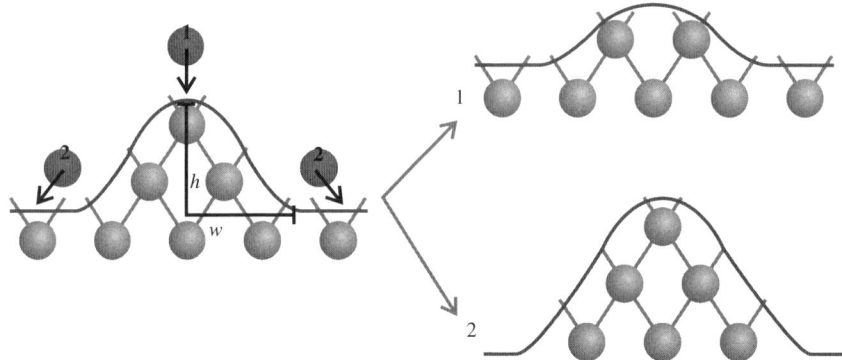

FIGURE 4.26 Two paths (1) and (2) that oxidation can reduce curvature of a surface asperity shown initially at left with a height h and width $2w$. Top right shows h decreasing and w constant and bottom right shows h and w increasing. [Adapted from Lai and Irene (9), Figure 6.]

above. Figure 4.25 shows a similar sequence but starting with an initially smooth surface (the top panel in Fig. 4.25) and then progressive oxidation of the smooth surface at two different temperatures for the left and right sides below the initial sample. The left side images in Fig. 4.25 show oxidation at 1000°C as for Figure 4.24, and the right side shows oxidation at 800°C in pure O_2. In this case of the initially smooth surface, the rms increases systematically with oxidation, but D_F decreases overall for both oxidation temperatures, although the extent of change is different. The size of the roughness features increase, but the surface complexity decreases. To explain these results, a careful consideration of how a surface feature can change to reduce the local free energy is required, and this is done with the use of Fig. 4.26. At the left of Fig. 4.26 is shown a surface feature with a width and height. On the right two pathways ways are shown by which this asperity can change, to reduce its radius of curvature and thus reduce its energy. In process 1, oxidation can occur at the top of the asperity, thus reducing the height h while the width w remains constant. In this way a tall, perhaps sharp, feature becomes rounded and shorter. In process 2 the width can increase by oxidation at the sides of the asperity. In this way, w increases and so does h. Now, if we start with a surface that has widely separated asperities, that is, a relatively smooth surface, both processes 1 and 2 can occur. Process 2 will lead to taller asperities after oxidation and hence an increased rms. If on the other hand we start with a surface that has many closely spaced asperities, that is, a roughened surface, then only process 1 can occur because process 2 can widen one asperity but sharpen the adjacent ones, and so reduce the energy for one asperity while increasing the energy of the adjacent ones by sharpening them. Thus, if an initially smooth surface statistically develops a vacancy or cluster of vacancies on the surface, a rough spot, this spot can lead to an asperity with increasing height through further oxidation and thus contribute to the rms. However, the complexity of the surface will remain low, that is, D_F will not significantly increase.

In summary the Kelvin equation is found to control the dominant surface processes, but careful attention to allowed or favored pathways is necessary to properly apply the Kelvin equation.

REFERENCES

1. Song JF, Vorburger TV. ASM handbook. Volume 18, p. 334.
2. Mandelbrot BB. The fractal geometry of nature. New York: W. H. Freeman; 1982.
3. Spanos L, et al. J Vac Sci Technol A 1994;12:2646.
4. Spanos L, et al. J Vac Sci Technol A 1994;12:2653.
5. Lewis TJ. J Appl Phys 1995;26:1405.
6. Liu Q, Spanos L, Zhao C, Irene EA. J Vac Sci Technol A 1995;13:1977.
7. Zhao C, Lefebvre PR, Irene EA. Thin Solid Films 1998;313–414:286.
8. Zhao C, Hu YZ, Labayen T, Lai L, Irene EA. J Vac Sci Technol A 1998;16:57.
9. Lai L, Irene EA. J Appl Phys 1999;86:1729.
10. Lai L, Irene EA. J Appl Phys 2000;87:1159.

5

SURFACE ELECTRONIC STATES

5.1 INTRODUCTION

In previous chapters the nature of surfaces has been examined from the viewpoint of surface sites, namely how a surface varies from point to point, by virtue of changes in structure and thermodynamics. Insights in terms of overall surface structure and reactivity were obtained without considering the specific electronic interactions, that is, without quantum mechanics. It is the purpose of this chapter to consider the electronic configurations of solid surfaces and arrive at a deeper understanding of surface via an extension of the bulk solid electronic structure, the so-called electronic energy band structure. It is clear that the electronic energy band structure for solids yields a level of understanding that not only enables electronic devices to be analyzed but also enables a more complete understanding of the properties and reactivity of solids. Now the powerful ideas of bulk solid electronic energy band structure are extended using the notions of solids developed in earlier chapters. For example, it is clear from Chapter 1 that the bonding at and near surfaces is different from the bulk. Nearest-neighbor considerations led to the TLK model discussed in Chapter 1 that enables visualization of the differing surface sites. Specifically, the binding potential for atoms is necessarily weaker at and near surfaces. The consequences of this prevailing situation can be evaluated starting from the bulk electronic energy band structure. For this purpose the Kronig–Penney (KP) model is presented (1) to provide both a methodology for the extension to surfaces as well as a framework for other viewpoints of surface electronic states. The KP model was published in the 1930s near the

---

Surfaces, Interfaces, and Thin Films for Microelectronics. By Eugene A. Irene
Copyright © 2008 John Wiley & Sons, Inc.

beginning of quantum mechanics, and it is therefore not competitive with the calculation power of more modern treatments. However, the KP model contains the essential if not complete physics of the quantum mechanics of solids and leads directly to allowed energy bands and gaps without overbearing mathematics. On the negative side the KP model does not permit accurate calculations of band energies.

5.2 THE KRONIG–PENNEY MODEL

5.2.1 The KP Model for Infinite Solids

The KP model considers an array of atomic potentials that has the order or periodicity of the crystal lattice. An example of one kind of ordered one-dimensional (1D) potential is displayed in Fig. 5.1a. This crystal potential is periodic and has a complex shape

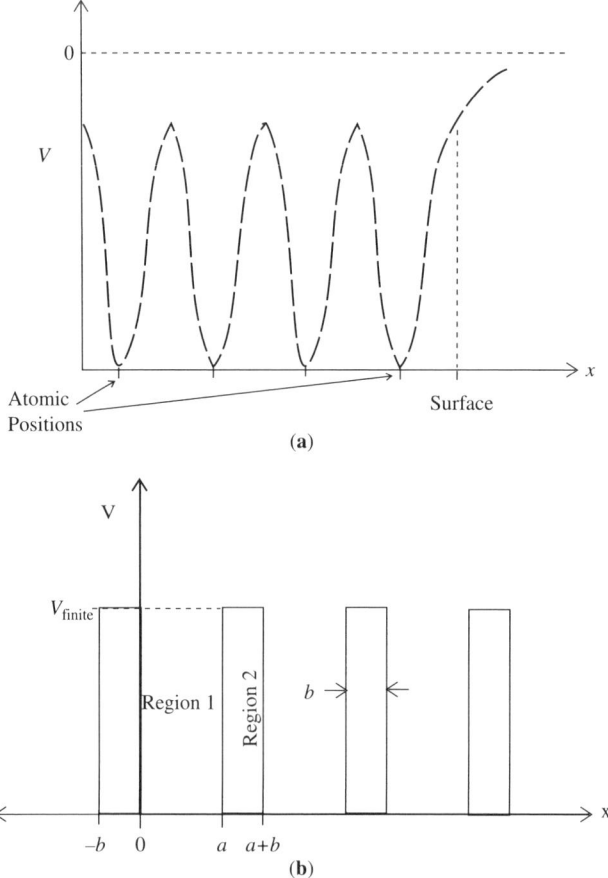

FIGURE 5.1 (a) One-dimensional periodic binding potential for a solid and (b) Kronig–Penney potential for a periodic solid.

5.2 THE KRONIG–PENNEY MODEL

that is essentially Coulombic and that depends on the reciprocal of the square of the atomic separation. The binding is stronger closer to the atomic core and drops sharply away from the core. It is noticed that Fig. 5.1a is sketched as a binding potential, and thus the zero of energy is at the top with the binding energies below zero with the maximum negative energies at the atomic cores. Of course, different atoms would display different shaped potentials for a monatomic solid, and thus the solution of the Shrödinger equation (SE) for one such potential would require modification for another kind of atom. Also, for materials with different kinds of atoms, different structures and morphologies, the situation becomes even more complicated. Yet in those cases in which potentials are available, precise calculations of the electronic structure have been made. The point is that procedures are presently known to solve complicated SEs. For the present purpose of understanding the origin of surface electronic states, a simple model for a solid that yields physically correct albeit inaccurate numerical results is useful. This idea was successfully pursued in the early 1930s and is called the Kronig–Penney (KP) model and this model will be outlined below.

The KP model commences with a simplification of the potential for a periodic solid shown in Fig. 5.1a. Rather than attempting to approximate complex shaped potentials, the KP model uses the conjoining of two regions and repeating these regions with the crystal periodicity. Figure 5.1b shows the KP periodic potential as made up of a potential free region ($V = 0$), region 1 with a width a as the separation between atoms (similar to the lattice parameter); and region 2 having a finite binding potential (V_{finite}) and width b. Notice that in this figure the reference zero is at the bottom and the bonding potentials are represented as positive energies as was done in the original literature (1). As mentioned above this simple-minded potential cannot yield correct numerical results, but as we will see below, when the results from the KP model are compared with real electronic structures, the KP model will be found to reveal the correct overall physics, but without some details. Furthermore, the KP model is amenable to algebraic analytical solution and follows directly from the solutions above for the free and bound electrons.

The procedure to obtain a solution of the SE using the KP model commences with Fig. 5.1b and the two relevant SEs written for region 1 and region 2 as follows:

$$\text{Region 1} \quad \frac{d^2\psi}{dx^2} + \frac{2m_e}{\hbar^2} E\psi = 0 \tag{5.1}$$

$$\text{Region 2} \quad \frac{d^2\psi}{dx^2} + \frac{2m_e}{\hbar^2} (E - V_{finite})\psi = 0 \tag{5.2}$$

The energies for regions 1 and 2, respectively, can be expressed as

$$\alpha = \sqrt{\frac{2m_e E}{\hbar^2}} \quad \text{and} \quad \beta = \sqrt{\frac{2m_e (V_{finite} - E)}{\hbar^2}} \tag{5.3}$$

For the KP model the potential is periodic throughout the crystal, and the solution for the two regions must, therefore, be a simultaneous solution extending over the entire crystal. Solutions for this kind of problem exist, and the solutions are given by the so-called Bloch theorem to yield solutions in one dimension of the form:

$$\psi(x) = u(x)e^{ikx} \tag{5.4}$$

where $u(x)$ is a periodic function with the same period as the barriers $(a + b)$ and k has the magnitude $2\pi/L$ where L is the wavelength or period. Notice that this solution is essentially a modulated wave where $u(x)$ is the modulating function. These waves are often referred to as Bloch waves. Of course, our discussion will be stated in only one dimension, but it should be understood that the periodicity can be different in different crystal directions and thus the solution in three dimensions is more complicated.

The Bloch theorem is used to provide the correct solution for our periodic potential problem. The appropriate derivatives of ψ are taken and the SE is reformatted with these derivatives. The first derivative of the solution $\psi(x)$ yields two terms:

$$\frac{d\psi(x)}{dx} = u(x)ike^{ikx} + e^{ikx}\frac{du(x)}{dx} \tag{5.5}$$

The second derivative yields four terms:

$$\frac{d^2\psi(x)}{dx^2} = u(x)i^2 k^2 e^{ikx} + ike^{ikx}\frac{du(x)}{dx} + e^{ikx}\frac{d^2u(x)}{dx^2} + ike^{ikx}\frac{du(x)}{dx} \tag{5.6}$$

and with terms collected we obtain

$$\frac{d^2\psi(x)}{dx^2} = e^{ikx}\left[\frac{d^2u(x)}{dx^2} + 2ik\frac{du(x)}{dx} - k^2 u(x)\right] \tag{5.7}$$

These derivatives are now substituted in the region 1 and 2 SEs, and the following substitutions are also made using Equation 5.3:

$$E = \frac{\hbar^2\alpha^2}{2m_e} \quad \text{and} \quad V_{\text{finite}} - E = \frac{\hbar^2\beta^2}{2m_e} \tag{5.8}$$

For region 1 the SE is as follows:

$$e^{ikx}\left[\frac{d^2u(x)}{dx^2} + 2ik\frac{du(x)}{dx} - k^2 u(x)\right] + \alpha^2\psi(x) = 0 \tag{5.9}$$

5.2 THE KRONIG–PENNEY MODEL

Also substitute $\psi(x) = u(x)e^{ikx}$ the final equation for region 1 is as follows:

$$\text{Region 1} \qquad \frac{d^2u(x)}{dx^2} + 2ik\frac{du(x)}{dx} - (k^2 - \alpha^2)u(x) = 0 \tag{5.10}$$

For region 2 the SE is as follows:

$$e^{ikx}\left[\frac{d^2u(x)}{dx^2} + 2ik\frac{du(x)}{dx} - k^2u(x)\right] - \beta^2\psi(x) = 0 \tag{5.11}$$

Also with a substitution for $\psi(x) = u(x)e^{ikx}$ as was done for region 1 above, we obtain for region 2:

$$\text{Region 2} \qquad \frac{d^2u(x)}{dx^2} + 2ik\frac{du(x)}{dx} - (k^2 + \beta^2)u(x) = 0 \tag{5.12}$$

The region 1 and 2 SEs can be recognized as having the same form as the differential equation for a damped vibration in classical physics. If $u(x)$ is defined as the displacement for the vibration, then the form is as follows:

$$\frac{d^2u(x)}{dx^2} + A\frac{du(x)}{dx} + Bu(x) = 0 \tag{5.13}$$

This equation has a general solution of the form:

$$u(x) = e^{-Ax/2}\left(Ce^{i\xi x} + De^{-i\xi x}\right) \quad \text{where } \xi = \sqrt{B - \frac{A^2}{4}} \tag{5.14}$$

For regions 1 and 2, $A = 2ik$; for region 1, $B = -(k^2 - \alpha^2)$ and for region 2, $B = -(k^2 - \beta^2)$. Putting this together, the following solutions are obtained:

$$\text{Region 1} \qquad u(x) = e^{-ikx}\left(Ce^{i\alpha x} + De^{-i\alpha x}\right) \tag{5.15}$$

$$\text{Region 2} \qquad u(x) = e^{-ikx}\left(Ae^{i\beta x} + Be^{-i\beta x}\right) \tag{5.16}$$

where A, B, C, and D are constants that need to be determined by the specific conditions of the problem, namely that the solutions and their derivatives are continuous at the boundary ($x = 0$) and at all periods ($x = a + b$). This will yield four equations in the four unknowns A, B, C, and D.

At $x = 0$ the solutions given by Equations 5.15 and 5.16 and the derivatives are equated (subscripts 1 and 2 are used to indicate the regions) to yield the first two equations as follows:

For $u_1(x) = u_2(x)$ at $x = 0$ $\quad C + D = A + B$ (5.17)

For $\dfrac{du_1(x)}{dx} = \dfrac{du_2(x)}{dx}$ at $x = 0$ $\quad Ci(\alpha - k) + Di(-\alpha - k)$
$= Ai(\beta - k) + Bi(-\beta - k)$ (5.18)

Also two equations can be obtained for the periodic boundary condition $x = a + b$. These equations are most easily obtained by using Fig. 5.1b and realizing that $u_1(x)$ at a must equal $u_2(x)$ at $-b$, and likewise for the derivatives, and this requirement yields the following two equations:

For $u_1(x)$ (at $x = a$) $= u_2(x)$ (at $x = -b$)
$Ce^{(i\alpha - ik)a} + De^{(-i\alpha - ik)a} = Ae^{(ik - i\beta)b} + Be^{(ik + i\beta)b}$ (5.19)

For $\dfrac{du_1(x)}{dx}$ (at $x = a$) $= \dfrac{du_2(x)}{dx}$ (at $x = -b$)
$Ci(\alpha - k)e^{ia(\alpha - k)} - Di(\alpha + k)e^{-ia(\alpha + k)} = Ai(\beta - k)e^{-ib(\beta - k)} - Bi(\beta + k)e^{ib(\beta + k)}$
(5.20)

Now there are four independent equations in the four unknowns A, B, C, and D, namely Equations 5.17 to 5.20. A simultaneous solution is obtained by forming a determinant of the coefficients C, D, A, B and setting the determinant equal to zero. The determinant has to form:

$$\begin{vmatrix} 1 & 1 & 1 & 1 \\ \alpha - k & -(\alpha + k) & \beta - k & -(\beta + k) \\ e^{ia(\alpha - k)} & e^{-ia(\alpha + k)} & e^{-ib(\beta - k)} & e^{ib(\beta + k)} \\ (\alpha - k)e^{ia(\alpha - k)} & -(\alpha + k)e^{-ia(\alpha + k)} & (\beta - k)e^{-ib(\beta - k)} & -(\beta + k)e^{ib(\beta + k)} \end{vmatrix}$$

The result after setting the determinant equal to 0, considerable algebra, and applying Euler's formulas is as follows:

$$\dfrac{\beta^2 - \alpha^2}{2\alpha\beta} \sinh(\beta b) \sin(\alpha a) + \cosh(\beta b) \cos(\alpha a) = \cos[k(a + b)] \quad (5.21)$$

This result can be further simplified using several physically relevant assumptions. The first is to assume that V_{finite} is large compared with the kinetic energy for the

5.2 THE KRONIG–PENNEY MODEL

electron, so that from Equation 5.3 the expression for β above becomes

$$\beta = \sqrt{\frac{2m_e V_{\text{finite}}}{\hbar^2}} \quad \text{and} \quad \beta b = \sqrt{\frac{2m_e V_{\text{finite}} b^2}{\hbar^2}} \tag{5.22}$$

Also, since $\beta \propto V_{\text{finite}}^{1/2}$ and $\alpha \propto E^{1/2}$, then $\beta > \alpha$ and $\beta^2 \gg \alpha^2$. Thus, α^2 can be removed from the first term in the result above. Now we assume that the electron binding potential drops off sharply at the atomic cores. Effectively, this assumes a narrow barrier. Thus, as $b \to 0$, βb becomes small. The cosh of a small argument is 1 and the sinh of a small argument is the argument. Including all these assertions simplifies the KP result in Equation 5.21 to the following:

$$\frac{\beta^2}{2\alpha\beta}(\beta b)\sin(\alpha a) + \cos(\alpha a) = \cos(ka) \tag{5.23}$$

which upon substitution using Equation 5.22 becomes the following:

$$\frac{m_e b V_{\text{finite}}}{\hbar^2 \alpha}\sin(\alpha a) + \cos(\alpha a) = \cos(ka) \tag{5.24}$$

It is typical to define a term P as follows:

$$P = \frac{a m_e b V_{\text{finite}}}{\hbar^2} \tag{5.25}$$

so that the final form from the KP analysis is as follows:

$$\frac{P}{\alpha a}\sin(\alpha a) + \cos(\alpha a) = \cos(ka) \tag{5.26}$$

This final simplified form is of great interest because from it important revelations about electronic structure of solids are directly derived, as we will see below. First, we should notice that the right-hand side is a cosine function with the argument (ka). Thus, the only permissible values for the left-hand side of the final form lie between 1 and -1. The first term on the left is essentially $\sin(\alpha a)/\alpha a$ where P is a composite constant defined above in Equation 5.25. This function, $\sin(\alpha a)/\alpha a$, is called a sinc function, and it is characterized as a periodic function with decreasing amplitude as is shown in Fig. 5.2a. Also keep in mind that α is an expression of the energy for the electron from Equation 5.3. Thus, the left-hand side of the final KP formula defines all the energies for electrons. When all these energies are bound by the right-hand side of Equation 5.26, the allowed energies lie between values of $+1$ and -1 for the left-hand side of Equation 5.16. These allowed energy regions are called allowed energy bands. Figure 5.2b shows a plot of the left-hand side of the KP formula versus αa (or energy) for two values of P; one value of P

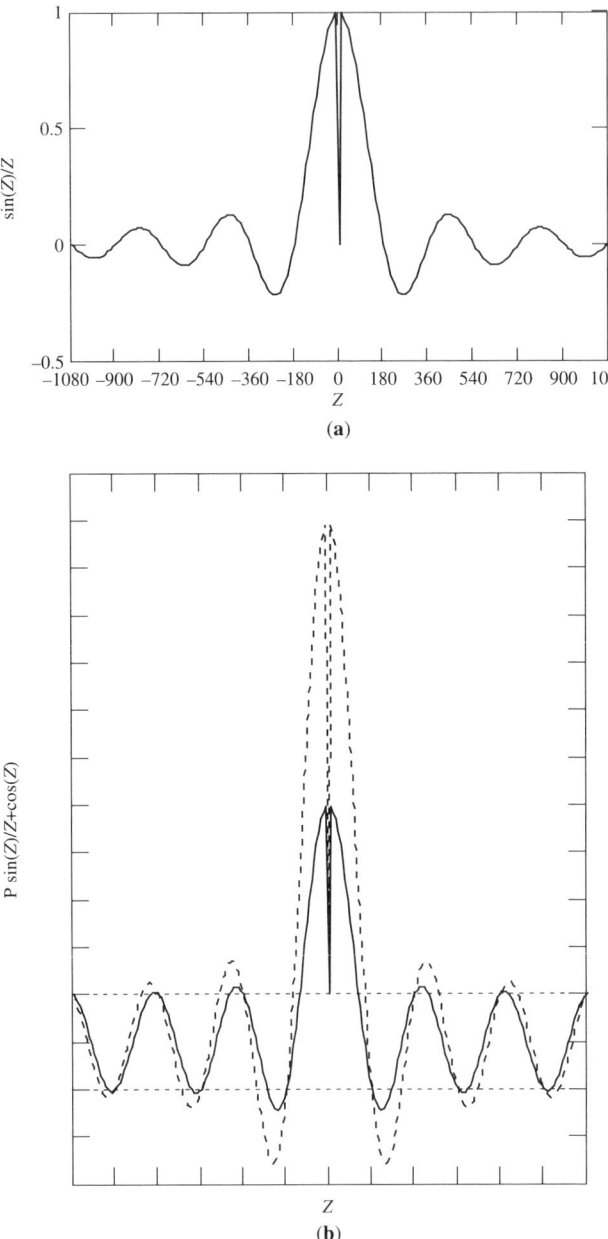

FIGURE 5.2 (a) Plot of sinc (Z) function where Z is an angle θ; (b) plot of left side of KP formula Equation 5.26 for two values of P (large dashed and small solid) where the argument Z is αa and the top and bottom horizontal lines are $+1$ and -1, respectively, indicate the limits of the sinc (Z) or cos(ka) from Equation 5.26.

5.2 THE KRONIG-PENNEY MODEL

is 2.5 times the other for comparison. Also plotted are the horizontal lines at values $+1$ and -1, which as was mentioned above form the boundaries for allowed electron energy solutions as dictated by the right-hand side of the KP formula (Equation 5.26). The scales are conveniently chosen and do not reflect real numbers. However, the shape of the solution in terms of allowed energy bands is revealed.

The KP formula plotted in Fig. 5.2b shows several important features. From the zero of energy at the center the KP function, there is a sharp rise and then the function decreases traversing the $+1$ to -1 region in its decent; and it repeats with ever decreasing amplitude until all the values lie between $+1$ and -1. The values of energy that the KP function (f_{KP}) takes in the region $+1 \geq f_{KP} \geq -1$ are the allowed energies. This region is called an allowed energy band. Notice that both before and after an allowed band there is a region of energies that lie above or below the allowed region. These regions are obviously not allowed energies and are therefore called energy band gaps. The electronic structure is therefore composed of allowed energy bands separated by disallowed energy gaps, or more simply of bands and gaps. With this picture and simple KP model in mind we extend the ideas to finite solids.

5.2.2 The KP Model Extended for Finite Solids

For the KP model as was applied above to infinite solids, the binding potential was strong in the neighborhood of the atom core, with width b in Fig. 5.1b. However, this strong binding potential dropped sharply to zero away from the atom core, as is depicted by the region of width a in Fig. 5.1b with $V = 0$ binding potential. This drop in potential away from the core is more realistically depicted in Fig. 5.1a, where the potential approaches zero near the surface and drops to zero beyond the surface. So the potential is not zero at the surface but less than the bulk potential. With this notion we can then reconsider Fig. 5.2b that shows that as the binding potential is reduced, the allowed energy bands widen, and the forbidden electron gaps (FEGs) become narrow. Keeping this picture for the infinite solid in mind, and now considering the formation of a surface as the cleavage of an infinite crystal, new surfaces are obtained that have the substantive crystal bonding in one direction toward the bulk solid and no bonding in the other direction. Needless to say in the direction away from the newly created surfaces there are no atoms and thus no bonding. However, the binding potential V is not zero at the surface and even exists some distance away from the surface, as seen in Fig. 5.1a. Because the near surface and surface atoms have fewer neighbors, the potential at and near the surface is smaller than the uniform potential for the bulk material. According to the KP model, this reduced potential at and near the surface will lead to a narrowing of the FEG. In effect the allowed bands include more states near the surface. These new states that are associated with altered (reduced) bulk potential near the surface are referred to as *surface electronic states*. Because these states arise from the intrinsic properties of surfaces, they are also referred to as *intrinsic surface electronic states*. The unsatisfied chemical bonds that result from the cleavage and the formation of a new surface are referred to as *dangling bonds*.

In the traditional KP treatment above, the continuity of the appropriate wave functions across a boundary where the binding potential abruptly changed from $V = 0$ to some finite value, V_{finite} was considered. Also, a similar situation will be discussed for scanning tunneling microscopy (STM) in Chapter 6 where in the region of high potential, the so-called tunneling barrier, the wave function decays exponentially into the barrier. This exponential decay arises from the requirement that the wave function be well behaved, that is, finite, as the coordinate x approaches infinity. If this were not the case, then the wave function would approach infinity and so would the probability for the electron, which is nonphysical.

A similar line of reasoning is now pursued to match the appropriate wave functions across the vacuum–crystal interface labeled as $x = 0$ in Fig. 5.3a. Figure 5.3b shows the appropriate wave functions discussed below for the vacuum (ψ_v) and crystal (ψ_c)

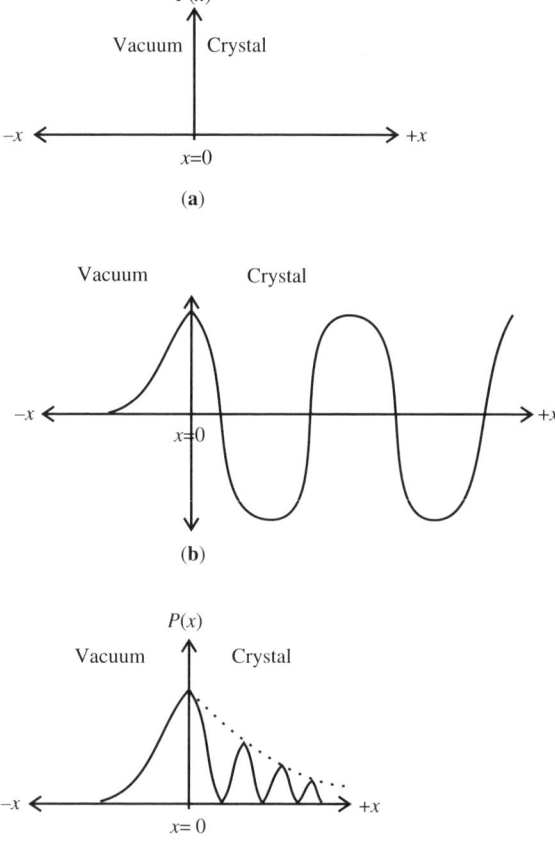

FIGURE 5.3 (a) Vacuum-crystal interface ($x = 0$) coordinates; (b) with appropriate wave functions matched at the interface, and (c) the relative probability $P(x)$ of an electronic state near the interface in the x direction.

5.2 THE KRONIG–PENNEY MODEL

sides of the interface. From our KP treatment of an infinite crystal and the Block functions given as Equation 5.4, we can obtain the wave function inside the crystal (ψ_c) of the following form:

$$\psi_c = Au_{\alpha_1}(x)e^{i\alpha_1 x} + Bu_{-\alpha_1}(x)e^{-i\alpha_1 x} \tag{5.27}$$

where from Equation 5.3

$$\alpha_1 = \sqrt{\frac{2m_e E}{\hbar^2}}$$

For the vacuum side of the interface, $x < 0$, as $x \to -\infty$, ψ_v (vacuum) is obtained from a solution for a finite barrier as is discussed in Chapter 6 and given as Equation 6.21. Here this result ψ_v is written for the positive and negative x coordinate in Fig. 5.3a as follows:

$$\psi_v = De^{i\alpha_2 x} + Ce^{-i\alpha_2 x} \tag{5.28}$$

From Equation 5.3 for α_2 for $-x$ for the vacuum side of the interface, we set $C = 0$ to make ψ_v finite and obtain

$$\psi_v = De^{\alpha_2 x} \tag{5.29}$$

From Equation 5.3 for α_2 for $V_{\text{finite}} > E$, α_2 is real while for $V_{\text{finite}} < E$, α_2 is imaginary.

The wave functions ψ and derivatives ψ' are matched at the surface at $x = 0$, from which the following relations are obtained:

$$D = Au_{\alpha_1}(0) + Bu_{-\alpha_1}(0) \tag{5.30}$$

$$\alpha_2 D = A[\alpha_1 u'_{\alpha_1}(0) + i\alpha_1 u_{\alpha_1}(0)] + B[\alpha_1 u'_{-\alpha_1}(0) - i\alpha_1 u_{-\alpha_1}(0)] \tag{5.31}$$

There are two possibilities for α_1: α_1 real and α_1 complex, and these are considered in terms of the resulting electronic energy states.

Possibility α_1 real means that the electron energies at or near the surface ($x = 0$) lie within an allowed band for the crystal. The matching conditions given as Equations 5.30 and 5.31 are linear equations in A, B, and D with solutions for any E in the allowed band. ψ_c is finite for any A and B and the matching conditions equations yield the same allowed states for the semi-infinite crystal as for the infinite crystal. In other words the same electronic energy levels allowed for an infinite crystal are also allowed for the crystal with a terminating surface.

The α_1 complex means that the electron energies in the crystal are not allowed at the surface and the new states lie in the FEG. A complex energy corresponds to the KP function $\cos ka$ exceeding ± 1. For the semi-infinite crystal ψ_c remains finite as $x \to \infty$ if $A = 0$. This yields a modulated wave as

$$\psi_c = Bu_{-\alpha_1}(x)e^{-i\alpha_1 x} \tag{5.32}$$

This wave function is also periodic in the crystal [from the modulating function $u(x)$], but it decreases exponentially as $x \to \infty$ and it is matched with ψ_v, which also decreases exponentially into the vacuum ($x \to -\infty$), and this result is shown in Fig. 5.3b.

To examine the location for the new states, the surface states, we can calculate the relative probability P for the location of the surface states relative to $x = 0$ where P is defined as

$$P = \frac{|\psi(x)^2|}{|\psi(0)^2|} \tag{5.33}$$

First we test $\psi_v = De^{\alpha_2 x}$ for negative x (in vacuum):

$$P = \frac{|\psi(-x)^2|}{|\psi(0)^2|} = \frac{D^2 e^{-2\alpha_2 x}}{D^2} = \frac{1}{e^{2\alpha_2 x}} \tag{5.34}$$

which is a maximum at $x = 0$.

Then we test $\psi_c = Bu_{-\alpha_1}(x)e^{-i\alpha_1 x}$ for positive x (in the crystal):

$$P = \frac{|\psi(x)^2|}{|\psi(0)^2|} = \frac{B^2 u_{-\alpha_1}^2(x)e^{-2i\alpha_1 x}}{B^2} = \frac{u_{-\alpha}^2(x)}{e^{2i\alpha_1 x}} \tag{5.35}$$

Recalling from the discussion of Equation 5.4 that u is a periodic function (a modulated wave) with the period of the lattice (in the x direction) and, for example, can be given as

$$u(x) = e^{-Ax/2}\left(Ce^{i\xi x} + De^{-i\xi x}\right) \tag{5.36}$$

The numerator in P is also a periodic function albeit squared, but the denominator forces the expression for P to be a maximum at $x = 0$ as well; $P(x)$ is shown in Fig. 5.3c.

Thus, the matching conditions for the wave functions yield states in the FEG that are localized near the surface. Consequently, these states are called surface states.

5.3 OTHER MODELS FOR SURFACE STATES

Another way to look at the formation of surface states is to consider the formation of energy bands and gaps from the splitting of atomic orbitals as atoms come together from infinity to their minimum energy configuration in the solid state. Figure 5.4a shows how bulk electron bands are formed from the splitting of atomic levels as atoms come from infinity (far right) to the appropriate spacing range (near a_0) for the formation of the solid-state lattice. The amount of splitting that occurs for a particular material is a strong function of the binding potential. From our previous discussion about binding potentials, it should be remembered that in the near surface region the potential is different from the bulk potential that was used to form the bulk bands in Fig. 5.4a. In fact, the binding potential in the near surface region is

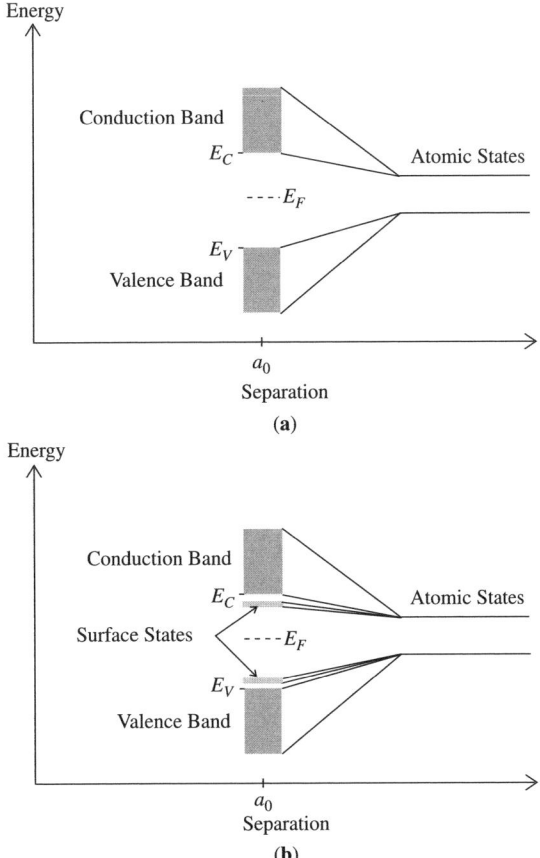

FIGURE 5.4 (a) Energy versus atomic separation for bulk atomic states arising from the bulk binding potential yielding energy bands at the equilibrium separation and (b) the surface states created as a result of the different surface binding potential.

smaller than that for the bulk due to the reduction in interactions among fewer neighbors for near surface and surface atoms. As we have learned before, the smaller potential for the near surface atoms leads to a smaller FEG, that is, less splitting. Thus, the situation is that we need to simultaneously consider the splitting of the bulk atomic levels with the associated bulk potential along with the near surface atoms that will have smaller potentials. Also there are fewer near surface atoms to come together relative to the number of bulk atoms. This situation will lead to a large number of levels to form bulk electron energy bands resulting from the bulk atoms, and in the FEG will be a smaller number of narrow discrete states called surface states, since they are associated with the near surface atomic splitting. In Fig. 5.4b the splitting associated with near surface atoms is shown along with the bulk bands. Because of the small number of surface states, they are largely localized states.

Another useful model for surface states arises from the binding potential derived from the positive ion cores that are immersed in a sea of free electrons. These electrons arrange themselves so as to reduce or screen the charge from the positive cores. This kind of solid is called a jellium, and this problem has been solved self-consistently for metals with a sampling of the results in Fig. 5.5 that displays the charge density. These results indicate that the electronic charge density is maximum near but below the surface and varies periodically into the solid and decays rapidly beyond the positive ion cores away from the surface. This kind of distribution of electronic charge yields a dipole at the surface that can affect the work function and chemical reactions at the surface that will be discussed later.

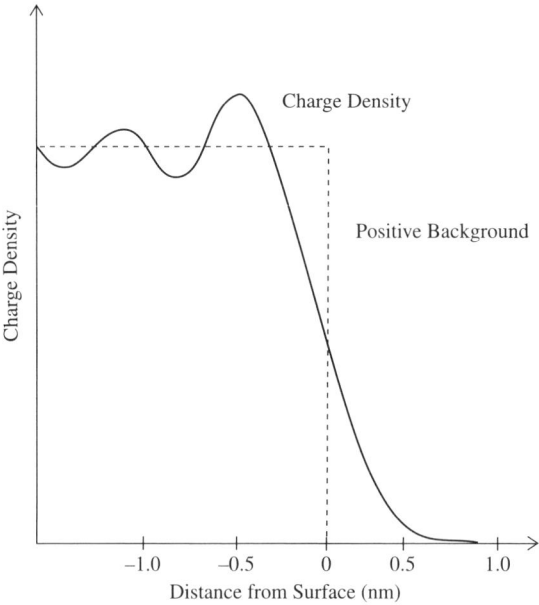

FIGURE 5.5 Jellium model of charge distribution. [Adapted from Lang and Kohn (2), Figure 2.]

5.3.1 Extrinsic Surface States

The discussion above about the origin and location of surface electronic states originated with the electronic band structure for materials and how that band structure is modified near a surface. The changes in the binding potential near the surface gave rise to new states associated with the surface potential. These surface electronic states arose from a consideration of the intrinsic nature of the material and are called intrinsic surface electronic states. Other changes in the interatomic potential can also occur near the surface that can yield new surface electronic states. Foreign atoms or molecules adsorbed or reacted at the surface and/or defects can perturb local chemical bonding and therefore alter the near surface interatomic potentials. Consequently, surface electronic states can arise from these extrinsic sources and such states are called extrinsic surface electronic states. Figure 5.6 summarizes several extrinsic surface electronic states as well as intrinsic states in cartoon form. The material shown is composed of two different atoms depicted as squares and circles. Near the surface of this solid the so-called dangling bonds seen in Fig. 5.6 are merely the result of the termination of the solid, and these unsatisfied bonds yield the lower binding potential for surface and near surface atoms and consequently the intrinsic surface electronic states discussed above. In this figure different kinds of unsatisfied bonds are seen arising from the different kind of surface atoms in this solid. In addition, an impurity atom is seen bonded to the surface (triangle). This foreign atom, whether it is chemically or physically bound to the surface, will alter the local potential through its unique bonding to the surface and through dangling bonds specific to its nature and yield extrinsic surface electronic states. Also shown is the absence of one kind of surface atom, a vacancy that can also alter the local potential and give rise to extrinsic states. Other kinds of defects such as dislocations and grain boundaries can also intersect the surface, alter the local potential, and give rise to extrinsic surface states. Stresses acting on a solid can cause a displacement in atoms, and often thin films on a surface can produce strain in the surface and lead to additional extrinsic surface states and/or change the energy of existing states. The diversity in the kinds of surface states can sometimes lead to a wide energy distribution of surface states. Whether any of the states will be localized or extended will depend on the interaction among the states. Owing to the low

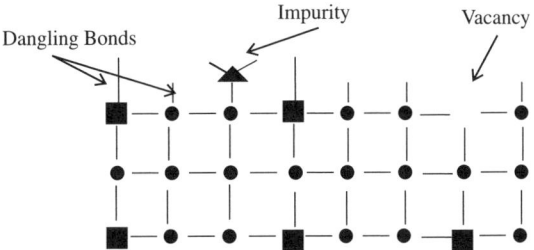

FIGURE 5.6 Illustration of some surface conditions that can result in new surface states.

number of the surface states, intrinsic or extrinsic and hence their separation, most surface electronic states are localized.

5.3.2 Band Bending

The surface electronic states that lie near the Fermi level of metals or in the FEG for insulators and semiconductors are the most important since they can alter the

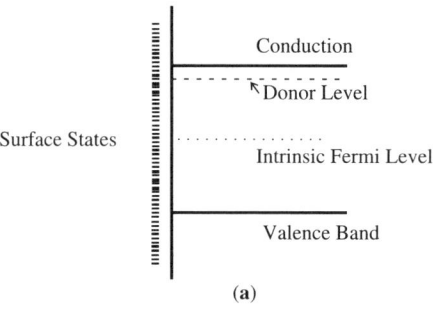

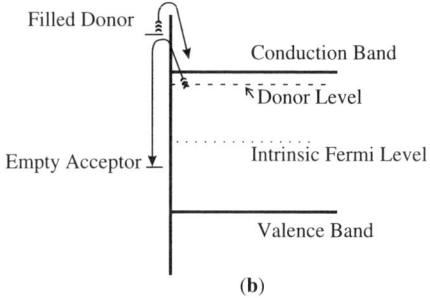

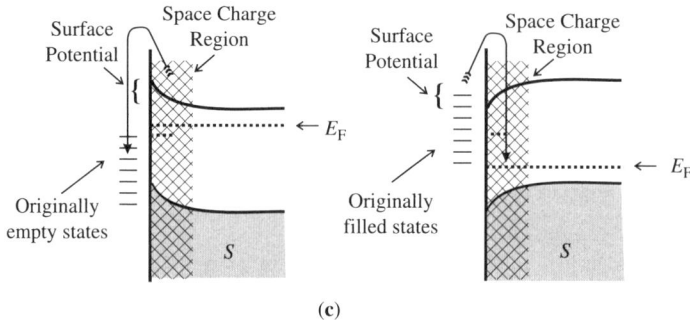

FIGURE 5.7 (a) Surface states and bulk electron energy bands for a semiconductor; (b) shows charge exchange between surface states and bulk energy bands; and (c) band bending resulting from charge exchange.

electronic behavior of the material. In particular electrons can flow from filled bulk states to empty surface states and vice versa and in so doing alter the surface potential of the material. This is best illustrated with a semiconductor or insulator that can support a space charge region. Figure 5.7a shows a semiconductor with a wide distribution of electronic surface state energy levels originating from both extrinsic and intrinsic sources. Filled surface states that lie in energy above empty conduction band states of the semiconductor can transfer their charge to the semiconductor. Likewise empty surface state levels that lie below occupied semiconductor levels can accept electrons from the semiconductor. These two cases are illustrated in Fig. 5.7b using an N-type semiconductor and selected energy surface states. The flow of electrons (indicated by arrows) from the semiconductor will result in a positively charged semiconductor, and the flow to the semiconductor will result in a negatively charged semiconductor. This charging will alter the way further electrons can flow and to illustrate this, the semiconductor bands are bent at the surface to indicate the development of a surface potential in response to the charge flow as is shown in Fig. 5.7c for the cases of charge flow from (left) and to the semiconductor (right). As surface states are filled and semiconductor bands are emptied (or the opposite), a surface dipole develops that can influence electronic device operation and even surface chemistry. We will return to this point many times in the following discussions and chapters.

5.4 MEASUREMENT OF SURFACE ELECTRONIC STATES

From this chapter onward a number of techniques will be discussed that can directly measure surface or interface electronic states and other technique that measure the after effects of the surface or interfaces states, that is, that indirectly measure the states. In keeping with the theme of this chapter, namely electronic structure, surface-sensitive techniques from both categories are discussed below. Furthermore, from here forward surface and interface states are not distinguished since for the most part they have the same origins in the altered surface potential as was discussed above. The first three techniques discussed below—thermionic emission, field emission, and the Kelvin probe—measure the charged state or charges at a surface. The last technique in this chapter, photoemission, can directly probe surface and interfaces electronic states.

5.4.1 Thermionic and Field Emission

Thermionic emission refers to the emission of thermally excited electrons from a solid. For an electron to be emitted from a solid, the emitted electron must attain enough energy to exceed the work function for the solid. Figures 5.8a and 5.8b show work functions for a metal ϕ_M and semiconductor ϕ_S (or insulator), respectively. Function ϕ is seen as the energy to elevate an electron from the Fermi level of the material to infinity. To calculate the flux of electrons emitted from a heated

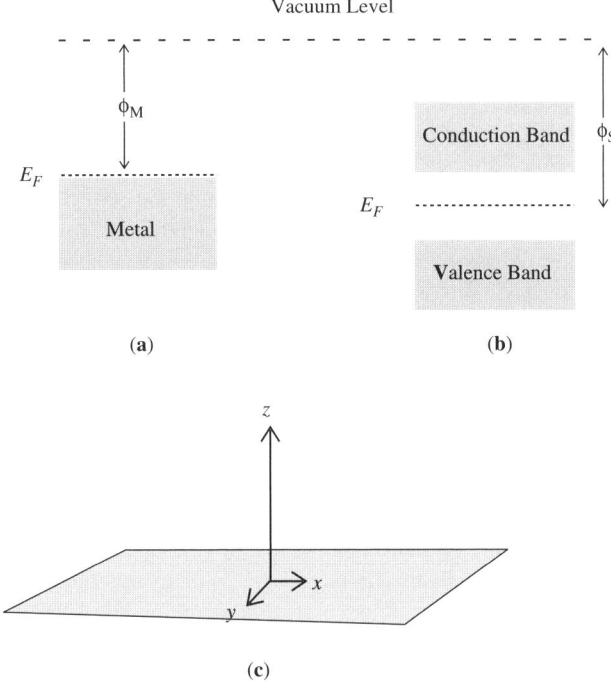

FIGURE 5.8 Energy band showing (**a**) the metal work function ϕ_M, (**b**) the semiconductor work function ϕ_S, and (**c**) the coordinates for thermionic electron emission.

surface, we consider electrons near a surface, which is shown with coordinates in Fig. 5.8c. The energy E is given in terms of the momentum $\mathbf{p}$ as

$$E = \frac{\mathbf{p}^2}{2m} \qquad (5.37)$$

It is desirable to obtain the flux of electrons emitted normal to the surface, $\mathbf{J_z}$. The momentum in the z direction, the direction for emitted electrons, is given as

$$\mathbf{p_z} = [2m_e(E_F + \phi)]^{1/2} \qquad (5.38)$$

The electron flux in the z direction is given by the product of the number of electrons, n, multiplied by the electronic charge e multiplied by the velocity in the z direction, $\mathbf{v_z}$, and can be written in terms of the momentum in the z direction as follows:

$$\mathbf{J_z} = ne\mathbf{v_z} = \frac{nem_e\mathbf{v_z}}{m_e} = \frac{ne\mathbf{p_z}}{m_e} \qquad (5.39)$$

5.4 MEASUREMENT OF SURFACE ELECTRONIC STATES

The number of electrons n with momentum in the range $\mathbf{p_z}$ to $\mathbf{p_z} + \mathbf{dp_z}$ and $\mathbf{p_y}$ to $\mathbf{p_y} + \mathbf{dp_y}$ and $\mathbf{p_x}$ to $\mathbf{p_x} + \mathbf{dp_x}$ is obtained from the density of states dn written in terms of momentum multiplied by the probability that the states are occupied and given by the Fermi–Dirac distribution function. In terms of momentum dn is given as

$$dn = \frac{2}{h^3} \mathbf{dp_z} \, \mathbf{dp_y} \, \mathbf{dp_x} \quad (5.40)$$

where h^3 is the volume of momentum space. The Fermi–Dirac distribution function $F(E)$ is given as

$$F(E) = \frac{1}{1 + e^{(E-E_F)/kT}} \quad (5.41)$$

For energies greater than kT ($E \gg kT$), $F(E)$ can be approximated by the Boltzman distribution $P(E)$:

$$P(E) = e^{E_F/kT} e^{-E/kT} = e^{E_F/kT} e^{-\left(p_z^2 + p_y^2 + p_x^2 / 2m_e kT\right)} \quad (5.42)$$

Then the electron flux emitted is given as:

$$\mathbf{J_z} = \frac{e \mathbf{p_z}}{4\pi^3 \, m_e h^3} P(E) \, \mathbf{dp_z} \, \mathbf{dp_y} \, \mathbf{dp_x}$$

$$= \frac{e}{4\pi^3 \, m_e h^3} \int_{-\infty}^{\infty} e^{-p_z^2/2m_e kT} \mathbf{dp_z} \int_{-\infty}^{\infty} e^{-p_y^2/2m_e kT} \mathbf{dp_y} \int_{p_z}^{\infty} p_z e^{em_e E_F - p_z^2/2m_e kT} \mathbf{dp_x} \quad (5.43)$$

This yields the so-called Richardson–Dushman equation:

$$\mathbf{J_z} = A_0 T^2 e^{-\phi/kT} \quad (5.44)$$

where A_0 is called the Richardson constant and is $1.2 \times 10^6 \, \text{A/m}^2 \, \text{K}$. From a measurement of the thermionic emission current as a function of T, a plot of $\ln(\mathbf{J_z}/T^2)$ versus $1/T$ will yield a slope from which ϕ can be obtained. Some work functions are given in Table 5.1.

The simple thermionic emission model needs some refinement to account for factors that affect the barrier for electron emission such as electrostatic image forces and then for an applied electric field. From these considerations several important measurement techniques arise.

Prior to a discussion of image forces, a brief review of useful formulas from electrostatics is in order. Coulomb's law for the force $\mathbf{F}$ between two charges Q_1 and Q_2

TABLE 5.1 Selected Work Functions

Element	Work Function ϕ (eV)
Cs	1.8
K	2.2
Na	2.3
Ba	2.5
W	4.5
Pt	5.3

separated by a distance r is

$$\mathbf{F} = \frac{KQ_1Q_2}{r^2} \tag{5.45}$$

where K is $\frac{1}{4\pi\varepsilon_0}$ and ε_0 is the permittivity of free space. The electric field $\mathbf{E}$ is given as

$$\mathbf{E} = \frac{\mathbf{F}}{Q} = \frac{KQ}{r^2} \tag{5.46}$$

and the electric potential V between two points a and b that are r apart is given by

$$V_{ab} = -\int_a^b \mathbf{F}\, dr = -\mathbf{F}r = -\frac{KQ_1Q_2}{r} \tag{5.47}$$

where $\mathbf{F}$ is in the r direction. From the thermionic emission case discussed above, when an electron leaves the surface of a metal and is at a distance z above the surface, the remaining charge needs to redistribute to maintain a uniform surface potential. It is as if a positive charge exists in the metal a distance z below the surface of the metal. This charge is called the image charge. The force between these two charges is given as

$$\mathbf{F} = \frac{Ke^2}{(2z)^2} = \frac{e^2}{16\pi\varepsilon_0 z^2} \tag{5.48}$$

The potential due to the image charge, V_i, is given by

$$V_i = -\frac{e^2}{16\pi\varepsilon_0 z} \tag{5.49}$$

The effect of the image charge would be to change the shape of the barrier potential from a step potential for an electron to go from the metal to vacuum as shown in

5.4 MEASUREMENT OF SURFACE ELECTRONIC STATES   143

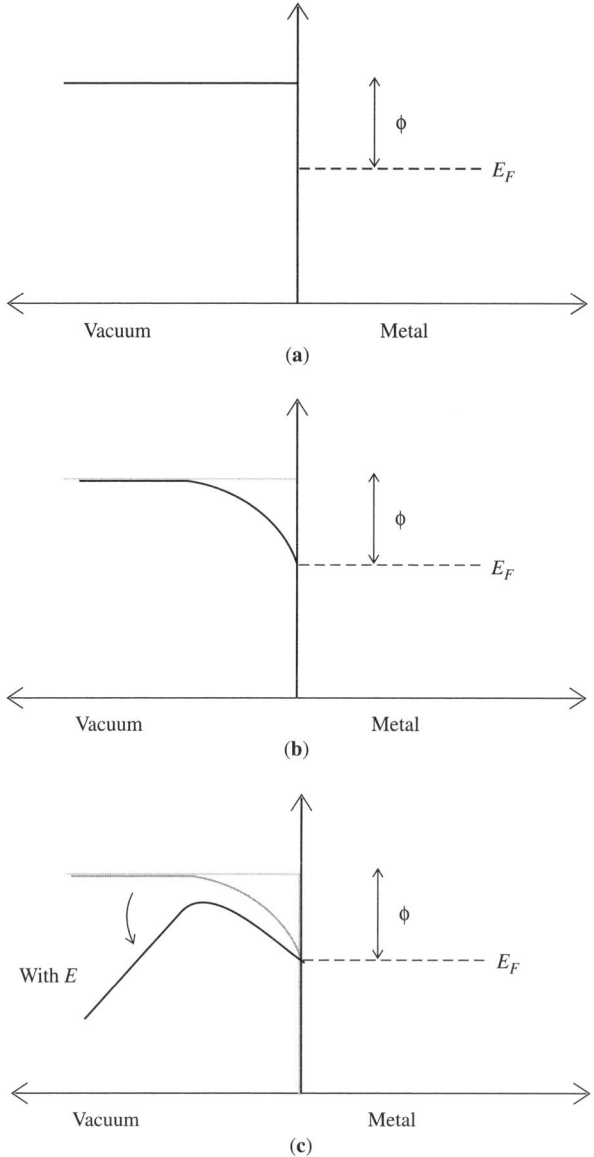

FIGURE 5.9 (a) Metal–vacuum interface with work function ϕ, (b) with image charge considered, and (c) with both image charge and an external electric field **E** that lowers the barrier.

Fig. 5.9a to a potential that varies as $1/z$ as shown in Fig. 5.9b. Now if an electric field **E** were applied with a direction normal to the surface and pointing such that the field enhances the emission of electrons, the barrier to electron emission would be lowered as is shown in Fig. 5.9c, which depicts the effects of both the image charge and an electric filed $\mathbf{E}_z$ with positive direction facing away from the surface. This electric field that effectively lowers the barrier in the z direction yields a potential of $e\mathbf{E}_z z$. The overall potential barrier V_b can be written as

$$V_b = E_F + \phi - \frac{e^2}{16\pi\varepsilon_o z} - e\mathbf{E}_z z \qquad (5.50)$$

The electric-field-assisted electron emission via the barrier lowering is called the Schottky effect.

The field emission microscope (FEM) uses the emission of electrons from a sharpened metal tip (often W) that is suspended in a vacuum chamber as shown in Fig. 5.10 and heated to emit electrons. The region around the tip is typically cooled so that only the tip is hot. A large electric field in the range of 10 to 100×10^6 V/cm is applied between the tip ($-$) and the electrode ($+$) that is coated with a conducting fluorescent screen. Electrons will be emitted from the heated tip with a flux proportional to $e^{-\phi}$ as discussed above. The emitted electrons are accelerated to the screen where an image is formed. Emission can take place from anywhere on the tip, which would expose different planes on different sides of the tip. Thus, the image on the screen would be a greatly magnified image of the tip because of the divergence in the emitted electron beam from the surface to the screen, and the image will vary in brightness according to the ϕ corresponding to the different planes on the tip that are emitting. Changes in the patterns of the clean tip can be observed when an adsorbate is added to the surface such as O to a W tip. A sequence of events of O on a W surface using FEM is shown in Fig. 5.11. Figure 5.11a shows the image from the fluorescent screen of a clean single-crystal W tip. Note the cubic symmetry of the image. Figure 5.11b shows that same tip coated with more than a monolayer of O but at 4 K. The bright regions indicating the W surface (significant electron

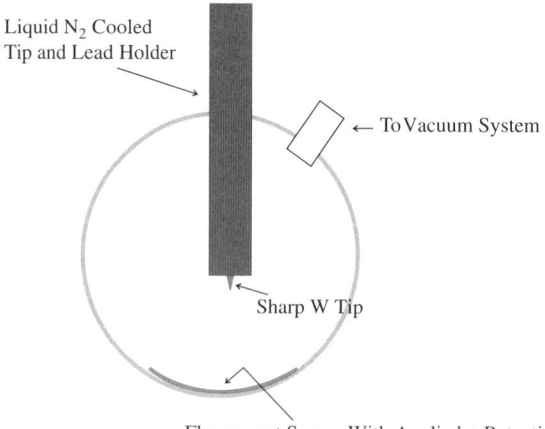

FIGURE 5.10 Sketch of a field emission microscope.

5.4 MEASUREMENT OF SURFACE ELECTRONIC STATES

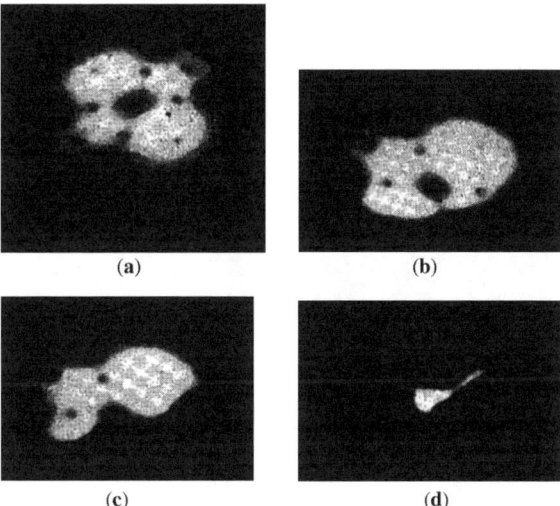

FIGURE 5.11 Sequence of observations with the field emission microscope of O on a W surface: (**a**) clear W surface, (**b**) W tip with O at 4.2 K, (**c**) same tip as (**b**) but at 27 K after 1.28 s, and (**d**) same tip as (**c**) after 12.6 s. [Adapted from Gomer and Hulm (3), Figs. 1, 2, 4, and 9, respectively.]

emission) is smaller due to the O coating. Figures 5.11c and 5.11d show the same O-coated tip at 27 K after progressive time intervals. The O is diffused across the surface leaving progressively smaller uncoated bright regions.

A similar and even better instrument that takes advantage of ϕ variations on a surface is the field ion microscope (FIM). Just like the FEM, the FIM uses a sharp tip of the metal to be studied with a large electric field applied to the tip but with the tip positive (+). An inert imaging gas, typically He, is leaked into the vacuum chamber in the milliTorr pressure range. The He gas atoms become ionized by losing an electron to the highly + tip and the He$^-$ ions are then accelerated in the field and strike the fluorescent screen. Single-crystal tips are used with radii of curvature less than 100 nm. Atom positions are seen with resolutions of several nanometers. Figure 5.12a shows a Re tip with a Re dimer on the surface. Atomic resolution is possible. Figure 5.12b shows a sequence in which a Re atom moves across a W surface. While the FIM is similar to the FEM, the FIM depends less on the surface ϕ. However, the FIM can also be used to study local variations in chemical bonding that depends on the surface electronic properties such as ϕ.

The methods above based on thermionic and electric field emission yield images of surfaces resulting from the emission. If there are adsorbates on the surface, either chemically or physically adsorbed (discussed in Chapter 8), the emission process will be altered through a change in ϕ and thus the image contrast will be altered. Surface electronic states can alter the adsorption process and hence indirectly the image. Because the specific relationship between the surface states and adsorbates and the

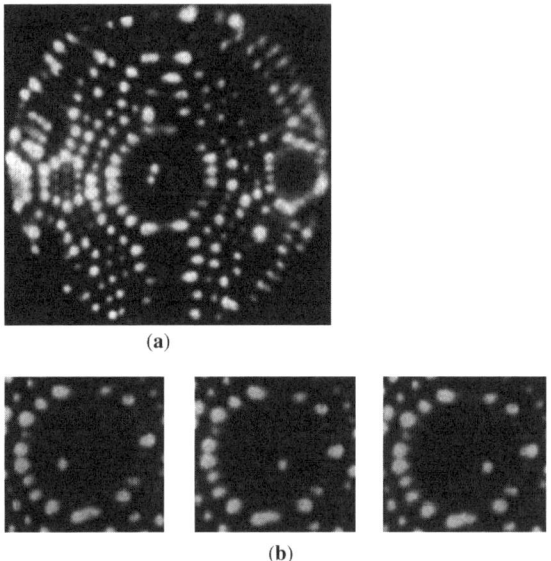

FIGURE 5.12 Field ion microscope (FIM) observations on (**a**) FIM image of a Re[211] tip with a dimer at the center of the tip, and (**b**) movement of a single Re atom across a W(211) surface at 30-s intervals. The observations were made at 20 K. [Adapted from Ehrlich (4), Figures 1 and part of 2, respectively.]

resulting image is usually unknown, the interpretation in terms of surface electronic states and the quantification of surface electronic states is not made. Rather the images are interpreted in terms of the specific reaction sites.

5.4.2 Kelvin Probe

As discussed above, a reaction at a surface or the adsorption of foreign atoms at a surface can alter ϕ. Thus, the sensitive measurement of the change in ϕ under different surface exposures can provide a means to investigate the kind and extent of surface interactions. The Kelvin method provides a sensitive measure of $\Delta\phi$ and also ϕ for an unknown surface, if the reference surface ϕ is known as will be discussed below. This method uses a reference metal electrode that is positioned close to the sample surface to be evaluated as is shown in Fig. 5.13 to form a capacitor structure and all within a vacuum enclosure to preserve clean surfaces. Typically, the work function for the reference electrode ϕ_{Ref} is known, and changes in the work function for the surface ϕ_{Sur} is to be determined. If, at first, the reference metal and sample touch, electronic current will flow that is driven by the differences between the Fermi levels of the reference metal and the surface under investigation. Once the Fermi levels equilibrate, however, the current will cease and an internal potential, the contact potential, develops that gives rise to an internal electric field.

5.4 MEASUREMENT OF SURFACE ELECTRONIC STATES

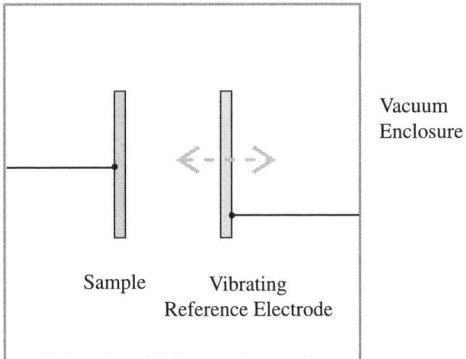

FIGURE 5.13 Sketch of Kelvin probe technique.

The contact potential is the $\Delta\phi$ given as

$$\Delta\phi = \phi_{Ref} - \phi_{Sur} \tag{5.51}$$

Because the Fermi levels of the reference electrode and surface are equal after contact and equilibration, this contact potential is difficult to measure with an external circuit. However, if the experimental arrangement enables the reference electrode to be moved away from the surface, a capacitance C will be developed between the reference and sample, and C will change with changing potentials. If C changes so does the charge Q because

$$C = \frac{Q}{V} \tag{5.52}$$

where V is the potential that is given by the contact potential $\Delta\phi$. The Kelvin probe technique moves the reference electrode at some frequency using magnetic solenoids or piezoelectric materials. In this way both C and Q are changing with the frequency or time of the vibration, and the changing charge gives rise to an alternating current I_{ac} as

$$\frac{dC}{dt} V = \frac{dC}{dt} \Delta\phi = \frac{dQ}{dt} = I_{ac} \tag{5.53}$$

Now if simultaneously and external direct current potential is applied, V_{dc}, in such a way to exactly cancel $\Delta\phi$, then the ac current I_{ac} would go to zero, and the following relationship obtains:

$$-V_{dc} = \Delta\phi = \phi_{Ref} - \phi_{Sur} \tag{5.54}$$

Thus, with V_{dc} known, $\Delta\phi$ is determined, and, if ϕ_{Ref} is known, then ϕ_{Sur} is also determined. The Kelvin probe can be operated in vacuum and then under gas adsorption pressure conditions in order to follow changes in ϕ.

The ϕ measured using the Kelvin probe is a composite of the ϕ for the surface, the changes due to adsorption of various substances, and band bending that can occur as a result of the charging of surface electronic states. In some circumstances adsorption can be prevented (ultra-high vacuum, certain materials, certain ambients, precleaning of surfaces), and, if ϕ for the material is already known, then the Kelvin probe can yield information about the surface electronic states. However, these circumstances are both difficult to achieve and to know if achieved, and so Kelvin probe measurements are usually performed to follow the course of adsorption processes.

5.4.3 Photoemission

As discussed above electron emission can yield information about surfaces, in particular direct surface site information and indirect surface state information. In a previous section electron emission by means of thermionic emission was discussed in which electrons are emitted by heating the surface and field emission in which thermionic electron emission is enhanced in strong electric fields. These simple methods for electron emission led to important techniques for determining the work function and enabled imaging of atomic processes that occur on metal surfaces. However, in all cases above only indirect information about surface states could be obtained. Now we discuss a more powerful tool for the study of electronic surface states, namely the emission of electrons from a surface using incident photons. This kind of emission process is best understood by commencing with a consideration of the photoelectric effect.

As shown in Fig. 5.14a, the photoelectric effect involves incident photons of energy $E = h\nu$. If the incident photon energy is greater than ϕ, then free electrons in a metal will be emitted from the surface with a kinetic energy equal to the amount that $h\nu > \phi$ and given as

$$E_{kin} = h\nu - \phi \qquad (5.55)$$

Using the circuitry shown in Fig. 5.14, the emitted electrons are collected and the current measured by the ammeter A. A potential is applied with a polarity to stop the emitted electrons from striking the collecting plate. The potential that causes the electron current to go to zero is called the stopping potential V_S. Then we can write Equation 5.55 in terms of V_S:

$$V_S = h\nu - \phi \qquad (5.56)$$

It follows that a plot of V_S versus the incident photon frequency ν should be a straight line with slope h and an intercept equal to the work function divided by Planck's constant, ϕ/h.

5.4 MEASUREMENT OF SURFACE ELECTRONIC STATES

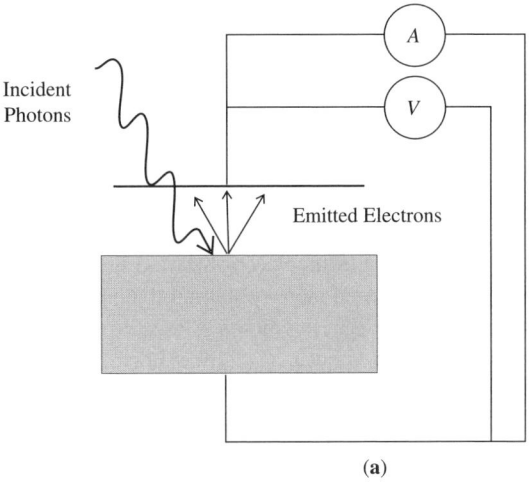

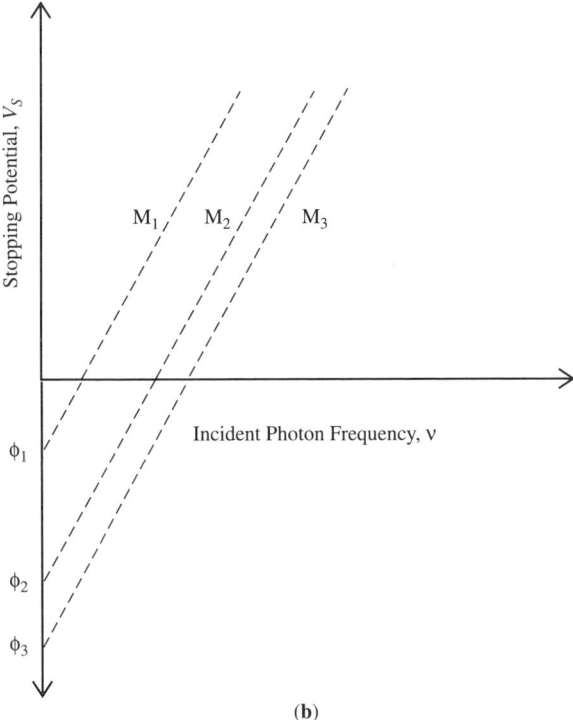

FIGURE 5.14 (a) Basic photoemission experiment and (b) photoelectric effect plot for three metals of the potential needed to preclude emitted electrons from striking the detector versus the incident photon frequency (energy). The intercept yields the work function ϕ.

For metals some electrons are not free and for semiconductors or insulators most of the electrons are strongly bound. Equation 5.55 can be rewritten to include the binding energy for electrons E_B and the work function for the surface ϕ_{Sur} as

$$E_{kin} = h\nu - \phi_{Sur} - E_B \tag{5.57}$$

In an actual photoemission or photoelectron spectrometer the Fermi level of a metal sample will equilibrate with that of the spectrometer (Fig. 5.15). An electron emitted from the sample surface will experience a potential that is the difference between the sample and spectrometers work function given as $\phi_{Sur} - \phi_{Spec}$. Equation 5.57 can be rewritten to express the measurement of E_{kin} in a spectrometer:

$$E_{kin} = h\nu - \phi_{Sur} - E_B + \phi_{Sur} - \phi_{Spec} = h\nu - E_B - \phi_{Spec} \tag{5.58}$$

The reported binding energies are referenced to the work function of the spectrometer. Furthermore, for semiconductors and insulators ejected electrons will cause the buildup of a positive space charge at the surface of the material that will effect the apparent ϕ_{Spec} and shift the measured kinetic energies. Thus, charging must be accounted for in reporting binding energies.

For most solids incident photons in the near-ultraviolet (UV) to x-ray regions are found suitable for electron emission. The resulting spectroscopy, namely the study of the intensity of emitted electrons versus the incident photon energy, is called photoelectron spectroscopy (PES). In the UV incident photon range PES is often called ultraviolet photoelectron spectroscopy, or UPS, and in the x-ray incident photon range the spectroscopy is called x-ray photoelectron spectroscopy, or XPS. XPS is also sometimes referred to as ESCA, which stands for electron spectroscopy for chemical analysis, because XPS yields significant information about chemical

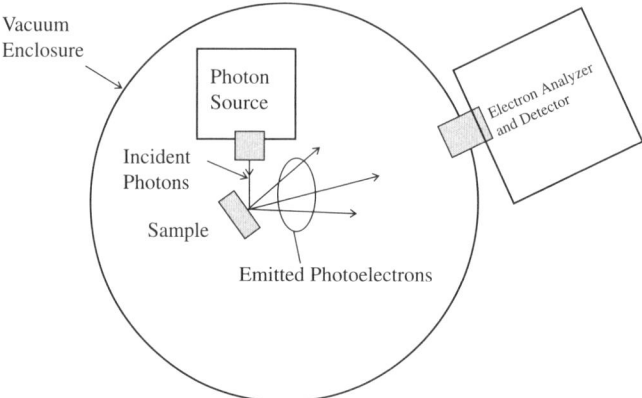

FIGURE 5.15 Photoemission spectrometer.

5.4 MEASUREMENT OF SURFACE ELECTRONIC STATES

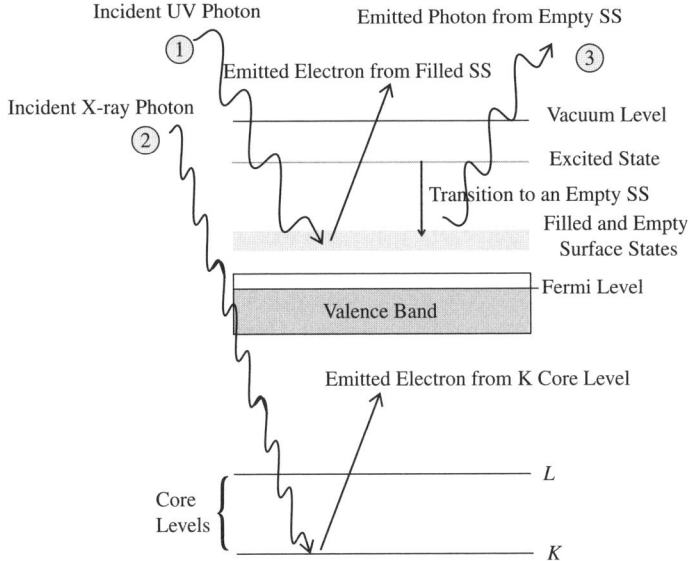

FIGURE 5.16 Energy band diagram with incident photons that give rise to three processes. Process 1 is UV photoemission process, PES of UPS; 2 is x-ray photoemission, XPS, and 3 is inverse photoemission, IPES.

environments as we will discuss later. It should also be noted that the processes in Fig. 5.16 and discussed below involve elastic electron processes and thus have an electron escape depth similar to that for LEED as was discussed in Chapter 2 of some 2 to 3 nm; the photoemission information is derived from atoms in the near surface region. For UPS studies the source of photons has often been He discharge lamps that yield emission lines at 21.2 and 40.8 eV and for XPS, x-rays commonly from Mg and Al are often used that have K_α emission lines of 1253.6 and 1486.6 eV, respectively.

5.4.3.1 Ultraviolet Photoemission Spectroscopy (UPS) and Inverse Photoelectron Spectroscopy (IPES)

As discussed above, UPS involves the emission of photoelectrons from a material. In the UV range of energy for the incident photons, electrons from the energy vicinity of the valence band of the material are typically probed. Thus, UPS is sometimes referred to as valence band spectroscopy. Of present importance is that this energy range is also similar to the energy range for surface electronic states. In its simplest manifestation, as shown in Fig. 5.15, monochromatic incident photons from a photon source as discussed above are impinged upon a surface under study, and the ejected electrons, the photoelectrons, are collected and analyzed in terms of energy and intensity. The source and detector for UV photons as well as the sample are contained within or connected to an ultra-high vacuum chamber in order to maintain the pristine surface for examination and to enable efficiency with both the incident photons that are absorbed by air and emitted photoelectrons that scatter in air. Using incident UV photons filled electron states are probed. To probe initially empty states, the empty states of the material are filled with electrons from higher lying excited states that were excited from an external source. When the excited electrons drop into the empty surface states, the energy release is detected. This latter spectroscopy is called inverse photoelectron spectroscopy, or IPES.

Figure 5.16 shows a schematic of the photoemission and inverse photoemission processes for both UV and x-ray incident photons. In process 1 we see an incident UV photon emitting a photoelectron from a filled surface state that is higher in energy than the valence band of the material. Process 2 shows a more energetic incident x-ray photon ejecting a core level photoelectron from the K shell of the material. Process 3 shows that a transition from an electron in an excited state into an empty surface state emits a photon in the energy loss process. Process 3 is inverse photoemission. These three processes yield different information relative to the electronic structure of surfaces and surface states, and we discuss each process separately with examples.

Figure 5.17a shows the energy level diagram for Ni with adsorbed O. In the top left part of the figure the shaded p and d orbitals represent the normal electronic structure at and below the Fermi level. Ni $3p$ and $3d$ and O $2p$ orbitals are seen as filled states in the spectrum. To the right of the top of Fig. 5.17a and in the inset below are the results from using an incident photon of energy hv. The filled p and d Ni orbital electrons and the p O orbital electrons are excited in energy above the vacuum level and are therefore emitted and collected for energy analysis (indicated by "measure" in the figure). From this analysis of the energy and intensity of the emitted electrons and the knowledge of the exciting photons, hv, the valence band of Ni with adsorbed O can be mapped. The inset shows that the emitted valence band electrons are included on a background of secondary electrons also emitted from the Ni surface. The secondary electrons are emitted as a result of inelastic electron–electron interactions between the excited electrons. Some electrons near the surface do not experience these inelastic interactions, and thus those emitted electrons carry the information about the energy levels near the Ni surface while the inelastically scattered electrons form the broad background intensity. The valence band at and below E_F are occupied valence band states.

Similar to the case for Ni above, Fig. 5.17b displays the PES spectra for both a clean and an oxidized Si surface. The spectra were taken from energies below E_F to near E_F. The most interesting part of these spectra is near the top of E_V where surface states if present would be discernable from valence band states (at the right of Fig. 5.17b). The clean Si surface spectrum shows higher emission in this energy region than the Si that has been saturated with O and oxidized. The difference in the clean and oxidized Si surface in this energy region is shown by the shaded feature. These filled states on the clean surface are surface electronic states that are removed via oxidation. In microelectronics this process of removing unwanted surface or interface electronic states is called electronic surface passivation and is discussed at length in Chapter 11. The ability to passivate the Si surface via the formation of a dielectric oxide is the single most important reason that Si has risen to be the most important semiconductor in modern electronic devices. This important result will be referred to again in Chapters 7, 10, and 11.

5.4 MEASUREMENT OF SURFACE ELECTRONIC STATES

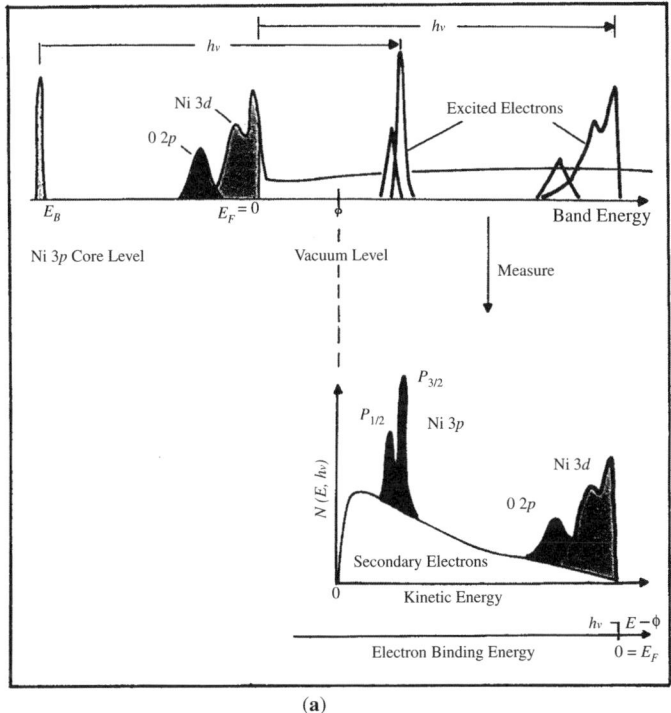

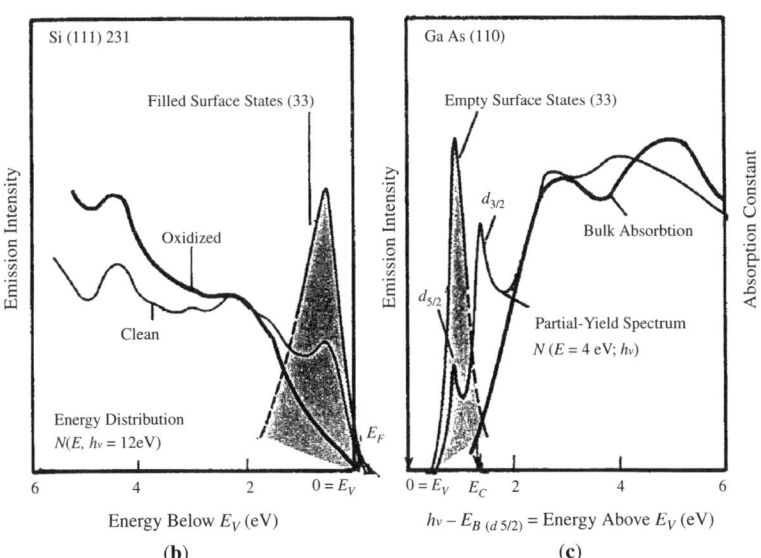

FIGURE 5.17 (a) UPS spectrum of a Ni surface with adsorbed O; (b) a comparison of the Si surface with an oxidized Si surface; and (c) a comparison of the UPS spectrum for GaAs surface with bulk GaAs optical spectrum. [Adapted from Eastman and Nathan (5), Figures 3 and 4.]

The observation of empty surface states can also sometimes be accomplished. As shown in Fig. 5.16, this corresponds to process 3 in which an excited state is first filled, and then via deexcitation of the filled excited state to an empty surface state, a photon is emitted. An example of this more complex process is shown in Fig. 5.17c for the GaAs surface. This figure shows the IPES spectrum along with an optical spectrum for bulk GaAs, which is not sensitive to surface states (thicker line). The result of the difference is the shaded region that shows the empty surface states in the energy region from E_V to E_C or empty states in the forbidden energy gap.

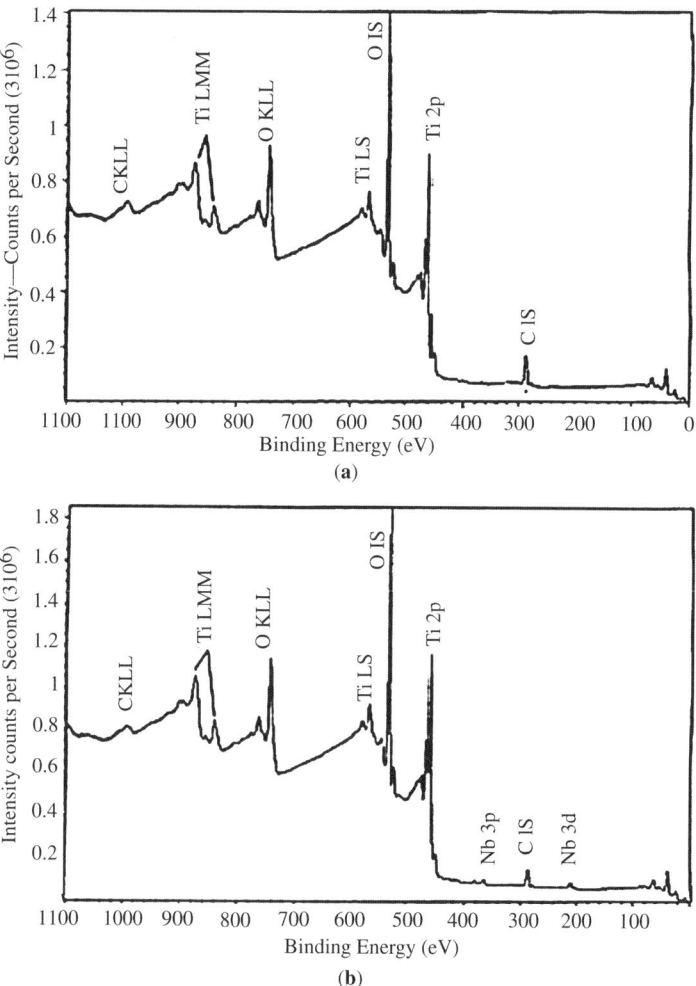

FIGURE 5.18 XPS survey spectra of (**a**) undoped and (**b**) Nb-doped TiO_2 on a Si wafer. [Adapted from Atashbar et al. (6), Figure 4.]

5.4 MEASUREMENT OF SURFACE ELECTRONIC STATES

The classical experiments discussed above show that PES and IPES can directly access surface electronic states. In Chapter 7 electronic measurements will be discussed that can also quantify surface electronic states and in Chapters 10 and 11 examples will be discussed.

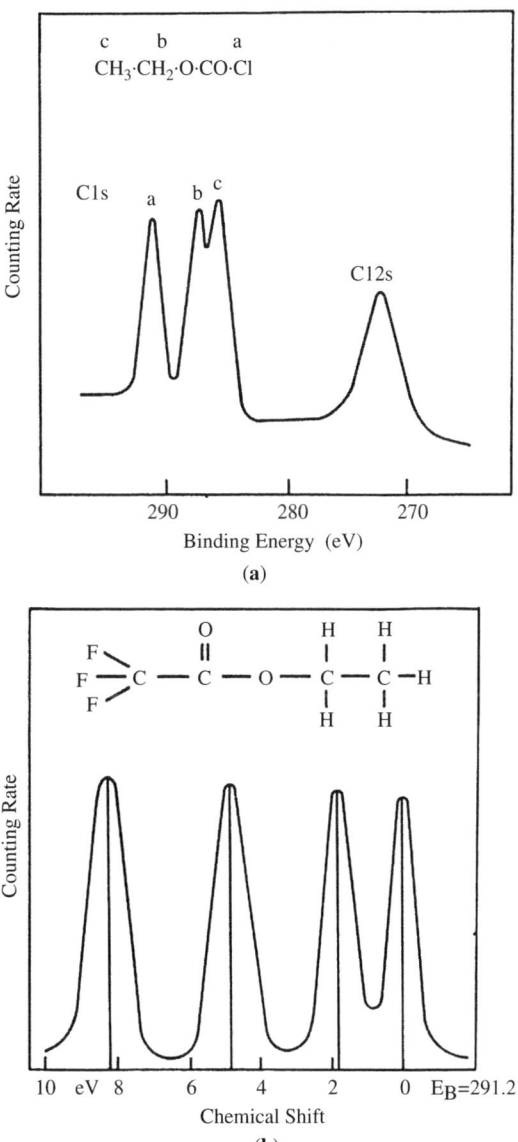

FIGURE 5.19 XPS spectra showing chemical shifts: (a) binding energy changes with the position of the C atom in a molecule of chloroacetic acid and (b) the chemical shifts in a similar molecule with CF_3 replacing Cl. [Adapted from Ghosh (7), Fig. 1.9, p. 12.]

5.4.3.2 X-ray Photoelectron Spectroscopy (XPS) As shown in Fig. 5.16 for process 2, XPS involves an incident x-ray photon that has sufficient energy to eject an electron from the atomic core levels. This process requires more energy than UPS, which ejects the higher lying valence band electrons. The core level electron spectra, the XPS spectra, yield information about the identity of the atoms from which electron emission takes place, and thus XPS can be used to identify constituent atoms on the surface. Since the XPS process also involves elastic electron processes, and thus has an electron escape depth of some 2 to 3 nm, the XPS information is derived from atoms in the near surface region. The core level energies are less affected by potential changes at the surface and thus less affected by surface electronic states. Figure 5.18 shows core level XPS survey spectra for TiO_2 deposited on Si single-crystal wafers. In Fig. 5.18a the peaks associated with Ti and O are seen and C is also seen as a surface contaminant. In Fig. 5.18b the same peaks are seen, but in addition Nb peaks are observed. The Nb dopant was at about the 1% level in the TiO_2. This figure shows that atoms of different elements can be distinguished with great sensitivity.

In addition, the electron environment for a particular kind of atom is effected by neighboring atoms. Adjacent atoms will cause an increase or decrease in the binding energies of predominantly the outer valence band electrons. Figure 5.19a shows the XPS spectrum for chloroacetic acid, CH_3—CH_2—O—CO—Cl. In this molecule the three C atoms are each in a different chemical environment as is indicated by the letter *a*, *b*, and *c* and seen in the corresponding XPS spectrum. The XPS peaks show that C in the *a* position has the largest binding energy. This can be explained by considering that this C atom has two electronegative O atoms nearby that can withdraw some of the electronic charge associated with the C at the *a* position. The decrease in charge associated with the *a* position C atom enables the core level to be bound more tightly and thereby increase the binding of the electron to the C atom. This can be thought of as a screening effect that is reduced when charge is withdrawn by adjacent electronegative atoms or groups. An XPS spectrum of the same molecule but with the Cl replaced by CF_3 is shown in Fig. 5.19b. This representation shows the spectral intensity plotted versus the relative change in the binding energy from the pure C. This change is called the chemical shift, and it varies with the environment of the C. Again one sees the largest chemical shift for the C atom with the most electronegative atoms nearby, namely 8 eV for the C in the CF_3 group. The next largest chemical shift of 5 eV is seen for the C with adjacent O's.

In Chapter 11 it will be shown that film thickness measurements can also be done using XPS, and XPS has been influential in determining the chemical environment of the important Si–SiO_2 interface.

REFERENCES

1. Kronig R de L, Penney WG. Proc Roy Soc London Series A 1931;130(814):499.
2. Lang ND, Kohn W. Phys Rev B 1970;1(12):4555.

3. Gomer R, Hulm JK. J Chem Phys 1957;27:1363.
4. Ehrich G. J Vac Sci Technol 1980;17:9.
5. Eastman DE Nathan MI. Phys Today 1975;April:44.
6. Atashbar MZ, Sun HT, Gong B, Woldarski W, Lamb R. Thin Solid Films 1998;326:238–244.
7. Ghosh PK. Introduction to photoelectron spectroscopy. New York: Wiley; 1983.

SUGGESTED READING

S. C. Davison and M. Steslicka, 1996. *Basic Theory of Surface States*, Oxford. A theoretical discussion of a wide variety of electronic surface states with an interesting and useful historical chapter.

H. Luth, 1995. *Surfaces and Interfaces of Solid Materials*, Springer. This text covers many of the same topics as in the present book. Some topics are covered at a higher level but well worth reading. The best feature of this book is that there are many experimental results and studies described.

6

OTHER SURFACE PROBES

6.1 SURFACE TOPOLOGY OR MORPHOLOGY: SCANNING PROBE MICROSCOPY

In Chapter 2 surface structure was discussed and examples of surface morphology analyses were presented using both transmission (TEM) and scanning (SEM) electron microscopy. It was shown that TEM can yield both diffraction and images of surfaces and for surface characterization was particularly useful in observing profiles in cross section (XTEM), and SEM was used only for surface imaging. In Chapter 3 roughness was introduced through a discussion of the effect of surface feature curvature on surface thermodynamic properties. Chapter 4 was dedicated entirely to the notion of rough surfaces and how to quantify roughness. In Chapter 4 studies of roughness were cited where the rough surfaces were imaged using atomic force microscopy (AFM) that produced detailed surface images, but the AFM technique was not discussed. Now we return to that powerful surface technique and start not with AFM, but rather with a brief discussion of the generic set of techniques labeled scanning probe microscopies (SPM). In reality SEM that uses a rastered electron beam could be considered a SPM technique, although it is not usually included in the SPM techniques. The SPM techniques include those techniques that use a sharp tip, often fabricated of metal or a hard substance such as silicon nitride that is positioned close to the surface being probed and moved or rastered across the surface. The family of techniques labeled SPM are differentiated by the nature of the interaction between the sharp tip and the surface. For example, the most popular SPM techniques and those selected for discussion below are scanning tunneling microscopy (STM) and

---

Surfaces, Interfaces, and Thin Films for Microelectronics. By Eugene A. Irene
Copyright © 2008 John Wiley & Sons, Inc.

AFM, with the former using the tunneling current that occurs between a conducting tip and conducting substrate surface while AFM uses the mechanical interaction. However, there are other interactions that have been shown to yield useful information such as electrochemical, thermal, spin, superconducting, capacitance, magnetic, and optical, all of which have resulted in SPM. The similarity is the fact that sharp probes are brought near to the surface under study. An important advantages of the SPM techniques is that the resolution is not limited by diffraction but rather by the size of the probe–sample interaction volume, which can be translate into subnanometer resolution in many cases. One disadvantage of SPM techniques is that most are slow in acquiring images, due to the scanning or rastering process.

Historically, the SPM revolution in surface imaging commenced with STM, which was invented in the early 1980s and with the award of the Nobel Prize in Physics to G. Binnig and H. Rohrer in 1986. The forerunners to STM were the surface profilometers with an example of the essential features of a profilometer shown in Fig. 6.1, which uses a sharp hard probe or stylus in mechanical contact with a surface. The stylus attached to a magnetic bar in the profilometer in Fig. 6.1 is rastered across the surface. The magnetic bar has a wire coil around it. As the stylus moves up and down in response to the hills and valleys (the roughness) on the surface, a current is induced that changes magnitude and direction with the convolutions of the surface. The small ac signal resulting from the rough surface is amplified, and then it can be digitized and input to a computer along with the associated xy surface positions. In that way a surface image can be obtained. This method of imaging is similar to that of the SEM, but the method for obtaining a signal is different. It should be noted that the profilometer depicted in Fig. 6.1 is similar to a record player (if anyone is old enough to recall such a device!), which uses a sharp stylus that vibrates in grooves in a plastic disk, the musical recording. Also the side surface of the grooves in the recording are undulating (are rough) at the frequency of the recorded speech and/or music. The stylus tracks in the concentric grooves and the undulations cause the tip of the stylus to vibrate at the intelligence frequency on the recording. That mechanical vibration can be converted with a magnetic transducer to an electronic signal that can be amplified and sent to audio speakers, which are essentially transducers that create air vibrations or sound waves at the electrical frequency, which is the music frequency. STM and later AFM are clever manifestations of the same general

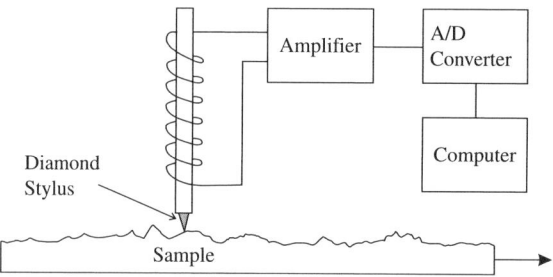

FIGURE 6.1 Schematic illustration of a stylus profiler.

6.1 SURFACE TOPOLOGY OR MORPHOLOGY: SCANNING PROBE MICROSCOPY

idea as the profilometer or record player but with greatly improved sensitivity. In addition both the profilometer and the record player have the stylus in contact with the surface. The force of the contact can damage the surface and also in the case of soft materials can even alter the surface morphology. In both cases the accuracy and sensitivity are reduced. For record players the tracking force was adjusted to be sufficient to obtain the music but light enough to ensure a long (but not infinite) lifetime for the recordings. The STM and AFM and virtually all the modern SPM techniques either have no contact or reduce the contact force so as not to damage the surface or alter the measurement.

6.1.1 Scanning Tunneling Microscopy

Like all the SPM techniques, STM uses a sharp tip that is brought near the surface to be probed and rastered above the surface. For STM the means to obtain signal is to tunnel electrons from the tip to the surface or the reverse and to measure the tunnel current and report the current at each position on the surface. From a brief review of electron tunneling theory to be presented below, the important factors for the STM experiment are elucidated, namely the electronic structure of the tip and surface and the distance between the tip and surface. From these factors and a measurement of the tunneling current, information about the surface topology and the nature of surface sites can be obtained.

Since STM depends on the flow of electronic current, the STM tip or probe needs to be highly conducting and thus STM tips are made of metal. The metal should not be easily deformed and so tungsten is often chosen for the tip material. Ideally, the end of the tip closest to the surface should be a single atom. In this way virtually all of the tunnel current comes from (or goes to) a single point, and this renders image interpretation straightforward. After a consideration of electron tunneling, the experimental setup for STM and some examples are discussed.

6.1.1.1 Electron Tunneling A brief review of elementary quantum mechanics is useful to understand how STM operates, the major parameters that are measured, and their interpretation. We commence with a simple model for strong chemical bonding, namely the "particle in a box" problem. This problem, while not specific for STM, provides a starting point to review the method for the solution of the Schrödinger equation (SE). Then from this classical quantum mechanics problem the requirement for an infinitely strong binding potential is relaxed and thus permits the wave functions to "leak" out of the box. This reduction in the binding potential is in essence a lowering of the barrier that electrons must overcome to be free, and this barrier lowering from infinite to finite enables quantum mechanical tunneling. Quantum mechanical tunneling refers to the fact that electrons can exist within a barrier and on the other side of a barrier, even if the barrier height is greater than the kinetic energy for the electron.

We start with the model for strongly bound electrons by assuming that the electrons as represented by wave functions (ψ) are contained within a potential well of infinite depth. This is the so-called particle (electron) in a box (infinitely deep potential well) formulation in which the walls of the one-dimensional (1D) box are infinitely high. Under this assumption of infinitely high barriers, there is zero probability for the electron to be outside the box, and this model applies for strongly bound electrons where the electrons reside only on an atom. We commence with the one-dimensional time-independent SE as

$$\frac{d^2\psi}{dx^2} + \frac{2m_e}{\hbar^2}(E-V)\psi = 0 \qquad (6.1)$$

where, as in Chapter 5, ψ is the time-independent wave function used to represent the electron, m_e is the electron mass, E is the total electron energy, and V the potential energy or binding potential. In this case $V = \infty$ outside the box but inside $V = 0$ (recall that for free electrons $V = 0$). This problem and the differential equation above is analogous to the classical vibrating string problem with the string held tightly (infinitely tightly!) at two positions along its length. Thus, when the string is plucked in between the pinning points, it will vibrate and sustain those vibrations that do not destructively interfere. Furthermore, because of the infinitely tight pinning, the vibrations cannot propagate beyond the pinned points. Also, within the pinned points the waves can propagate to the left and right. The general solution to the differential Equation 6.1 is as follows:

$$\psi(x) = Ae^{i\alpha x} + Be^{-i\alpha x} \qquad (6.2)$$

with $\alpha = 2m_e E/\hbar^2$ [obtained from Equation 6.1 as the square root of the quotient of the coefficient of the second term on the left $(2m_e E/\hbar^2)$ divided by the coefficient of the first term (1) and keeping in mind that $V = 0$ inside the well or box] and recalling that $E = p^2/2m_e$, $p = h/\lambda$, and $\hbar = h/2\pi$, we obtain α as follows:

$$\alpha = \sqrt{\frac{2m_e E}{\hbar^2}} = \sqrt{\frac{2m_e p^2}{\hbar^2 2m_e}} = \frac{p}{\hbar} = \frac{2\pi p}{h} = \frac{2\pi}{\lambda} = k \qquad (6.3)$$

The barriers (pinning points) are shown in Fig. 6.2 at 0 and l. Now this general solution, Equation 6.2, needs to be tailored to be in conformity with the specific

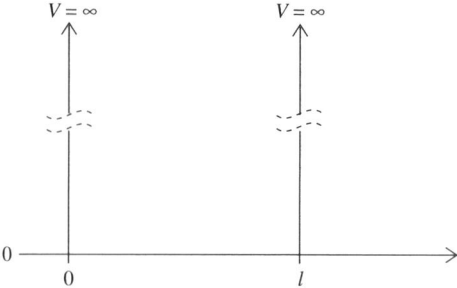

FIGURE 6.2 One-dimensional potential well of length l and with infinite potential barriers.

6.1 SURFACE TOPOLOGY OR MORPHOLOGY: SCANNING PROBE MICROSCOPY

conditions of the problem, the so-called boundary conditions. With reference to Fig. 6.2 the boundary conditions are:

1. At 0 and l (the pinning points), $V = \infty$.
2. $\psi = 0$ at $x \leq 0$, and $x \geq l$.

These boundary conditions mean that the vibrations do not exist to the right or left of the barriers but do exist in the region between 0 and l, namely between the pinned points. The condition $\psi = 0$ at $x = 0$ in Equation 6.2 yields the following:

$$Ae^{i\alpha 0} + Be^{-i\alpha 0} = 0 \tag{6.4}$$

The solution for Equation 6.4 is that $A = -B$. The condition $\psi = 0$ at $x = l$ in Equation 6.2 yields the following:

$$Ae^{i\alpha l} + Be^{-i\alpha l} = 0 \tag{6.5}$$

To simplify the analysis of this latter result we can make use of an Euler relationship:

$$\sin \rho = \frac{1}{2i}\left(e^{i\rho} - e^{-i\rho}\right) \tag{6.6}$$

For the case of Equation 6.5 $\rho = \alpha l$ and $A = -B$ we obtain:

$$A\left(e^{i\alpha l} - e^{-i\alpha l}\right) = 2Ai \sin \alpha l = 0 \tag{6.7}$$

This equation will hold true if and only if $\alpha l = n\pi$ where n is an integer, $n = 0, 1, 2, 3, \ldots$. Therefore, we can express this condition as

$$\alpha = \frac{n\pi}{l} = \sqrt{\frac{2m_e E}{\hbar^2}} \tag{6.8}$$

and solve for the energy:

$$E = \frac{\hbar^2}{2m_e} \frac{n^2 \pi^2}{l^2} \tag{6.9}$$

The implications of this solution are quite important. For the tightly bound electrons only certain values of the energy are permitted, specifically only those values of E are allowed when n is an integer. Thus, the energy is quantized in units of n^2. The quantization of the energy is a result of the boundary conditions imposed on the solution of the SE.

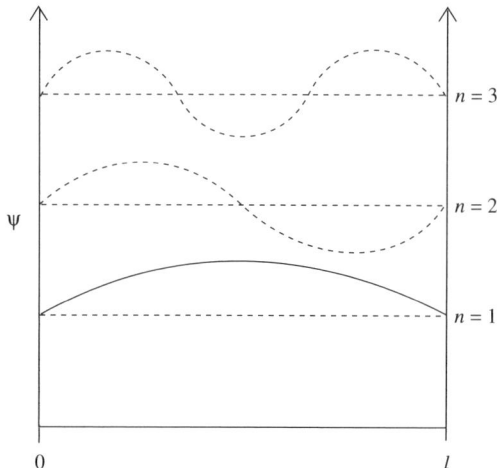

FIGURE 6.3 Three solutions ($n = 1$, 2, and 3) for the bound electron problem.

Returning to the plucked string analogy used above, we ask what wavelengths for the vibrations are permitted that do not suffer from destructive interference. It is shown in Fig. 6.3 (for $n = 1$, 2, and 3) that the wavelengths that survive are those that "fit" integrally in the length of the string (the length of the box, l). These allowed waves constructively interfere and lead to standing waves. All the others destructively interfere and die out.

We now modify this picture and release the infinitely strong pinning of the string. This means that the requirement that $V = \infty$ is changed, and we now set V to a finite value and solve the governing SE. For this problem we consider an electron wave propagating in one dimension from left to right ($+x$ direction) as is shown in Fig. 6.4. Figure 6.4 also shows that the wave is traveling in two different regions with respect to the binding potential. In region 1 the electron is free and $V = 0$, and in region 2 the

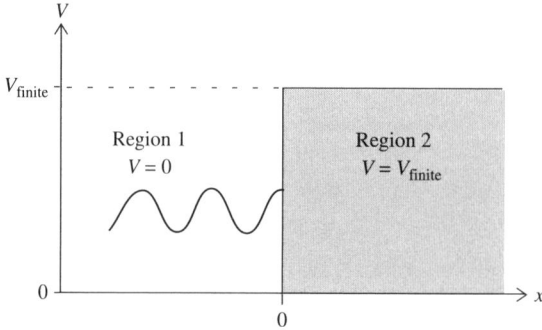

FIGURE 6.4 Electron in two regions: Region 1 is the free electron region, and region 2 has a finite binding potential.

6.1 SURFACE TOPOLOGY OR MORPHOLOGY: SCANNING PROBE MICROSCOPY

electron experiences a finite binding potential, $V = V_{\text{finite}}$. The finite binding potential is greater than the kinetic energy of the electron wave, $V_{\text{finite}} > E_{\text{kin}}$, but it is not infinite as was the case previously (see Fig. 6.2). The origin $x = 0$ is set at the border of regions 1 ($x < 0$) and 2 ($x > 0$). To solve for the allowed energies, we first write separate SEs for the two regions and then solve the SEs so that the solutions match at the boundary of the regions, at $x = 0$.

The SE for region 1 is the SE at $V = 0$ and is as follows:

$$\text{Region 1} \quad \frac{d^2\psi}{dx^2} + \frac{2m_e}{\hbar^2} E\psi = 0 \tag{6.10}$$

For region 2 where $V = V_{\text{finite}}$ the SE is as follows:

$$\text{Region 2} \quad \frac{d^2\psi}{dx^2} + \frac{2m_e}{\hbar^2}(E - V_{\text{finite}})\psi = 0 \tag{6.11}$$

The solutions can be written as follows:

$$\text{Region 1} \quad \psi_1(x) = Ae^{i\alpha_1 x} + Be^{-i\alpha_1 x} \tag{6.12}$$

where

$$\alpha_1 = \sqrt{\frac{2m_e E}{\hbar^2}} \tag{6.13}$$

$$\text{Region 2} \quad \psi_2(x) = Ce^{i\alpha_2 x} + De^{-i\alpha_2 x} \tag{6.14}$$

where

$$\alpha_2 = \sqrt{\frac{2m_e(E - V_{\text{finite}})}{\hbar^2}} \tag{6.15}$$

If $|V_{\text{finite}}| > |E|$, then α_2 will be imaginary. The complications arising from this can be avoided by making the following substitution:

$$\gamma = i\alpha_2 \tag{6.16}$$

With this substitution γ can be written as follows:

$$\gamma = \sqrt{\frac{2m_e(V_{\text{finite}} - E)}{\hbar^2}} \tag{6.17}$$

and for the solution in region 2 we obtain the following:

$$\psi_2(x) = Ce^{\gamma x} + De^{-\gamma x} \tag{6.18}$$

To complete the solution the constants A, B, C, and D are obtained from the details of the problem. We consider the behavior of $\psi_2(x)$ as x approaches ∞ and impose the continuity conditions that $\psi_1(x) = \psi_2(x)$ at the border $x = 0$, and that the derivatives $d\psi_1/dx$ and $d\psi_2/dx$ are also equal at $x = 0$. As $x \to \infty$ we obtain the following:

$$\psi_2(x) = Ce^{\gamma \infty} + 0 \tag{6.19}$$

However, this result is not physical because the probability over all space must be 1 so that

$$\int_{-\infty}^{\infty} \psi \psi^* \, dV = 1 \tag{6.20}$$

The way to ensure this condition is for $C = 0$ in Equation 6.18 and then $\psi_2(x)$ becomes

$$\psi_2(x) = De^{-\gamma x} \tag{6.21}$$

For the condition $\psi_1(x) = \psi_2(x)$ at $x = 0$ we obtain

$$Ae^{i\alpha_1 x} + Be^{-i\alpha_1 x} = De^{-\gamma x} \tag{6.22}$$

which at $x = 0$ becomes

$$A + B = D \tag{6.23}$$

For the condition $d\psi_1/dx$ and $d\psi_2/dx$ at $x = 0$ we obtain

$$Ai\alpha_1 e^{i\alpha_1 x} - Bi\alpha_1 e^{-i\alpha_1 x} = -D\beta e^{-\gamma x} \tag{6.24}$$

which at $x = 0$ becomes

$$Ai\alpha_1 - Bi\alpha_1 = -D\gamma \tag{6.25}$$

6.1 SURFACE TOPOLOGY OR MORPHOLOGY: SCANNING PROBE MICROSCOPY

From the results in Equations 6.23 and 6.25, A and B in terms of D can be obtained as follows:

$$A = \frac{D}{2}\left(1 + \frac{i\gamma}{\alpha_1}\right) \quad \text{and} \quad B = \frac{D}{2}\left(1 - \frac{i\gamma}{2}\right) \tag{6.26}$$

For our present purposes to describe tunneling between tip and surface across the barrier or gap, the most interesting result is the solution above in Equation 6.21 where $\psi_2(x) = De^{-\gamma x}$ and where $\psi_2(x)$ is shown versus x in Fig. 6.5a in region 2. Figure 6.5a shows the shape of the wave function ψ_2 in region 2 to be an exponentially decaying wave function. Thus, the wave function penetrates into region 2 but it decays rapidly. This penetration of the wave function into the barrier is called quantum mechanical tunneling and results from changing the barrier height from infinitely high to

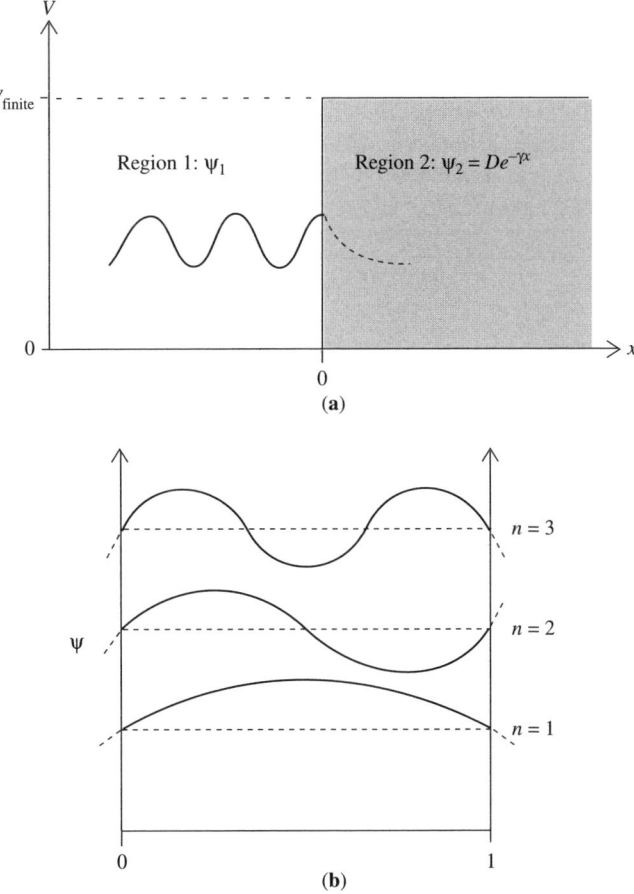

FIGURE 6.5 (a) Solutions to the SE in two regions from Fig. 6.4 as ψ_1 and ψ_2; (b) other solutions with dashed portion indicating leaking into adjacent regions.

finite. Likewise Fig. 6.5b shows the result for other wave functions corresponding to different n's as was shown in Fig. 6.3 for the infinite barrier. Tunneling is an important result of quantum mechanics since it predicts a probability for an electron to exist in a region that is classically disallowed. One consequence of tunneling is that if an electronic detector was placed close to the barrier in region 2, the electron could be detected (albeit weakly), and this is the underlying idea for scanning tunneling microscopy.

For STM we rename x as the tip to surface distance d and E_F is the Fermi energy of the metal so that the quantity $(V_{finite} - E_F)$ is the tunneling barrier and is named ϕ. Then the tunneling current I_{tun} is given as

$$I_{tun} \propto \psi^2 = D^2 e^{-2\gamma d} \tag{6.27}$$

γ is given as

$$\gamma = \frac{1}{\hbar}\sqrt{2m_e \phi} \tag{6.28}$$

6.1.1.2 Scanning Tunneling Microscopy Operation

It is seen directly from Equation 6.27 that the tunneling current is exponentially related to the tip to sample separation d. A simple estimation of the ratio I_{tun}/D^2 can yield insight about the effect of tip to sample separation on the tunneling current. The ratio I_{tun}/D^2 can be written from Equation 6.27 as follows:

$$\frac{I_{tun}}{D^2} \propto e^{-2\gamma d} \tag{6.29}$$

From Equation 6.28 γ is estimated to be $0.1\,\text{nm}^{-1}$ using a conveniently (and realistically) chosen barrier $\phi = 4$ eV (1 eV = 1.602×10^{-12} g cm^2/s), $m_e = 9.11 \times 10^{-28}$ g, $\hbar = 1.054 \times 10^{-27}$ g cm^2/s. Table 6.1 summarizes the calculation according to Equation 6.29.

From Table 6.1 it is seen that beyond a separation of 1 nm the tunnel current drops rapidly, which indicates that for STM usable tip to sample separations are not more than several nanometers. Furthermore, there is about a 10× decrease in the current for each 0.1-nm increase in separation, which indicates great sensitivity in the vertical

TABLE 6.1 Estimation of Ratio I_{tun}/D^2 for Various Tip–Sample Separations

d (nm)	I_{tun}/D^2
0.1	0.14
0.2	0.02
0.3	0.002
0.5	4.5×10^{-5}
1.0	2.0×10^{-9}
10.0	1.4×10^{-87}

6.1 SURFACE TOPOLOGY OR MORPHOLOGY: SCANNING PROBE MICROSCOPY

direction. In fact, it is straightforward to measure changes in current of several percent, which would yield vertical resolution of below 0.001 nm. Horizontal resolution is not as good and is limited by the tip geometry, which we will discuss later. However, lateral or horizontal resolution of a few tenths of a nanometer is achievable. Thus STM has atomic resolution.

With atomic resolution of STM as mentioned above, the next issue is the methodology required to position the tip relative to the sample in terms of both the tip to sample separation and the ability to raster across the surface. Specifically, three-dimensional motion of both tip and sample are required, and two broad ranges of motion are required. The first range is relatively coarse motion to allow for sample and/or tip management (changing, altering, etc.) This coarse motion capability should also be able to position the tip and sample to be within the fine adjustment range of motion to achieve the desired tip to sample distance and attitude, but this coarse motion does not require the capability to raster. The coarse motion can be readily obtained using mechanical screw and slide hardware developed for optics. The fine motion used for final tip to surface positioning within several nanometers and the rastering capability can be performed using piezoelectric materials. A piezoelectric material is a material that displays a potential difference across the material, a polarization charge, when a mechanical stress is applied. There are many materials that exhibit this property: quartz, Rochelle salt (sodium potassium tartrate tetrahydrate) $AlPO_4$, many perovskite structures such as $BaTiO_3$, and the most popular $Pb(ZrTi)O_3$ or PZT, and many other materials. In fact SPM techniques use converse piezoelectricity in which a potential applied to a bar of PZT causes a change in the dimension of the bar. Nanometer step motion can be obtained with nominal voltages of less than hundreds of volts. Figure 6.6 shows a schematic of an STM that could be similar for

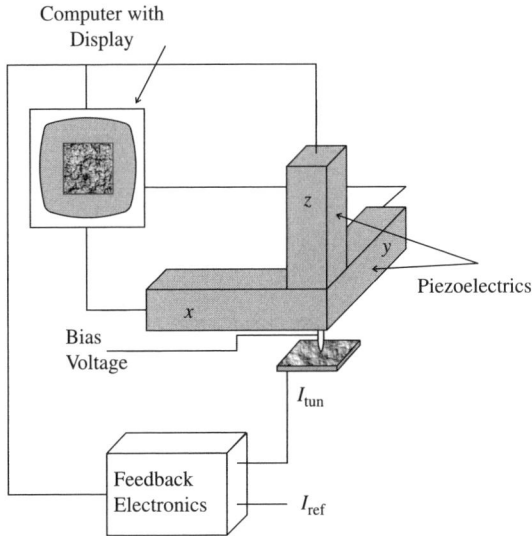

FIGURE 6.6 Schematic of a scanning tunneling microscope.

virtually all the SPM techniques mentioned above. The piezoelectric elements (piezos) used for fine motion are shown in an *xyz* arrangement. In this arrangement the *z* direction, or the tip to sample distance, can be separately controlled from the *xy* rastering across the sample surface. Based on a separate calibration of the piezoelectric elements, the voltage applied to the piezoelectrics can be converted to distance and fed to a computer for storage and display. The tunneling current that is controlled by the bias voltage applied between the tip and sample and the tip to sample distance can be obtained from a separate circuit and also fed to the computer. The sample image is obtained from the position and current information. For example, a high current at a particular sample position can be represented by a bright spot on the display. Of course, with modern imaging electronics and software, colors and shading as well as data-enhancing routines can produce dramatic images of surfaces.

The feedback electronics included in Fig. 6.6 play a central role in STM and other SPM techniques as well. To elucidate the purpose and mechanism for feedback circuitry, we consider that for STM there are two usual modes of operation shown in Fig. 6.7: constant current mode in Fig. 6.7a or constant tip position or height mode

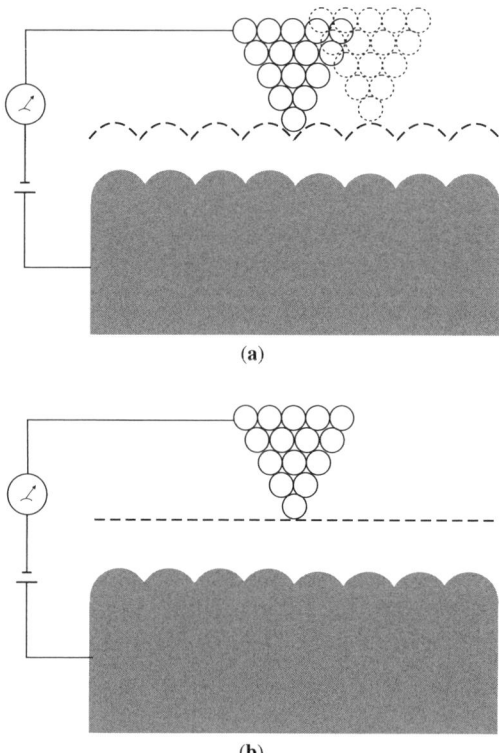

FIGURE 6.7 Two modes of STM operation: (**a**) constant current mode and (**b**) constant height or separation mode.

shown in Fig. 6.7b. In constant current mode the tunneling current I_{tun} is held constant as the tip at the top is scanned over the surface that shows corrugations. With the tip traveling across the corrugated surface the tip to sample distance would vary if the tip were not raised or lowered. Consequently, I_{tun}, which depends exponentially on the separation, would be varying in step with the corrugations. However, if the varying current is fed into the piezo control circuitry for the z direction piezo, when I_{tun} increased above a preset value the piezo would increase the tip–sample separation, thus reducing I_{tun} to the preset value. Likewise, when I_{tun} dropped during rastering, the tip to sample distance would be reduced to again achieve the preset value for I_{tun}. The path of the tip is shown in Fig. 6.7a by the dashed line, and it is seen to follow the surface corrugations. The variations in the z direction as well as the position of the tip (x, y) are recorded, and this information is used for image construction. Alternatively, in the constant height mode the tip is held in a fixed separation at some position on the surface at the start of a surface scan. This position of the tip or the tip height is held constant throughout the surface scan. At this starting position I_{tun} is adjusted by adjusting z to a suitable value for measurement. Once the rastering begins, the I_{tun} will change due again to the corrugation of the surface relative to the fixed height of the tip. The level path of the tip in the constant separation is shown in Fig. 6.7b by the dashed line. The changing I_{tun} at each surface position (x, y) is then used for image construction. In the constant current mode the tip must be adjusted in accordance with the feedback circuit signal that activates the z piezo. While the electronics are fast, the response of the piezo is not, and for the large number of positions samples on a surface scan the constant current mode is slow. On the other had the constant height mode is faster because it does not require the adjustment of the tip to sample distance. However, if the surface has corrugations higher then the preset tip to sample distance, the tip will collide with the surface and typically cause tip failure. Thus, the experimenter needs to make an informed decision based on the sample characteristics about which mode to use. For example, often a scan in the constant current mode is performed to learn about the surface. Then, if it is necessary to observe the changes with various surface exposures, the experimenter can do faster constant height scans over selected regions of the surface.

As was mentioned above the material of choice for STM tips are hard durable metals such as W and Mo and Ir tips are also used. Tip making is an art with simple processes leading to reproducibly sharp tips with radii of 10 nm and even smaller. Among the methods developed over the years are electrochemical methods and ion milling. Often electrochemical methods will be used first to produce a sharp tip, and ion milling will be used to remove any surface films such as oxides or impurity layers. The ion milling often improves the sharpness as well as cleans the tips. Once fabricated, tips are usually characterized using SEM and sometimes TEM. Field ion microscopy (FIM) can also be used to characterize the tip shape and structure (see Chapter 5). The ideal is to have a single atom tip with a small radius of curvature so that most of the tunneling current is from (or to) the very end of the tip. In that way the sample area participating in the tunneling event is minimized, and the lateral resolution is optimized.

6.1.1.3 Applications Before the results from specific STM studies can be understood, it is important to understand exactly what is represented in an STM image. The previous electron tunneling discussion indicates that electrons are tunneling from either the tip to the surface or the surface to the tip. This process requires electrons in allowed states to tunnel through a barrier and through a distance and then into allowed empty states. The resulting current I_{tun} is measured and then reported as a function of the position on the surface where the tip is positioned. For the purpose of analyzing this kind of problem, it is useful to know the number of electrons in allowed states in the material from which the tunneling originates, and the number of allowed states is called the density of states (DOS). The DOS represents the amount of electron states at specific allowed values of energy. When the DOS can vary over a surface and the tunneling experiment takes place at a small point on the surface, it is usual to use a local density of states (LDOS) and thus I_{tun} is proportional to the LDOS. Thus, what is being mapped in an STM experiment is a representation of the LDOS. Now we can consider two metals separated by a small distance as shown in Fig. 6.8a. In this particular case M_1 has a higher Fermi level and smaller work function as compared to M_2. So, if these metals were placed in

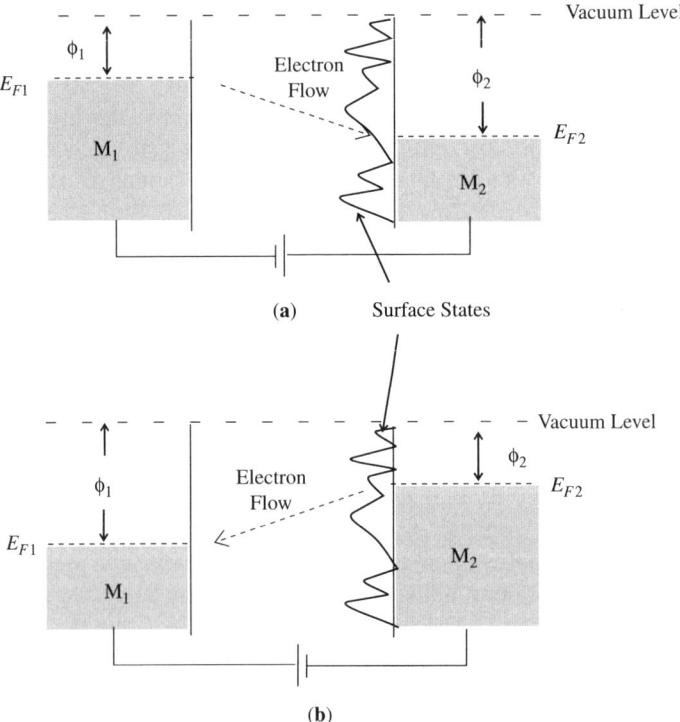

FIGURE 6.8 (a) Two separated different metals with Fermi levels (E_F) work functions (ϕ) and a distribution of surface electronic states on M_2 indicated; (b) the same metals as in (a) but after joining and at equilibrium.

6.1 SURFACE TOPOLOGY OR MORPHOLOGY: SCANNING PROBE MICROSCOPY

contact, electrons would flow from M_1 to M_2 and establish the contact potential at equilibrium ($\phi_2 - \phi_1$). When this occurs, the Fermi levels would equilibrate and current flow would cease. After equilibration the metals can again be separated and a potential (bias) applied as is shown in Fig. 6.8a. In this case M_1 Fermi level would rise and then tunneling could take place across the gap since there are filled states on M_1 at E_{F1} and available empty states on M above E_{F2} that due to the bias is below E_{F1}. The bias enables current flow and enhances the tunneling probability, and the direction of the current is from filled to empty states. One could imagine the tip metal to be M_1 and the sample surface to be M_2, and then in Fig. 6.8a I_{tun} flows from filled states in the tip to empty allowed states in the sample. Figure 6.8b shows the same situation but with the bias reversed where after equilibrium the bias causes I_{tun} to flow from filled states in the sample to empty states in the tip. In addition Fig. 6.8 shows a distribution of surface electronic states on M_2. As the bias on the metal changes say in both situations in Fig. 6.8, tunneling will take place to different unoccupied states in M_2 in Fig. 6.8a or from occupied states in M_2 in Fig. 6.8b. Therefore, a surface electronic state spectroscopy of filled and/or empty states can be performed using the bias to probe the surface states at various energetic levels.

In summary, there are two broad kinds of information obtained from STM. The first is the location of occupied or unoccupied allowed electronic states relative to the position of the tip and sample. Atoms and molecules have associated electronic states; and, therefore, with the high spatial resolution STM can locate the relative positions of atoms and/or molecules from the positions of the LDOS. In addition, surface spectroscopy can be performed. With the tip at a fixed surface position (xy) and for a fixed separation (z), the tip to sample bias can scan through the surface state distribution. Examples of this kind of information are given in the following paragraphs.

Scanning tunneling microscopy has contributed greatly to the understanding of surface structures and reconstruction. Figure 6.9 displays a series of theoretical calculations (6.9a to 6.9e) along with STM results in panel 6.9f for Si(111) that reconstructs to Si(111)-7 × 7. At the time of the publication of these results, there was controversy about the exact structure of the Si(111)-7 × 7 surface. The names in Figs. 6.9a to 6.9e indicate the primary author of the structural model depicted. First notice that in Fig. 6.9f the experimental STM results indicate the order and relative atomic positions for the Si atoms on the Si(111)-7 × 7. The reconstruction is clearly apparent in this representation of the electronic structure of the Si surface. Then, when the experimental intensities are compared with the various models, the best-fit model is that of Takayanagi in Fig. 6.9e. This is an impressive result because to the eye several of the models in the top row of Fig. 6.9 look the same, and it is not until the details are compared that discrimination can be achieved. Thus, both the qualitative and quantitative aspects of STM are in evidence.

Figure 6.10 shows an example of the use of tunneling direction. Figure 6.10a shows tunneling to unoccupied states in Si (from tip to Si), and Fig. 6.10b shows tunneling from occupied Si states (from Si to tip). The intensity is easily noticed in the right-hand portion of the unit cells (one unit cell is indicated by the

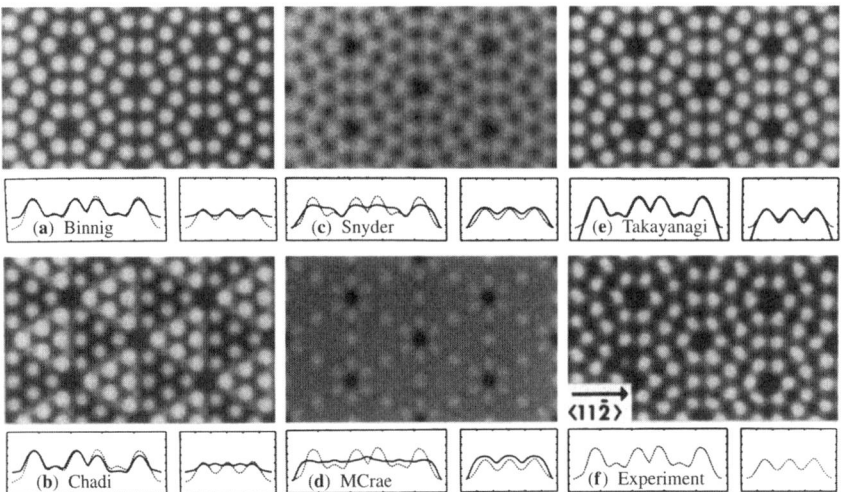

FIGURE 6.9 Panels (**a**)–(**e**) display various theoretical models for Si(111)-7 × 7 with images and the line scans below the images are line scans across various parts of the image with the solid line being theory and dotted line not being experimental data from (**f**). Panel (**f**) is experimental STM results for the Si(111)-7 × 7. [Adapted from Tromp et al. (1), Figure 2.]

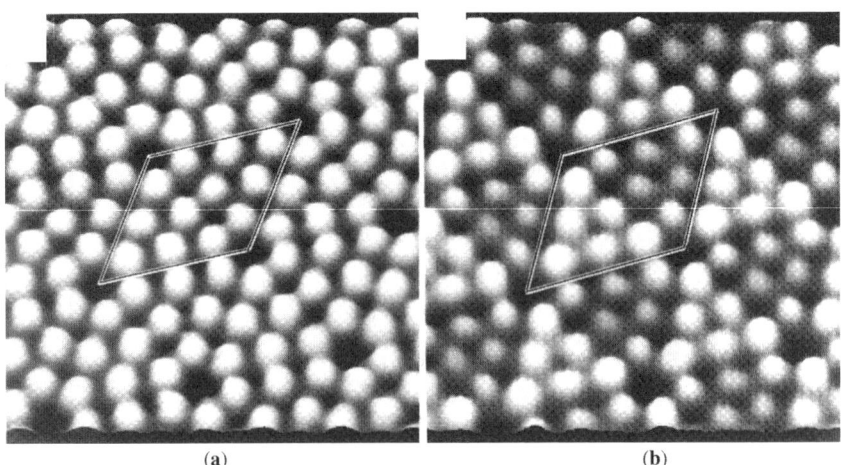

FIGURE 6.10 STM results for the Si(111)-7 × 7: (**a**) displays unoccupied Si electronic states where the sample bias was +1.5 V so that tunneling took place from occupied states in the tip to unoccupied states in the sample and (**b**) displays occupied Si electronic states where the sample bias was −1.5 V so that tunneling took place from occupied Si states to unoccupied tip states. The 7 × 7 unit cell with 12 atoms is identified by the lines figures. The intensity differences in the occupied states image indicate different positions relative to the surface adatoms. [Adapted from Avouris and Wolkow (2), Figure 2.]

6.1 SURFACE TOPOLOGY OR MORPHOLOGY: SCANNING PROBE MICROSCOPY   175

diamond-shaped figure). This indicates that several darker atoms are at some distance below the surface adatoms, which appear bright.

Figure 6.11 compares STM tunneling results for Si(111)-7 × 7 with photoelectron spectroscopy (PES) and inverse PES (IPES) to identify the distribution of surface electronic states on a surface. Figure 6.11a displays the tunneling conductance current divided by bias (I/V) versus bias at various positions in the unit cell. The inset is the Si(111)-7 × 7 unit cell with the positions that were measured marked with the symbols used in the plot. The left side of Fig. 6.11a corresponds to tunneling from occupied surface states and the right side from unoccupied states as is indicated at the bias that determines the energy of the states probed. Figures 6.11b and 6.11c compare, respectively, the photoemission results: PES left and IPES right, with

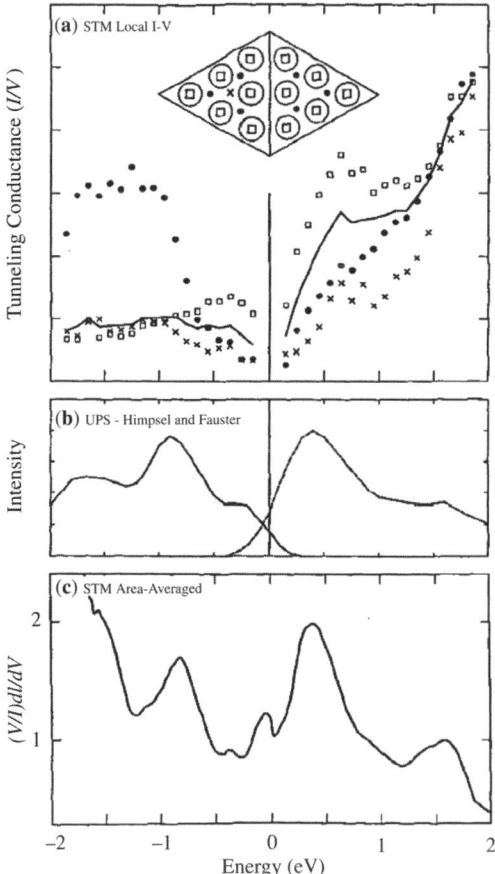

FIGURE 6.11 (a) Plot of the tunneling current divided by the applied bias (I_{tun}/V) on a Si(111) surface (reconstructed 7 × 7 surface); (b) shows the photoemission PES and inverse photoemission IPES spectra for the same surface; and (c) displays the tunneling current spectrum dI/dV. [Adapted from Hamers (3), Figure 7.]

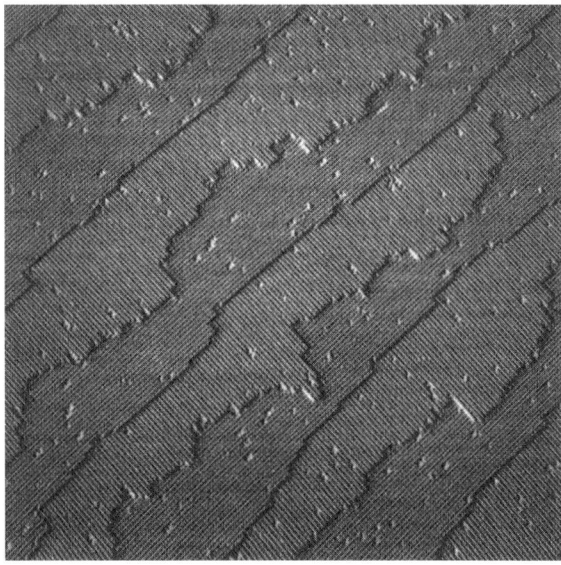

FIGURE 6.12 STM of vicinal Si(100)-2 × 1. [Adapted from Kariotis and Legally (4).]

tunneling spectroscopy results plotted in a way that corresponds to the density of the states probed, namely $d(\ln I)/d(\ln V)$. Notice that the same distribution shape is obtained indicating that the interpretation in terms of LDOS being probed is correct.

With the availability of high resolution it is tempting to always use the highest magnification for a given technique. However, it is sometimes useful to raster over larger areas of a new sample, especially at first to gain perspective about the sample and the features that are worthy to observe at higher magnifications (using smaller raster areas). Figure 6.12 shows an STM image of vicinal Si(100) that reconstructs to Si(100)-2 × 1. First, the vicinal surface shows the expected step surfaces (see Chapter 1) that are low index planes, Si(100)-2 × 1. In addition, the rows of atoms on each step alternate direction. This indicates nearly equal amounts of the 2 × 1 and 1 × 2 reconstructions. It is interesting to note that all the features discussed about the terrace–ledge–kink (TLK) model in Chapter 1 are seen in this image.

6.1.2 Atomic Force Microscopy

The atomic force microscope (AFM), another in the SPM group, was developed by Binnig and co-workers in 1986 (5). It allows atomic-scale imaging of surface topography on both conducting and nonconducting surfaces. For STM, the tunneling current between the tip and the sample is measured, whereas for AFM, the force between the atoms at the apex of the tip and the surface is determined. While AFM usually has lower resolution than STM, the ability to operate AFM with any sample and in many gas or fluid ambients has proven to be a more versatile technique. Also the force interaction used for AFM usually renders image interpretation simpler that for STM.

6.1 SURFACE TOPOLOGY OR MORPHOLOGY: SCANNING PROBE MICROSCOPY 177

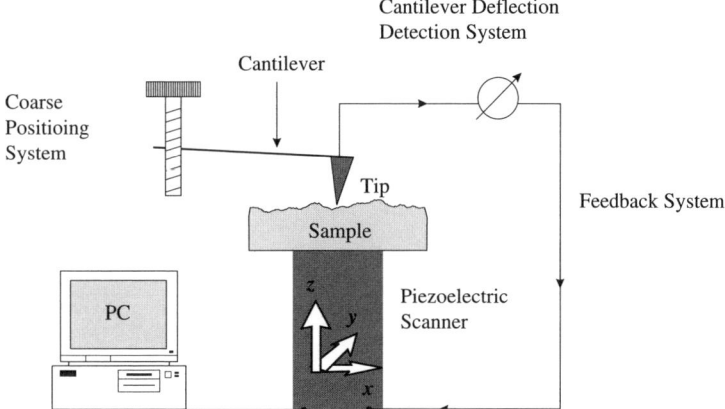

FIGURE 6.13 Schematic diagram of AFM.

6.1.2.1 Atomic Force Microscopy Operation The operation of a typical AFM is described using Fig. 6.13 in which a sample is affixed to a piezoelectric tube scanner and is rastered under a stationary sharp tip that is attached to a flexible cantilever. As the sample surface is rastered, the changes of the surface morphology cause the cantilever that is in mechanical contact (discussed below) with the surface to deflect accordingly. The length and the material of the cantilever determine the sensitivity. The cantilever deflection is detected, and a feedback system is used to maintain a constant cantilever deflection by controlling the displacement of the piezoelectric scanner tube in the z direction to move the sample up or down. This displacement is recorded and along with the xy position is used to produce a topographic map of the sample surface. AFM tips are commercially available and have approximate dimensions of about 2 μm long and less than 20 nm in diameter while the cantilever is on the order of 100 to 200 μm long. A popular material for AFM tips is Si_3N_4 because the material is hard and durable and tips can be mass produced using microelectronics lithography and film deposition techniques (specifically chemical vapor deposition discussed in Chapter 10).

Atomic force microscopy uses the interatomic force between the atoms at the apex of a sharp tip and those on the sample surface as the control interaction. This interatomic force can be either attractive or repulsive as described by the Lennard-Jones potential (discussed further in Chapter 8) of the form:

$$U(r) = \frac{A}{r^{12}} - \frac{B}{r^6} \qquad (6.30)$$

where r is the distance between two atoms, and A, B are constants. Figure 6.14 shows the interatomic force between two atoms as a function of their separation distance. As the two atoms approach each other from afar, they experience an attractive van der Waals force resulting mainly from the induced dipole–dipole interaction. The r^{-6} term from Equation 6.30 prevails. When the atoms are brought closer, their electron orbitals repel each other electrostatically, and, consequently, the atoms experience an

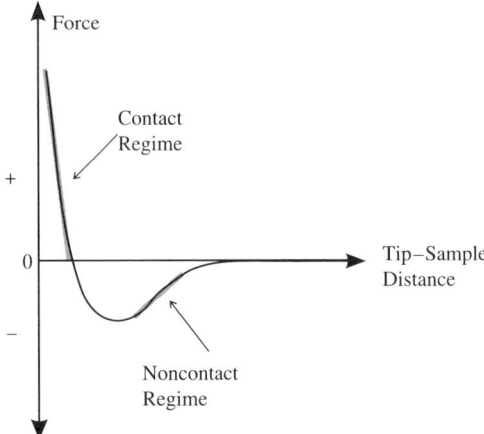

FIGURE 6.14 Force versus tip to sample distance showing the major force interaction regimes: contact and noncontact.

even stronger interatomic repulsion approximated by the r^{-12} term in Equation 6.30. Because the mechanical interatomic forces depend strongly on the separation distance between the atoms at the sharp tip and those on the surface, atomic scale imaging is possible. The Lennard-Jones potential represents the simple interatomic force interaction. Under actual AFM imaging conditions usually in laboratory ambient, other factors such as the condition of the surface and the shape of the tip complicate the imaging through the addition of other forces such as capillary forces (see Chapter 3) and cantilever forces during scanning. The capillary forces occur largely because in laboratory ambient the sample surface is covered with a thin layer of moisture of the order of a few nanometers. When the tip is brought close to the surface, it is subjected to a strong attractive capillary force that depends on the shape of the tip. Figure 6.15 shows that for a tip with low aspect ratio, the water or contamination

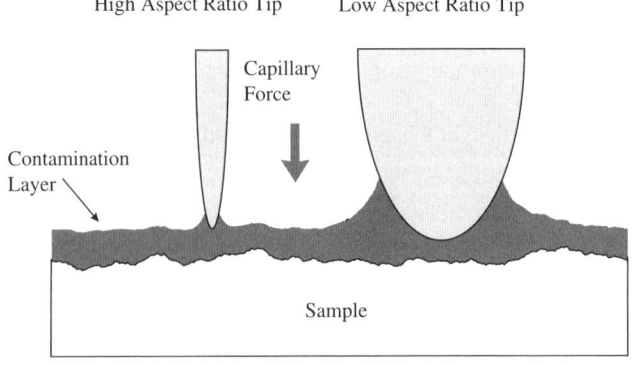

FIGURE 6.15 High aspect ratio (sharp) and lower aspect ratio (rounder) tips contacting a surface with a contamination layer (usually H_2O).

6.1 SURFACE TOPOLOGY OR MORPHOLOGY: SCANNING PROBE MICROSCOPY

layer will wet a larger contact area as compared to a tip with a high aspect ratio, and as a result the attractive capillary force is smaller for a sharper tip. As the tip is being held onto the surface by this attractive capillary force from the moisture layer while the sample is moving away, the cantilever is deflected and results in a cantilever force that has stored energy similar to that of a stretched spring. The magnitude of this force depends on the deflection and the spring constant of the cantilever. At some distance from the surface, the tip will retract and cause the cantilever to oscillate in order to release the stored energy.

The interatomic force (F) between the atoms at the apex of the tip and those on the surface causes the cantilever to deflect according (x) to Hooke's law:

$$F = -cx \quad (6.31)$$

where c is the spring constant of the cantilever. For this cantilever-type spring to achieve high sensitivity, a large deflection range is desirable, which means that the cantilever needs to be soft with small spring constant. However, it also needs to be sufficiently stiff such that it has a high resonant frequency so as to minimize the tip sensitivity to mechanical vibrations. According to the following relationship of the resonant frequency in the spring system:

$$\omega_0 = \left(\frac{c}{m}\right)^2 \quad (6.32)$$

where m is the effective mass of the spring; in order to have a large ω_0 and a small c, m should be kept as small as possible. This is accomplished by fabricating small but rigid cantilevers.

There are many techniques available for the detection of cantilever deflection such as electron tunneling, capacitance, optical interferometry, and optical deflection. Historically, the first AFMs used electron tunneling since it was found to be sensitive to atomic dimension changes for the newly developed STM technology. Figure 6.16

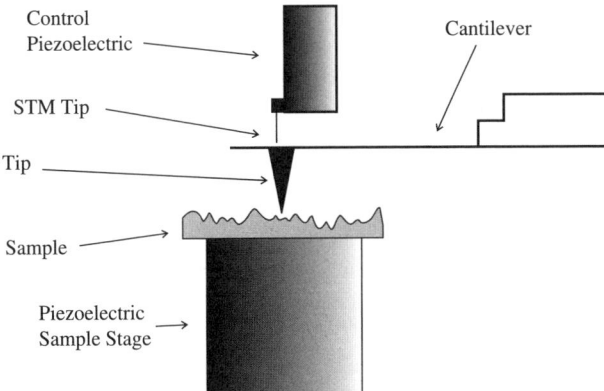

FIGURE 6.16 AFM that employs the tunneling current to detect the cantilever deflection.

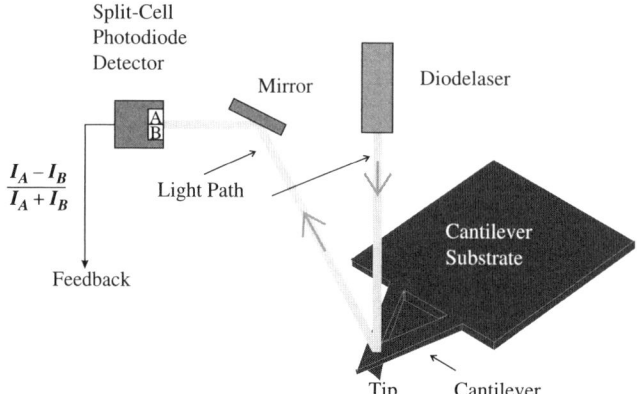

FIGURE 6.17 AFM with light reflection sensing method.

shows an AFM using electron tunneling where the deflection of the cantilever causes changes in the tunneling distance that is sensed as a large change in I_{tun}. While this method has been found to operate, the adjustment of two tips and the calibration of the tunneling current has been found to be cumbersome, and the laser beam deflection method shown in Fig. 6.17 has evolved as the detection scheme of choice in virtually all modern AFM designs. The laser beam deflection method employs a laser diode as the source of light and a split-cell (A and B) photodiode detector. With this method the top side of the cantilever needs to have a sufficiently large mirrorlike surface such that light can be reflected with minimal scattering. The light beam is emitted from the laser diode and reflected from the cantilever onto the split-cell photodiode detector where the light produces I_A and I_B. The detector signal is amplified and the ratio $(I_A - I_B)/(I_A + I_B)$ is electronically produced. The relative magnitudes of I_A and I_B depend on the position of the laser beam spot on the detector that is dependent of the deflection of the cantilever. This output is used to maintain a constant distance between the tip and the sample during scanning via a feedback control loop by altering the piezoelectric tube scanner in the z direction.

There are basically three operation modes of AFM—contact, noncontact, and tapping modes—that are determined by different interatomic forces displayed in Fig. 6.14 that are used in monitoring the tip–sample interaction.

The contact mode exists where the interatomic forces are strongly repulsive and thus where the tip is very close to the surface so that the separation distance cannot be further decreased without damaging the tip or the sample surface. In this operation mode, the interaction force is highly localized so that atomic resolution is possible. However, this mode is limited to hard surfaces such as Si or ceramics and some metals. Organic and biological materials would likely be damaged in the contact mode.

The noncontact mode is operative at larger tip to sample distances, and the interaction force is predominantly the much weaker van der Waals attractive force. This weaker force minimizes tip damage to the sample, but the large separation yields some instability of the tip-to-surface engagement, and the result is that the resolution of AFM images is usually lower than those taken in contact mode. In general, for a small feature to be resolved, its lateral size has to be larger than both the effective tip radius and the tip–sample separation distance.

The third mode, the tapping mode, combines the advantages of the noncontact and the contact modes of operation such as the minimal sample surface damage and the high resolution imaging, respectively. In this mode, a piezoelectric crystal is used to drive the tip in and out of the surface region at near the resonant frequency of the cantilever. Therefore, the tip is brought into contact with the surface for imaging with the high resolution repulsive force range and pulled out of contact when the surface is moved to a different position such that the tip is not being dragged across the surface and damaging the sample.

Examples of semiconductor studies using AFM to determine surface morphology are given in Chapter 4. Since all the samples were hard, the contact mode was used exclusively.

6.2 SURFACE COMPOSITION: AUGER ELECTRON SPECTROSCOPY AND ION SCATTERING AND RECOIL SPECTROSCOPY

In previous chapters several surface-sensitive techniques were discussed that can yield information about surface composition. Now two more techniques are added that have been used extensively for surface composition, namely Auger electron spectroscopy, and another that shows great promise, namely ion scattering and recoil spectrometry. The underlying physics of these techniques are discussed along with several relevant applications.

6.2.1 Auger Electron Spectroscopy

Auger electron spectroscopy (AES) is a surface-specific technique that uses the emission of low energy electrons by the Auger process discussed below and is one of the commonly employed surface analytical techniques for determining the composition of the surface layers of a sample.

To effect Auger emission, electrons in the energy range of about 2 to 20 keV are incident upon a sample and cause core electrons to be ejected from atoms in the near surface region of the sample. Incident x-ray photons can also cause the ejection of a core-level electron. In addition to an ejected photoelectron the surface atom is left with a core hole. Figure 6.18a shows the emission of a core-level electron (filled circles) by means of an incident photon in an energy range ($h\nu$) commensurate with the core levels, namely an incident x-ray photon, and the creation of a core hole (open circle). The atom now in a perturbed state relaxes via electrons with a lower binding energy dropping into and filling the core hole. The energy released in this relaxation can be converted into an emitted x-ray photon or an electron can be emitted. Figure 6.18b shows the emission of an x-ray photon as a valence band electron drops down in energy to fill the core hole. This process is also useful for elemental analysis because the emitted photon is characteristic of the atom, and the

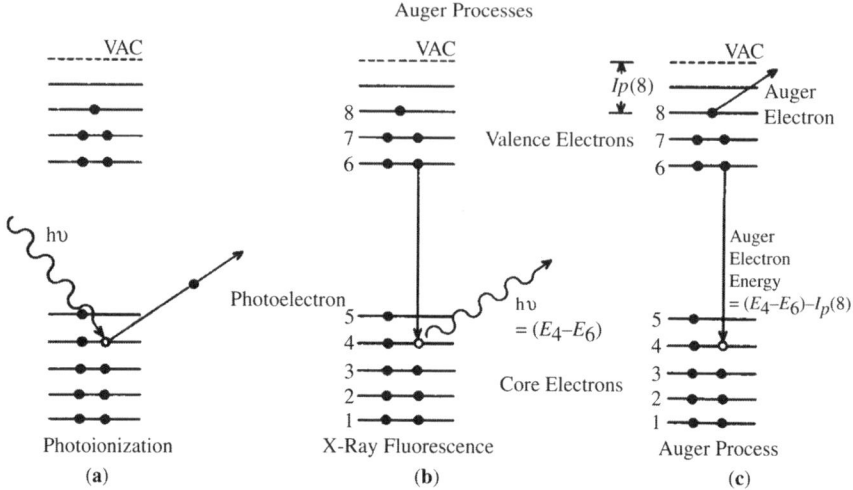

FIGURE 6.18 Comparison of (**a**) photoionization, (**b**) x-ray fluorescence, and (**c**) Auger processes. [Adapted from Ghosh (6), Figure 1.3.]

process is referred to as energy dispersive x-ray analysis, or EDAX. TEMs and SEMs can be equipped with EDAX hardware and use the electron beam to excite the x-ray photoemission. The emission of an electron is called the Auger process and is shown in Fig. 6.18c, and the electron emitted is called the Auger electron after Pierre Auger who has been credited with the discovery of this relaxation process. Notice that after the Auger electron emission, the atom is left in a doubly ionized state. The energy of the Auger electron is characteristic of the element that emitted it and can thus be used to identify the element. The Auger electrons emitted are typically in the range of 50 eV to 3 keV, and in this range the electrons cannot escape from more than a few nanometers of the surface (recall Fig. 2.14). The short inelastic mean free path of Auger electrons in solids yields surface sensitivity similar to that of photoemission discussed in Chapter 5, and therefore AES is popular for determining the composition of the top few layers of a surface. It cannot detect hydrogen or helium but is sensitive to all other elements and is most sensitive to the low atomic number elements.

Like XPS and PES, AES must be carried out in ultra-high vacuum (UHV), which is usually considered to be below 10^{-8} Torr. Often AES is used to characterize the composition of buried layers and is therefore used in conjunction with a technique that can cross section the sample. One such technique is called sputtering and will be discussed in Chapter 10 and was discussed briefly in Chapter 2 under TEM specimen preparation. Sputtering is essentially the use of a high energy heavy ion beam such as Ar^+ ions that are made to impinge upon the surface to be tested and thereby erode the sample and expose layers beneath the surface for AES analysis. Also, when a sample is brought into the UHV environment from laboratory ambient, it is coated with carbon and oxygen and this unwanted layer(s) can be removed by sputtering. The sputtering process is used extensively with AES and other surface techniques. However, it leads to damage to the sample and often

6.2 SURFACE COMPOSITION: AUGER ELECTRON SPECTROSCOPY

changes the surface composition due to preferential sputtering of some constituents. Thus great care must be taken in interpreting AES results from ion-sputtered surfaces.

The use of AES profiles obtained from sputtering through interfaces in thin-film structures is of great importance in microelectronics. This is the case because interfaces are crucial in microelectronics devices and often interfaces are broadened and otherwise altered by various microelectronics processes, especially heat treatments. In recent years there has been a quest in microelectronics for dielectric films with high static dielectric constant K (discussed further in Chapter 7). The materials selected as candidate high K materials are often complex oxides, and one of these (Ba, Sr)TiO$_3$ (BST) was a serious candidate. One problem with this material was that it reacted with Si and with SiO$_2$, producing unwanted products and degraded properties including a significant lowering of K. Some of the details of this study will be presented below and again in Chapter 11. Figure 6.19 displays AES sputter profile results for BST that had been sputter ion deposited on Si in 6.19a and SiO$_2$

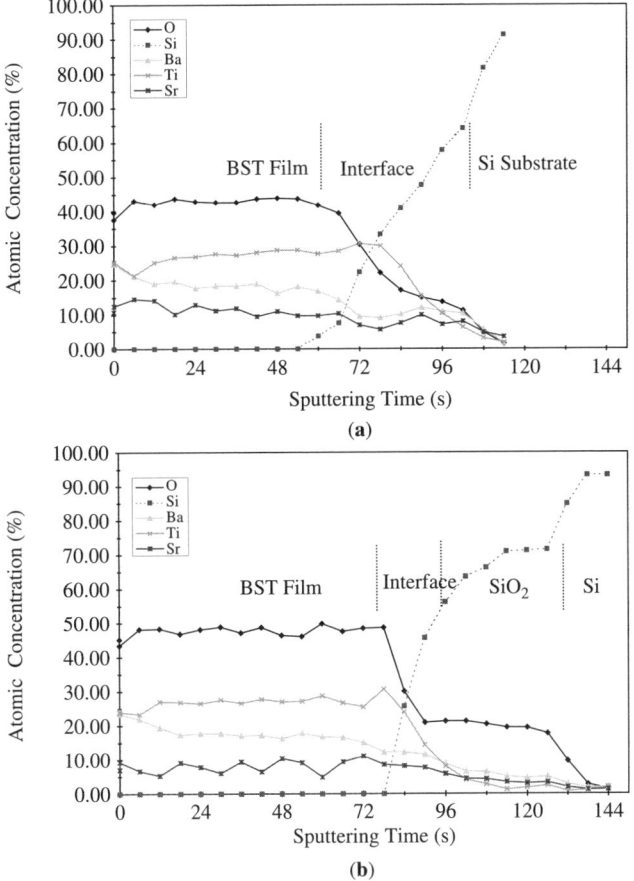

FIGURE 6.19 AES sputter profiles results for BST on (**a**) Si and (**b**) SiO$_2$ on Si. [Adapted from Gao et al. (7), Figure 4.]

in 6.19b. After deposition the samples were then inserted into an AES UHV chamber that was also equipped with a sputter ion gun, and the samples were etched and remeasured periodically during the etching process. The results in Fig. 6.19 are reported in terms of atomic concentration of the BST film constituents (O, Si, Ba, Ti, Sr) versus sputter time. For the same materials and sputter conditions the sputtering time is proportional to the film thickness. It is often not necessary to calibrate the sputtering time in terms of actual thickness. The results show that the interface region where there is maximum intermixing and reaction is wider on Si that on SiO_2, and those results were confirmed with other measurements. This kind of interface profile is difficult to obtain with other methods and very important in thin-film research in particular in the field of microelectronics. Consequently, AES has been widely used for this purpose.

6.2.2 Ion Scattering

There are many very useful material characterization techniques that use ions and ion scattering. However, the discussion here is limited to those techniques that afford the maximum in surface sensitivity, and this usually involves the use of low energy ions. Low energy (<15 keV) ion-scattering spectroscopy (ISS), direct recoil spectroscopy (DRS), and mass spectroscopy of recoiled ions (MSRI) are all surface analytical methods that are enhanced with the use of time of-flight detection of ions and can be collectively referred to as time-of-flight ion scattering and recoil spectroscopy (ToF-ISARS). ToF-ISARS can provide a wide range of surface-specific information such as surface composition, atomic structure of the first few monolayers, lattice defect density, and trace element analysis. In addition ToF-ISARS can be used in real time while a film grows and thus can analyze from the starting surface to a thick film. In this way compositional profiles are obtained as was done above using AES along with sputtering, but with ToF-ISARS the problems with sputtering (damage and preferential sputtering of some elements) are avoided. Most important for surface analyses ToF-ISARS provides surface-specific information from about 1 to 2 monolayers. In addition, this technique is less sensitive to multiple-scattering effects compared to other conventional surface analysis techniques such as low energy electron diffraction (LEED) and Auger electron spectroscopy (AES). By using differential pumping, ToF-ISARS becomes even more compatible with a moderate vacuum processing environment, up to hundreds of milliTorr. The energy of the incident primary ion beam for ToF-ISARS is generally in the 10- to 15-keV range. With the application of a pulsed primary ion beam and time-of-flight (ToF) detection, the requirement of beam dose (typically 10^{10} to 10^{12} ions/cm^2) is 2 to 3 orders less than for continuous electrostatic analysis, which renders ToF-ISARS an essentially nondestructive technique. Moreover, the data acquisition rate is also dramatically increased by using ToF detection, which is a key factor for *in situ* monitoring of thin-film growth.

The powerful surface techniques already discussed such as LEED, AES, UPS, and XPS detect 100 to 2000 eV electrons, which have a typical range of about 5 to 40 Å in solids. However, the very short mean free path of electrons at high background pressures degrades the energy information and limits these electron-based analytical techniques usefulness to high and ultra-high vacuum environments because electrons undergo significant gas-phase scattering.

Methods that detect higher energy (Mega electron Volts) ions such as Rutherford backscattering spectroscopy (RBS) and elastic recoil detection (ERD) are less subject to gas-phase scattering and may be used at pressures up to 1 atm. However, the sampling depth increases to 0.5 to 2 μm, which makes them much less surface sensitive, although ion channeling techniques for single-crystal materials can enhance surface sensitivity. At the same time, the high energy incident ion beam increases the possibility of film damage during exposure of sample to the test probe.

In summary, the use of low energy ions enables sensitive surface analyses that can access surface structure and composition and that can operate with low damage and in real time in many process environments.

6.2.2.1 Time-of-Flight Ion Scattering and Recoil Spectrometry Time-of-flight ion scattering and recoil spectrometry (ToF-ISARS) refers to a collection of ion-scattering and recoil spectroscopy techniques that use time-of-flight mass spectrometry instrumentation. Energy analysis of atomic signals arising from binary collisions between an incident probe beam at low incident energies typically in the range of 5 to 15 kV using a pulsed Ar^+ beam yields information about the samples outermost surface layers. Each of the included techniques differ whether the signal is derived from the scattered primary ion beam, known as scattering techniques, or ions ejected from the surface by a binary collision process, referred to as recoil techniques.

The use of a low energy primary beam bombarding a surface to elicit signal has been thoroughly studied for analytical techniques such as secondary ion mass spectrometry (SIMS) and fast atom bombardment (FAB) with particular attention being paid to the damage caused to the surface of the analyte by the incident beam. The consensus solution to reduce damage is to minimize the incident ion flux. The time-of-flight mass spectrometry instrumentation is directed at reducing the incident ion flux needed for analysis. Energy analysis is performed on the signal detected from fast moving particles colliding with the surface of an electron multiplier-type detector, by timing the particles flight over a known distance, and calculating the kinetic energy of the particle. Peak assignments for recoil spectra are then made by comparing the resulting particle energies to the theoretical energy possessed by each element recoiling at the angle of analysis. Scatter peak assignments are made by considering the theoretical kinetic energy remaining with the scattered primary ions after redirection by elements acting as scattering centers. Time analysis such as this requires a well-defined time zero and thus cannot be performed with the sample under constant beam exposure. A short (<5 ns) pulsed event is required to determine the detected particle energies precisely enough to make peak assignments with atomic differentiation. Thus, the nature of the instrumentation itself limits the flux incident upon the sample to approximately 2×10^5 ions/mm^2/pulse). With a pulse rate of 20 kHz and a generous acquisition time of 5 s, this results in an exposure of 2×10^{10} ions/mm^2, anywhere from 10^3 to 10^5 times less than the atomic surface concentration.

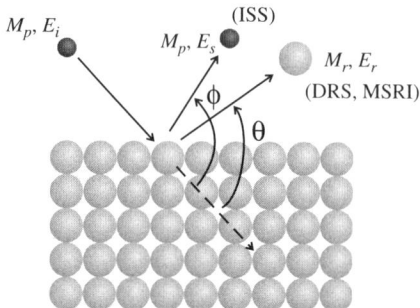

FIGURE 6.20 Ion scattering (ISS) and recoil events (DRS, MSRI) at a solid surface.

Even taking into account the theoretical value of 250 atoms displaced by each 10 keV Ar$^+$ ion (for Si)^2, the damage to the surface should be negligible.

Since low energy ion scattering and recoil spectrometry employs primary ions in the range of 5 to 15 keV, the collision between the primary ion and the surface is well described using classical two-body elastic collision kinematics as shown in Fig. 6.20. In this figure a primary ion of mass M_p and energy E_i impinges upon a surface. In one case the primary ion is scattered through angle ϕ relative to the trajectory (dashed arrow) and its energy is changed to E_s. The other case, called recoil, is that a surface atom M_r is ejected at energy E_r and angle θ. Conservation of kinetic energy (E) and linear momentum (**P**) imply that the sum of energy and momentum of the scattered primary (E_s, **P**$_s$, respectively) and recoiled scattering center (E_r, **P**$_r$) must sum to equal the energy and momentum initially possessed by the primary ion (E_i, **P**$_i$):

$$E_i = E_s + E_r$$
$$\mathbf{P_i} = \mathbf{P_s} + \mathbf{P_r} \tag{6.33}$$

Considering the latitudinal and longitudinal components for the momentum of these particles in two-dimensional, and using $E = \frac{1}{2}m\mathbf{v}^2$, these equations may be rewritten to yield:

$$\frac{1}{2}m_p\mathbf{v_s}^2 + \frac{1}{2}m_r\mathbf{v_r}^2 = \frac{1}{2}m_p\mathbf{v_i}^2$$
$$m_r\mathbf{v_r}\cos\theta + m_p\mathbf{v_s}\cos\phi = m_p\mathbf{v_i} \tag{6.34}$$
$$m_r\mathbf{v_r}\sin\theta + m_p\mathbf{v_s}\sin\phi = 0$$

where the subscripts p and r refer to the primary and recoil atoms, and the angles θ and ϕ to the recoil and scatter angles, as indicated in Fig. 6.20. It should be noted that the recoil atom and the target atom is the same with mass m_r or m_t since the recoil atom is the ejected target atom. Equations 6.34 illustrate that as the impinging particle approaches at a sufficiently close distance (called the impact parameter distance), the trajectory is altered by interaction with the surface atom(s), also referred to as

6.2 SURFACE COMPOSITION: AUGER ELECTRON SPECTROSCOPY

the recoil atom. During this interaction, the target or surface atom is recoiled with some kinetic energy at some angle to directly offset the momentum gained by the primary atom in the direction normal to the incident beam path. Thus, at grazing angles of incidence, primary particles scattering away from the sample bulk will produce recoils into the bulk, while particles scattering into the bulk of the sample will yield recoiled atoms that escape from the sample surface and may be analyzed.

Solving Equations 6.33 and 6.34 for the kinetic energies possessed by the scattered (E_s) and recoiled (E_r) particles, and summing various mechanisms of energy loss as Q, with introduction of a reduced energy term f^2, the following energy ratios are obtained:

$$\frac{E_s}{E_i} = \left[\frac{\cos\phi \pm (A^2 f^2 - \sin^2\phi)^{1/2}}{1 + A}\right]^2 \tag{6.35}$$

and

$$\frac{E_r}{E_i} = \frac{A}{(1+A)^2}\left[\cos\theta + (f^2 - \sin^2\theta)^{1/2}\right]^2 \tag{6.36}$$

where

$$f^2 = \frac{E_i - Q}{E_i} \quad \text{and} \quad A = \frac{m_t \text{ or } m_r}{m_p} \tag{6.37}$$

For most practical cases of ToF-ISARS $Q \ll E_i$, so the factor f^2 is approximately 1, yielding the following simplified equations:

$$\frac{E_s}{E_i} = (1+A)^{-2}\left(\cos\phi \pm \sqrt{A^2 - \sin^2\phi}\right)^2 \tag{6.38}$$

and

$$\frac{E_r}{E_i} = 4A(1+A)^{-2}\cos^2\theta \tag{6.39}$$

where for Equation 6.38, the case $A \leq 1$ is interesting. Recall that A is the ratio of the mass of the target atom m_t (or recoil atom m_r) to the incident or primary ion m_p (Equation 6.37). Thus, for the case of m_r or the target atom mass being less than that of the impinging particle m_p, it can be shown that the scattering angle ϕ will be limited by the requirement that $\phi_{max} \leq \arcsin A$, that in turn imposes a mass limit for detecting backscattered ($\phi > 90°$) primary particles. If $A > 1$, that is m_p is less than m_r, the energy of the scattered primary is double-valued according to Equation 6.38. These equations enable an estimation of the particle energies and thus the ability to identify signal peaks. However, the formulas developed herein are based on classical collision theory and for precise identification it is often necessary to perform a simulation that includes the attractive potentials of the atoms and the energy losses included in the f^2 term. Energy losses that can cause deviation from the ideal elastically scattered and recoiled particles arise from inelastic processes such as electronic excitation or ionization of the target atoms during collision. Also surface charging for dielectric surfaces can alter the field-free region traversed by the incoming primary ion and consequently reduces the energy of the incident beam.

Time-of-flight spectra collected without electrostatic or magnetic mass selection requires assigning the spectral features according to the speed of the particle detected. Equations 6.40 show the time of flight for scattered particles t_s and recoil particles t_r in terms of the masses and angles from Fig. 6.20 and the path L in the system from sample to detector.

$$t_s = \frac{L(m_p + m_s)}{\sqrt{2m_p E_p}\left[\cos\phi + \sqrt{(m_s/m_p)^2 - \sin^2\phi}\right]}$$

$$t_r = \frac{L(m_p + m_r)}{\sqrt{8m_p E_p}\,\cos\theta} \quad (6.40)$$

For the three analytical methods—ISS, DRS, and MSRI—the mass of the surface or target atoms m_t can be determined by measuring the kinetic energies E_s and E_r or the times of flight t_s and t_r with detectors placed at different angles, and by varying the incidence angle of primary ion beam relative to the sample normal. For ISS, the mass of surface atom is determined by detecting the kinetic energy loss of primary ions in the backscattered direction when the primary ions are incident close to the normal to the sample surface. DRS and MSRI employ relatively grazing entrance and exit angles in order to measure the kinetic energy of forward ejected surface atoms (recoiled atoms) via a single collision with primary ions. As was noted above, Equation 6.38 always has real solutions if $m_t > m_p$. Therefore, ISS is generally sensitive to target atoms heavier than primary ions. According to Equation 6.39, DRS can detect all atomic species, and the primary advantage of DRS is that it has sensitivity to light elements including hydrogen and helium. Additionally, MSRI detects only ion fractions of the direct recoil spectrum, and therefore provides a high mass resolution technique for surface composition and isotope analysis.

The kinetic energy of the backscattered primary ions or forward recoiled surface atoms are measured using either an electrostatic energy analyzer (ESA) or a pulsed primary beam and ToF detection scheme. The ESA only detects the ion fraction, which is typically 10^{-2} to 10^{-4} of the scattered primary beam. In addition, ESA only transmits a narrow portion of the energy distribution of these ions. As a result of low detection efficiency of the ESA, large ion doses are required to obtain sufficient signal. In contrast, the ToF detection scheme simultaneously collects particles including both ions and neutrals of all energies. Consequently, the required primary ion doses for ToF detection are about 3 to 5 orders of magnitude smaller than that for ESA. A typical ion dose for ToF analysis is about 10^{11} to 10^{12} ions/cm^2, which is one part in 10^3 to 10^4 of the average surface atoms. Thus, ToF detection scheme renders ISARS an essentially nondestructive technique. Moreover, ToF detection significantly reduces data acquisition time.

More details for each of the three complementary low energy ion techniques—ISS, DRS, and MSRI—are presented below.

6.2.2.2 Ion Scattering Spectroscopy

In ISS, the primary ion beam incident close to the sample normal and the mass of surface atoms can be determined using Equation 6.38 by measuring the kinetic energy of the backscattered primary ions that have known mass. ISS can be considered as a low energy version of the widely used Rutherford backscattering spectroscopy (RBS), which is sensitive to long-range bulk structure by using the incident ions with energy of Mega electron Volts. ISS with ion energy in the kiloelectron volt range has much higher sampling depth sensitivity than RBS. Among all surface analysis techniques, ISS is the most surface-specific surface analytical method with one atomic layer sensitivity. From an analysis of the ISS spectra from crystals, information on composition and structure can be provided.

By varying the incidence angle of the primary beam, ISS can be used to determine the surface atomic structure and defect configuration of the sample surface, and this method is called angular resolved ion scattering spectroscopy (ARISS). The basic concept of the shadowing effect is illustrated using Fig. 6.21. As a result of scattering from the topmost atoms, trajectory calculations indicate that there is a region behind each target atom called the "shadow cone" from which the primary beam is excluded. Target atoms (B) located inside this shadow cone are therefore invisible to the

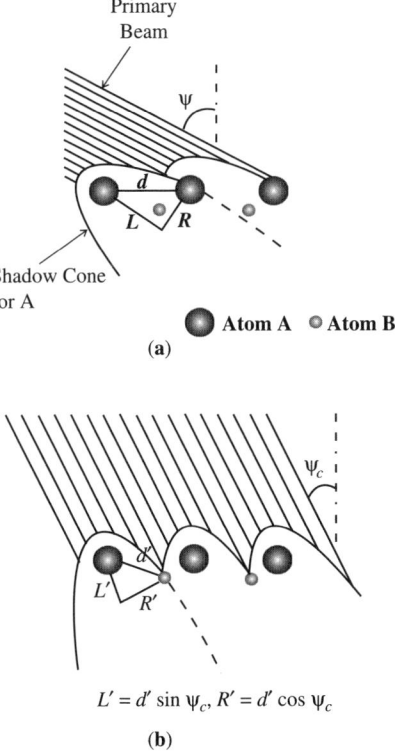

FIGURE 6.21 (a) Shadow effect of an ion beam where atom B is invisible and (b) by changing the incident angle atom B is accessed by the incident beam.

primary ions and do not contribute to the ISS signal. At the incident angle ψ for Fig. 6.21a only signal from top layer atoms A are expected to appear in the ISS spectrum. As the incidence angle approaches the sample normal, there is a critical angle ψ_c at which the shadowed B atoms become visible to the incident beam as shown in Fig. 6.21b, and the ISS signal from the B atoms increases abruptly. The sharpness of the onset of the B atoms signal provides a measure of the degree of surface disorder, and the value of the critical angle can be quantitatively related to the bond distance and bond angle for specific atomic species within the first few atomic layers on the surface. Therefore, ISS provides a direct measure of interatomic distances in the first and subsurface layers in real space.

6.2.2.3 Direct Recoil Spectroscopy For DRS the primary ion beam impinges on the surface at a glancing angle (large ψ in Fig. 6.21) and the surface atoms recoiled into a forward scattering angle are detected. One of the early advantages identified for DRS is the ability to detect all surface species including H and He that are difficult by most techniques.

Direct recoil spectroscopy is particularly suitable for surface elemental analysis. The large scattering and recoil cross sections in the kiloelectron volt energy range used for DRS along with the simultaneous detection of both neutrals and ions by ToF detection result in high surface sensitivity for DRS. Qualitative elemental analysis can be obtained with primary ion doses of only $\sim 10^{11}$ ions/cm^2, which corresponds to 10^{-4} monolayer and therefore provides both a nondestructive and surface-sensitive technique. DRS is especially important for detecting light impurities such as H, C, and O on surfaces, although heavy elements can also be detected by DRS. This method is particularly sensitive to surface H since the recoil cross section for H is large.

The direct recoiled atoms are those species ejected from the surface as a result of a single collision of the primary ion and without losing energy through collisions with neighboring atoms. Thus, the kinetic energy of direct recoil particles can be calculated from Equation 6.39, which gives one energy value for a specific surface element m_t and produces a sharp peak in the DRS spectrum. However, multiple collisions are common in these processes and may cause energy loss of recoiled particles, which results in a broad and low intensity structure lying beneath and extending toward longer flight time in the DRS spectrum. The broad distribution of peaks from the energy spread degrades the resolution of DRS and also prevents the detection of species present in small amount. The technique using a time-refocusing analyzer to improve resolution and sensitivity of DRS is referred as to MSRI.

6.2.2.4 Mass Spectroscopy of Recoiled Ions For MSRI, the ionized fraction of recoiled particles are diverted into a time-refocusing analyzer called a reflectron. For ions with a given mass, those with higher kinetic energy travel far into the reflectron before being turned around, while ions with lower energy do not penetrate as deeply and therefore travel a shorter distance. By properly adjusting the reflectron potentials and experimental geometry, all ions with a given mass are forced to arrive at the detector simultaneously. In this way, the resolution of recoil spectrum is dramatically enhanced. An isotopic resolution of all surface species can be

6.2 SURFACE COMPOSITION: AUGER ELECTRON SPECTROSCOPY

obtained, and MSRI has a sensitivity of better than one part per million (ppm). Mass resolution values $M/\Delta M$ as high as 400 have been achieved.

With isotopic resolution and ppm sensitivity, MSRI has been successfully applied in a variety of surface analyses such as detection of dopants in semiconductor materials, detection of surface hydrogen, and detection of isotopes.

The hardware used for MSRI is similar to that used for ToF secondary ion mass spectroscopy (SIMS): a high intensity ion source, ion beam line, sample holder, and ToF detector. However, the techniques are significantly different and the major differences can be understood using Fig. 6.22. First, MSRI geometry in Fig. 6.22a top panel emphasizes single-collision ejection events with a large angle of incidence (relative to the surface normal), rather than the multiple-collision cascade mechanism associated with the SIMS process with lower incidence angles relative to the surface normal, as shown in Fig. 6.22b top panel. In MSRI the more energetic single-collision processes leads to mainly ejected atoms, as is seen in Fig. 6.22a bottom panel. In SIMS the incident beam causes numerous small-angle collisions between the recoiling target species, as is indicated in Fig. 6.22b top panel with the collision cascade paths in the solid surface. Consequently, the ejected particles are often not fully dissociated, as seen in Fig. 6.22b bottom panel, and leave the surface with energy of only a few electron volts as a result of the collision cascade. Because of the low energy of ejected particles from the surface in SIMS, there is enough time for charge exchange between the surface and departing particles to occur, which leads to a high degree of neutralization of sputtered particles. Thus, in SIMS, the ion fraction is typically as low as only 10^{-2} to 10^{-5}. Since the extent of neutralization strongly depends on the surface composition, the ion yield varies up to 5 orders of magnitude, as the surface composition changes. In analytical chemistry this is called a *matrix effect*, which renders SIMS difficult to quantify. In contrast, the MSRI signal corresponds to direct recoiled particles ejected from the surface with much higher kinetic energy (a few hundred of electon volts to thousands of electron

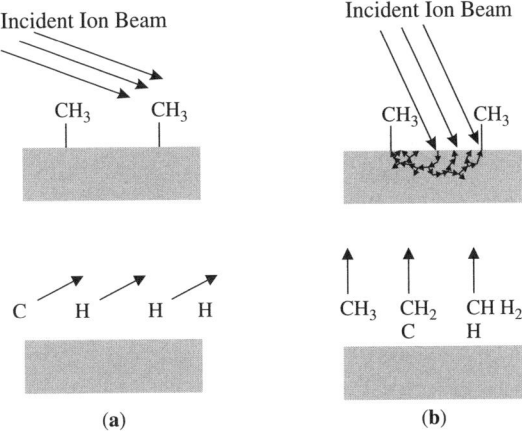

FIGURE 6.22 Comparison of (a) MSRI/DRS collision processes with that of (b) SIMS.

volts) than those comprising the SIMS signal. Therefore, the MSRI ions are less subject to neutralization by charge exchange with the surface than SIMS ions, which makes the MSRI signal much less sensitive to the matrix effect. The ion fraction of the direct recoil particles often exceeds 10%. In addition, both elemental species and molecular fragments are detected in SIMS, which makes SIMS spectra difficult to interpret. While in MSRI, the energy of collision completely breaks the molecular bonds and result in complete decomposition. Only the elemental species appear in MSRI spectrum, which minimize mass overlaps that would otherwise hinder the detection of minority species.

6.2.2.5 Mass Spectroscopy of Recoiled Ions Applications For surface and interface studies MSRI has shown to be a very sensitive and therefore useful technique for exploring the formation of intermixed/interdiffused interfaces between a film and substrate. Furthermore MSRI has been found to be operable in vacuum that is also amenable to various deposition (discussed in Chapter 10) and characterization techniques such as ellipsometry (see Chapter 9). The ability of a characterization technique to operate in moderate vacuum and while a surface process is ongoing enables real-time observations of surface changes and interface formation, and such techniques are both powerful and desirable for surface and interface studies. An example of a real-time system in which a surface can be treated in various ways and surface reactions can occur such as film deposition, and in which the evolution of the surface can be monitored in real time is shown in Fig. 6.23. The system

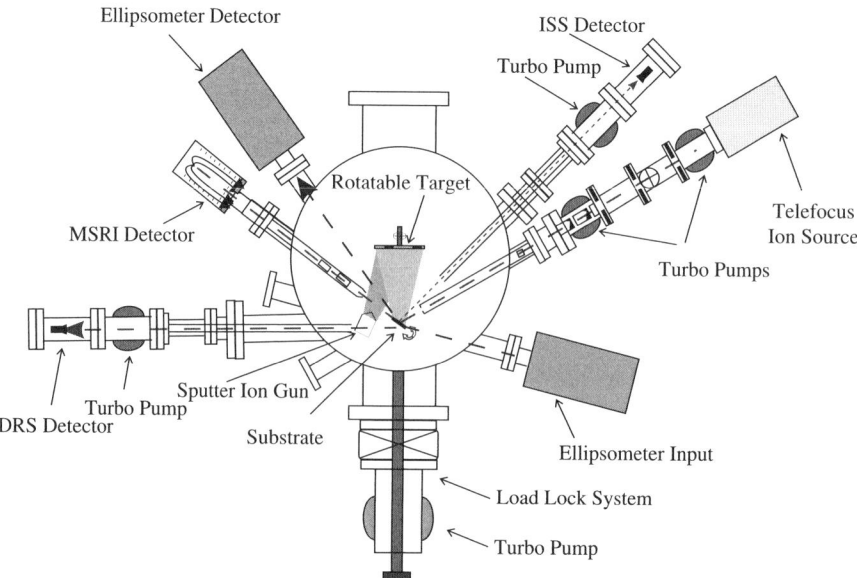

FIGURE 6.23 Integrated ion sputter deposition, spectroscopic ellipsometer, and time-of-flight ion scattering and recoil (ToF-ISARS) spectrometry system.

depicted in Fig. 6.23 is equipped with all the ToF-ISARS techniques mounted on a vacuum process chamber. In addition to ToF-ISARS, there is a spectroscopic ellipsometer (SE) (discussed in Chapter 9) and an ion beam sputter deposition system (discussed in Chapter 10) for preparing thin films on a substrate that is mounted in the vacuum chamber. The system is configured so that during deposition optical properties of the film being deposited can be obtained using SE and ToF-ISARS techniques, in particular MSRI can also be used in real time to obtain surface composition during deposition.

Before presenting some experimental results from the system shown in Fig. 6.23, it is useful to discuss the system itself as one example of the recent use of multifunctional systems to deposit and analyze surfaces, interfaces, and films. Figure 6.23 shows a top view schematic of a cylindrical stainless steel vacuum chamber. The chamber is evacuated using turbomolecular vacuum pumps that are backed up using conventional mechanical vane pumps. In addition each subsystem to be discussed separately is also evacuated separately using turbomolecular pumps that are backed up using conventional mechanical vane pumps, and each subsystem can be isolated from the other subsystems. As was pointed out in previous chapters, in particular Chapter 5, the use of vacuum in surface and interface science is very important and worthy of separate study for the main reasons that vacuum is crucial to preserve surfaces and interfaces for characterization, and many surface characterization techniques (e.g., TEM, SEM, STM, and XPS) require vacuum to operate properly. The ability to isolate the various subsystems enables more efficient maintenance and operation when one system is inoperable.

The first step in any experiment with such a system is to provide a sample for processing and study and to enter that sample into the system. For this purpose a separately pumped load lock subsystem is used. This subsystem is first isolated from the main chamber using a large gate valve. The load lock has a door to the laboratory that can open to accept the sample. The sample loading mechanisms vary, but typically there is a mechanical apparatus that can hold the sample in the load lock chamber while the sample is brought to vacuum. Then with the load lock chamber brought to proper vacuum, the gate valve is opened and the sample is inserted into the main chamber. There is a push rod and bellows assembly to move the sample into the main chamber and a sample stage to which the sample can be affixed. Often the sample is affixed to the sample stage using a screw or clip mechanism. Once the sample is on the sample stage in the main chamber, the push rod assembly is withdrawn and the gate valve is closed. The sample stage is capable of xyz motion and tilting. Typically, the sample stage can be heated and in some instances cooled as well. The rotation of the sample stage first enables positioning to accept the sample, and then it is rotated for sample processing and/or characterization.

The next step is to commence processing and real-time characterization. The system in Fig. 6.23 has several processing options. First, there is an ion beam sputter deposition subsystem. This system has an ion gun that can be aimed at a target that contains the desired atoms for film formation. The sputter ion beam is generated using heavy inert atoms, usually Ar, Kr, or Xe that are ionized by electron bombardment, and then the ions such as Ar^+ are accelerated by negative grids and electrostatically focused upon the target. As discussed in earlier chapters, the impinging ion beam erodes the target, creating atoms and excited atomic species. With the sample placed facing the target, the ejected (sputtered) target atoms can impinge upon the sample surface reacting and or depositing upon the sample surface. Often there are several targets of different materials so that different atoms can be selected for deposition by simply rotating the appropriate target to the sputter beam and the process can be readily automated. In addition, the ambient inside the vacuum system can be controlled. In particular, a small amount of oxygen or other desired gaseous reactant can be bled into the system for reaction with the sample surface. In most systems the sample could be heated to stimulate reaction.

Before, during, and after sample processing, it is usually desirable to perform surface, interface, and film characterizations. The system shown in Fig. 6.23 has two main surface characterization subsystems that are capable of characterizations as mention above: ToF-ISARS and SE. Several MSRI results will be shown below to illustrate the use of MSRI to obtain surface chemical composition. In Chapter 9 ellipsometry will be discussed and an example of ellipsometry result given and also in Chapters 6, 10, 11, and 12, where further results using the system in Fig. 6.23 will be shown.

As was mentioned previously in Chapter 4, in the practice of microelectronics a gate dielectric film is required to support the electric field that is necessary to operate the metal–oxide–semiconductor field-effect transistor (MOSFET). Furthermore in Chapter 7 it will be pointed out that there is an ongoing search to replace the presently used SiO_2 film (on Si) as the gate dielectric film with a dielectric having a higher static dielectric constant K. One of the candidate materials for that purpose is barium strontium titanate (BST). Figure 6.24 displays a typical MSRI spectrum of BST that was taken using the system depicted in Fig. 6.23. BST was sputtered onto an MgO substrate. MgO was found to be inert to BST and thereby provides a sharp interface. The spectrum shows peaks corresponding to the elemental composition of the BST film, namely Ba, Sr, Ti, and O. A doubly charge Ba peak is also seen as well as an Ar peak. Ar^+ was used to sputter a BST target, and thus Ar

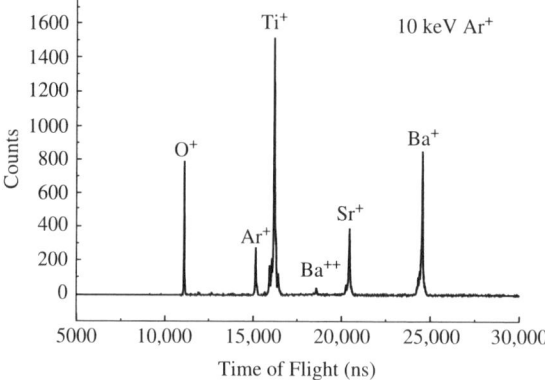

FIGURE 6.24 MSRI spectrum of barium strontium nitrate BST film sputter deposited on an MgO substrate.

6.2 SURFACE COMPOSITION: AUGER ELECTRON SPECTROSCOPY   195

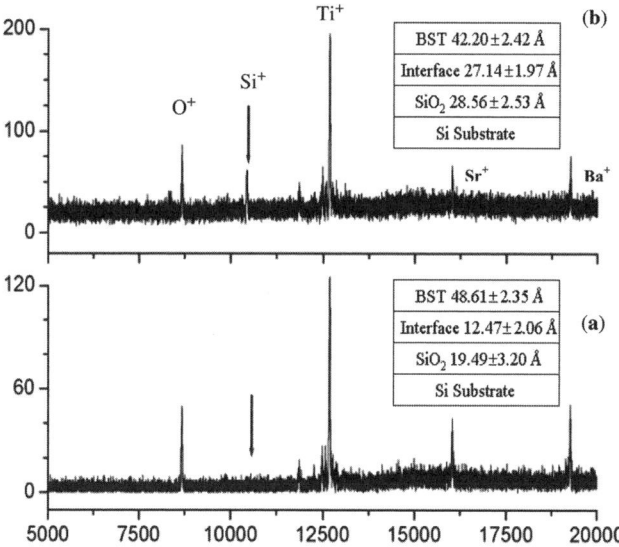

FIGURE 6.25 MSRI spectra of BST sputter deposited on Si with native SiO_2 layer: (**a**) initial MSRI spectrum with on Si and (**b**) result with Si present after less than 4 min anneal at 650°C. [Adapted from Mueller et al. (8), Figure 4.]

ions were on the sample surface of found their way to the detector. While not resolved well in Fig. 6.24, the base of the Ti peak shows some of the isotopic spread in Ti. Because of the different ion interaction efficiencies for different atoms and the different atom positions at the surface, the spectrum does not yield a direct

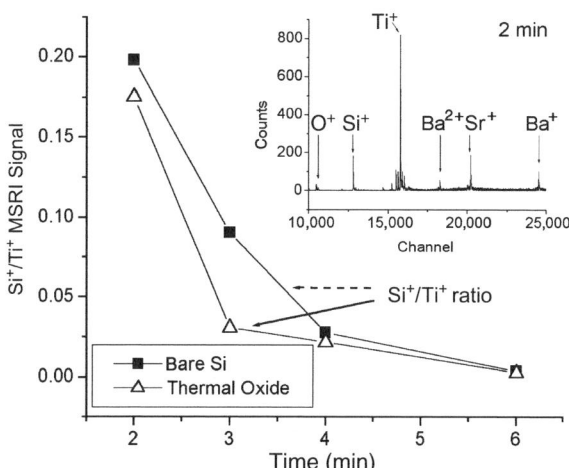

FIGURE 6.26 Ratio of Si/Ti as obtained using MSRI for BST deposited on bare Si (with native SiO_2 about 2 nm thick) and thicker SiO_2 (about 20 nm thick). [Adapted from Mueller et al. (8), Figure 2.]

quantitative assessment of the sample surface. Either the use of a known sample as reference or careful calibration is required to render the MSRI spectra quantitative. To study the intermixing of the BST with Si, BST was sputter deposited on Si that has a native oxide layer of about 2 nm after deposition. Actually, the native oxide on Si is closer to 1 nm or less after Si cleaning, but the sputtering of BST takes place in O_2 (at pressures around 10^{-4} Torr) to ensure O stoichiometry. During the sputter deposition of BST, the oxide on Si grows about 1 nm in the low pressure O_2. The results (8) are shown in Fig. 6.25, where it is seen that upon deposition of BST, the Si signal from the underlying SiO_2 is not seen, but after a brief anneal the Si signal appears, indicating that Si diffused to the top of the BST. Also shown in the inset of the figure is the layer structure for the films present and the film thicknesses as was obtained using spectroscopic ellipsometry (SE is discussed in Chapter 9). The inset indicates considerable intermixing of BST with the oxide and with Si at the surface confirmed by MSRI. In fact this intermixing and reaction of the BST

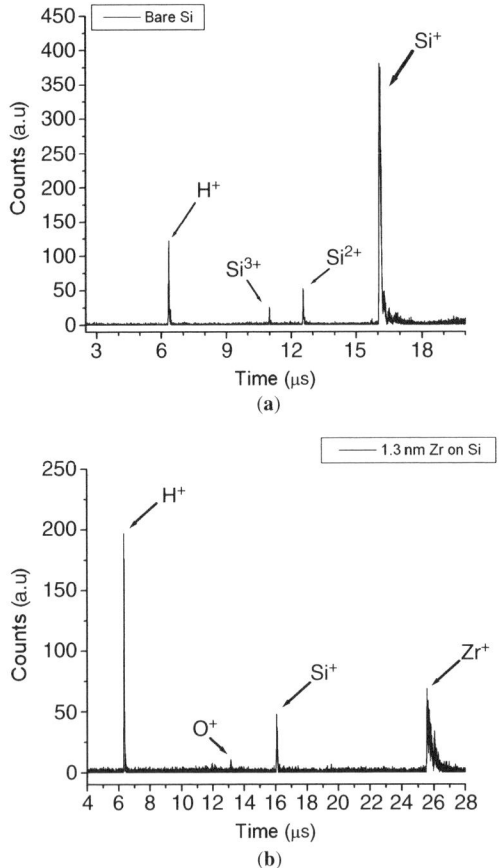

FIGURE 6.27 MSRI spectra for (**a**) bare Si and (**b**) 1.3 nm Zr on Si. [Adapted from Lopez et al. (9), Figure 5.]

degrades the BST and renders it not usable as a high K dielectric. Figure 6.26 shows the Si/Ti ratio at the surface of BST deposited on bare Si and Si with 20 nm SiO_2. The ratios were obtained at various times during the sputter deposition of BST with a 2 min MSRI spectrum shown in the inset. The results indicate that Si at the BST surface can derive both from Si and from SiO_2 with more Si coming from the thinner underlayer of SiO_2 and therefore displaying a higher Si/Ti ratio at given deposition time.

Mass spectroscopy of recoiled ions was also used to determine the extent of intermixing between Zr/Si and Zr/SiO_2 (9). In these experiments approximately 0.5-nm increments of Zr were deposited on the substrate (Si or SiO_2) and probed *in situ* after each increment. Figures 6.27a and 6.27b show the MSRI spectra for a bare Si surface and after a deposition of 1.3 nm Zr, respectively. It is seen that Si is present at the surface after a 1.3-nm deposition of Zr. Features near 26 µs are the peaks for the naturally occurring isotopes of Zr. The Si^+ peak is completely attenuated after depositing 2.0 nm Zr for both the Si and SiO_2 substrates. After oxidation of the 2.0 nm Zr film for both Si and SiO_2 substrates, the Si peak was still not detected. Though, when the samples had an initial layer of 1.3 nm Zr, the Si^+ peak intensity did increase after oxidation relative to the unoxidized sample. To eliminate the possibility that the MSRI Si signals were due to Zr film discontinuities, AFM and SEM measurements were done and led to the conclusion that the thin Zr films were continuous.

In Chapter 11 further MSRI examples will be discussed along with surface and interface results using other characterization techniques.

REFERENCES

1. Tromp RM, Hamers RJ, Demuth JE. Phys Rev B 1986;34:1388–1391.
2. Avouris Ph, Wolkaw R. Phys Rev B 1989;39:5091.
3. Hamers RJ. Annu Rev Phys Chem 1989;40:531.
4. Kariotis R, Lagally MG. Surf Sci 1991;248:295.
5. Bining G, Quate CF, Gerber C. Phys Rev Lett 1983;56:930.
6. Ghosh PK. Introduction to photoelectron spectroscopy. New York: Wiley; 1983.
7. Gao Y, Mueller AH, Irene EA, Auciello O, Krauss A. J Vac Sci Technol A 1999;17:1880.
8. Mueller AH, Suvorova NA, Irene EA, Auciello O, Schultz JA. Appl Phys Lett 2002; 80:3796.
9. Lopez CM, Suvorova AA, Saunders M, Irene EA. J Appl Phys 2005;98:033506.

SUGGESTED READING

L. C. Feldman and J. W. Mayer, 1986. *Fundamentals of Surface and Thin Film Analysis*, North Holland. An excellent book on surface science techniques where theory, hardware, and results are presented in a readable manner. This book was often used and referred to in the surface science courses at UNC.

J. M. Walls (ed.), 1990. *Methods of Surface Analysis: Techniques and Applications*, Cambridge University Press. A collection of chapters on various surface science techniques.

J. M. Walls and R. Smith (eds.), 1994. *Surface Science Techniques*, Pergamon. A collection of short articles on important surface science techniques.

R. Wiesendanger, 1994. *Scanning Probe Microscopy and Spectroscopy*, Cambridge University Press. A thorough coverage of scanning probe microscopy.

7
CHARGED SURFACES

7.1 INTRODUCTION

As pointed out in Chapter 5 both intrinsic and extrinsic surface electronic states can give rise to charge at a surface. In particular, using Fig. 5.7, it is seen that a double layer of charge can form as a result of charge exchange between the surface electronic states and the bulk energy bands of the material. The double layer can be closely spaced thin layers of positive and negative charges, as shown in Fig. 7.1a, that arise from charge transfer from the metal to an adsorbed oxygen layer that is relatively more electronegative and therefore can withdraw electronic charge from the metal. Figure 7.1b shows the potential distribution after charge exchange. For this case of a metal, the metal cannot support an electric field or space charge so that the distribution of potential is sharp. Alternatively, for the case of a semiconductor or an insulator, the charge exchange with the surface can result in a relatively thick layer in which the charge is distributed in the layer. Figure 7.2a shows this situation for an adsorbed electronegative oxygen layer similar to Fig. 7.1a but now upon a semiconductor rather than a metal, as seen in Fig. 7.2a. The charge is now distributed in the semiconductor near the dopants (donors), which can readily donate electrons to the surface layer. This results in a space charge layer as shown in Fig. 7.2b that is not sharp but rather extends into the semiconductor. Both cases shown in Figs. 7.1 and 7.2 consider extrinsic states derived from an adsorbate on a surface, but the same scenario could obtain for intrinsic surface states. This chapter will first set out to understand the nature of this charge, starting from a review of electrostatics culminating with the Poisson equation, whose

---

Surfaces, Interfaces, and Thin Films for Microelectronics. By Eugene A. Irene
Copyright © 2008 John Wiley & Sons, Inc.

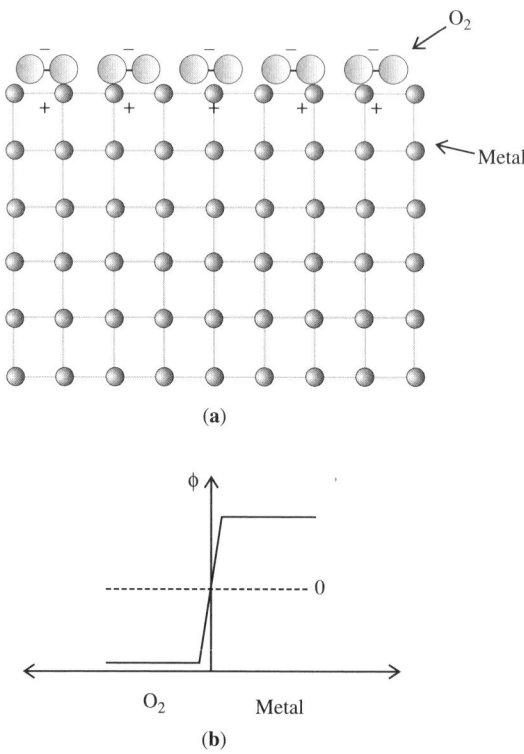

FIGURE 7.1 (a) Metal with adsorbed O_2 on the surface and (b) the resulting potential distribution across the metal–adsorbate interface.

solution yields the potential or charge distributions shown in Figs. 7.1b and 7.2b. Then to make this analysis somewhat realistic surface states will be discussed in relation to a device in which the interfacial charge is crucial, the metal–oxide–semiconductor field-effect transistor, or MOSFET. Finally, methods to measure interface charge and interface electronic states will be discussed.

7.2 ELECTROSTATICS AND THE POISSON EQUATION

To arrive at the Poisson equation, we commence with Coulomb's law and the definition of the electric field and then consider electric flux and Gauss's law. The Poisson equation is readily obtained from Gauss's law by a mathematical manipulation. The first integration of the Poisson equation yields the electric field, and the second integration yields the electric potential shown in Figs. 7.1b and 7.2b.

Coulomb's law indicates that an electric charge q_1 exerts a force **F** on another charge q_2, acting through a separation distance r_{12} as follows:

$$\mathbf{F} = \frac{kq_1q_2}{r_{12}^2}$$

7.2 ELECTROSTATICS AND THE POISSON EQUATION

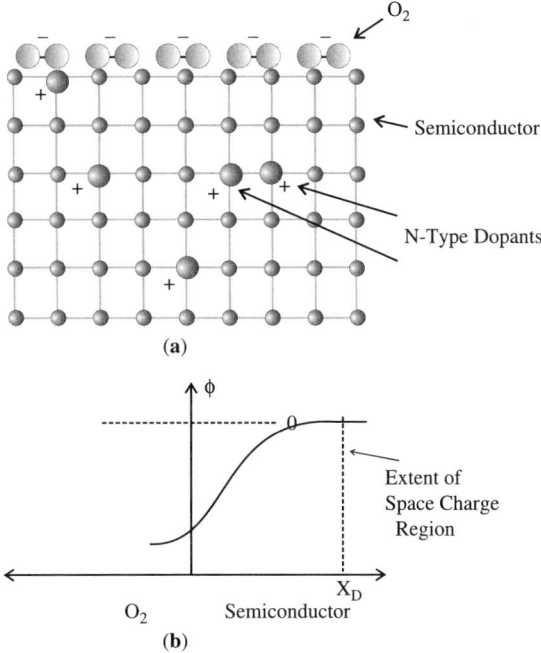

FIGURE 7.2 (a) Semiconductor with adsorbed O_2 on the surface and (b) the resulting potential distribution across the semiconductor–adsorbate interface.

where

$$k = \frac{1}{4\pi\varepsilon_0} \quad (7.1)$$

where ε_0 is the permittivity of free space. The electric field **E** at a distance r is the force per charge and is given as

$$\mathbf{E} = \frac{\mathbf{F}}{q} = \frac{kq}{r^2} \quad (7.2)$$

Also the resultant electric field can be given as the vector sum of i individual fields as

$$\mathbf{E} = \sum_i \mathbf{E}_i \quad (7.3)$$

The force on a charge q in an electric field **E** is as follows:

$$\mathbf{F} = q\mathbf{E} \quad (7.4)$$

Now we consider a good conductor in an electric field. Because charges in the conductor can move easily and will do so to reduce the force due to the field to zero, when $\mathbf{F} = 0$, $\mathbf{E} = 0$, as a result of the rearrangement of the free electrons in a conductor. Electric field $\mathbf{E}$ for a conductor exists only at the surface of a conductor and is normal to the surface. If $\mathbf{E}$ were not normal to the surface, then a force would be exerted; $\mathbf{E}$ in a conductor is zero.

Figure 7.3a shows a surface area A penetrated by an electric field as is indicated by the field lines. The electric flux Φ_E can be calculated for a planar surface:

$$\Phi_E = \mathbf{E} \cdot A = EA \cos\theta \tag{7.5}$$

For $\mathbf{E}$ perpendicular to A as was the case discussed above for a good conductor:

$$\Phi_E = EA \tag{7.6}$$

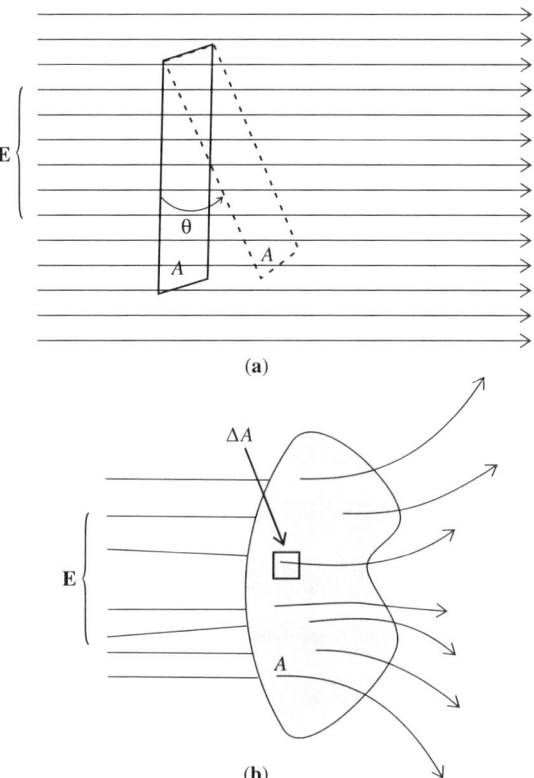

FIGURE 7.3 (a) Electric field (field lines) orthogonal to a rectangular area A and at angle θ to the field and (b) irregular area A penetrated by an electric field (field lines).

7.2 ELECTROSTATICS AND THE POISSON EQUATION

However, for the general case shown in Fig. 7.3b the flux can be evaluated by the following integral:

$$\Phi_E = \int \mathbf{E} \, dA \tag{7.7}$$

where dA is the incremental area shown in Fig. 7.3b as ΔA and converted to a differential. Gauss's law is obtained by considering a closed surface that is penetrated by an electric field, as shown in Fig. 7.4. Figure 7.4a shows that the number of field lines entering the closed surface equals the number exiting, yielding a net flux of zero:

$$\Phi_E = 0 \tag{7.8}$$

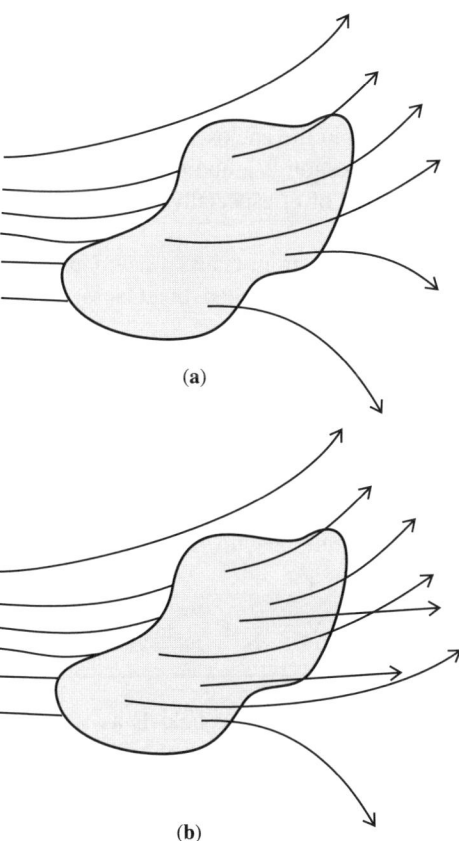

FIGURE 7.4 (a) Electric field (field lines) penetrating a closed volume where the same number of lines enter and exit the volume and (b) electric field (field lines) penetrating a closed volume where more field lines exit the volume than enter it.

For this case Equation 7.7 becomes

$$\int \mathbf{E}\, dA = 0 \tag{7.9}$$

Figure 7.4b shows that there are two field lines that originate in the closed surface while all the others enter and leave the closed surface as for Fig. 7.4a. Thus, for Fig. 7.4b:

$$\Phi_E \neq 0 \tag{7.10}$$

and this situation implies that charges must exist within the closed volume and act as the origin of the electric field difference between the number of incoming and outgoing flux lines. For this case Equation 7.7 can be rewritten and yields Gauss's law:

$$\int \mathbf{E}\, dA = \frac{Q}{\varepsilon} \tag{7.11}$$

where Q is the net charge within the enclosed volume and $\varepsilon_r \varepsilon_0 = \varepsilon$. The dielectric constant for the material in question is ε and ε_r is called the relative dielectric constant. Often K is used in place of ε_r especially for the static dielectric constant for a material.

To obtain the Poisson equation from Gauss's law, Green's theorem is applied to Gauss's law. Green's theorem (also called the Gauss divergence theorem) from vector calculus can be expressed as follows:

$$\iint \mathbf{F}\, dS = \iiint \nabla \cdot \mathbf{F}\, dV \tag{7.12}$$

where $\mathbf{F}$ is any vector function such as force $\mathbf{F}$ or electric field $\mathbf{E}$, S is surface area that bounds the volume V, and ∇ is the del operator given as

$$\nabla = \frac{\partial}{\partial x} + \frac{\partial}{\partial y} + \frac{\partial}{\partial z} \tag{7.13}$$

Now applying this theorem to the electric field $\mathbf{E}$ as was given in Equation 7.11, which is an integral over area dA and could be written with the double integrals, yields the following:

$$\iint \mathbf{E}\, dA = \frac{Q}{\varepsilon} = \iiint \nabla \cdot \mathbf{E}\, dV \tag{7.14}$$

7.3 TWO SIMPLE SOLUTIONS TO THE POISSON EQUATION

The charge Q results from the integral of the charge density ρ over volume as follows:

$$Q = \int \rho \, dV \tag{7.15}$$

Using Equation 7.15 in 7.14, we obtain the following result:

$$\iiint \nabla \cdot \mathbf{E} \, dV = \frac{1}{\varepsilon} \int \rho \, dV \tag{7.16}$$

Since both sides of Equation 7.16 are volume (triple) integrals, the arguments can be equated:

$$\nabla \cdot \mathbf{E} = \frac{\rho}{\varepsilon} \tag{7.17}$$

The electric potential ϕ between two points a and b is defined as

$$\phi_{ab} = -\int \mathbf{E} \, dx \tag{7.18}$$

Then, the derivative of the potential is the electric field $\mathbf{E}$ as follows:

$$\frac{d\phi}{dx} = -\mathbf{E} \tag{7.19}$$

Finally, we can put this together to yield the Poisson equation in one dimension:

$$\nabla \cdot \mathbf{E} = \frac{d^2\phi}{dx^2} = -\frac{d\mathbf{E}}{dx} = -\frac{\rho}{\varepsilon_r \varepsilon_0} \tag{7.20}$$

For the Poisson equation it is seen that the first integration yields the electric field $\mathbf{E}$ and the second integration yields the potential ϕ.

7.3 TWO SIMPLE SOLUTIONS TO THE POISSON EQUATION

Two appropriate and relatively simple solutions to the Poisson equation are discussed below. These are the solutions shown in Figs. 7.1 and 7.2 for sheets of charge and distributed charges, respectively.

The first solution is for a metal surface with an adsorbate, as was shown in Fig.7.1a. This situation leads to a double layer of charge that can be approximated by two sheets of charge with no charge density between the two layers, $\rho = 0$. For this situation Equation 7.20 is written as follows:

$$\frac{d^2\phi}{dx^2} = -\frac{d\mathbf{E}}{dx} = -\frac{\rho}{\varepsilon_r \varepsilon_0} = 0 \qquad (7.21)$$

The first integration of Equation 7.21 yields $\mathbf{E}$ as follows:

$$\mathbf{E} = \frac{\sigma}{\varepsilon_r \varepsilon_0} \qquad (7.22)$$

where σ is the number of charges per area on the metal surface; σ is obtained by integrating the charge per volume ρ multiplied by the dx dimension, which yields charge per area.

The second integration yields the potential and starts with $\mathbf{E}$ from Equations 7.19 and 7.22 as follows:

$$\mathbf{E} = -\frac{d\phi}{dx} = \frac{\sigma}{\varepsilon_r \varepsilon_0} \qquad (7.23)$$

The integral equation to solve is given as

$$-\int_{\phi_s}^{0} d\phi = \int_{0}^{x_0} \frac{\sigma}{\varepsilon_r \varepsilon_0} dx \qquad (7.24)$$

where the conditions or limits of integration are $\phi = \phi_s$ at $x = 0$ (at the metal surface where the charge resides) and $\phi = 0$ at the position of the second sheet of charge at $x = x_0$. Upon reversing the limits and sign, the solution is

$$\phi_s = \frac{\sigma x_0}{\varepsilon_r \varepsilon_0} + c \qquad (7.25)$$

where c is a constant of integration, and because at x_0, $\sigma = 0$ and $\phi = 0$, then $c = 0$. The final solution is as follows:

$$\phi_s = \frac{\sigma x_0}{\varepsilon_r \varepsilon_0} \qquad (7.26)$$

It should be recognized that Equation 7.26 is the formula for a parallel-plate capacitor. From the basic capacitor relationship where C is the capacitance and V the potential or

7.3 TWO SIMPLE SOLUTIONS TO THE POISSON EQUATION

voltage applied to the capacitor:

$$CV = Q \tag{7.27}$$

C is given as

$$C = \frac{K\varepsilon_0 A}{x} \tag{7.28}$$

where K is the static dielectric constant for low frequencies, A is the area of the capacitor plates, and x the plate separation. Then V or ϕ is given as

$$\phi = \frac{Q}{C} = \frac{Qx}{K\varepsilon_0 A} \tag{7.29}$$

where Q/A is σ, so the parallel-plate capacitor formula is the solution to the Poisson equation for the double layer on a metal surface, and the solution is depicted in Fig. 7.1b.

The second useful solution for our present purposes is the case where the surface is not a metal. This means that there can be a charge density in the bulk of the material away from the surface, and the result is a double layer for distributed charge. For this case we assume $\rho \neq 0$ in the material or for our case above of the parallel-plate capacitor between plates. The Poisson equation (Equation 7.20) is written as follows:

$$\frac{d^2\phi}{dx^2} = -\frac{d\mathbf{E}}{dx} = -\frac{\rho}{\varepsilon_r \varepsilon_0} \neq 0 \tag{7.30}$$

For typical semiconductors that are doped, we can consider ionized immobile dopants that give rise to positive charges that are uniformly distributed in the semiconductor lattice via substitution, as shown in Fig. 7.2a. The electrons are delocalized in either the conduction or valence bands. These ideas can be used to construct a model to solve the Poisson equation as written in Equation 7.30. This model is called the Schottky model. In this model N_D is the number of fixed donors that give rise to positive charges $+Q$, and N_A is the number of acceptors that give rise to negative charges $-Q$. The number of electrons in the conduction band is n and the number of holes in the valence band is p. Charge neutrality requires the following:

$$N_D - N_A = n - p \quad \text{and} \quad N_D + p = N_A + n \tag{7.31}$$

At the surface majority carriers in the semiconductor are captured by surface states and the relatively small number of minority carriers are ignored. Therefore, from the surface inward there are no or few carriers, and this region is called a depletion distance or depletion width extending to a distance x_D. For an N-type semiconductor

$p = 0$ and Equation 7.31 becomes

$$N_D - N_A = n \tag{7.32}$$

With electrons as the majority carriers and captured by surface states, a positive charge develops in the semiconductor through a distance $x < x_D$. The charge density is given as

$$\rho = e(N_D - N_A) \quad \text{for } x < x_D \quad \text{and} \quad \rho = 0 \quad \text{for } x > x_D \tag{7.33}$$

Now ρ is substituted into the Poisson Equation 7.30 to yield

$$\frac{d^2\phi}{dx^2} = -\frac{d\mathbf{E}}{dx} = -\frac{\rho}{\varepsilon_r \varepsilon_0} = \frac{-e(N_D - N_A)}{\varepsilon_r \varepsilon_0} \tag{7.34}$$

The first integration of Equation 7.34 yields $\mathbf{E}$ and is performed from x_D in the bulk to x:

$$\int_{x_D}^{x} \frac{d^2\phi}{dx^2} = -\mathbf{E} = \frac{d\phi}{dx} = -\frac{e(N_D - N_A)(x_D - x_A)}{\varepsilon_r \varepsilon_0} \tag{7.35}$$

The second integration to yield the potential ϕ is performed from the bulk where the potential is 0 at x_D to $x = 0$ at the surface where a surface potential ψ_s obtains and is given as follows:

$$\int_0^{\psi_s} d\phi = \int_{x_D}^{x} -\frac{e(N_D - N_A)(x_D - x)}{\varepsilon_r \varepsilon_0} dx \tag{7.36}$$

The result is as follows:

$$\psi_s = \frac{e(N_D - N_A)(x_D - x)^2}{2\varepsilon_r \varepsilon_0} \tag{7.37}$$

where x at the surface has the value $x = 0$, so the final result is

$$\psi_s = \frac{e(N_D - N_A)x_D^2}{2\varepsilon_r \varepsilon_0} \tag{7.38}$$

Figure 7.2b shows the parabolic profile of ψ_s (or ϕ) as is given by Equation 7.37. Often it is useful to express the surface potential in terms of the surface charge or the charge per area σ_s, where ψ_s by the product of the charge per volume and the thickness of the depletion width as follows:

$$\sigma_s = (N_D - N_A)x_D \quad (7.39)$$

With this substitution for x_D Equation 7.38 becomes

$$\psi_s = \frac{e\sigma_s^2}{2(N_D - N_A)\varepsilon_r\varepsilon_0} \quad (7.40)$$

7.4 METAL–OXIDE–SEMICONDUCTOR FIELD-EFFECT TRANSISTOR AND FERMI LEVEL PINNING

Many semiconductor devices operate based on the ability to alter the surface potential of the semiconductor, and the most prevalent device in this category and in all of microelectronics is the MOSFET. The MOSFET and related MOSFET surface and interface issues were mentioned in Chapters 1 to 6. Here with the background thus far discussed in Chapter 7, we expand on the MOSFET ideas already introduced from the perspective of charges at the interface and the effect of these charges on the operation of the device. Figure 4.16a displays a sketch of a N-channel MOSFET that is fabricated on P-type single-crystal Si, and the associated discussion of this figure in Chapter 4 outlines how an N-channel device operates. Figure 7.5a shows a P-channel MOSFET device that is fabricated on N-type Si and operates similar to the N-channel device in an on–off current mode, as shown in Fig. 7.5b, to create the binary 1 and 0 states used in computer switches and memory storage chips. As can be seen from Fig. 4.16 and 7.5, the MOSFET is a device fabricated using various thin films deposited upon a Si substrate. A real MOSFET device

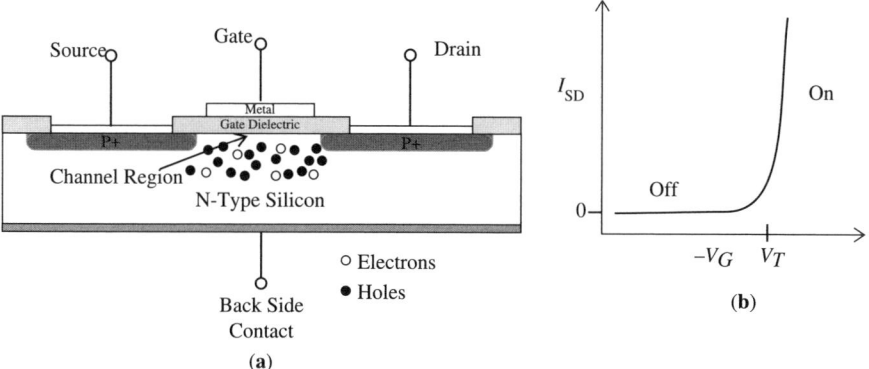

FIGURE 7.5 (a) P-channel MOSFET fabricated on N-type Si and (b) ideal operating characteristic of the MOSFET.

typically contains many more films with different functions than depicted in Fig. 4.16 and 7.5. The formation of thin films is discussed in Chapter 10. Furthermore present device structures include a complimentary MOSFET (CMOS) geometry that is comprised of both a P-Channel and an N-Channel MOSFET usually connected in a source-to-drain configuration. The CMOS configuration draws less power than individual devices and hence is advantageous in present-day high device density applications where power consumption and heating are crucial design considerations.

The Si substrate for MOSFETs is single-crystal Si with usually the Si<100> surface exposed for the device. The reason for this choice of Si orientation will be discussed below and is related to the lower level of surface electronic states on this orientation of Si. In Chapter 1 (Table 1.1) it was mentioned that the Si<100> has the lowest Si atom density of the low index planes. In Chapter 5 the origin of surface electronic states was discussed as related to the altered bonding potential near the surface of a material as compared to the uniform bulk potential for a single-crystal material. For those surfaces with fewer atoms on the surface, the smaller is the alteration of the potential near the surface and usually the lower is the intrinsic surface state density. This line of reasoning has been experimentally confirmed for the major orientations of Si. Hence the choice of the Si<100> is made on this basis. In addition Chapter 5 included methods to measure surface electronic states. In this context Fig. 5.17b shows that the intrinsic surface states on Si are greatly reduced (around 4 to 5 orders of magnitude) as a result of the thermal oxidation of the Si single-crystal surface. The Si oxidation process is discussed in Chapter 10, and this process is crucial to the manufacture of MOSFET devices. Also in Chapter 5 using Fig. 5.7 charge exchange between the bulk semiconductor and surface electronic states was shown to alter the surface potential. The surface or interface potential can also be altered using an externally applied voltage, as was seen for the operation of the MOSFET in Chapter 4 and more specifically again in the following discussion.

As the gate potential V_G goes positive for the MOSFET in Fig. 7.5a, electrons that are the majority carriers in N-type Si are attracted to the interfacial region where Si and SiO_2 meet. The electrons accumulate in this region, called the channel region of the MOSFET. The condition of the device is called *accumulation* and refers to majority carriers. In accumulation the device cannot conduct current from the P-type heavily doped source to the drain since the holes (the majority carriers in the source and drain regions) injected into the channel will recombine with accumulated electrons in the channel. Furthermore, without injection of carriers the adjacent P and N regions at the source and drain will form high resistance depletion regions when electron and hole recombination occurs. Therefore, in accumulation the MOSFET is "off." As V_G goes negative, the majority carrier electrons in the substrate will be repelled from the channel, and the condition is called *depletion* and also refers to the depletion of majority carriers. With a further increase of $-V_G$ the minority carrier holes from the N-type Si (the intrinsic number) form a P-type layer at the interface of the gate oxide (typically SiO_2) and Si connecting the source and drain regions and permitting current flow via holes. With the source-to-drain current flowing, the MOSFET is "on." The gate potential at which the MOSFET turns on is called the

7.4 METAL–OXIDE–SEMICONDUCTOR FIELD-EFFECT TRANSISTOR

threshold voltage and labeled as V_T in Fig. 7.5b. This phenomenon of changing the carrier type from electrons (the majority carriers in the N-type Si) to holes or vice versa in the channel region between the source and drain is called *inversion*. The negative gate potential used to operate the device via carrier inversion causes a change in the Si surface potential ψ_s, and this is indicated by band bending. In this example of the operation of the MOSFET, it was assumed that there are no surface electronic states and that the applied potential causes charges to move that results in band bending sufficient to invert the channel region. Figure 7.6a displays N-type Si with the condition that the majority carrier electrons have been removed from the bulk material to the surface and creating a space charge region extending into the Si. This situation can occur with a $+V_G$ on the metal (M) electrode.

If there are empty surface electronic states at the Si–SiO$_2$ interface, as is shown in Fig. 7.6b, then charge exchange between the semiconductor and the surface states also causes band bending even without the application of an external potential V_G. If there are sufficient numbers of surface states below E_F of the semiconductor, the filled surface state levels will increase in energy as these states fill and ultimately equilibrate with E_F. For the N-type Si E_F is near the conduction band, as indicated by the dotted line on the Si side of the interface. As electrons fill lower lying empty surface states, E_F in the Si drops as is indicated by the dashed line. At this point the bands are already bent in a way similar to that which occurred with a negative gate potential used to operate the device. The maximum swing of the gate potential to effect operation of the device is about 1 V since beyond that value for Si the valence band crosses E_F. Assuming a 1-V potential, then one can calculate the number of extrinsic (or intrinsic) states that must be present to swing the ψ_s as far as it can go. The first step is to calculate the surface charge σ_s using Equation 7.40 rewritten as follows:

$$\sigma_s = \left(\frac{2\psi_s \varepsilon_r \varepsilon_0 (N_D - N_A)}{e}\right)^{1/2} \qquad (7.41)$$

For Si the static dielectric constant is $\varepsilon_r \varepsilon_0$ (or $K\varepsilon_0$) = 12, $\psi_s = 1$ V (approximately the band gap) and with two assumptions are made about doping: $N_D - N_A = 10^{14}\,\text{cm}^{-3}$ for lightly doped Si and $= 10^{19}\,\text{cm}^{-3}$ for heavily doped Si. From Equation 7.40 we write

$$\sigma_s = \{[2 \times 12 \times 8.85 \times 10^{-12}\,\text{C}^2/\text{Nm}^2(10^{14}\,\text{cm}^{-3} \text{ or } 10^{19}\,\text{cm}^{-3})$$
$$\times 1\,\text{V/m}(1\,\text{N/C})]/1.6 \times 10^{-19}\,\text{C}\}^{1/2}$$

and

$$\sigma_s = 3.6 \times 10^{10}\,\text{cm}^{-2} \quad \text{or} \quad 1.2 \times 10^{13}\,\text{cm}^{-2} \qquad (7.42)$$

Now this approximate calculation yields some information about the operation of the MOSFET device that is controlled by adjusting the surface potential, ψ_s. In terms of intrinsic surface states a number that would yield about 10^{10} charges on a one for one

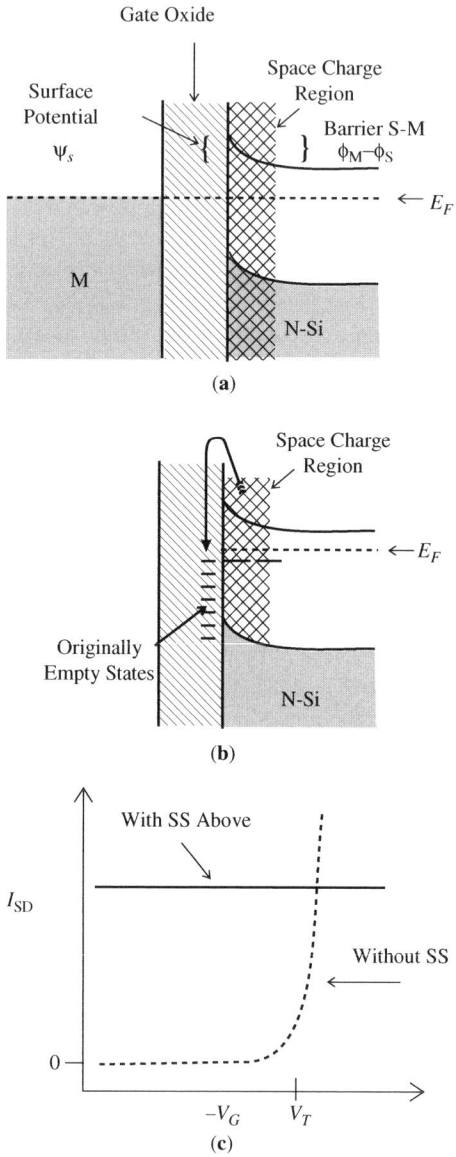

FIGURE 7.6 (a) MOS capacitor on N-type Si at equilibrium; (b) MOS in (a) after charge exchange with interface states; and (c) operating characteristics with and without interface states.

basis would be about 10^{10} surface electronic states for lightly doped Si and 10^{13} states for heavily doped Si. In fact, it is now well established that less than 10^{12} interface electronic states must be achieved to merely get a MOSFET to marginally operate, but less that 10^{10} is required for consistent reliable operation. Thus, the crude calculation yields some insight to what one must do to a semiconductor surface to enable MOSFET operation. To examine what this means in terms of extrinsic states, we first consider that there are about 10^{15} atoms/cm^2 in a monolayer. Thus, the amount of charge to change ψ_s by 1 V for Si corresponds to only 4×10^{-5} or 1×10^{-2} monolayers for lightly and heavily doped Si, respectively. As we recall from Chapter 5, the extrinsic states can be caused by impurities. Thus, to change the surface potential by the amount of the gap can be accomplished by small amounts of surface contamination. That is why most microelectronics processing is performed in expensive clean rooms and with exhaustive cleanliness procedures.

If there are more than sufficient surface states to swing ψ_s without an external potential, the external potential is rendered ineffective for device operation. This condition is known as Fermi level pinning. The E_F equilibration with the surface state level(s) effectively prevents the surface potential from changing unless large potentials are used that would be able to first of all saturate all the surface states and would likely also cause other device failure mechanisms. Thus, under conditions of Fermi level pinning the device is inoperable. Figure 7.6c shows source-to-drain current I_{SD} for a MOSFET with a normal turn-on characteristic (dotted line) and a well-defined threshold voltage for turn-on V_T. Such a device would ideally have no interface electronic states at or below a level of 10^{10} cm^{-2}. In addition the flat I_{SD} for the case with high levels of interface states (solid line) is indicative of no device control with the V_G and thus Fermi level pinning.

The interface charges as a result of surface state charging and other possible sources can be measured using a MOSFET. For many of the important measurements all one needs is the MOS part of the MOSFET. That is a capacitor structure with one plate being Si, the other being metal and with SiO$_2$ in between. This simple MOS structure does not require the heavily doped and well-defined source and drain regions and thus its simplicity renders the MOS capacitor as the main test vehicle to evaluate interface charges and other processing and materials issues without complex and costly processing, and some key ideas about MOS measurements are discussed below.

The opposite of Fermi level pinning is called electronic passivation where surface or interface electronic states are reduced to below the level necessary to operate the device. Chapter 11 discusses passivation processes on a variety of semiconductor interfaces.

7.5 METAL–OXIDE–SEMICONDUCTOR MEASUREMENTS FOR INTERFACE CHARGE

The basic MOS capacitor structure is shown in Fig. 7.5. It consists of an Si substrate on which an SiO$_2$ film that is an insulator has been grown usually using thermal oxidation. A metal contact (the gate contact for an FET) is then deposited onto the oxide using evaporation (thermal oxidation and evaporation are discussed in Chapter 10) or other metal deposition processes. Figure 7.7 shows the energy band structure for an MOS capacitor where equilibrium has occurred so that the metal and semiconductor

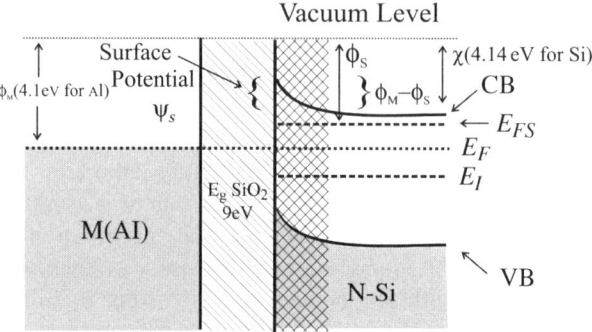

FIGURE 7.7 MOS structure with Al metal as gate contact on SiO_2 on N-type Si.

Fermi levels align, and where the Fermi levels for the N-type Si and the metal (Al) were initially different. For the case shown in Fig. 7.7 electrons flowed from the Si to the metal until equilibrium E_F was achieved. The amount of charge that flowed was proportional to the work function difference ϕ_{MS}, which represents the internal potential after equilibrium. The value for ϕ_{MS}, depends on the specific metal and semiconductor (through their Fermi levels) and to a lesser extent the doping level in the semiconductor. At equilibrium the charge distribution as a result of electron flow to the metal is shown in Fig. 7.8a where Si is positive and the Al metal is negative. Since electrons are the free carriers in the N-type silicon, the electrons are depleted from the semiconductor–oxide interface. The result is an MOS capacitor that is charged to a potential given as

$$\phi_{MS} = \phi_M - \phi_S \tag{7.43}$$

An externally applied potential is called a bias and can effect the situation shown in Figs. 7.7 and 7.8a. The application of a positive bias to the metal will reduce the band bending and the charge. The bent bands in Fig. 7.7 can be flattened at the interface, if the positive bias is equal to ϕ_{MS}, and this condition is known as *flat bands* and the applied voltage as the *flat band potential* ϕ_{FB}:

$$\phi_{FB} = \phi_{MS} = \phi_M - \phi_S \tag{7.44}$$

The measurement of the voltage applied to achieve the flat band condition is then a measure of ϕ_{MS} in the absence of any other charges at the interface. Thus, from a knowledge of the flat band voltage and of ϕ_{MS} the charge at the interface can be determined. More will be said below about the measurement of charges and the flat band potential ϕ_{FB} or flat band voltage V_{FB}.

By applying a more positive gate voltage, the majority carrier electrons in the Si will be attracted to the Si surface and accumulate near the $Si-SiO_2$ interface of the MOS. In this case the bands bend downward and the condition is called accumulation

7.5 METAL–OXIDE–SEMICONDUCTOR MEASUREMENTS

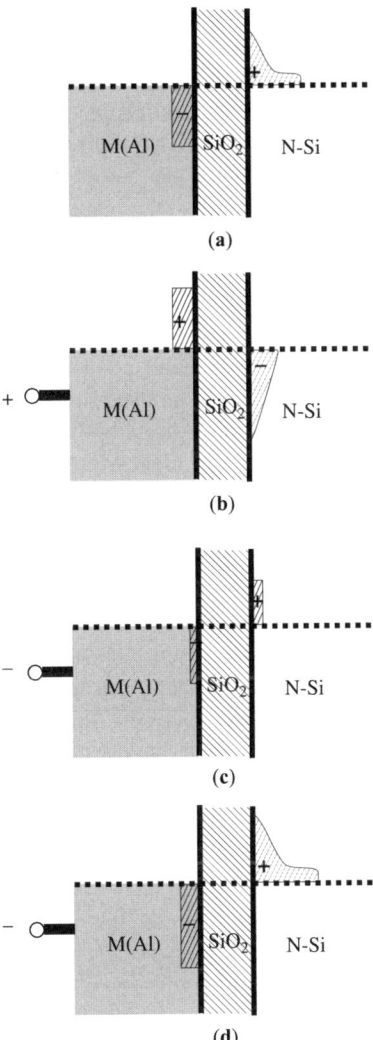

FIGURE 7.8 MOS structures showing the charge distribution: (a) at equilibrium, (b) with positive (+) gate bias on Al; (c) with negative (−) gate bias, and (d) with larger negative (−) gate bias.

with reference to the majority carriers. The charge distribution in accumulation for the MOS in Fig. 7.7 is given in Fig. 7.8b. With a negative-going V_G the majority electrons in the Si are repelled from the Si surface and minority holes are attracted. The Si surface is said to be in a depleted condition relative to majority carriers, and the charge distribution is shown in Fig. 7.8c. There are few holes in the N-type Si so the number of charges in the distribution is small. This negative applied bias adds to the built-in potential.

An even more negative gate bias further attracts minority holes to the Si surface and leads to a thin layer at the originally N-type Si surface that now appears to be P-type as a result of the large concentration of holes. This condition is called inversion and is shown in Fig. 7.8d. As discussed above, this MOS capacitor can be converted to a MOSFET by the addition of two adjacent source and drain regions that are heavily P-doped as was shown in Fig. 7.5a. Now in inversion these P-regions are connected via the P-channel created by the gate bias enabling source-to-drain current to flow.

The measurement of interfacial charges, some of which may be associated with surface electronic states, is performed using the MOS capacitor, and one important measurement is based on the measurement of the capacitance at various applied gate voltages. These measurements are called $C-V$ measurements. For $C-V$ measurements we first consider the capacitance of an ideal MOS capacitor where there are no interfacial charges associated with surface states or other external sources nor is there any work function difference. After the ideal MOS capacitor is understood, then the effects of other sources of charge will be considered. The total capacitance of the MOS structure with no interface traps, oxide charge, or work function difference (ideal MOS) is the series combination of the silicon capacitance per unit area, C_{Si}, and the oxide capacitance C_{ox} as shown in Fig. 7.9a. The silicon capacitance C_{Si} is displayed as a variable capacitor because, as discussed above, the depletion width is varied by the applied voltage. The extent of the depletion width determines the capacitance of the Si. For example, if the Si is in accumulation, the MOS capacitor appears to be a parallel-plate capacitor where the only dielectric is the SiO_2 layer in between the Si and the metal. On the other hand with an external voltage applied to create a depletion width, there are two capacitors in series as discussed above. The C_{Si} is varied by the applied voltage. For series capacitors the total capacitance C_T is given as

$$\frac{1}{C_T} = \frac{1}{C_{Si}} + \frac{1}{C_{ox}} \tag{7.45}$$

This yields the following result:

$$C_T = \frac{C_{Si} C_{ox}}{C_{Si} + C_{ox}} \tag{7.46}$$

where C_{ox} and C_{Si} are given as follows:

$$C_{ox} = \frac{K \varepsilon_0 A}{L_{ox}} \quad \text{and} \quad C_{Si} = \frac{\partial Q}{\partial \phi_{Si}} \tag{7.47}$$

In the formula for C_{ox}, $K\varepsilon_0$ is the static dielectric constant ($\varepsilon_r \varepsilon_0$), A is the area of the capacitor, and L_{ox} is the SiO_2 film thickness. The formula for C_{Si} derives first from the relationship that $CV = Q$ and then the variability of C_{Si} with the potential.

7.5 METAL–OXIDE–SEMICONDUCTOR MEASUREMENTS

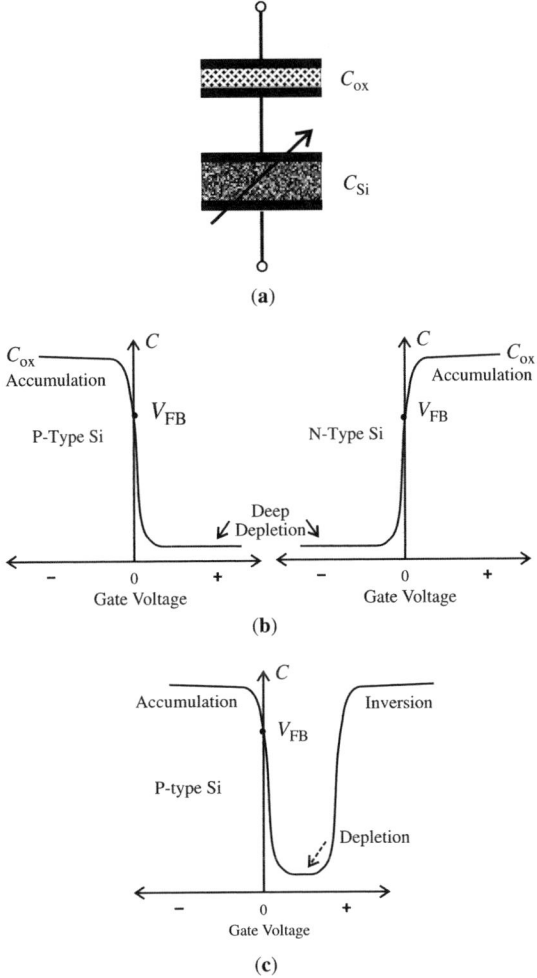

FIGURE 7.9 (a) MOS structure approximated as two capacitors in series, one fixed and one variable; (b) high test frequency C versus V characteristics for P and N-type MOS structures; and (c) low frequency C–V characteristic for a P-type MOS capacitor.

This variability is understood by considering that majority carriers can be accumulated at the Si–SiO$_2$ interface yielding a metallic-like Si and thus a parallel-plate capacitor structure where the capacitance measured is that of the SiO$_2$ in between metal plates, C_{ox}. Alternatively, the majority carriers can be depleted by a change in surface potential, forming a depletion width that is nearly devoid of carriers, and hence this depletion region acts as a capacitor in series with C_{ox} as given by Equation 7.45. Furthermore, with still further changes in the Si surface potential ϕ_{Si}, minority carriers can be accumulated at the interface yielding again a parallel-plate capacitor. Since the term accumulation refers to majority carriers, the accumulation of minority carriers at the interface is called inversion. For this latter effect of minority carrier response, the sensing of the inversion effect is greatly dependent on the test frequency. Capacitance is usually measured using a high frequency alternating test signal (ac) superimposed on a dc bias. The dc bias applied to the gate changes the carrier quantity and type at the Si surface adjacent to the Si–SiO$_2$ interface. The ac signal usually in the range of 10 to 20 mV rms is used to determine the capacitance by measuring the amplitude of the phase-shifted current. For majority carriers the high frequency ac presents no issue because there are so many available the generation rate is not crucial. However, this is not the case for the relatively few minority carriers that need to be generated at a rate greater than the test frequency to be sensed. At high test frequencies (1 MHz for Si) the minority carriers are not sensed since the test frequency is higher than the generation rate. Consequently, the parallel-plate single-capacitor scenario does not materialize and Equation 7.45 is used. Figure 7.9b shows the high frequency C–V curves for both P-type (left panel) and N-type (right panel) Si with a SiO$_2$ film. For the P-type curve one observes accumulation of majority carriers (holes) when the gate potential goes negative, thereby pulling the holes to the Si surface at the Si–SiO$_2$ interface. For this case the depletion width does not exist and the measured C is very close to the capacitance of the oxide C_{ox}. As the gate potential swings positive, minority carriers (electrons) are attracted to the Si surface while holes are pushed away. However, with the high frequency alternating test voltage the holes are not sensed and thus there appears to be two capacitors in series as given by Equation 7.45. A minimum capacitance is reached. The same scenario applies to the N-type Si as shown in the right panel of Fig. 7.9b, albeit at the opposite gate potentials. At low test frequencies under small signal ac excitation, the minority carriers can respond to variations in the ac test signal electric field and therefore inversion will be sensed. This results in both accumulation and inversion conditions with appropriate gate potentials, and the resulting C–V curve is a V-shaped characteristic shown in Fig. 7.9c.

For an ideal MOS capacitor (no work function difference and no charges in the device) the applied gate voltage (V_G) will appear across the oxide (V_{ox}) and across the semiconductor (ψ_{Si}) as follows:

$$V_G = V_{ox} + \psi_{Si} \tag{7.48}$$

where V_{ox} is given as

$$V_{ox} = E_{ox} L_{ox} = \frac{Q_{Si} L_{ox}}{K_{ox} \varepsilon_0 A} = \frac{Q_{Si}}{C_{ox}} \tag{7.49}$$

where Q_{Si} is the charge per unit area in the semiconductor that includes both depletion layer and inversion layer charge. The applied voltage V_G is known as the threshold voltage V_T when the following condition holds:

$$\psi_{Si} = \psi_{Siinv} = 2\psi_B \tag{7.50}$$

7.5 METAL-OXIDE-SEMICONDUCTOR MEASUREMENTS

where ψ_{Siinv} is the potential across the Si inversion layer and ψ_B is the potential difference between the Si Fermi level and the intrinsic level; $2\psi_B$ is approximately the potential needed to produce strong inversion and operate the device. Then from Equations 7.48 and 7.50 V_T is given as

$$V_T = V_{ox} + 2\psi_B \tag{7.51}$$

The Si–SiO$_2$ interface can contain charges and traps for charge that can have a detrimental effect on MOSFET device parameters. The most effective methods available to characterize the charges and traps at the Si–SiO$_2$ interface make use of the MOS capacitor shown in Fig. 7.9a. Many methods have been developed over the years, and two have been selected for discussion below: capacitance–voltage (C–V) method and the conductance method $G(\omega)$. The C–V measurement can in principle determine all charges at the Si–SiO$_2$ interface, but in reality it is sometimes difficult to determine interface traps resulting from surface electronic states. For that reason and the importance of interface traps on MOSFET quality, the $G(\omega)$ method that is particularly good for interface traps is described. Before the methods are discussed, a digression is useful to discuss the nature of charges to be measured.

7.5.1 Oxide Charges

There are four types of charges associated with the Si–SiO$_2$ system: fixed oxide charges (Q_f), mobile oxide charges (Q_m), oxide trapped charges (Q_{ot}), and interface trapped charges Q_{it}. Fixed oxide charges (Q_f) in the Si–SiO$_2$ system are often positive charges associated with incomplete oxidation of Si at the interface. The Si crystal orientation, the oxidation temperature, and ambient are the factors that determine the value of Q_f. Using an MOS capacitor, Q_f can be obtained from the shift of the flat band voltage V_{FB} that is located on a high frequency C–V curve. V_{FB} for the case where only Q_f is present is given as follows:

$$\Delta V_{FB} = (V_{FB})_{meas} - (V_{FB})_{ideal} = \phi_{MS} - \frac{Q_f}{C_{ox}} \tag{7.52}$$

where ϕ_{MS} is metal–semiconductor work function difference, and C_{ox} is the oxide capacitance in accumulation. The flat band voltage V_{FB} shown in Figs. 7.9b and 7.9c is obtained by determining the capacitance C_{FB} that should obtain at V_{FB} and while not derived here is given for P-type Si by the following equation:

$$C_{FB} = \frac{K_{ox}\varepsilon_0}{L_{ox} + (K_{ox}/K_{Si})\sqrt{kTK_{Si}\varepsilon_0/p_0 q^2}} \tag{7.53}$$

where p_0 is the equilibrium number of holes in the bulk Si, and q is the electronic charge. Equation 7.53 shows that besides the static dielectric constants of the dielectric and semiconductor, C_{FB} depends on the thickness of the dielectric film L_{ox} and the doping level of the semiconductor yielding p_0. A convenient form of Equation 7.53

is a normalized form where C_{FB} is divided by C_{ox} and given as follows:

$$\frac{C_{FB}}{C_{ox}} = \frac{1}{1 + \frac{136 C_{ox}}{K_{ox}\varepsilon_0 A \sqrt{N_A \text{ or } N_D}}} = \frac{1}{1 + \frac{136}{L_{ox}\sqrt{N_A \text{ or } N_D}}} \qquad (7.54)$$

where the oxide thickness L_{ox} is in centimeters and N_A (or N_D) is the acceptor (or donor) concentration in reciprocal cubic centimeters (cm^{-3}) for P-type (or N-type) Si. Often a normalized curve is obtained by first determining C_{ox} and then dividing each measured C by C_{ox} as V_G is changed. From the normalized plot C_{FB} is obtained for the Si doping level; $V_{FB} = 0$ for an ideal MOS capacitor where $\phi_{MS} = 0$, and there are no charges in the system as shown in Figs. 7.9b and 7.9c. If $\phi_{MS} \neq 0$ but there are no other charges in the system, then $V_{FB} = \phi_{MS}$. Of course, with $\phi_{MS} \neq 0$ and fixed oxide charges present, then Equation 7.52 obtains. As a practical matter the entire high frequency C–V curve will shift due to charges at the interface. Figure 7.10a shows an applied potential $-V$ on the metal (M), causing an electric field as represented by the electric field lines that originate at positive charges and terminate at negative charges. If no charges are at the Si–SiO$_2$ interface, all the field lines would originate or terminate in the Si and not on charges at the interface, and thereby the entire applied V would alter the Si potential. The field as shown would drive P-type Si potential toward accumulation. If positive charges are present, then a higher applied potential would be needed to achieve the same Si potential since some charges near the interface would originate (or terminate) field lines. Consequently, the high frequency C–V curve would shift in the $-V_G$ direction for positive interface charges and in the $+V_G$ direction for negative interface charges as shown in Fig. 7.10b. Under some circumstances such as high levels of Q_{it}, the C–V curve will distort. Thus, the parallel shift method to determine the charge levels will not be accurate, and the method to determine V_{FB} and the shift in V_{FB} should be used.

In addition to Q_f it is possible to have mobile ionic charges Q_m from Na$^+$, Li$^+$, and K$^+$ with Na$^+$ being the most common one derived from human proximity; Q_m can drift under applied field to the Si–SiO$_2$ interface and is usually a function of the cleanliness of the process. To determine Q_m a bias–temperature test is done. After the MOS sample is heated to around 50°C and with $+V_G$ applied, the sample is cooled with the bias applied. The C–V curve is then recorded. The procedure is repeated but using $-V_G$ so as to move the charge to the M–SiO$_2$ interface where it will have little influence on the C–V characteristic. The C–V curve is obtained again and compared to the first C–V curve done with $+V_G$. Any shift can be used to calculate Q_m using Equation 7.52 or the parallel shift method; Q_{ot} is a relatively immobile charge associated with trapping centers that are defects. The interface is a natural place for defects between the different materials, but bulk oxide traps can be caused by OH groups in SiO$_2$ that disrupt the network bonding. Often these traps are neutral but can be charged by trapping holes or electrons during device operation. OH-related traps will be distributed in the SiO$_2$ film, and thus not all charges

7.5 METAL–OXIDE–SEMICONDUCTOR MEASUREMENTS

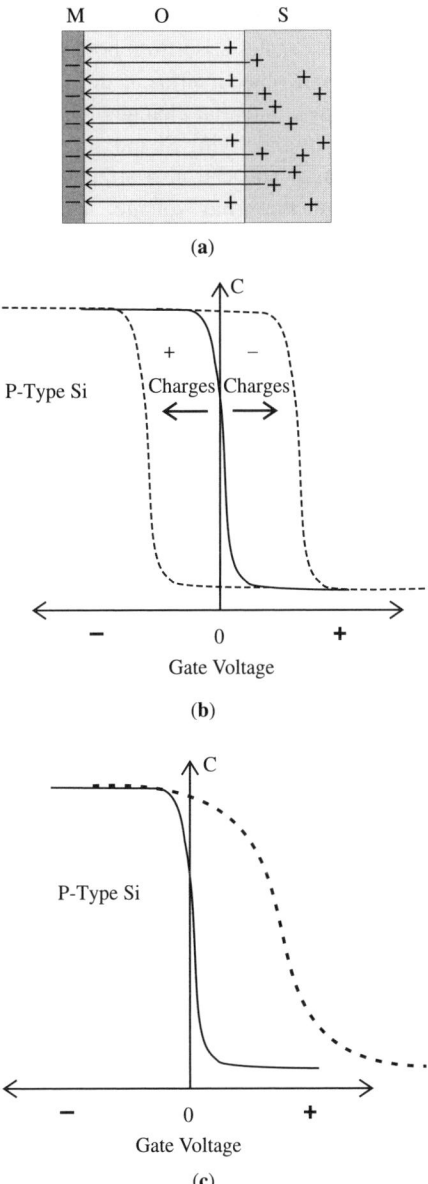

FIGURE 7.10 (a) Electric field on an MOS structure; (b) shift in the high frequency C–V characteristic with positive (+) and negative (−) charges at the interface; and (c) effect of interface-trapped charge in interface states on the high frequency C–V characteristic.

are at the interface. The effect on the C–V curve could be small relative to the charges present at the interface for the charges that are distant from the Si–SiO$_2$ interface. Other methods are preferable when high densities of bulk-trapped charges are present.

Interface-trapped charges, Q_{it}, are the charges trapped at interface states that were discussed in Chapter 5, and these states arise from both the termination of the semiconductor lattice or network and other interfacial defects, and the interface states are the result of local interatomic potential changes. Unlike the oxide charges, Q_{it} can rapidly exchange charge with underlying Si and, therefore, charge and discharge according to the surface potential and can also affect MOSFET device characteristics such as carrier mobility, threshold voltage, and reliability. To determine the value of Q_{it}, the electrical characteristics of MOS capacitor structures are frequently used. For C–V measurements of MOS structures, the presence of interface traps causes the high frequency C–V curve to stretch and to shift along the voltage axis, as shown in Fig. 7.10c, which compares an ideal high frequency C–V characteristic (solid line) with a stretched-out C–V curve (dashed line). The stretch-out and shift is due to the fact that in the presence of interface-trapped charges more potential is necessary on the metal to achieve a given Si surface potential, as was discussed using Fig. 7.10a. The stretch-out is indicative of a sample with a high density of interface-trapped charge D_{it}, which results from charge trapping Q_{it} in interface (or surface) electronic states, and the appearance of the stretched-out C–V curve is usually not perfectly symmetrical due to the distribution of interface traps. Note that Q_{it} is an interface-trapped charge and D_{it} is the density of that charge. While the observation of stretch-out is straightforward, it is difficult to quantify and only noticeable with very high D_{it}. Another more sensitive C–V method for interface electronic states and the charge therein Q_{it} is to use both high frequency and low frequency C–V characteristics. This technique derives from the fact that Q_{it} is not sensed by the high frequency test frequencies (typically 1 MHz) used for high frequency C–V measurements, but Q_{it} is sensed by the low test frequencies (below 10 Hz). All the charges are sensed at the low test frequencies, so the low frequency C–V as compared with the high frequency C–V can yield Q_{it}.

A typical scenario to obtain a good quantitative measure of the charges in MOS structures using C–V measurements is to first perform a high frequency C–V measurement and compare that curve with the ideal C–V curve for the particular ϕ_{MS} for the MOS structure. This comparison yields all the charges near the Si–SiO$_2$ interface except Q_{it}, namely Q_f, Q_{ot}, and Q_m. Then the bias–stress–temperature test described above is performed to determine Q_m. Next a low frequency C–V curve is determined and the comparison yields Q_{it}. This measurement scenario enables distinguishing Q_{it}, Q_m, and the sum $Q_f + Q_{ot}$.

To obtain high frequency C–V curves, the measuring circuit consists of the MOS device, which can be approximated using Fig. 7.11. Figure 7.11a shows the equivalent circuit for an ideal MOS capacitor. In this case the device acts like a capacitor with capacitance C_{ox} in series with the variable capacitance of the depletion layer in the Si; C_{Si}, which varies with gate bias. In this way the Z-shaped C–V characteristic is obtained. The MOS device can also be represented by a parallel conductance–capacitance (G–C) circuit, with G being the conductance of the gate bias induced space-charge region of the semiconductor and C its capacitance. The output

7.5 METAL–OXIDE–SEMICONDUCTOR MEASUREMENTS

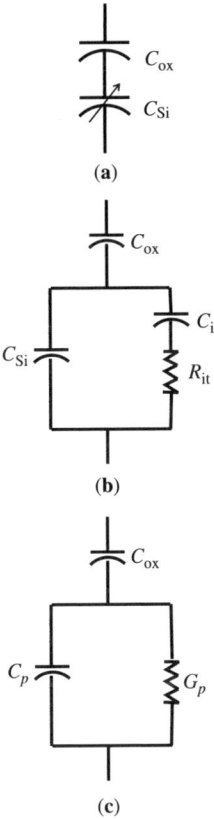

FIGURE 7.11 Approximate equivalent circuits for (a) an ideal MOS capacitor on a Si surface and (b) with interface electronic states and (c) a simplified version of (b).

voltage V_o can be related to the input voltage V_i as follows:

$$V_o = (RG + j\omega RC)V_i \qquad (7.55)$$

It is seen that V_o has an in-phase component given as RG ($V_o = RGV_i$ for 0° phase); and an out-of-phase component $j\omega RC$ ($V_o = j\omega RCV_i$ for 90° phase). With known R and $\omega = 2\pi f$, a phase-sensitive detector is used to determine G or C versus gate voltage.

The use of C–V measurements to determine all four charges in the Si–SiO$_2$ system has been widespread in the field of microelectronics. However, for Q_{it} or D_{it} measurements, a superior technique is the conductance (G) method, which is considered more accurate and sensitive than the C–V method. Essentially, the G method is a loss method that measures the loss due to interface state charge capture and emission. The loss mechanism from capture and emission of carriers from interface traps is represented by the resistance R_{it} in the equivalent circuit of an MOS capacitor, as shown in Fig. 7.11b where the interface traps are included as R_{it} and C_{it} in series, and where the time constant for capture of charge in interface states (or interface trap lifetime) is given as $\tau_{it} = R_{it}C_{it}$. It is usual to simplify this circuit by combining the C's in the parallel branch to yield the equivalent circuit in Fig. 7.11c where C_{si} and C_{it} in series with R_{it} are replaced by parallel capacitance C_p and parallel conductance G_p and given as

$$C_p = C_{si} + \frac{C_{it}}{1 + (\omega\tau_{it})^2} \tag{7.56}$$

and

$$\frac{G_p}{\omega} = \frac{q\omega\tau_{it}D_{it}}{1 + (\omega\tau_{it})^2} \tag{7.57}$$

where D_{it} is the interface trap density per unit energy, $C_{it} = qD_{it}$, $\omega = 2\pi f$ (f is the measurement frequency), and τ_{it} is given above as interface trap lifetime. The conductance G_p is only dependent on the interface trap response, while C_p requires C_{Si}. For interface traps continuously distributed in the band gap G_p becomes

$$\frac{G_p}{\omega} = \frac{qD_{it}}{2\omega\tau_{it}} \ln[1 + (\omega\tau_{it})^2] \tag{7.58}$$

The conductance as a function of frequency at each bias voltage is plotted as G_p/ω versus $\log \omega$ with the maximum at $\omega \approx 2/\tau_{it}$, so that $D_{it} = 2.5G_p/\omega$.

7.5.2 Measurement of Charges in the Si–SiO$_2$ System

As discussed in Chapter 1, the different low Miller index planes of single-crystal Si have different surface atom densities (see Table 1.1). Also in Chapter 1 it was pointed out that the different Si surface orientations yielded different oxidation rates for the thermal oxidation of Si in O$_2$ and different mechanical properties were also mentioned (interfacial stress at the Si–SiO$_2$ interface will be discussed further in Chapter 12). For microelectronics the electronic properties are most relevant as is the fact discussed below that different Si orientations display different Si–SiO$_2$ interface electronic properties. Both fixed oxide charge Q_f and interface-trapped charge density D_{it} display strong dependencies on oxidation processing variables. Figure 7.12 shows first that both Q_f and D_{it} are lower for higher oxidation temperatures. Furthermore, for an oxide grown at a specific temperature Q_f decreases (as indicated by the arrow in Fig. 7.12a), if the oxide is annealed after oxidation in an inert ambient (typically Ar or N$_2$) at the oxidation temperature and eventually reaches a minimum Q_f level, as shown by the dashed line in Fig. 7.12a. While D_{it} is smaller for higher Si oxidation temperatures, it decreases less than Q_f upon inert gas annealing. However, if H$_2$ is present in the anneal ambient, then D_{it} decreases by as much as an order of magnitude depending on specific anneal conditions. It is also seen in Fig. 7.12b that Si(100) displays a lower D_{it} than Si(111) at any oxidation

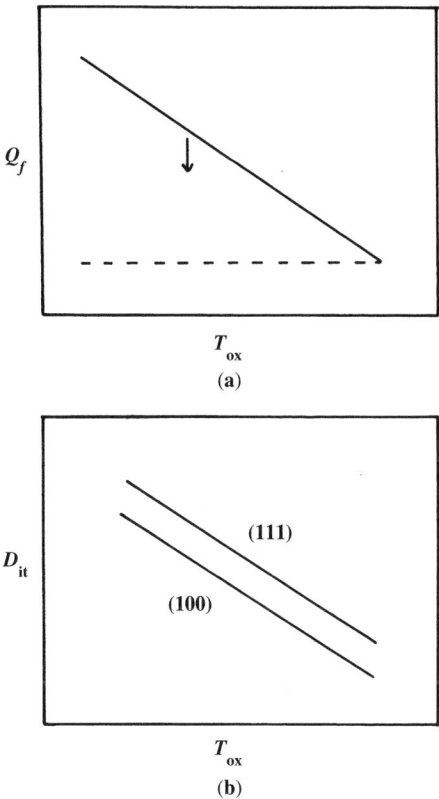

FIGURE 7.12 (a) Fixed oxide charge Q_f and (b) interface-trapped charge D_{it} as a function of oxidation temperature T_{ox} and annealing. [Adapted from Irene (1), Figure 5.]

temperature. In addition SiO_2 film stress, density, and refractive index are all smaller at higher Si oxidation temperatures, and all decrease upon inert gas annealing, and these properties along with Q_f and D_{it} are summarized in Fig. 7.13. That all these film and interfacial properties behave similarly suggests a relationship among the properties, and these results provide a backdrop to other more relevant surface and interface studies. Of specific interest here is a study by Vitkavage et al. (2) that focuses on Si substrate orientation driven trends in fixed oxide charge, Q_f, and interface-trapped charge, D_{it}, for four Si orientations: (100), (110), (111), and (511), for oxidation temperatures from 750 to 1100°C, with and without hydrogen containing postmetal anneals, and for processing within and without a cleanroom.

Fixed charge was calculated by rearranging Equation 7.52 as follows:

$$Q_f = C_{ox}(\phi_{MS} - V_{FB}) \tag{7.59}$$

where as before C_{ox} is the oxide capacitance, V_{FB} is the measured flat band voltage, and ϕ_{MS} is the metal–semiconductor work function. Fixed charge was obtained from

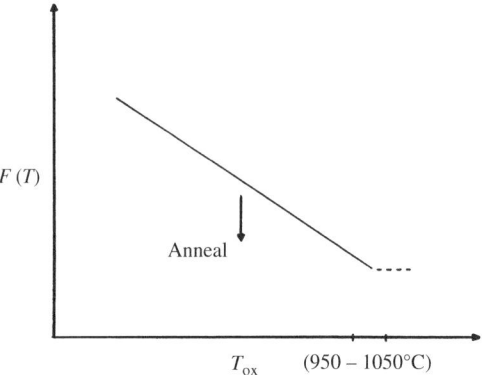

FIGURE 7.13 Summary of SiO$_2$ properties where $F(T)$ is refractive index n, density ρ stress σ, fixed oxide charge Q_f, and interface-trapped charge D_{it} as a function of oxidation temperature T_{ox} and annealing. [Adapted from Irene (1), Figure 6.]

high frequency (1 MHz) C–V curves and the flat band voltage measurements made on MOS samples that received a post-metal-annealing treatment (PMA), that minimizes any flat band voltage shift due to the presence of interface traps. The PMA is typically performed at 400 to 500°C for 5 to 15 min using forming gas that is 10% H$_2$ in N$_2$ or Ar and is a gas mixture that does not detonate but provides sufficient H$_2$ to effect the electronic properties. For comparison Q_f measurements were also reported from samples that did not receive postmetal anneal, NPMA. Rather than the more reliable $G(\omega)$ measurements described above, D_{it} was obtained using the quasi-static capacitance voltage, C–V, technique where a low frequency C–V curve was compared with the high frequency C–V result as discussed above. This technique uses two experimental curves, one a high frequency (1 MHz) C–V curve and the other a low frequency current voltage, I–V (voltage ramp rate less than 100 mV/s), curve to obtain the interface state density according to

$$D_{it} = \left[\left(\frac{C_{LF}}{C_{ox} - C_{LF}} \right) - \left(\frac{C_{HF}}{C_{ox} - C_{HF}} \right) \right] \frac{C_{ox}}{qA} \tag{7.60}$$

where C_{LF} and C_{HF} are, respectively, the low and high frequency capacitances at a given voltage, C_{ox} is the oxide capacitance, A is the area of the electrode, and q is the elementary charge. The high frequency C–V curve was corrected for the presence of a series resistance by measuring the capacitance and equivalent parallel conductance in strong accumulation from a conductance–voltage (G–V) curve. The Si surface potential was obtained by numerical integration of the low frequency C–V curve.

Bias–temperature–stress high frequency C–V measurements were performed to determine Q_m. For the noncleanroom samples flat band voltage shifts due to mobile ionic contamination, Q_m, of from -0.25 to -1.0 V were found that correspond to positive Q_m values of 1 to $5 \times 10^{11} \text{cm}^{-2}$, and cleanroom samples had flat band shifts of less than -0.04 V, which corresponds to a positive Q_m of about $1.5 \times 10^{10} \text{cm}^{-2}$.

Table 7.1 contains a sampling of Q_f and D_{it} data for MOS samples with PMA. From this table several facts emerge. First, the use of cleanrooms and higher oxidation temperatures typically produce samples with lower Q_f and D_{it} and as discussed above lower Q_m. These results confirm earlier studies in the literature and show that similar trends are seen for both cleanroom and noncleanroom samples, albeit with different levels of charge.

Table 7.2 compares D_{it} levels for different Si orientations. For NPMA samples where the H_2 is not present and thus cannot lower or thereby mask the number of interface states that can exchange charge, the highest levels of D_{it} corresponds to those Si orientations with the greatest number of Si atoms per area on the Si surface, namely the Si(110) (see Table 1.1). On the other hand where H_2 is present the Si(111) orientation displays the highest level of D_{it}. The reason for this is not elucidated in this study.

Table 7.3 summarizes the D_{it} results by comparing the D_{it} at the same oxidation temperature with the lowest values as obtained on Si(100). In addition Table 7.3 shows the ratios of the Si atom densities as well. This table shows the amazing similarity with D_{it} and the Si surface atom density. The same startling similarity is not

TABLE 7.1 Comparison of Fixed Oxide Charge and Interface-Trapped Charge for Samples with PMA

Oxidation Temperature (°C)	Fixed Oxide Charge (10^{11}cm^{-2})		
	(100)	(110)	(111)
Cleanroom			
750	0.74	10.80	5.89
800	1.60	1.92	4.99
900	1.27	1.71	4.86
1000	1.73	1.08	4.33
Noncleanroom			
750	6.30	8.20	11.80
800	3.12	3.35	6.73
900	2.87	3.45	6.78
1000	2.43	2.90	6.60
1100	1.42	2.02	3.92

Oxidation Temperature (°C)	Interface-Trapped Charge ($10^{11} \text{cm}^{-2} \text{eV}^{-1}$) at Midgap for Si (111)	
	Cleanroom	Noncleanroom
750	1.3	3.2
800	1.3	2.8
900	1.0	1.6
1000	1.2	4.9

TABLE 7.2 Interface-Trapped Charge Density for Cleanroom Post-Metal-Annealed, PMA, and Non-Post-Metal-Annealed, NPMA, Samples

Oxidation Temperature (°C)	Interface-Trapped Charge Density ($cm^{-2} eV^{-1} 10^{11}$)			
	(100)	(511)	(110)	(111)
PMA				
750	12	7.5	16	13
800	9.0	8.0	9.5	13
900	6.0	5.6	9.3	10
1000	5.0	5.7	7.9	12
NPMA				
750	8.2	9.5	35[a]	22[a]
800	5.4	6.7	22[a]	16[a]
900	4.1	6.8	36	14[a]
1000	2.6	3.2	8.0	11[a]

[a] Values reported at 0.3 eV above the valence band where a minimum occurs.

TABLE 7.3 Ratios of Measured Interface-Trapped Charge Density D_{it} and Silicon Surface Atom Density C_{Si} for Various Orientations Oxidized at 750°C (with PMA)

	(110)	(111)	(511)	(111)
	(100)	(100)	(100)	(110)
D_{it}/D_{it}	1.3	1.1	0.63	0.81
C_{Si}/C_{Si}	1.40	1.15	0.63	0.86

apparent for higher temperatures, but the general trend is also the same and indicates a strong and expected correlation of the available electronic interface states with the number of atoms on the surface.

REFERENCES

1. Irene EA. Phil Mag 1987;55:131.
2. Vitkavage SC, Irene EA, Massoud HZ. J Appl Phys 1990;68:6262.

SUGGESTED READING

E. H. Nicollian and J. R. Brews, 2003. *MOS Physics and Technology*, A classic and authoritative treatment of the basic ideas about MOSFET physics and measurements. Another must have book in microelectronics.

S. M. Sze, 1981. *Physics of Semiconductor Devices*, Wiley. A classic and authoritative treatment of the basic ideas about semiconductor devices. Another must have book in microelectronics.

8

ADSORPTION

8.1 INTRODUCTION

The process of adsorption considered here is one result of the collision of gas atoms or molecules with a solid surface. Numerous outcomes of this collision are possible, and some of the more common results are discussed with the use of Fig. 8.1. First, the impinging particles (atoms and/or molecules) can reflect from the surface either elastically without a change in energy or inelastically upon energy exchange with the surface and shown as process 1 in Fig. 8.1. Second, the particles may become adsorbed either strongly, called chemisorption, or relatively weakly adsorbed, called physisorption, as depicted by process 2. As the name implies, chemisorption involves the formation of chemical bonds with typical bonding energies on the order of 100 kcal/mol (about 400 kJ/mol). Physisorption involves weak forces of physical attraction such as van der Waals forces with energies typically 10 times smaller than chemisorption. Process 3 shows that a chemical reaction can occur to yield a new chemical moiety as a result of collision. There are many kinds of reactions that can occur. For example, the incoming particle can provide a reactant along with atoms/molecules from the solid surface (indicated in Fig. 8.1) or either the incoming molecule or surface molecule can undergo chemical change such as decomposition or disproportionation. Another possible outcome, process 4, is desorption of a species already on the surface: incoming adsorbed particle, surface species (indicated in Fig. 8.1), or reaction product.

There are many more kinds of interactions not discussed here, but the four common processes discussed above can each or in combination determine many

---

Surfaces, Interfaces, and Thin Films for Microelectronics. By Eugene A. Irene
Copyright © 2008 John Wiley & Sons, Inc.

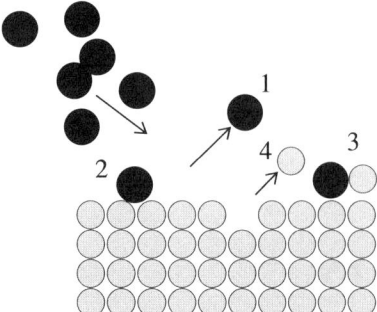

FIGURE 8.1 Four process occurring as a result of collisions with a surface: (1) reflection of incident particles, (2) adsorption, (3) reaction, and (4) desorption of adsorbed particle.

surface phenomena such as sputtering if the incoming particles are ions, nucleation of a new phase as species accumulate at a surface, film formation as nuclei grow and coalesce, catalysis, surface modification, implantation, and the like. In this chapter some of the basic ideas in the broad field of adsorption, process 2, are discussed.

8.2 PHYSISORPTION

Physisorption derives from the weak attraction between nonreacting atoms or molecules, called van der Waals forces with energies around 10 kcal/mol, and consists mainly of attractive dispersion or London forces and dipole forces and repulsive forces at close distances. The dispersion forces arise from instantaneous fluctuations of the electronic configurations of atoms or molecules that give rise to fluctuating dipoles. These rapidly changing dipoles can induce dipoles in adjacent atoms or molecules. The mutual interaction of the fluctuating dipoles is the origin of these weak forces. The attractive negative ($-$) energy from these forces E_d is obtained from perturbation theory and has the form

$$E_d = -\frac{C}{r^6} \qquad (8.1)$$

where r is the separation of the atoms or molecules and C is a constant that depends largely on the polarizability, which is a measure of the ability for the electronic configuration to deform. If some of the molecules are dipoles with dipole moment μ, then when these approach another molecule, whether dipolar or nonpolar, there will be a force of attraction called the dipole force. In the case of two polar molecules there is dipole–dipole interaction yielding a interaction energy E_{dd} as follows:

$$E_{dd} \propto \frac{\mu_1 \mu_2}{r^6} \qquad (8.2)$$

For the case of one dipole approaching a nonpolar molecule or atom, the dipole will induce a dipole in the nonpolar species with strength proportional to the species

8.2 PHYSISORPTION

polarizability α and the resulting energy E_{dn} is given as follows:

$$E_{dn} \propto \frac{\mu_1 \alpha_2}{r^6} \tag{8.3}$$

Figure 8.2a shows a plot of the interaction energy for physisorption with attractive interaction energy on the right-hand side. It is seen that as the separation between species decreases from infinity, the physisorption interaction energy decreases to a

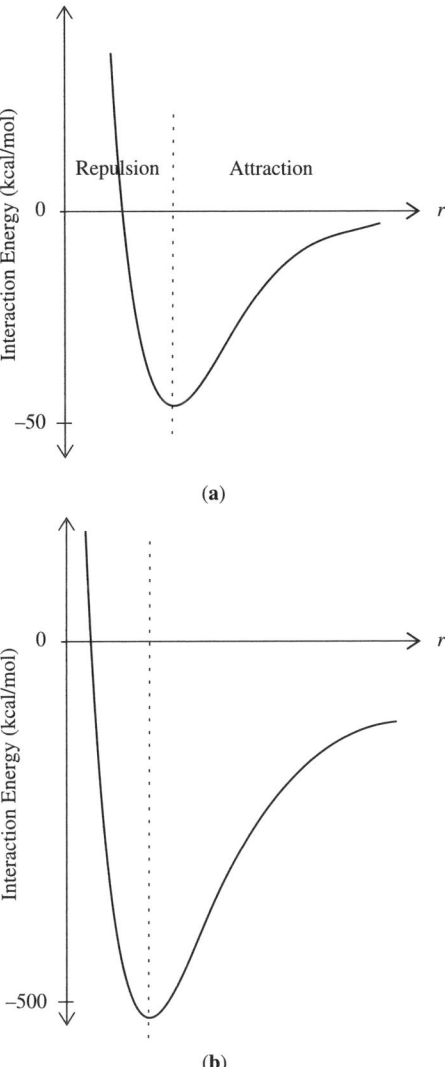

FIGURE 8.2 Interaction energy versus separation distance r for atoms or molecules approaching each other or a surface: (**a**) is physisorption and (**b**) is chemisorption.

minimum at some separation, and, thereafter, as the interacting species come closer, there is a strong repulsion as electron distributions repel. There are many theories about the details of the repulsions, and one empirical but well-respected analysis shows that the repulsion energy is proportional to r^{-12}, which when combined with the attractive interaction yields the Lennard-Jones potential E_{LJ} as follows:

$$E_{LJ} = 4\varepsilon \left[\left(\frac{\sigma}{r}\right)^{12} - \left(\frac{\sigma}{r}\right)^{6} \right] \tag{8.4}$$

where σ is separation r at $E_{LJ} = 0$ and ε is the minimum energy. The exponents 6 and 12 have also given this potential the name "6–12 potential," and it has been found to apply well to gas-phase interactions. However, for a species approaching a solid surface, for example, the interactions are more complex with cooperative effects on neighboring species in the solid. For interactions with solids the exponents have been empirically found to be closer to 3 and 9 but with approximately the same shape.

It should be noted that the above discussion of physisorption was material independent, and therefore physisorption is a phenomenon that applies to virtually all materials and with about the same interaction energies. On the other hand chemisorption discussed below is far more materials system specific and much stronger when it occurs.

Another important factor for solids is the concept of screening that can be understood by considering the case of a charge (left) and a dipole (right) approaching a surface as is shown in Fig. 8.3. Both the charge and the dipole induce an image charge of opposite polarity that is screened (diminished) by the medium (solid material) with dielectric constant ε as

$$q = e \frac{1 - \varepsilon}{1 + \varepsilon} \tag{8.5}$$

where a charge $-e$ induces q. Notice that the image produces an attractive interaction with the approaching charge or dipole. The level of the attraction is a materials

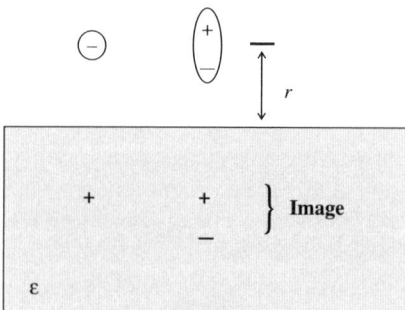

FIGURE 8.3 Negative charge (left) and dipole (right) a distance r from a surface producing image charges in the material with dielectric constant ε.

property that depends on the dielectric response function ε for the material. For example, for a perfect metal surface $\varepsilon = \infty$, so using Equation 8.5 for the charge $q = -e$ (as obtained from derivatives of numerator and denominator), and the attraction from the image charge is strong. For a material with a smaller dielectric constant of 4 the polarization of the material is less and the material would screen a portion of the charge and diminish the attraction. Again using Equation 8.5 for a charge with $\varepsilon = 4$, the material would screen the charge by $-3e/5 = -0.6e$. The approaching atom or molecule need not be charged or a dipole. An induced dipole also causes similar attraction albeit weaker. One implication of the image force interaction and screening is that the physisorption for a species will vary according to its ability to screen the charge.

8.3 CHEMISORPTION

A Morse potential used to describe chemical bonding mostly for diatomic molecules can also be used for chemisorption in that for chemisorption the formation of a chemical bond between surface and adsorbate is the central issue. The Morse potential for the chemical bong energy E_M is given by an expression of the form

$$E_M = D\{1 - \exp[-C(r - r_0)]\}^2 \tag{8.6}$$

where D is the minimum energy in the potential well, r_0 is the separation at the minimum in E_M, and C is a materials constant. The overall shape shown in Fig. 8.2b is similar to the Lennard-Jones potential curve for physisorption discussed above and shown in Fig. 8.2a, but the energy scales are much different with chemisorption larger by an order of magnitude, and the separation at minimum energy is smaller for the stronger chemical bonding.

A typical case is the interaction of a reactive gas such as O_2 on a metal surface. First, the adsorption is physisorption of the diatomic O_2 gas on a metal. The physisorbed molecular species dissociates via the input of energy that exceeds the dissociation energy D_0, and the resulting reactive atomic species now on the surface can readily react with the metal, forming strong exothermic chemical bonds. This oxidation reaction is often overall thermodynamically favored since the gain in energy for the formation of the strong chemical bonds overcomes the required energy input for dissociation of the diatomic gas species. This is the case despite the fact that the entropy change is negative as the gas molecule in three dimensions is constrained to the two-dimensional surface.

For the case of oxidation of metals one plausible explanation for the favorability of the overall oxidation reaction is obtained by realizing that it is energetically favorable for O_2, a strongly electronegative moiety, to gain an electron. Using Fig. 8.4, which shows Morse potential diagrams for O_2 (solid line) and O_2^- (dashed lines for ground and excited states), it is seen that the molecular ion O_2^- forms by the electronegative O_2 acquiring an additional electron (1). The additional electron gained by O_2 pairs an electron in the next available orbital, which is a π^* antibonding orbital in the

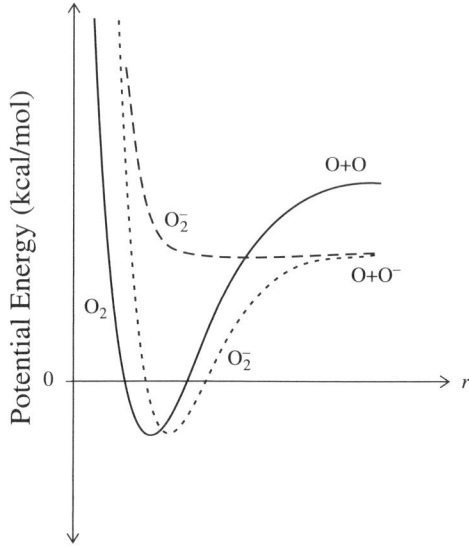

FIGURE 8.4 Potential energy versus atomic separation r for the O_2 molecule (solid line) and the O_2^- molecular ion ground and excited state (dashed lines). [Adapted from Chanin et al. (1), Figure 1.]

originally neutral O_2 to form the O_2^- ion that is equally as stable as O_2 (notice the same minima in Fig. 8.4). Once formed, the molecular ion requires a smaller dissociation energy to produce atomic species $O + O^-$ then for O_2 to produce atomic species $O + O$, as is seen in Fig. 8.4.

This idea about the formation of an unstable O_2^- ion along with the notion of surface electronic states has been used to understand the oxidation of metals and semiconductors. From Chapter 5 one can readily imagine a metal surface with dangling bonds, some of which can be occupied with electrons. These occupied surface states can give up electrons to the highly electronegative adsorbed O_2 molecules. This transfer of an electron from metal to molecule is thermodynamically favored. The formed oxygen molecular ion as discussed above is favored for dissociation into active species that reacts with the metal to form two metal–oxygen bonds. Overall a thermodynamically favored process results starting from physisorption and that evolves to chemisorption. In some cases additional external energy may be required to overcome potential barriers.

This oxidation scenario has been applied to the technologically relevant oxidation of Si surfaces to form SiO_2 films, especially in the very initial regime where oxidant transport through an already formed SiO_2 film is not significantly affecting the oxidation rate (2). For this system the energy band diagram in Fig. 8.5 is used to determine the energies required to render electrons available to the oxidation mechanism. From Fig. 8.5 it is seen that several different energetic routes are available for electrons from Si to be available for oxidation. From the Si valence band about 4 eV are required, while for the Si conduction band, about 3 eV are necessary. Intermediate

8.3 CHEMISORPTION

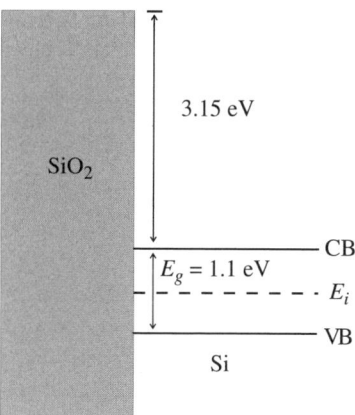

FIGURE 8.5 Energy band diagram for the Si–SiO$_2$ system. E_g is the Si band gap, E_i is the intrinsic Fermi level, and CB and VB are the conduction and valence band edges, respectively.

in energy are the defect levels (for N- or P-type Si) and the intrinsic Fermi level at which there are no allowed electron states, only a probability. To determine which, if any, of these barriers may be implicated in Si oxidation, the electron flux, J_e, using the Richardson–Dushman thermionic emission equation (Equation 5.44) is compared with the flux of O$_2$, which is derived from the experimental oxidation rates. This treatment assumes that one electron per O$_2$ molecule is required for oxidation, which is reasonable from the previous discussion of the molecular O$_2^-$ and further justified below when a specific mechanism is proposed. Table 8.1 shows the calculated barrier heights, ϕ, such that the J_e equals the oxygen flux J_{O_2} at various oxidation temperatures. It is seen that barriers of the order of 3 eV are appropriate. This is the energy barrier value for the Si conduction band electrons. Also a calculation confirmed that there are sufficient conduction band electrons for oxidation, by a factor of 10 or more, at any temperature above room temperature at which the numbers are also sufficient but marginal.

With these results a mechanism was formulated in which there exists a rapid relative (to the consumption of O$_2$) flux of O$_2$ to the Si surface on the SiO$_2$ side of the Si–SiO$_2$ interface, and also a rapid flux of electrons on the Si side, with the flux of electrons over the barrier to be rate limiting with an activation energy of 3 eV. Once an $e-$ goes over the barrier, it attaches by a favorable reaction as was discussed above to O$_2$ forming O$_2^-$, which decomposes to O and O$^-$ more readily than O$_2$ (by 25% or more). Oxidation then proceeds readily by reaction of Si with O atoms. Oxidation can also occur more slowly by reaction with O$_2$. Another interesting aspect is that this electron-based mechanism yields insight into the formation of the 1-nm native oxide, which forms virtually instantly on a fresh Si surface even at room temperature, yet virtually ceases to grow after 2 nm unless the temperature is raised. If we consider the approximately 10^{15} Si surface electronic states, most of which have eventually captured an electron from the bulk Si, then there are about 10^{15} electrons available for Si oxidation. These electrons exist in closely spaced

TABLE 8.1 Barrier Heights That Yield Equivalent Thermionic Electron and Experimental Oxygen Fluxes for Different Oxidation Temperatures, Si Orientations, and Thickness Ranges

Oxidation Temperature (°C)	Si Orientation	Oxidation Rate (nm/min)	SiO$_2$ Thickness Range (nm)	Barrier Height (eV)
600	<100>	0.0004	2–3	2.87
	<110>	0.0012	2.5–6	2.79
650	<100>	0.014	2.5–7	2.95
	<110>	0.041	2.5–10	2.86
700	<100>	0.014	2.5	2.92
		0.0057	10	3.00
		0.0043	20	3.02
	<110>	0.033	2.5	2.85
		0.014	10	2.92
		0.0094	20	2.96
750	<100>	0.028	5	3.02
		0.024	10	3.03
		0.019	20	3.06
	<110>	0.057	5	2.96
		0.035	10	3.00
		0.024	20	3.04
1000	<100>	1.79	7	3.35
		0.9	20	3.43
		0.27	100	3.56
	<110>	3.45	9	3.28
		1.74	20	3.35
		0.4	100	3.52

levels and require little energy for promotion. Thus, these electrons are available in addition to the thermionically produced electrons. The 10^{15} electrons could yield just about 1 nm SiO$_2$ for one electron per O$_2$ molecule, which is the native oxide thickness. Once the states are removed via oxidation, however, this native oxide can no longer form, and the thermionic or otherwise excited electrons are required for further oxidation. This mechanism explains the rapid formation of about 1 to 2 nm of native oxide on Si at room temperature, and, after this initial oxide formation, high temperatures are needed to produce thicker SiO$_2$ films.

8.4 VAPOR–SOLID EQUILIBRIUM AT SURFACES

Vapor–solid equilibrium requires that the chemical potentials μ for each phase be equivalent:

$$\mu_V(P, T) = \mu_S(P, T) \tag{8.7}$$

8.4 VAPOR–SOLID EQUILIBRIUM AT SURFACES

With $d\mu = -S\,dT + V\,dP$ we obtain

$$-S_V\,dT + V_V\,dP = -S_S\,dT + V_S\,dP \tag{8.8}$$

Upon rearrangement we obtain

$$(V_V - V_S)\,dP = (S_V - S_S)\,dT \tag{8.9}$$

Assuming that $V_V \gg V_S$ and that $V_V = RT/P$ for a mole, namely that the vapor is ideal, and that at constant pressure $\Delta S = \Delta H/T$, we obtain the Clausius–Clapeyron equation:

$$\frac{dP}{P\,dT} = \frac{\Delta H_{\text{vap}}}{RT^2} \tag{8.10}$$

where ΔH_{vap} is the heat of vaporization and a more usable form is

$$\frac{d\ln P}{dT} = \frac{\Delta H_{\text{vap}}}{RT^2} \tag{8.11}$$

In this development we have assumed that ΔH_{vap} was not dependent on the coverage θ at the surface where coverage is defined as

$$\theta = \frac{\text{Amount absorbed}}{\text{Amount for monolayer}} \tag{8.12}$$

However, in surface science this is not always a good assumption in that there are a variety of sites and defects with different configurations and hence energies, and also that the electronic distribution at the surface is altered as each species is adsorbed. This is especially true for chemisorption where interaction and therefore bonding is strong, but even sometimes significant for physisoption. If ΔH_{vap} is a function of θ, a different designation indicating constant coverage is used, namely ΔH_{ist} where the subscript ist is for isosteric. For this case the Clausius–Clapeyron equation is written as

$$\left(\frac{d\ln P}{dT}\right)_\theta = \frac{\Delta H_{\text{ist}}}{RT^2} \tag{8.13}$$

For ease of application this formula can be written as follows:

$$\ln\left(\frac{P_1}{P_2}\right) = -\frac{2\Delta H_{\text{ist}}}{R}\left(\frac{1}{T_1} - \frac{1}{T_2}\right) \quad \text{where } T_1 > T_2 \tag{8.14}$$

This formula can be illustrated using Fig. 8.6, where it is seen that at any constant pressure (isobar) the coverage is larger at lower temperatures. Furthermore, to

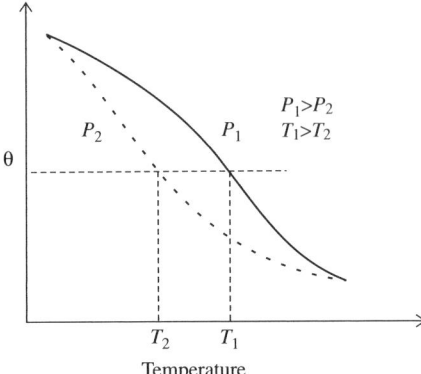

FIGURE 8.6 Surface coverage θ versus temperature at two pressures.

achieve the same coverage at a lower pressure, a lower temperature is necessary. From this kind of data for a narrow region of coverage, a value for ΔH_{ist} can be estimated. Likewise at a fixed temperature (isotherm) higher pressure yields a higher coverage.

8.5 ADSORPTION ISOTHERMS

Empirical observations indicate that there are two main kinds of adsorption isotherms that are illustrated in Fig. 8.7: (1) θ levels off with pressure as shown in Fig. 8.7a and (2) θ increases continually with pressure as shown in Fig. 8.7b. It should be understood that each of these broad classifications in Fig. 8.7 can have many subgroups with somewhat different shapes.

For both of these classes of isotherms the coverage increases approximately linearly with increasing pressure at low pressures. In this low pressure regime the number of moles of absorbate is directly proportional to the pressure. After this initial stage one kind of isotherm shown in Fig. 8.7a levels off to a limiting value of coverage, while Fig. 8.7b displays a knee of coverage but then an increase with higher pressure. The usual interpretation of these classes is that Fig. 8.7a corresponds to monolayer coverage, while isotherms like that shown in Fig. 8.7b are indicative of multilayer coverage. Monolayer coverage implies that there is strong attraction between absorbate and substrate so that when a surface is exposed to absorbate vapor, θ increases with time. However, after the surface attains θ = 1 or full coverage, there are no longer any surface sites available, and the interaction among the absorbate species is small so that another layer does not form. Thus, θ saturates, indicating monolayer coverage. For multilayer coverage after a monolayer forms, there is sufficient interaction among absorbate species so that another layer forms. However, there is no reason to believe that the second layer forms similar to the first. After a number of layers form, the bonding becomes more uniform as the underlayer is further from the influence of the original surface.

8.5 ADSORPTION ISOTHERMS

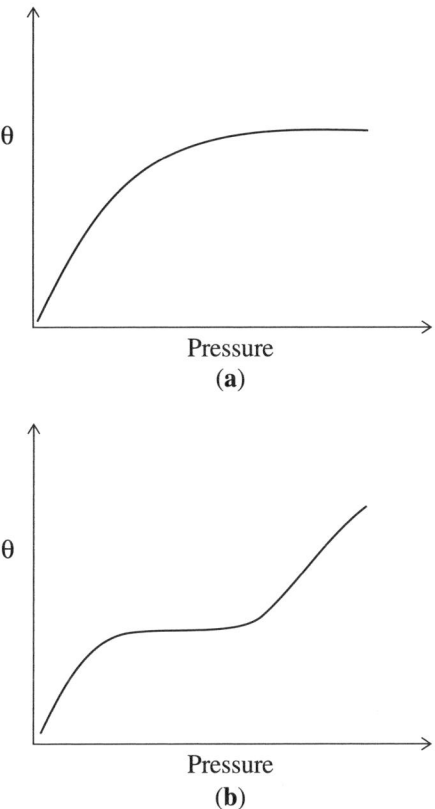

FIGURE 8.7 Surface coverage θ versus pressure showing two different isotherms: (a) monolayer coverage and (b) multilayer coverage.

For monolayer coverage the Langmuir isotherm is often evoked. The derivation of this isotherm considers a homogenous surface where all sites are equivalent and independent and that only one monolayer forms. A simple surface interaction where a gas species A chemisorbs on a metal surface M as follows:

$$A_g + M_{sur} = AM_{sur} \tag{8.15}$$

For this reaction k_a is the forward rate constant for adsorption, and k_r is the reverse rate constant for desorption. The fraction of surface covered is given as θ as usual, and $1-θ$ is the vacant fraction; N is the total number of surface sites. For adsorption the change in coverage is proportional to the number of vacant sites, which is the total number multiplied by the faction vacant as follows:

$$\frac{dθ}{dt} \propto N(1-θ) \tag{8.16}$$

Also the rate of adsorption is given by a second-order rate expression as the product of the number of vacant sites and the gas pressure P as follows:

$$\frac{d\theta}{dt} = k_a N(1 - \theta)P \tag{8.17}$$

For desorption the change in coverage is proportional to the number occupied as follows:

$$\frac{d\theta}{dt} \propto N\theta \tag{8.18}$$

And, similarly, the rate of desorption is given by a first-order expression as the product of the desorption rate constant k_d and the number of filled sites as follows:

$$\frac{d\theta}{dt} = k_d N\theta \tag{8.19}$$

At equilibrium the adsorption and desorption rates are equal:

$$k_a N(1 - \theta)P = k_d N\theta \tag{8.20}$$

For all equivalent sites and for $K = k_a/k_d$ the Langmuir isotherm expression is obtained:

$$\theta = \frac{k_a P}{k_d + k_a P} = \frac{KP}{1 + KP} \tag{8.21}$$

From this final expression it is seen that for high P, θ approaches 1 or full monolayer coverage. At low pressures θ is proportional to P.

It is useful to approach this derivation of the Langmuir isotherm from a different perspective. We commence with the collision flux formula from the kinetic theory of gases where the flux J (number of particles/area × time) of particles impinging upon a surface is proportional to the pressure P:

$$J = \frac{P}{(2m\pi kT)^{1/2}} \tag{8.22}$$

where m is the mass of the particles and T the absolute temperature (see Chapter 10 on evaporation for a derivation of this formula). We also consider that not all the collisions predicted by the flux lead to adsorption, and thus we need to consider the probability for fruitful collisions Γ. This probability is proportional to the energy for adsorption E_{ad} in a Boltzmann factor multiplied by the number of empty sites as

$$\Gamma \propto e^{-E_{ad}}(1 - \theta) \tag{8.23}$$

8.5 ADSORPTION ISOTHERMS

This expression can be converted to an equality using a factor called the condensation coefficient S^* as follows:

$$\Gamma = S^* e^{-E_{ad}/kT}(1 - \theta) \tag{8.24}$$

Thus, the rate of adsorption is given as $J\Gamma$:

$$\frac{P}{(2m\pi kT)^{1/2}} S^* e^{-E_{ad}}(1 - \theta) \tag{8.25}$$

The rate of desorption is proportional to the number of occupied sites θ and a Boltzmann factor with the energy for desorption E_D and converted to an equality by forming a rate expression using k_d as follows:

$$k_d \theta e^{-E_D/kT} \tag{8.26}$$

These rate terms are equated at equilibrium and solved for P to yield

$$P = (2m\pi kT)^{1/2} \frac{k_d}{S^*} \frac{\theta}{1 - \theta} e^{\Delta H_{ad}/kT} \tag{8.27}$$

where $\Delta H_{ad} = E_{ad} - E_D$. (It should be remembered that for solids H and E are often used interchangeably.) Using this expression Equation 8.27 can be written as

$$P = \frac{\theta}{b(1 - \theta)} \tag{8.28}$$

where

$$\frac{1}{b} = (2m\pi kT)^{1/2} \frac{k_d}{S^*} e^{\Delta H_{ad}/kT} \tag{8.29}$$

From this expression we can obtain the above Langmuir isotherm:

$$\theta = \frac{bP}{1 + bP} \quad \text{where } b = K = \frac{k_a}{k_d} \tag{8.30}$$

Now we can observe that K is given by the above expression for b, which embodies several parameters that define the adsorption process. There are many variations on the Langmuir isotherm that considers specific circumstances. For example, for dissociative adsorption that is common for O_2, N_2, and halogens, the fragments of the dissociation each occupy a single site. The rates of adsorption and desorption, respectively, are given as follows:

$$\frac{d\theta}{dt} = k_a N (1 - \theta)^2 P \qquad \frac{d\theta}{dt} = k_d N \theta^2 \tag{8.31}$$

Then the final expression for monolayer adsorption is given as

$$\theta = \frac{(KP)^{1/2}}{1 + (KP)^{1/2}} \tag{8.32}$$

which reveals a weaker pressure dependence. Other important monolayer isotherms consider that some sites are more favorable for adsorption, and these sites will fill first. This means that ΔH_{ad} will decrease as θ increases. When the pressure dependence follows a power law, the Freundlich isotherm results with the form:

$$\theta = cP^{1/2} \tag{8.33}$$

where c is a constant. When the decrease in ΔH_{ad} is linear, the Temkin isotherm results:

$$\theta = c_1 \ln(c_2 P) \tag{8.34}$$

The most often used multilayer adsorption isotherm, the BET isotherm (for Brunauer, Emmett, and Teller, the scientists who discovered this isotherm) is derived by considering that the above-derived Langmuir isotherm obtains for each monolayer formed. The first layer formed has a different ΔH_{ad} than all the succeeding layers that have the same ΔH_{ad}, and all the layers are in equilibrium. This is reasoned based on the fact that the first monolayer is deposited upon the substrate surface with an associated ΔH_{ad}. All succeeding layers are deposited on an already formed layer of itself and consequently have the same ΔH that is equal to the latent heat of vaporization ΔH_{vap} and that is different from the values for the first monolayer. While this is a compelling assumption for multilayer adsorption, it is not accurate because the bonding is altered near the interface, and so it is more likely that there is a more gradual change in ΔH_{ad} away from the interface and the first monolayer. The BET isotherm is often used to estimate surface area, and the form best suited for this purpose is as follows:

$$\frac{P}{(P^\circ - P)V} = \frac{1}{V_m C} + \frac{C-1}{V_m C} \frac{P}{P^\circ} \tag{8.35}$$

where P is the pressure of absorbate that has an equilibrium pressure P° at the temperature of the experiment, V is the volume of the gas adsorbed, and V_M is the volume of a monolayer of adsorbate; C is given as

$$C = e^{-(\Delta H_D - \Delta H_{vap})/RT} \tag{8.36}$$

To measure the surface area of a sample, the amount of an inert gas that is adsorbed at a series of pressures P is measured at low temperature (usually liquid nitrogen temperature, -195°C). With the gas identity known P° is estimated at the temperature of the experiment $P - P^\circ$ and the ratio P/P° is calculated. With V measured from the pressure drop in a fixed volume of gas and assuming ideal inert gas, the left-hand

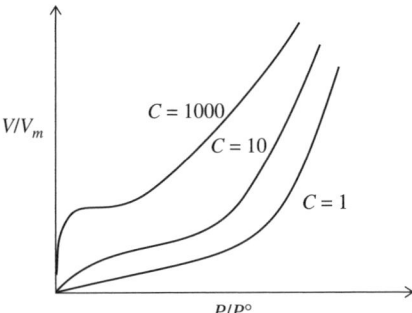

FIGURE 8.8 Ratio of volume V of gas adsorbed relative to the volume for a monolayer V_m versus the ratio of the pressure of the gas P relative to its equilibrium pressure $P°$ for three different monolayer stabilities C (see Equation 8.36).

side of the BET formula $P/(P° - P)V$ is plotted versus $P/P°$. A straight line with a slope of $(C - 1)/V_m C$ and an intercept $1/V_m C$ is obtained. With numerical values for the slope and intercept, there are two equations in two unknowns, C and V_m, and thus these values are readily obtained; V_m is the volume of the inert test gas in a monolayer, and from this value the number of gas molecules is calculated from the number of moles multiplied by Avogadro's number N_A. If we assume that one dense packed monolayer is forming from V_m and we know the area for each molecule of the test gas, then the area of surface is obtained. For example, typical test gases are N_2 and Kr, which have areas of 0.16 and 0.195 nm^2, respectively.

The BET isotherm method to measure surface areas is widely used. However, its applicability needs to be tested before use. The key issue is whether one monolayer is actually formed. This can be checked using a plot of the measured data in the form V/V_m versus $P/P°$ with an illustration of this plot in Fig. 8.8. Figure 8.8 shows three curves that have different values of C: C measures the relative stability of the layer formed, with higher C indicating that the first layer is more stable then succeeding layers. The curve with the highest C displays a distinct knee, indicating the formation of the first monolayer. Thus, the pressure range where this knee is observed should be the ideal pressure range to apply the BET isotherm and perform an area measurement. Furthermore, these curves imply that the BET measurement should not be used for systems (test gas and surface) that do not yield sufficiently large C values.

8.6 SURFACE REACTION MECHANISMS

Considerable research effort is often expended in determining the extent to which a surface influences a chemical reaction. For example, in the field of chemical catalysis the focus of study is often the influence on the reaction rate of various surfaces purposely introduced into a chemical reaction. Selected surfaces in specific reactions are known to improve reaction rates and yields while others do not. A surface that adsorbs a reactant restricts the reactant to two-dimensional motion rather than three-dimensional and thereby improves the statistics for reactants to collide. The attachment of a reactant at a surface site exposes a particular side, the unattached side, of a molecular reactant, hence potentially modifying the reactivity of the molecule to an incoming colliding species. Also, as discussed in Chapter 5, a surface has particular electronic states both filled and empty. Therefore, a surface can potentially enhance electron transfer among reactants. These surface states exist at various potentials, and if the potential matches that which a particular reaction species requires, the surface can greatly influence the rate of reaction and possibly the mechanism. A first step in understanding surface-mediated reactions is to determine whether the reaction takes place among or between adsorbed reactants or between an adsorbed reactant and a colliding reactant that is not yet adsorbed. These two possibilities comprise important surface reaction mechanism categories that explain a large amount of surface reaction data.

Figure 8.9a shows two of many reaction possibilities that can occur with either one or two kinds of species that are adsorbed on a surface. In Fig. 8.9a the circles represent atoms or molecules, with the left-side reaction being among like moieties that react and desorb while the right-side reaction is among unlike moieties that react and desorb. The assumptions made are that the surface reaction is rate limiting,

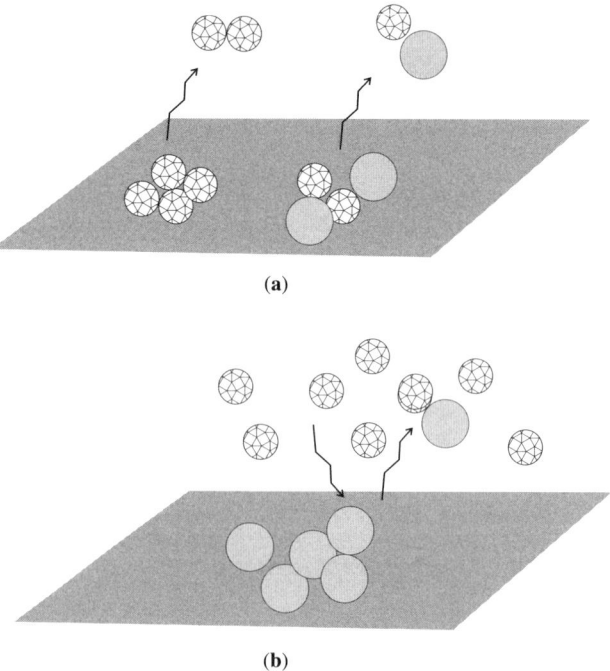

FIGURE 8.9 Representation of surface-mediated reactions where the various circles represent atoms or molecules: (**a**) the Langmuir–Hinshelwood mechanism among surface moeties yielding desorbed product and (**b**) the Ely–Rideal mechanism between adsorbed and colliding gas-phase moeties.

8.6 SURFACE REACTION MECHANISMS

that is, determines the kinetics, and that the initial monolayer formed is approximated by a Langmuir adsorption isotherm. This kind of mechanism between surface-adsorbed species is called a Langmuir–Hinshelwood (LH) mechanism. The simplest example of an LH mechanism is a unimolecular process where a single kind of adsorbed surface species changes to form a product (Fig. 8.9a, left side). This kind of reaction can take place where the products leave the surface, as in Fig. 8.9a, or where the products are adsorbed. Where products leave the surface the reaction is as follows:

$$A_{ad} \rightarrow \text{Product} \quad (8.37)$$

The rate of formation of product P_r is given as

$$\frac{d[P_r]}{dt} = k\theta_A \quad (8.38)$$

where $[P_r]$ indicates the concentration of P_r. With θ_A from the Langmuir adsorption isotherm above:

$$\theta_A = \frac{KP_A}{1 + KP_A} \quad (8.39)$$

The rate is given as

$$\frac{d[P_r]}{dt} = k\frac{KP_A}{1 + KP_A} \quad (8.40)$$

For the case where the product is adsorbed and competes for sites with the original adsorbate, the reaction scheme is as follows:

$$A_{ad} \rightarrow B_{ad} \quad (8.41)$$

where the product B can also desorb. For this case the Langmuir isotherm is recast to consider the possibility of two adsorbed species on the surface. For adsorption the change in coverage is proportional to vacant sites as a result of the joint probability of coverage by A and B species and the pressure of A as follows:

$$\frac{d\theta_A}{dt} = k_a N P_A (1 - \theta_A)(1 - \theta_B) \quad (8.42)$$

For desorption of A we have as above the equation:

$$\frac{d\theta_A}{dt} = k_d N \theta_A \quad (8.43)$$

As before we equate the adsorption and desorption rates and solve for θ_A to reinsert in the rate equation. Equating yields

$$\theta_A = K_A P_A (1 + \theta_A \theta_B - \theta_A - \theta_B) \tag{8.44}$$

where $K_A = k_a/k_d$. Multiplying through and bringing all θ_A terms together yields the following:

$$\theta_A (1 - K_A P_A \theta_B + K_A P_A) = K_A P_A (1 - \theta_B) \tag{8.45}$$

Now we consider that B_{ad} also behaves according a Langmuir isotherm; then θ_B is given as

$$\theta_B = \frac{KP_B}{1 + KP_B} \tag{8.46}$$

Substituting θ_B and solving for θ_A yields the following result:

$$\theta_A = \frac{K_A P_A}{1 + K_B P_B + K_A P_A} \tag{8.47}$$

where $K_B = k_a/k_d$ for B and similarly for θ_B:

$$\theta_B = \frac{K_B P_B}{1 + K_B P_B + K_A P_A} \tag{8.48}$$

Then the rate of adsorption of product is given as

$$\frac{d[P_r]}{dt} = \frac{kK_A P_A}{1 + K_B P_B + K_A P_A} \tag{8.49}$$

For bimolecular reactions there are also several cases. First, as before, we form products that are not adsorbed. This situation is as follows:

$$A_{ad} + B_{ad} \rightarrow \text{Product} \tag{8.50}$$

The rate for this reaction can be written as follows:

$$\frac{d[P_r]}{dt} = k\theta_A \theta_B \tag{8.51}$$

If both A and B reactants adsorb onto the surface according to a Langmuir isotherm and the reactants A and B compete for surface sites, then the overall rate is expressed as

$$\frac{d[P_r]}{dt} = \frac{kK_A K_B P_A P_B}{(1 + K_A P_A + K_B P_B)^2} \tag{8.52}$$

For the bimolecular reaction above it may be possible that product species P are also

8.6 SURFACE REACTION MECHANISMS

adsorbed and therefore also compete for adsorption sites:

$$\frac{d\theta_A}{dt} = k_a N P_A (1-\theta_A)(1-\theta_B)(1-\theta_P) \tag{8.53}$$

For desorption of A we have as above the equation:

$$\frac{d\theta_A}{dt} = k_d N \theta_A \tag{8.54}$$

Solving for θ_A as above yields the following result:

$$\frac{d[P_r]}{dt} = \frac{kK_A K_B P_A P_B}{(1 + K_A P_A + K_B P_B + K_P P_P)^2} \tag{8.55}$$

Figure 8.9b shows a surface with adsorbed A species (shaded circles) and gas-phase B species that can collide with the surface at a rate determined by the pressure P_B. We assume the reaction proceeds as follows:

$$A_{ad} + B_g \rightarrow \text{Product} \tag{8.56}$$

The formation of product P_r is written as a bimolecular reaction between gas-phase B species and A species that are absorbed, and the rate is given as

$$\frac{d[P_r]}{dt} = k P_B \theta_A \tag{8.57}$$

where k is the second-order rate constant and θ_A is the coverage of A as usual. If we now assume that the adsorption of A is described by the Langmuir isotherm as follows:

$$\theta_A = \frac{KP_A}{1 + KP_A} \tag{8.58}$$

The rate of reaction is given by the following expression in terms of the partial pressures of the reactants:

$$\frac{d[P_r]}{dt} = kK \frac{P_A P_B}{1 + KP_A} \tag{8.59}$$

This mechanism is called the Eley–Rideal reaction mechanism. From this result for $KP_A \gg 1$ the rate is proportional to P_B. Therefore, a plot of the rate of reaction versus P_A at constant P_B would display a limiting value at high P_A values.

8.7 TEMPERATURE-PROGRAMMED DESORPTION

The process shown in Fig. 8.9a in which adsorbed species leave the surface and enter the gas phase is a temperature-activated process, and the energy dependence can be summarized using a so-called Boltzmann factor. An expression for the Boltzmann factor can be readily obtained by considering a two-state situation where the change in the population of the higher energy state δN relative to the lower energy or ground state N_0 is proportional to the negative of the energy difference in the states $-\Delta E$ as follows:

$$\frac{\delta N}{N_0} \propto -\Delta E \qquad (8.60)$$

Changing differences to differentials and the proportionality to an equality with a constant and then integrating yields the probability for the occupancy for the upper state as proportional to $e^{-\Delta E/kT}$, which is the Boltzmann factor and where $1/kT$ is used to define a suitable energy scale. For the case of desorption the desorption rate $-d\theta/dt$ is proportional to the coverage multiplied by the probability to desorb and is given as

$$-\frac{d\theta}{dt} \propto \theta e^{-\Delta H_{des}/kT} \qquad (8.61)$$

where the energy is the heat of desorption ΔH_{des}.

This nature of adsorption and desorption processes can yield considerable information about the surface reactions. Measurement of the desorption processes is particularly noteworthy and will be developed below as an example of a powerful surface probe derived from the simple process of a species leaving a surface when the temperature is increased. The surface probe discussed below and based on desorption is called temperature-programmed desorption (TPD) and sometimes called thermal desorption spectroscopy (TDS).

To develop the central idea of TPD, the above proportionality expression (8.61) for desorption is converted to an equality by including the fact that the rate could depend on coverage to a higher order than 1 and hence replace θ by θ^n. This substitution expresses the fact that the exponent n represents the molecularity or kinetic order of the desorption process, that is, whether desorption depends on one or several species interacting. Figure 8.9a shows both a unimolecular process where A_{ad} directly desorbs and a bimolecular processes where A_{ad} reacts with B_{ad} to form a product that desorbs. Also there is another factor needed in converting the proportionality and that is a frequency ν (for the purpose of proper units since ν has units of $1/t$) and the form for the desorption rate is as follows:

$$-\frac{d\theta}{dt} = \nu \theta^n e^{\Delta H_{des}/kT} \qquad (8.62)$$

A typical TPD experiment shown in Fig. 8.10 involves heating a sample and observing the change in coverage with time. The experiment is enclosed in a

8.7 TEMPERATURE-PROGRAMMED DESORPTION

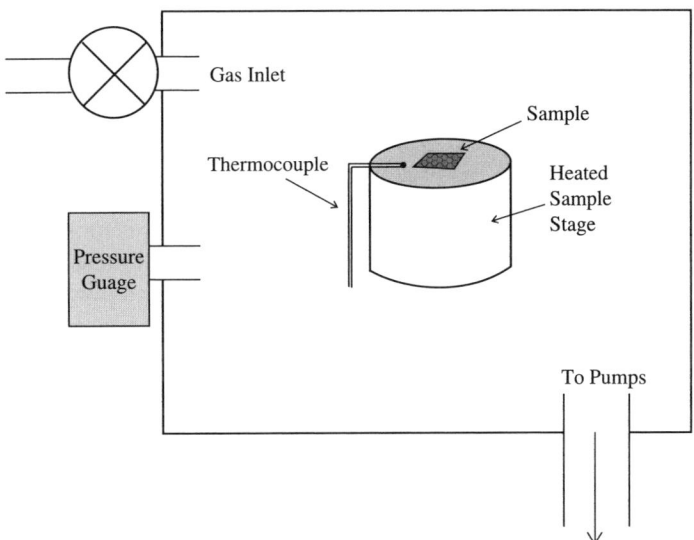

FIGURE 8.10 Schematic of a basic temperature-programmed desorption TPD system.

vacuum chamber, and there are usually some methods of surface and gas-phase analysis to determine surface coverage, surface species, gas composition, and pressure. A key aspect of a TPD experiment is temperature programming. The precise way in which T is varied yields interpretable results. For example, two common heating methods are to employ a linear sample heating rate or a parabolic heating rate and are given, respectively, as

$$T = T_0 + \alpha t \quad \text{and} \quad \frac{1}{T} = \frac{1}{T_0} + \beta t \tag{8.63}$$

where T is temperature at time t and T_0 is the initial temperature (at $t=0$), and α and β are constants that define the linear or parabolic T programming, respectively. In a static vacuum system, as the sample is heated and adsorbate is vaporized, the pressure in the vacuum system will rise, and the change in pressure would be due entirely to the change in coverage as follows:

$$\frac{dP}{dt} = -\frac{d\theta}{dt} \tag{8.64}$$

However, a practical vacuum system is not static and is maintained at low pressure using pumps. Thus, as the pressure tends to rise as a result of desorption, the new vapor will be pumped away at a rate determined by the pumping speed and efficiency. Also no vacuum system is entirely without leaks, so there is a source of gas other than the desorbed material that adds to the chamber pressure. Neglecting the desorbed gas, a given vacuum system with a set of pumps will reach a steady-state pressure depending on the speed and efficiency of the pumps and the size of the leaks. Thus, without any other sources the change in pressure in a vacuum system can be expressed as follows:

$$\left(\frac{dP}{dt}\right)_{sys} = k(P_{ss} - P_p) \tag{8.65}$$

where P_{ss} is the steady-state pressure from a source of gas that is for now a leak (and below also desorption), and P_p is the ultimate pressure achievable by the pumps without leaks. In the development below the $+$ and $-$ signs are for gas in and out, respectively. Now we include the effect of a sample that desorbs, which as mentioned above will increase the pressure, so it is represented by a $+$ change in system pressure as represented in Equation 8.65. The sample alters the above system pressure change formula to yield

$$\left(\frac{dP}{dt}\right)_{sys} = k(P_{ss} - P_p) - \frac{A}{cV}\frac{d\theta}{dt} \tag{8.66}$$

where A is the sample area, V is the chamber volume, c is a constant, and from Equation 8.62 $d\theta/dt$ is negative $(-)$ so by making the right-hand term negative the desorbing material is adding to the chamber pressure. To use this formula we rearrange it as follows:

$$\left(\frac{dP}{dt}\right)_{sys} - k(P_{ss} - P_p) = -\frac{A}{cV}\frac{d\theta}{dt} \tag{8.67}$$

This form has the vacuum system terms on the left and the sample-related term on the right. We can consider two limiting cases for the vacuum system: The pumping is slow relative to the sample desorption or the pumping is fast. If pumping is slow, then the second term on the left is small and we will ignore it. If the pumping is fast, then the system only changes pressure when desorption occurs and then quickly returns to the original pressure and so we can ignore the left-most term on the left side of the equation.

For slow pumping the simplified version of Equation 8.67 is as follows:

$$\left(\frac{dP}{dt}\right)_{sys} = -\frac{A}{cV}\frac{d\theta}{dt} \tag{8.68}$$

If the sample is heated using the linear heating rate given by Equation 8.63 above, we can then transform t to T. Starting with the rearranged formula and its derivative:

$$T - T_0 = \alpha t \quad \text{and} \quad \frac{dT}{dt} = \alpha \tag{8.69}$$

8.7 TEMPERATURE-PROGRAMMED DESORPTION

We can transform t to T as follows:

$$\frac{d\theta}{dt} = \alpha \frac{d\theta}{dT} \tag{8.70}$$

This result can be used in Equation 8.62 to yield:

$$-\frac{d\theta}{dT} = \frac{\nu}{\alpha} \theta^n e^{\Delta H_{\text{des}}/kT} \tag{8.71}$$

For slow pumping typical P vs. time t results for a desorbing sample are shown in Fig. 8.11a. For the linear T ramp, t and T are linearly related, so the t scale in Fig. 8.11 could be directly converted to a T scale and the slope is also proportional

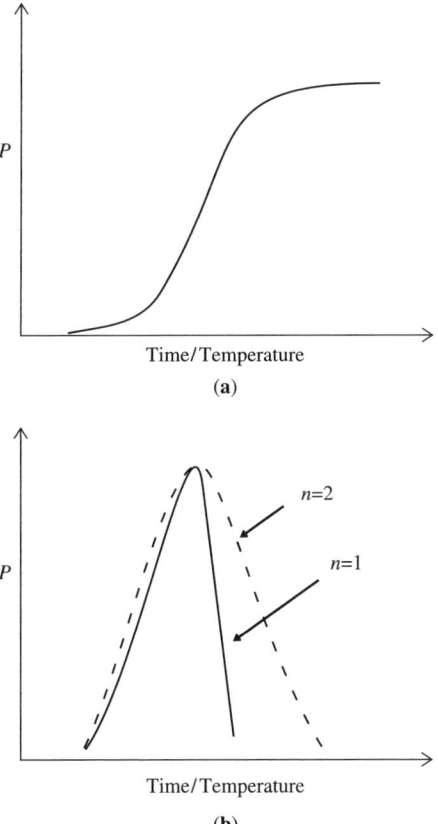

FIGURE 8.11 Linear temperature-programmed TPD experiments: (a) slow pumped TPD system pressure versus time or temperature and (b) fast pumped TPD system pressure versus time or temperature for first- ($n = 1$) and second- ($n = 2$) order desorption mechanisms.

to $d\theta/dT$. Then, from a determination of the coverage θ and an assumed value for n starting with $n = 1$, the following plot is made:

$$\ln\left\{\frac{1}{\theta^n}\frac{d\theta}{dT}\right\} \text{ versus } \frac{1}{T} \quad (8.72)$$

The finding of a straight line indicates that the choice of n is correct and the molecularity or order for the desorption is determined and the slope of the line yields a value for ΔH_{des}. Also with α known as an experimental parameter v can be determined.

For fast pumping dP/dt is small and Equation 8.66 above becomes

$$\frac{d\theta}{dt} = \frac{cV}{A}k(P_{ss} - P_p) \quad (8.73)$$

Figure 8.11b shows that desorption taking place with rapid pumping results in pressure transients. Note that the shapes are different with higher order desorption ($n > 1$) are symmetrical, while first-order desorption is not and displays a more rapidly decreasing trailing edge. The maximum in the transient can be obtained for a linear T ramp from the condition:

$$\frac{d^2\theta}{dt^2} = \frac{d^2\theta}{dT^2} = 0 \quad (8.74)$$

Using the result for $d\theta/dT$ in Equation 8.70, the following second derivative is obtained and equated to 0 to find the maximum:

$$0 = \frac{d^2\theta}{dT^2} = \frac{v}{\alpha}\left\{n\theta^{n-1}\frac{d\theta}{dT}e^{-\Delta H_{des}/kT} + \theta^n\frac{\Delta H_{des}}{kT^2}e^{-\Delta H_{des}/kT}\right\} \quad (8.75)$$

From this relationship, which yields the peak at $T = T_{max}$ and at $n = 1$, we obtain

$$\frac{d\theta}{dT} = -\theta\frac{\Delta H_{des}}{kT_{max}^2} \quad (8.76)$$

Now we can use this relationship in Equation 8.70 to yield the result:

$$\frac{\Delta H_{des}}{kT_{max}^2} = \frac{v}{\alpha}e^{-\Delta H_{des}/kT_{max}} \quad (8.77)$$

This result teaches that at $n = 1$ the peak temperature T_{max} is not a function of the coverage θ. Figure 8.12a shows a set of P vs. T desorption results for $n = 1$ but at various starting coverages. Note that T_{max} does not shift with θ.

8.7 TEMPERATURE-PROGRAMMED DESORPTION

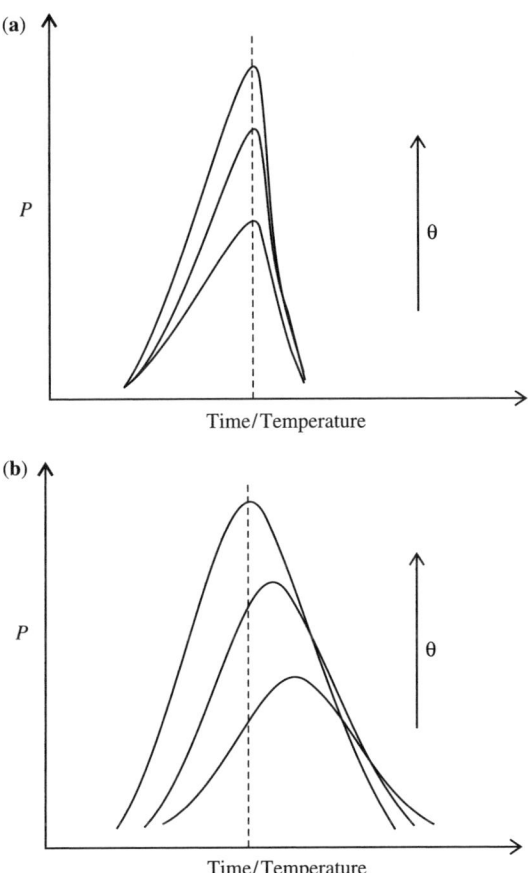

FIGURE 8.12 Fast pumped TPD experiments with linear temperature programming in terms of system pressure P versus time or temperature: (a) n = 1 with three coverages θ and (b) $n = 2$ at three θ.

For $n = 2$ we can obtain T_{max} using the second derivative at $n = 2$ to obtain

$$\frac{\Delta H_{des}}{kT_{max}^2} = \frac{2v\theta_{max}}{\alpha} e^{-\Delta H_{des}/kT_{max}} \tag{8.78}$$

Taking the logarithm we obtain

$$\ln\left(\frac{\alpha \Delta H_{des}}{2kv\theta_{max}}\right) = 2 \ln T_{max} - \frac{\Delta H_{des}}{kT_{max}} \tag{8.79}$$

It is seen that for this case T_{max} is a function of coverage and as θ increases T_{max} decreases. Figure 8.12b shows a set of P vs. T results for $n = 2$ at various coverages.

Notice that T_{max} decreases with larger θ. Thus, from TPD coverage studies as well as from the shape of the curves, surface desorption orders can be discerned. For values of ν of 10^{13} s^{-1} as obtained from $\nu = kT/h$ at 300 K, ΔH_{des} can also be obtained.

It is also useful to vary α, which determines the heating rate. Starting with Equation 8.77 for $n = 1$ the logarithm is as follows:

$$\ln\left(\frac{\Delta H_{des}}{k}\right) = \ln \nu + 2 \ln T_{max} - \ln \alpha - \frac{\Delta H_{des}}{kT_{max}} \tag{8.80}$$

From TPD studies using different α values the derivative is given as

$$\frac{\partial(\ln \alpha)}{\partial(\ln T_{max})} = 2 + \frac{\Delta H_{des}}{k} \tag{8.81}$$

Equation 8.81 teaches that a plot of $\ln \alpha$ vs. $\ln T_{max}$ yields a straight line with a slope proportional to ΔH_{des} if $n = 1$. Similarly, for $n = 2$, Equation 8.78 is used as well as the fact that coverage at T_{max}, θ_{max}, is one-half of the initial coverage θ_0 (from the symmetry of the P vs. T plot in Fig. 8.11) and the logarithm is as follows:

$$\ln(\theta_0 T_{max}^2) = \frac{\Delta H_{des}}{kT_{max}} - \ln\left(\frac{\nu k}{\alpha \Delta H_{des}}\right) \tag{8.82}$$

A plot of $\ln(\theta_0 T_{max}^2)$ vs. $1/T_{max}$ will yield a linear plot with a slope proportional to ΔH_{des} for $n = 2$.

In addition, Equation 8.75 can be integrated to yield for $n = 1$ and $n = 2$, respectively:

$$\ln\left[\frac{(d\theta/dT)_{max}}{\theta_{max}}\right] = \frac{\Delta H_{des}}{k}\left(\frac{1}{T} - \frac{1}{T_{max}}\right) + \left(\frac{T}{T_{max}}\right)^2$$

$$\times \exp\left(-\left(\frac{\Delta H_{des}}{k}\right)\left(\frac{1}{T} - \frac{1}{T_{max}}\right)^{-1}\right) \tag{8.83}$$

$$\frac{(d\theta/dT)_{max}}{\theta_{max}} = \frac{1}{4}\left\{\exp\frac{\Delta H_{des}}{k}\left(\frac{1}{T} - \frac{1}{T_{max}}\right) + \left(\frac{T}{T_{max}}\right)^2\right.$$

$$\left. \times \exp-\left(\frac{\Delta H_{des}}{2k}\right)\left(\frac{1}{T} - \frac{1}{T_{max}}\right)\right\}^2 \tag{8.84}$$

Figure 8.11b shows that this case for $n = 1$ yields an unsymmetrical curve while that for $n = 2$ is symmetrical about T_{max}.

In summary, TPD experiments can lead to an understanding of surface reaction mechanisms. In many cases, however, the experimental results are not readily interpreted and require other surface analyses before confident assignment of the surface mechanism can be made, but nonetheless TPD is a powerful and frequently used tool.

REFERENCES

1. Chanin LM, Phelphs AU, Biondi MA. Phys Rev 1962;128:219.
2. Irene EA, Lewis EA. Appl Phys Lett 1987;51:767.

SUGGESTED READING

R. P. H. Gasser, 1985. *An Introduction to Chemisorption and Catalysis by Metals*, Oxford Scientific. This book has many of the adsorption ideas discussed in this chapter and other chapters in this text. It may be out of print and hard to find but worth the effort on this topic. This book was used as a text in the early days of the UNC courses.

H. Luth, 1995. *Surfaces and Interfaces of Solid Materials*, Springer. This book has many useful sections on adsorption and measurements.

9

ELLIPSOMETRY AND OPTICAL PROPERTIES OF SURFACES, INTERFACES, AND FILMS

9.1 INTRODUCTION

Surfaces, interfaces, and thin films by their nature do not contain large numbers of atoms or molecules compared with the bulk materials that they bind. Because of this, most optical techniques are unable to discriminate surface or interface optical properties from bulk properties. However, reflection ellipsometry, which is the most surface-sensitive optical technique (as will be discussed below), is well suited to determine the optical properties of surfaces, interfaces, and thin films. Therefore, an understanding of ellipsometry will enable an appreciation for what optical properties of surfaces, interfaces, and films are obtainable and how to obtain and evaluate those important properties. Thus, this chapter commences with a discussion of ellipsometry and intertwined is the determination of optical properties of surfaces, interfaces, and films in this and other chapters throughout this text particularly in Chapters 10, 11, and 12.

Ellipsometry is an old technique dating to the 1800s (1), but with the advent of the technological importance of thin films, in particular, in microelectronics, and with the availability of computers to do the precise lengthy calculations, ellipsometry is experiencing a renaissance. Ellipsometry has been used in microelectronics since the early 1960s [see, e.g., an early authoritative study by Archer (2)], when it became necessary to measure the thickness of transparent films on highly reflective substrates. Ellipsometry is a simple technique (do not confuse simple with easy!) that relies primarily on the classical physical optics of stratified media. There are

Surfaces, Interfaces, and Thin Films for Microelectronics. By Eugene A. Irene
Copyright © 2008 John Wiley & Sons, Inc.

several in-depth treatments of ellipsometry and two introductory texts to which the reader is directed (see citations at the end of this chapter) for further study of the technique and applications. The approach of this chapter is to answer some basic questions about the ellipsometry technique: What is it? What and how does ellipsometry measure? How well does it measure? Following a brief discussion of the hardware, several illustrative applications are discussed that reinforce and enhance previous and later presented discussions about surfaces, interfaces, and thin films in which ellipsometry greatly contributed to the study with more in-depth applications in Chapters 10, 11, and 12.

9.2 WHAT IS ELLIPSOMETRY?

Essentially, ellipsometry is an optical technique that measures the changes that occur in light when polarized monochromatic light interacts with matter. For the purpose of characterizing surfaces and interfaces, we restrict the discussion to the reflection of light from matter (transmission ellipsometry follows a similar development but is far less useful for surfaces, interfaces, and films), and more specifically reflection from solid surfaces that may or may not be covered with a film or films. For a treatment of ellipsometry one needs to be reminded of the transverse wave nature of light and the definition of polarized light. Figure 9.1 shows a monochromatic light wave as a transverse wave with a single electric field **E** direction, $\mathbf{E}_x$. This is called linearly polarized light. The light wave is an electromagnetic wave with orthogonal electric **E** and magnetic **B** field vectors perpendicular to the direction of propagation z. The **E** and **B** fields are in phase from the source as shown. While both the magnetic and electric fields can be polarized and the polarization for both vectors can change upon interaction with matter, ellipsometry deals with the polarized electric field and how **E** interacts with matter. The effect on **E** is much larger than for **B**. Figure 9.2a shows linearly polarized monochromatic light propagating in the z direction from the top right toward the bottom left, as indicated with a single electric field vector **E** shown as the darker wave. Field vector **E** has orthogonal components projected in

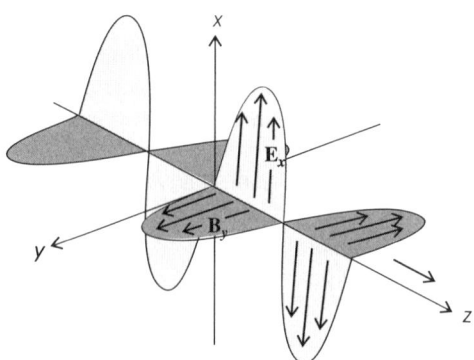

FIGURE 9.1 Transverse electromagnetic wave with electric **E** and magnetic **B** fields perpendicular to the propagation direction z.

9.2 WHAT IS ELLIPSOMETRY?

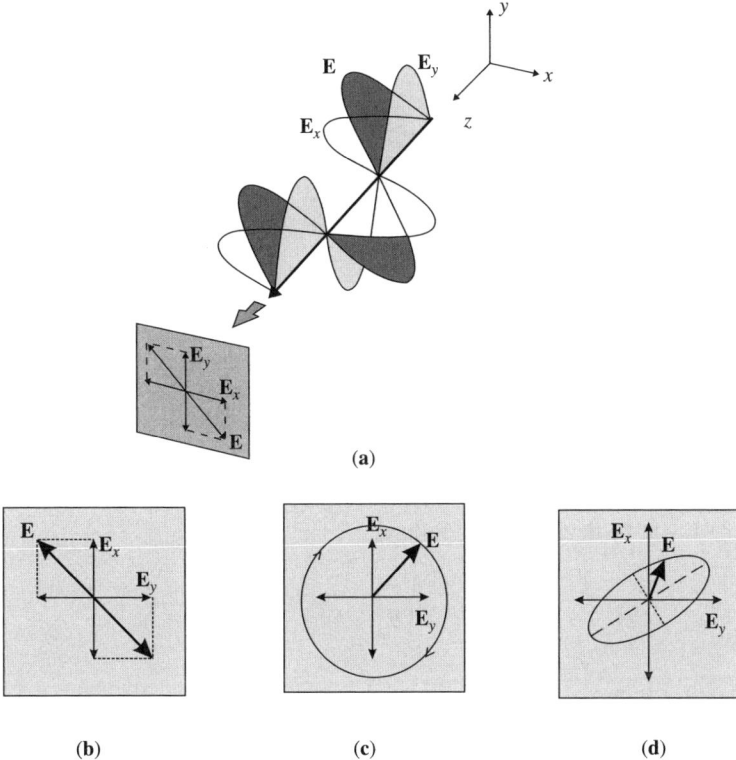

FIGURE 9.2 Various polarization states of a monochromatic light wave with **E** field projections in the x and y planes yielding $\mathbf{E}_x$ and $\mathbf{E}_y$: (a) linearly or plane polarized light, (b) projected linearly polarized light, (c) projected right circularly polarized light, and (d) projected elliptically polarized light.

the x and y directions and labeled $\mathbf{E}_x$ and $\mathbf{E}_y$ and is obtained from the square root of the sum of the squares of the two orthogonal components as

$$\mathbf{E} = \left(\mathbf{E}_x^2 + \mathbf{E}_y^2\right)^{1/2} \tag{9.1}$$

Figure 9.2a also shows the projection of the linearly polarized wave with **E**, $\mathbf{E}_x$, and $\mathbf{E}_y$ onto a plane. Figures 9.2b, 9.2c, and 9.2d show a comparison of the projections of linear, circular, and elliptically polarized light, respectively. The locus of the endpoint of the **E** vector traces a circle on the shaded plane for Fig. 9.2c when the $\mathbf{E}_x$ and $\mathbf{E}_y$ projections are equal and an ellipse when they are unequal as seen for Fig. 9.2d. The term ellipsometry refers to this latter more general case for unequal projections of **E**, and the ellipsometry technique is the determination of this polarization ellipse after reflection from a material or structure in comparison to the polarization of the incident light. For the reflection of light, the plane of incidence (POI) is defined by the lines of the incident light beam and the normal to the surface (N), as shown by the shaded plane in Fig. 9.3. In this representation one of the orthogonal components of the **E**

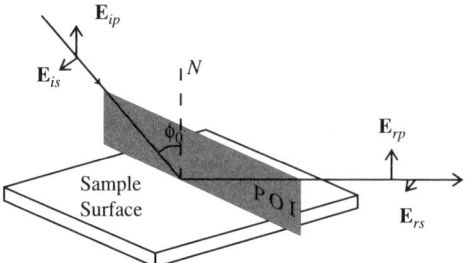

FIGURE 9.3 Incident linearly polarized light with orthogonal electric field components $\mathbf{E}_{ip}$ and $\mathbf{E}_{is}$ reflected at incidence and ϕ_0 yielding reflected components $\mathbf{E}_{rp}$ and $\mathbf{E}_{rs}$. The plane of incidence (POI) contains the surface normal N and the incident ray.

vector is called $\mathbf{E}_p$ and is parallel to the POI, and the other is $\mathbf{E}_s$ and is perpendicular to the POI (where the subscript s is from the German for perpendicular, *senkrecht*), and as before $\mathbf{E}$ can be determined as follows:

$$\mathbf{E} = \left(\mathbf{E}_p^2 + \mathbf{E}_s^2\right)^{1/2} \tag{9.2}$$

Recalling that light intensity I is given as

$$I = \mathbf{E}\mathbf{E}^* = \mathbf{E}^2 \quad \text{for real } \mathbf{E} \tag{9.3}$$

where the asterisk denotes the complex conjugate, and notice from Fig. 9.3 that the $\mathbf{E}_p$ and $\mathbf{E}_s$ components are in phase before reflection from the sample, that is, these components are in phase from the source. In general when polarized monochromatic light reflects from (or transmits through) a surface, both the phase and amplitude of the components describing the light can change. Whether both phase and amplitude will change and by how much is a function of the optical nature of the reflecting surface, and the quantification of the phase change called Δ, and the amplitude change called Ψ, comprise the essence of the ellipsometric measurement. Figure 9.3 shows that for the reflected light (subscript r) the amplitude $\mathbf{E}_{rp}$ is in general different from the incident (subscript i) p component $\mathbf{E}_{ip}$ and likewise for incident and reflected s components. Also the phase between $\mathbf{E}_{rp}$ and $\mathbf{E}_{rs}$ is different than the incident phase between $\mathbf{E}_{ip}$ and $\mathbf{E}_{is}$. Figure 9.4a shows the reflection ellipsometry experiment at a surface where all the light in the reflected beam derives from the reflecting surface. Figure 9.4b shows a film on a surface that can transmit light as well as reflect light, and, because of multiple reflections at the ambient–film interface (the 01 interface) and the film–substrate interface (the 12 interface), additional phase and amplitude changes can occur. In Fig. 9.4 all the reflection coefficients are denoted by r and transmission coefficients by t. The detected light is obtained from a sum of all the components that reach the detector with some terms indicated

9.2 WHAT IS ELLIPSOMETRY?

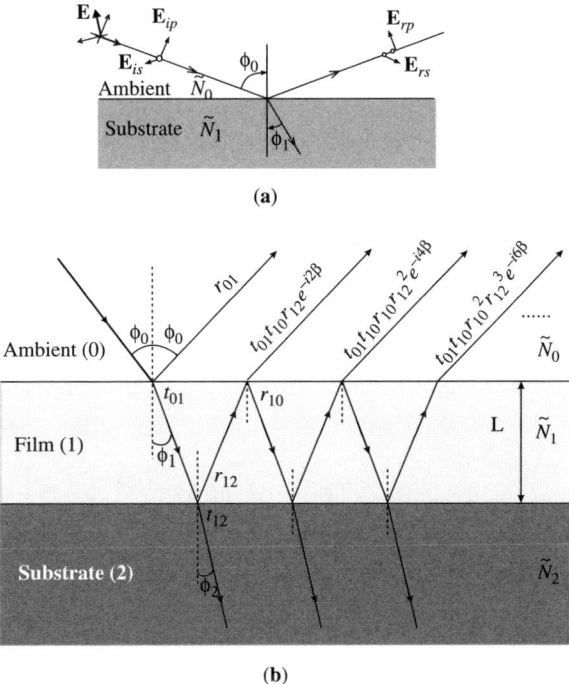

FIGURE 9.4 Reflection at (a) bare substrate and (b) film covered substrate showing incident and reflected electric field components **E**, angles (ϕ), and optical properties ($\tilde{N}$), r_{ij} and t_{ij} are the Fresnel reflection and transmission coefficients, respectively.

in Fig. 9.4b, and all the changes are measured as changes in Δ and Ψ and analyzed in terms of the optical properties of the reflecting materials. For a bare reflecting surface, the form for Δ and Ψ is as follows:

$$\Delta = \delta_p - \delta_s \quad \text{and} \quad \tan\psi = \frac{|r_p|}{|r_s|} \tag{9.4}$$

where δ_p is the phase change for the p component upon reflection and similarly for the s component as δ_s. The Fresnel coefficients r_p and r_s are obtained from matching the tangential **E** (and **B**) fields across an interface and have the form:

$$r_p = \frac{\mathbf{E}_{rp}}{\mathbf{E}_{ip}} = \frac{\tilde{N}_1 \cos\phi_0 - \tilde{N}_0 \cos\phi_1}{\tilde{N}_1 \cos\phi_0 + \tilde{N}_0 \cos\phi_1} \quad \text{and}$$

$$r_s = \frac{\mathbf{E}_{rs}}{\mathbf{E}_{is}} = \frac{\tilde{N}_0 \cos\phi_0 - \tilde{N}_1 \cos\phi_1}{\tilde{N}_0 \cos\phi_0 + \tilde{N}_1 \cos\phi_1} \tag{9.5}$$

where $\tilde{N}$ is the complex refractive index (to be defined below), ϕ is the reflection or refraction angle, and the subscripts refer to the interface at which the event occurs counting from the outer ambient containing interface designated with the 0 subscript as shown in Fig. 9.4b. Transmission coefficients are also defined similarly but in terms of the transmitted electric fields. For film-covered surfaces Fig. 9.4b shows that additional reflections and refractions occur at the film–film and film–substrate interfaces, and in the different media and as mentioned above the reflected output at the detector must be a sum of all reflected beams. The result, R, as obtained from a converging geometric series is as follows:

$$R_p = \frac{r_{01p} + r_{12p}\exp(-i2\beta)}{1 + r_{01p}r_{12p}\exp(-i2\beta)} \quad \text{and} \quad R_s = \frac{r_{01s} + r_{12s}\exp(-i2\beta)}{1 + r_{01s}r_{12s}\exp(-i2\beta)} \tag{9.6}$$

where

$$\beta = 2\pi \left[\frac{L\left(\tilde{N}_1^2 - \tilde{N}_0^2 \sin^2\phi_0\right)^{1/2}}{\lambda} \right] \tag{9.7}$$

and where the Fresnel coefficients r_{mnl} where mn identify the interface and l is either the p or s component of $\mathbf{E}$ are as follows:

$$r_{01p} = \frac{\tilde{N}_1 \cos\phi_0 - \tilde{N}_0 \cos\phi_1}{\tilde{N}_1 \cos\phi_0 + \tilde{N}_0 \cos\phi_1} \quad r_{12p} = \frac{\tilde{N}_2 \cos\phi_1 - \tilde{N}_1 \cos\phi_2}{\tilde{N}_2 \cos\phi_1 + \tilde{N}_1 \cos\phi_2}$$

$$r_{01s} = \frac{\tilde{N}_0 \cos\phi_0 - \tilde{N}_1 \cos\phi_1}{\tilde{N}_0 \cos\phi_0 + \tilde{N}_1 \cos\phi_1} \quad r_{12s} = \frac{\tilde{N}_1 \cos\phi_1 - \tilde{N}_2 \cos\phi_2}{\tilde{N}_1 \cos\phi_1 + \tilde{N}_2 \cos\phi_2} \tag{9.8}$$

The exponential factor β, often referred to as the phase factor, contains the usually desired information from an ellipsometry measurement, namely film thickness, L, and the complex refractive indices, $\tilde{N}$.

It is usual to define a complex reflection coefficient, ρ, for both bare and film-covered substrates:

$$\rho = \frac{r_p}{r_s} \quad \text{or} \quad \frac{R_p}{R_s} \tag{9.9}$$

where the reflection coefficients, the r's and R's, can also be expressed as complex exponentials:

$$r_p = |r_p|\exp(i\delta_p) \quad \text{and} \quad r_s = |r_s|\exp(i\delta_s) \tag{9.10}$$

$$R_p = |R_p|\exp(i\Delta_p) \quad \text{and} \quad R_s = |R_s|\exp(i\Delta_s) \tag{9.11}$$

9.3 WHAT DOES ELLIPSOMETRY MEASURE?

Combining the results above, the following is obtained for ρ:

$$\rho = \frac{E_{rp}E_{is}}{E_{ip}E_{rs}} = \frac{|r_p|\exp(i\delta_p)}{|r_s|\exp(i\delta_s)} \quad \text{or} \quad \frac{|R_p|\exp(i\Delta_p)}{|R_s|\exp(i\Delta_s)} = \tan\psi \, \exp(i\Delta) \quad (9.12)$$

where

$$\tan\psi = \frac{|r_p|}{|r_s|} \quad \text{or} \quad \frac{|R_p|}{|R_s|} \quad (9.13)$$

and $\Delta = \delta_p - \delta_s$ or $\Delta_p - \Delta_s$. From the development above the general formula for ellipsometry is obtained that relates measurables to properties as follows:

$$\rho = \tan\psi \, \exp(i\Delta) = \rho(\tilde{N}_0, \tilde{N}_1, \tilde{N}_2, \ldots, L_1, \ldots, \phi_0, \lambda) \quad (9.14)$$

where the last term on the right contains all the parameters of the measurement, namely film thicknesses, optical properties (discussed below), the wavelength of light, and the angle of incidence. Of course, the last two are known and the others are sought from the measurement. It must be remembered that, however complex is the film(s) upon a substrate, there are always only two measurables in ellipsometry, Δ and Ψ. Therefore, from a typical measurement of Δ and Ψ only two properties from the general ellipsometry equation can be obtained. Either the others need to be supplied from independent measurements or additional experimentally controlled hardware variables can be systematically varied in order to increase the number of independent equations. Often the angle of incidence and the wavelength are selected. For many *in situ* measurements the experimental situation precludes changes in angle of incidence, but scanning wavelength ellipsometry or spectroscopic ellipsometry is comparatively straightforward to implement, and we will return to spectroscopic ellipsometry below.

9.3 WHAT DOES ELLIPSOMETRY MEASURE?

This question is answered at once from Equation 9.14, which shows materials properties ($\tilde{N}$'s and L's) and experimental parameters (ϕ_0 and λ that are usually known *a priori*) on the right that are accessible from the measurement of Δ and Ψ. However, it is worthwhile to look at the important relationships more closely (optical properties and roughness effects on optical properties were briefly discussed in Chapter 4), and the first is the definition of the complex refractive index $\tilde{N}$ that was used in previous equations. $\tilde{N}$ is defined as

$$\tilde{N} = n + ik \quad (9.15)$$

where n is the real part of the complex refractive index and is the ratio of the speed of light in vacuum compared to the material in question, and k is the absorption index that is related to the absorption constant α; and α and k are related by the formula

$$\alpha = \frac{4\pi k}{\lambda_0} \qquad (9.16)$$

where λ_0 is the wavelength in vacuum. The dielectric response function, or simply the dielectric function ε, is a measure of the response of a material to the interaction with electromagnetic radiation and is given as

$$\varepsilon = \varepsilon_1 + i\varepsilon_2 \qquad (9.17)$$

$$\varepsilon_1 = n^2 - k^2 \quad \text{and} \quad \varepsilon_2 = 2nk \qquad (9.18)$$

It is important to realize that the dielectric function is defined for a pure homogenous material. The dielectric function for a simple homogenous surface or a pseudodielectric function for a complex film-covered surface, and denoted by brackets, can be written in terms of the ellipsometric variable ρ as follows:

$$\varepsilon \text{ or } \langle \varepsilon \rangle = \sin^2 \phi_0 + \sin^2 \phi_0 \tan^2 \phi_0 \left[\frac{1-\rho}{1+\rho} \right]^2 \qquad (9.19)$$

The use of formula 9.19 is straightforward for a pure material: Δ and Ψ are measured at some angle of incidence ϕ_0 and wavelength λ, and then ε is calculated using Equations 9.14 and 9.19. However, in many cases Δ, Ψ is measured for a complicated system such as a film-covered surface, possibly with many films and/or roughness and/or impurities. In all of these cases only a single Δ, Ψ is measured at ϕ_0 and λ. The measured Δ, Ψ corresponds to a complex system in which ε is undefined, and thus the use of Equation 9.19 will not yield a value for ε but rather simply a number. To indicate that this was the case for a complex system, angular brackets are used for the calculated quantity, $\langle \varepsilon \rangle$, and it is called the pseudodielectric function in that $\langle \varepsilon \rangle$ is obtained using the relationship for ε.

From the ellipsometric measurement of Δ and Ψ, values for ρ and then L and N can be obtained as well as ε. Later we introduce the notion of spectroscopic ellipsometry where the dispersion in $\tilde{N}$ and ε is measured.

9.4 HOW WELL CAN ELLIPSOMETRY MEASURE?

The great sensitivity of ellipsometry is derived from the fact that the measurement of Δ, Ψ is a relative measurement of the change in polarization imparted by the sample

9.4 HOW WELL CAN ELLIPSOMETRY MEASURE?

TABLE 9.1 Calculated Δ, Ψ for Various Transparent Film Thickness with Conditions: $n = 1.5$ on Si, $\lambda = 632.8$ nm, $\phi_0 = 70°$

Δ (degrees)	Ψ (degrees)	Film Thickness (nm)
179.257	10.448	0.0
178.957	10.448	0.1
178.657	10.449	0.2
178.356	10.450	0.3
178.056	10.451	0.4
177.756	10.453	0.5
176.257	10.462	1.0

upon the incident polarized light as is evidenced by Equations 9.12 and 9.14. Because ellipsometry is a relative measurement unlike absolute light measurement spectroscopies, ellipsometry is not especially sensitive to long-term drift in the light source or detector. Hence ellipsometry is sensitive to small surface changes. Indeed fractions of a monolayer are sensed by reflection ellipsometry. This level of sensitivity is achievable only after careful alignment, which will be discussed in the hardware section.

One way to quantify the sensitivity of ellipsometry is to first calculate the effect of the presence of overlayers or films on a substrate on the measurables Δ, Ψ and then to compare that result with the capability of an ellipsometer to measure Δ, Ψ. Table 9.1 shows calculated results for an Si surface (with $n = 3.085$ and $k = 0.018$ at $\lambda = 632.8$ nm) coated with a transparent film with $n = 1.5$ and $k = 0$. Under these conditions of the calculation for an imaginary film that is similar to SiO_2, it is seen that Δ changes by about $0.3°$ per 0.1 nm of film and Ψ by about $0.001°$ for 0.1 nm. Considering that a properly aligned ellipsometer with high quality optics is capable of precision of about 0.01 to $0.02°$ in Δ and Ψ, sensitivity approaching 0.01 nm or submonolayer sensitivity (considering that atomic diameters are of the order of 0.1 nm) is achievable with the determination of Δ. For other film thicknesses and other measurement conditions, there may be more Ψ sensitivity and we will return to this point below.

Automated ellipsometers often have lower sensitivity, but even with an order of magnitude worse sensitivity than calculated above, sensitivity to the presence of a film of the order of 0.1 nm thick is achievable. It must be kept in mind that while ellipsometry has great sensitivity, it is also a precision optical technique. To achieve the sensitivity and use it in terms of accurately measured properties, each optical component as well as the angle of incidence ϕ_0 must be carefully calibrated. To illustrate the level to which calibration must be done, Table 9.2 is used to show the effect of errors in ϕ_0 on film thickness L and refractive index n. The first column has film thickness of 10 and 100 nm, and column 2 shows the variation in ϕ_0 used for the calculation. Columns 3 and 4 display the corresponding calculated Δ and Ψ values. These Δ and Ψ values are then used to calculate L and n values but assuming that ϕ_0 is fixed at $70°$ rather than the value of ϕ_0 actually used and given in column 2.

TABLE 9.2 Calculation of Errors in Film Thickness L and Refractive Index n Due to Errors in ϕ_0^a

Film Thickness, L (nm)	ϕ_0 (°)	Δ (°)	Ψ (°)	L @ 70° (nm)	n @ 70°
10	70.00	150.815	11.404	9.84	1.52
10	70.01	150.770	11.390	9.58	1.56
10	70.02	150.726	11.376	9.36	1.60
10	70.03	150.682	11.362	9.18	1.64
10	70.05	150.593	11.334	8.93	1.71
100	70.00	76.026	43.541	100.01	1.500
100	70.01	75.989	43.540	99.97	1.500
100	70.02	75.952	43.539	99.93	1.501
100	70.03	75.915	43.538	99.89	1.501
100	70.05	75.841	43.536	99.82	1.502

^{a}Calculation parameters are $L = 10$ and 100 nm, $n = 1.5$ on Si, $\lambda = 632.8$ nm, various ϕ_0, Δ, and Ψ calculated results in columns 3 and 4, and L and n are calculated from Δ and Ψ assuming $\phi_0 = 70°$.

In all cases except for the first lines for the 10- and 100-nm films, this introduces an error due to ϕ_0 used.

The first entries in column 5 and 6 are the values for L and n that were recalculated from the column 3 and 4 Δ and Ψ values. One might expect exact agreement between columns 1 and 5 and for n in column 6 to be exactly 1.5 since all the parameters and input values are the same. Rather for the 10-nm film there is a 1.6% difference in L and 1.3% in n that is due to the truncation of the Δ and Ψ values in the thousandths decimal place. This error is barely noticed for the 100-nm film. Since there is a limit of about 0.01° in the measurement of Δ and Ψ, errors of this order can be expected as minimum errors in the 10-nm film thickness range. When erroneous ϕ_0 values are used for the remaining 4 entries, for the 10 and 4 entries for the 100-nm films larger errors are seen that are always larger for the thinner films. For example, an error of 0.05° in ϕ_0 yield an error of almost 11% in L and 14% in n for the 10-nm film, and about 2% in L and 0.1% in n for the 100-nm film.

It is clear that ellipsometry is a very sensitive technique, sensitive to fractions of a monolayer on a surface so long as the ellipsometers are carefully aligned and calibrated. In addition there are other factors such as truncation errors and thickness regimes that also can affect the accuracy of the measurement.

9.5 OPTICAL MODELS

To extract useful optical properties for films and surfaces from the two ellipsometric measurables, Δ and Ψ, the optical system (substrate, film, ambient) being investigated needs to be approximated, and this approximation of reality is called a

9.5 OPTICAL MODELS

model and for ellipsometry an optical model. The extent to which the model is correct will determine the physical meaningfulness of ellipsometry determined properties. For example, to accurately extract optical properties from an ellipsometric measurement, it must first be determined whether the system under study is a bare or film-covered surface with sharp interfaces as shown in Fig. 9.5a or whether the single film is inhomogeneous as shown in Fig. 9.5b where the single inhomogeneous film on a substrate has sharp interfaces, and/or whether the films and/or interfaces have significant roughness as is shown in Fig. 9.5c, or multiple films each with different characteristics as shown in Fig. 9.5d. In some instances the interfaces between films and substrate may not be sharp due to interaction or diffusion. Once a model for the system is obtained, an algorithm can be formulated that considers reflection and refraction at each interface with different optical properties for each film. The model-based algorithm is then used to invert Equation 9.14 to obtain the desired optical properties that are in the model. It should be noted that film thickness is also considered to be a film optical property in that thickness determines the path length for the optical wave in the material, and the optical thickness is the product of thickness and refractive index, nL, and this product is found in Equation 9.7.

One tried and true strategy is to derive an optical model based on an often lengthy materials science study that typically involves a number of independent measurements using a variety of techniques. Another strategy for studying a new film–substrate system is to commence with one of the established models that appears to fit the situation under investigation and then try to determine how well the physical parameters obtained using the model agree with independently determined values. In essence this method is to make an educated guess based on previous experience and then provide an independent test. For example, in microelectronics where devices are fabricated using various films on semiconductor surfaces, the dielectric film thickness L on a semiconductor surface is usually of great interest, in particular SiO_2 films on Si. For this system a single film model is used because it is known that there is little interaction between Si and SiO_2, and for films greater than 10 nm the sharp interface model is applicable. Thickness L can not only be measured by ellipsometry but also by angle resolved x-ray photoelectron spectroscopy (ARXPS) and transmission electron microscopy, on cross-sectioned samples (XTEM). One could use the results from these independent techniques to compare with the ellipsometric optical model results. The models shown in Fig. 9.5 are ones that have already been used successfully for a number of systems of interest and will be referred to later in this chapter.

Figure 9.5a is a simple single film on substrate model in which both the film and substrate are discrete, planar, and homogenous. The film thickness L can vary upward from 0 nm, which is indicative of a smooth bare substrate. This model works very well for grown and deposited dielectric films on Si and for many other film substrate situations where nearly perfect smooth substrates are used, and where uniform stoichiometric films are possible, that is, where there is little interaction between film and substrate during film formation, and for thicker films (>10 nm) where interfacial effects are relatively small. In this single film model there are possibly seven model

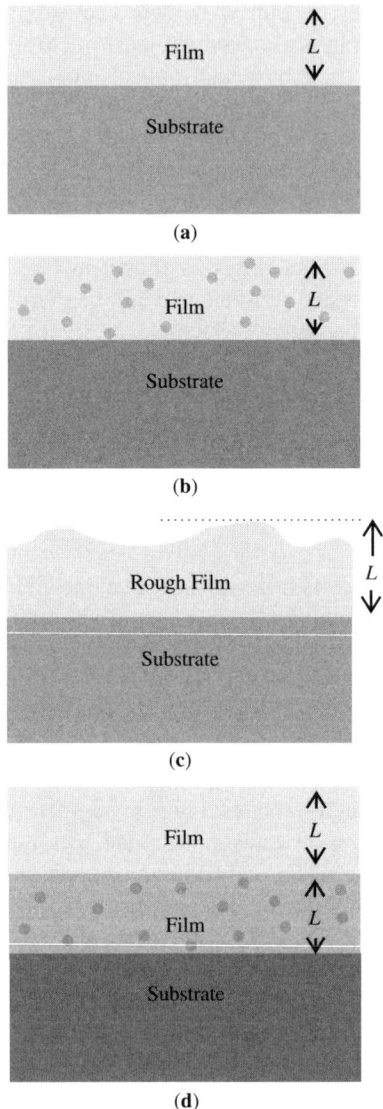

FIGURE 9.5 Optical models: (a) uniform single film with thickness L on a substrate, (b) inhomogeneous film on substrate, (c) rough film on substrate, and (d) multiple films on substrate.

parameters to be determined or supplied: n and k for ambient, film, and substrate and the film thickness L. With ambient air or vacuum the optical properties are known with $n = 1$ and $k = 0$ and can therefore be provided as input to the single film model algorithm. For other ambients, for example, liquids, the properties of the medium (n and k) would add another two unknowns. Typically, n and k for the substrate can be measured separately for a bare surface and then input the algorithm. With known substrate and ambient optical properties, a single ellipsometric measurement of Δ and Ψ would not yield the remaining three film parameters (n, k, L). To overcome this problem sometimes λ and a range of λ can be chosen so that k for the film is 0, that is, the film is optically transparent. This is possible and convenient for many dielectric films in the visible photon energy range. Another way around this problem of too many unknowns is to measure thickness by another technique and then use it as input to obtain n and k. Once found, n and k can be used as input to find L. Regression analyses can also be used that enable the best fit for underspecified systems, and this will be discussed below.

Figures 9.5b to 9.5d show the more complicated cases where the film(s) and/or the substrate are inhomogeneous. For Fig. 9.5b there are two different materials and therefore two different film compositions with commensurate properties. In this case there will be composition unknowns in addition to the unknowns in Fig. 9.5a. For many situations similar to Fig. 9.5b it has been shown that the Bruggeman effective medium approximation (BEMA) can yield excellent results for a variety of inhomogeneous film–substrate situations. Essentially, this model, which is discussed further below, considers that an inhomogeneous layer is discrete, that is, with sharp interfaces as for the simpler models, but with a dielectric response that is the composite of the dielectric responses of the individual components. The manner in which the individual components with their respective contributions to the total dielectric response, $<\varepsilon>$ are summed, varies according to assumptions made based about the state of aggregation. To understand this more clearly and without lengthy derivations, we proceed quickly to the final forms used for optical response calculations. Justification for the application of the various models is in the original literature as are derivations. The various effective medium approximations that can be used for a variety of applications are discussed after other useful approximations are introduced.

It is often useful to approximate the dispersion of ε (the changes in ε with photon energy or ν or λ) or changes in $\tilde{N}$ for a film or substrate. Typical spectra for a material in terms of ε and $\tilde{N}$ are shown in Fig. 9.6 for the visible photon energy range. Figure 9.6a shows a strong absorption near 5 eV with both ε_1 and ε_2 slowly varying at lower energies away from the absorption. Figure 9.6b shows the same spectrum but in terms of n and k. At energies lower than the adsorption k is close to zero and n is slowly varying. For the subabsorption photon energies where k can be safely approximated as 0, the Cauchy formula is often used to approximate $n(\lambda)$ as follows:

$$n(\lambda) = A + \frac{B}{\lambda^2} + \frac{C}{\lambda^3} \qquad (9.20)$$

where A, B, and C are Cauchy parameters. For materials or at photon energies where absorption is in evidence, the Cauchy formula is not a good approximation and a

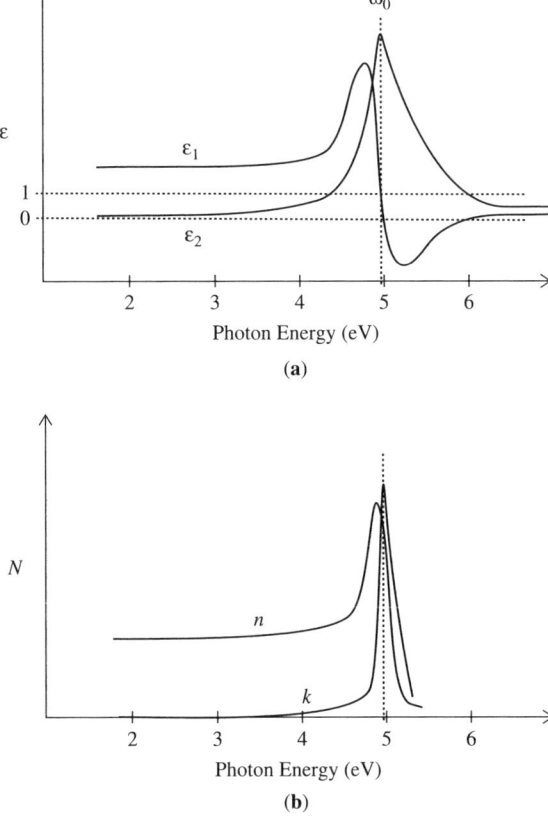

FIGURE 9.6 Optical properties of a typical dielectric in the visible light range: (a) in terms of the optical dielectric functions ε and real and imaginary components ε_1 and ε_2, respectively, and (b) in terms of the refractive index n and absorption constant k.

useful formula that includes optical absorption, the Lorentz oscillator formula, is used and for a single oscillator is given as

$$\varepsilon = 1 + \frac{4\pi e^2}{m\left(\omega_0^2 - \omega^2 - j\Gamma\omega\right)} \qquad (9.21)$$

where m is the oscillator reduced mass, ω_0 is a oscillator resonance frequency, ω is the probe frequency, and Γ is a broadening parameter for the resonance. For more than one oscillator the Lorentz formula is a sum as

$$\varepsilon = 1 + \frac{4\pi e^2}{m} \sum_i \frac{N_i}{\omega_i^2 - \omega^2 - j\Gamma_i\omega} \qquad (9.22)$$

9.5 OPTICAL MODELS

where N_i is the number of oscillators per volume with resonance at ω_i. For $\sum_i N_i = N$ and ω_0 is the resonance frequency, ε_1 and ε_2 are as follows:

$$\varepsilon_1 = 1 + \frac{4\pi Ne^2(\omega_0^2 - \omega^2)}{m\left[(\omega_0^2 - \omega^2)^2 + \Gamma^2\omega^2\right]} \quad \text{and}$$

$$\varepsilon_2 = \frac{4\pi Ne^2 \Gamma \omega}{m\left[(\omega_0^2 - \omega^2)^2 + \Gamma^2\omega^2\right]}$$

(9.23)

Figure 9.6a shows a single oscillator with Γ given by the width of the resonance at ω_0 at half maximum. The region near ω_0 where ε_1 decreases rapidly and even becomes negative and n goes through a maximum is called the anomalous dispersion region. The Cauchy and/or Lorentz formulas are used to approximate various films in single or multiple film models for nonabsorbing or absorbing films, respectively. The remaining issue is the modeling of inhomogeneous films and rough interfaces, and this can be done using effective medium approximations (EMAs).

The effective medium approximation formulas used for approximating inhomogeneous films are now discussed and are derived by considering how the incident electromagnetic light waves interact with the inhomogeneous material. Thus, different formulas are obtained for different materials and different materials circumstances such as mixtures. The starting point is the Clausius–Massotti equation, which connects a microscopic material property, the polarizability, α, to the macroscopic dielectric response, ε:

$$\frac{\varepsilon - 1}{\varepsilon + 2} = \frac{4\pi}{3} n\alpha$$

(9.24)

In this formula n is the number of polarizable species in the volume of material probed or the density. This equation obtains for a pure substance and is derived from a consideration of the local electric fields. For a material that is approximated as a heterogeneous mixture of polarizable points (atoms/molecules) a and b in vacuum and each with a different polarizability, α_a and α_b, then the Lorentz–Lorenz equation applies:

$$\frac{\varepsilon - 1}{\varepsilon + 2} = f_a \frac{\varepsilon_a - 1}{\varepsilon_a + 2} + f_b \frac{\varepsilon_b - 1}{\varepsilon_b + 2}$$

(9.25)

where the f's are the volume fractions of constituents.

If a and b are not points in a vacuum, but rather are in a host with a dielectric response ε_h, then the Maxwell–Garnet equation is obtained:

$$\frac{\varepsilon - \varepsilon_h}{\varepsilon + 2\varepsilon_h} = f_a \frac{\varepsilon_a - \varepsilon_h}{\varepsilon_a + 2\varepsilon_h} + f_b \frac{\varepsilon_b - \varepsilon_h}{\varepsilon_b + 2\varepsilon_h}$$

(9.26)

If a is considered as the host and thus we have a mixture of points of b in a, then one obtains

$$\frac{\varepsilon - \varepsilon_a}{\varepsilon + 2\varepsilon_a} = f_b \frac{\varepsilon_b - \varepsilon_a}{\varepsilon_b + 2\varepsilon_a} \qquad (9.27)$$

The Maxwell–Garnet relationship has found application in the field of cermets, which are ceramic composites composed of hard brittle ceramic particles in a connected ductile phase. If $f_a \approx f_b$, that is, there are ample amounts of both materials present, and we let $\varepsilon_h = \varepsilon$ with a as host the result is

$$f_a \frac{\varepsilon_a - \varepsilon}{\varepsilon_a + 2\varepsilon} + f_b \frac{\varepsilon_b - \varepsilon}{\varepsilon_b + 2\varepsilon} = 0 \qquad (9.28)$$

which is the Bruggeman effective medium approximation (BEMA), as was mentioned above, and this formula is generalized for i constituents as follows:

$$\sum_i f_i \frac{\varepsilon_i - \varepsilon}{\varepsilon_i + 2\varepsilon} = 0 \qquad (9.29)$$

The BEMA assumes mixtures on a scale smaller than the wavelength of light, but that each constituent retains its original dielectric response. One can imagine that this model might be appropriate for mixed-phase films, large amounts of impurities in substrates and damage, and roughness and indeed applications to these cases have been successful in many instances.

Now with these approximation tools in hand a regression analysis can be used to extract desired optical properties from Δ and Ψ measurements at numerous photon energies [spectroscopic ellipsometry (SE) will be discussed further below]. One recipe for this analysis is shown in Fig. 9.7 and is described as follows:

1. Measure Δ, Ψ at various λ's and from Equation 9.14 obtain ρ_{exp}. This experimentally determined quantity, ρ_{exp}, provides one input to the regression analysis and is displayed in the top box in Fig. 9.7.
2. As discussed above, the optical model is deduced from independent experiments and measurements and/or good guesses, and this is seen as the second box in Fig. 9.7. The identity of each film, its constituents, and state of aggregation (homogeneous, inhomogeneous, rough, etc.) are determined.
3. Measure or find in the literature the ε's ($\tilde{N}$'s) for all the constituents. Steps 2 and 3 are summarized in the second box from the top in Fig. 9.7.
4. The third box from the top indicates that from items 2 and 3 above ρ_{calc} is calculated.
5. The values for ρ_{exp} and ρ_{calc} are compared as shown in the fourth box from the top where N is the number of measurements (the number of λ's) and P is the number of parameters.

9.5 OPTICAL MODELS   273

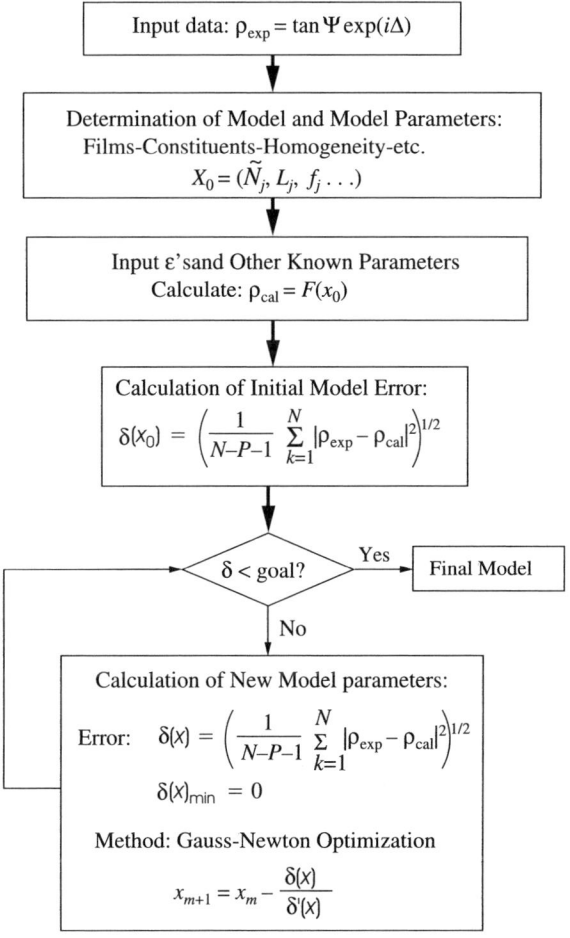

FIGURE 9.7 Regression analysis flowchart. Δ and Ψ are measured varibales and ρ is the complex reflection coefficient. $\tilde{N}_j$ is the complex index for the j component, L_i is film thicknesses, and f_i are the volume fractions of constituents, N is the number of measurements, and P the number of parameters and δ the tolerable error.

6. The difference is compared in the fifth box with a tolerable difference between the model and experimental results. If this result is satisfactory, the calculation ends and the parameter values obtained are deemed to be correct.
7. If not, then the process repeats with a change of parameters, as shown in the sixth box. This first comparison derives from the initial values for the parameters in the model. There are many mathematical methods that could be used for the minimization routines with one shown in the last box. Also, it should be remembered that the parameter values obtained are only as good as the model. It is possible to get a good fit to a physically or chemically

incorrect model, and correlation among parameters is also a source for error and often determined. With a well-substantiated optical model, a regression analysis has been found to be useful in obtaining desired materials parameters and properties.

9.6 MANUAL AND AUTOMATED ELLIPSOMETRY TECHNIQUES

9.6.1 Introduction

The objective of this section on technique is to present a sufficient amount of analysis and hardware information for the reader to obtain a basic understanding of how polarized light propagates through the ellipsometer optical system and what kind of apparatus is required for ellipsometric measurements. To this end it is instructive to begin with manual single wavelength ellipsometry and then to make appropriate changes for spectroscopic ellipsometry and automation. Figure 9.8 shows a manual ellipsometer in the polarizer (P), compensator (C), sample (S), analyzer (A), or PCSA, configuration. The light source is typically a low power laser such as a He–Ne laser at 632.8 nm or a wide-band source such as a Xe high pressure lamp with filters to pass only the desired wavelength. For spectroscopic ellipsometry discussed below the use of a wide bandwidth light source is required. For the visible spectrum (200 to 900 nm) it is usual to use Xe short-arc high pressure lamps with appropriate collimating optics. In the configuration shown in Fig. 9.8 the usual measurement is the "null" measurement, where for a given sample the light intensity at the detector is adjusted to zero by adjusting the P, C, A azimuths with a sample in place. This null condition plus knowledge of λ, ϕ_0, and the P, C, and A azimuths enable a deduction of what polarization the sample must have imparted to the light, which in combination with the optical elements yield zero light intensity at the detector, I_D. With knowledge of how much polarization has been imparted by the sample, and with an optical model for the sample as discussed above, the optical properties of the sample can be obtained. For null ellipsometry the following condition is sought:

$$I_D = 0 = G\mathbf{E}_{AO}\mathbf{E}^*_{AO} \tag{9.30}$$

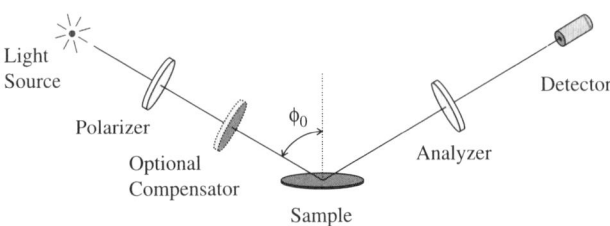

FIGURE 9.8 Polarizer, compensator, sample, and analyzer PCSA ellipsometer configuration.

9.6 MANUAL AND AUTOMATED ELLIPSOMETRY TECHNIQUES

where G is a constant and $\mathbf{E}_{AO}$ is the electric field after the analyzer and $\mathbf{E}_{AO}^*$ is the complex conjugate of $\mathbf{E}_{AO}$. To find expressions for the intensity we need to follow the light as it propagates from the source through P, C, and reflects from S and then through A.

When light interacts with an optical element, the polarization state of the light changes. If we are only interested in the change in polarization state of the light before and after it interacts with an (ideal) optical element (P, C, S, or A), the effect of an optical element on the polarization of light can be represented by a 2×2 matrix $\mathbf{T}$ called a Jones matrix, which can express the change as follows:

$$\mathbf{E}_o = \mathbf{T}\mathbf{E}_i \tag{9.31}$$

where $\mathbf{E}_i$ and $\mathbf{E}_o$ are the field vectors of the input and output waves, respectively, and the Jones matrix of the optical element is given as:

$$\mathbf{T} = \begin{pmatrix} T_{11} & T_{12} \\ T_{21} & T_{22} \end{pmatrix} \tag{9.32}$$

Figure 9.9a shows the interaction using an x, y reference frame. Figure 9.9b shows several optical elements, each represented by a different $\mathbf{T}$, and for a light wave propagating through all the elements and by using matrix algebra a combination matrix $\mathbf{T}_{comb}$ is generated and expressed with the following relationship:

$$\mathbf{E}_o = \mathbf{T}_N \mathbf{T}_{N-1} \cdots \mathbf{T}_{II} \mathbf{T}_I \mathbf{E}_i = \mathbf{T}_{comb} \mathbf{E}_i \tag{9.33}$$

The Jones matrices for some most frequently encountered optical elements in ellipsometry systems will be presented next.

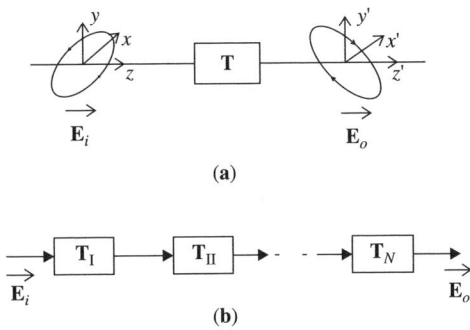

FIGURE 9.9 The effect of (a) one optical element or (b) several elements in series as represented by boxes (with T inside), on the polarization state of light where each box is represented by a charactcristic Jones matrix and with $\mathbf{E}_i$ and $\mathbf{E}_o$ the incident light and out light electric field, respectively.

276 ELLIPSOMETRY AND OPTICAL PROPERTIES

9.6.2 Isotropic Media

When light of wavelength λ propagates through an isotropic medium of thickness L and refractive index n (see Fig. 9.10a), the exiting **E** field, $\mathbf{E}_o$, is given by a Jones matrix operating on the incident $\mathbf{E}_i$ as follows:

$$\mathbf{E}_o = \begin{pmatrix} e^{-i2\pi nL/\lambda} & 0 \\ 0 & e^{-i2\pi nL/\lambda} \end{pmatrix} \mathbf{E}_i \qquad (9.34)$$

Notice the similarity of the exponential in Equation 9.34 with Equation 9.7, where the product nL appears, and this product is called the optical path for a material. If

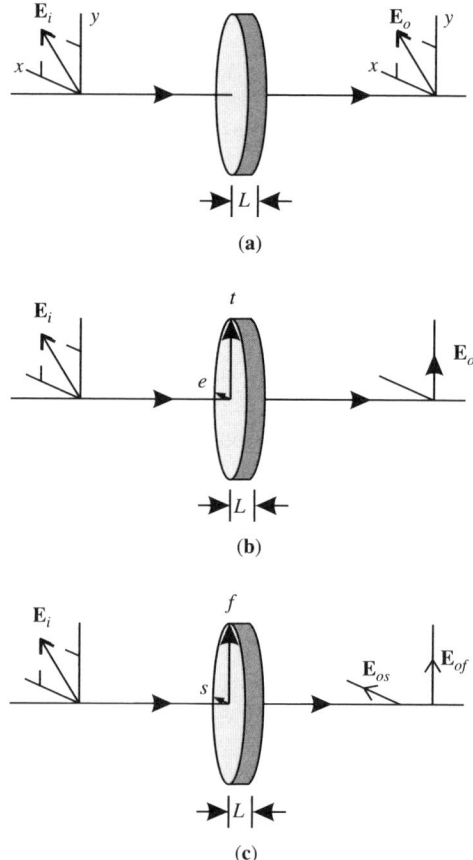

FIGURE 9.10 Effects of (**a**) an isotropic medium, (**b**) a linear polarizer, and (**c**) a compensator, on the polarization state of a linearly polarized light. $\mathbf{E}_i$ and $\mathbf{E}_o$ are the input and output electric fields, respectively, t is the transmission axis of the polarizer, and f the fast transmission axis of the compensator and L is the thickness of the optical element.

the medium is isotropically absorbing, the effect can be represented simply by replacing the refractive index n in the above matrix with the complex refractive index $\tilde{N}$ of the medium.

9.6.3 Linear Polarizer

A linear polarizer is the main element in an ellipsometer system, and it converts incident light of any polarization state into linearly polarized light at its output. Additionally, a polarizer is used to resolve the polarization state of light reflected from a sample and before the detector, and in this position the polarizer is usually called an analyzer. A linear polarizer has two orthogonal axes, that is, a transmission axis and an extinction axis, as indicated by t and e, respectively, in Fig. 9.10b. When unpolarized or, in general, elliptically polarized light passes through a linear polarizer, the light is transformed into linearly polarized light with an electric field vector that is parallel to the transmission axis of the polarizer, as shown in Fig. 9.10b. The effect of a linear polarizer (of thickness L and index n) can be represented by the following Jones matrix:

$$\mathbf{T} = K \begin{pmatrix} 1 & 0 \\ 0 & 0 \end{pmatrix} \tag{9.35}$$

where $K = e^{-i2\pi n L/\lambda}$.

A linear polarizer is characterized by a parameter called the extinction ratio, which is the ratio of intensity of light along the extinction and transmission directions. For ellipsometry high extinction ratios are required, and suitable polarizers are commercially available with extinction ratios of 10^{-6}. Most linear polarizers operate based on three different physical mechanisms: birefringence, dichroism, and reflection. Among them, birefringence prisms are the most practical polarizers, and these use the principle of double refraction in transparent uniaxially or biaxially anisotropic crystals. A variety of polarizers have been built based on the double refraction mechanism, including Wollaston, Rochon, Glan-Thompson, and Glan-Taylor prisms with a Glan-Taylor prism shown in Fig. 9.11. This prism consists of two sections of (usually) calcite crystals that are separated by a narrow air gap. The Glan-Taylor prism differs from the Glan-Thompson in that the latter uses a refractive index matching glue instead of an air gap. The result is a better extinction ratio but a narrower optical range due to absorption of light by the glue. The optical axes of the two crystals are parallel to each other and to the entrance plane and exit end face of the prism. The incident light perpendicular to one of the end faces of the prism can be resolved into two components $\mathbf{E}_e$ (extraordinary) and $\mathbf{E}_o$ (ordinary), which are parallel and perpendicular to the optical axis, respectively. Because the two components have different refraction indexes in calcite, the incident angle at the calcite/gap interface can be adjusted by the prism dimension design, so as to satisfy the conditions that $\mathbf{E}_o$ is totally reflected at an incident angle greater than the critical angle, while $\mathbf{E}_e$ is highly transmitted by striking the interface at Brewster's angle. The internally reflected light can be absorbed by blackening the interior side of prism, while the

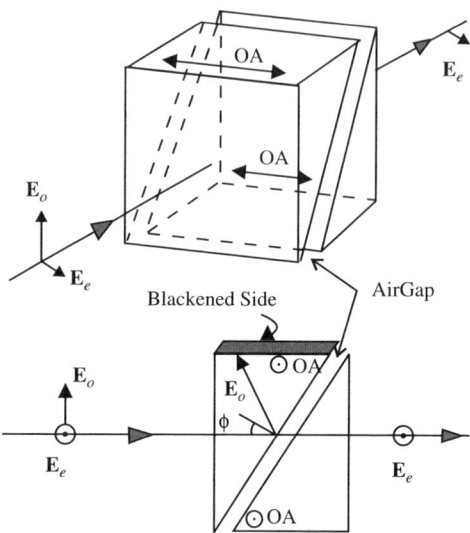

FIGURE 9.11 Glan–Taylor polarizing prism and with ordinary E_o and exraordinary E_e components indicated.

other linear polarized light component E_e exits the end face of prism with a slight deviation, but still nearly parallel to the incident beam.

The Brewster angle is important for the reflection of polarized light, and it is obtained from a consideration of Poynting's theorem where the optical power or rate of energy transmitted S is given by the cross product of the electric E and magnetic B field vectors as

$$\mathbf{E} \times \mathbf{B} = \mathbf{S} \qquad (9.36)$$

The direction of S is the propagation direction for the light. Thus, S is a vector orthogonal to the plane formed by E and B. In the plane of E and B there is no electromagnetic energy. If a component of the E field of light is in the direction of the propagation of the light, the magnitude of the component of E will be 0. Hence $I = \mathbf{EE}^* = 0$. The angle at which the E_p component goes to zero is called the Brewster angle ϕ_B and is given as

$$\phi_B = \tan^{-1} \tilde{N} \qquad (9.37)$$

where $\tilde{N}$ is the complex refractive index as before. For the case where $k = 0$, $n = \tilde{N}$ (transparent material), the calculation yields a true Brewster's. For $k \neq 0$, E will not be 0, hence $I = \mathbf{EE}^* \neq 0$ and the angle ϕ_B ($=\tan^{-1} \tilde{N}$) is called the pseudo-Brewster angle.

When light at some incident angle (i) is reflected from a surface, the difference in refractive index across the interface causes the direction of the transmitted light (t) to

9.6 MANUAL AND AUTOMATED ELLIPSOMETRY TECHNIQUES

be altered as given by Snell's law of refraction:

$$n_i \sin \theta_i = n_t \sin \theta_t \tag{9.38}$$

It should be noted that the incident angle θ_i and the refracted angle θ_t are measured from the surface normal, but ϕ_B is measured from the surface. Now considering the $\mathbf{E}_p$ and $\mathbf{E}_s$ components of polarized light where the $\mathbf{E}_p$ component is in the POI and the $\mathbf{E}_s$ is orthogonal to the POI, if the incident light is bent so that the $\mathbf{E}_p$ component is in the direction of the reflected ray, then from Poynting's theorem this component will be zero, that is, no $\mathbf{E}_p$ component. In effect the surface at the Brewster angle provides polarization.

9.6.4 Compensator

A compensator is an anisotropic element in which light travels at different speeds in different directions, thereby causing a phase change for the light exiting the compensator. The light traversing the compensator travels at different speeds due to different refractive indices along two axes, a fast (f) and a slow (s) axis that are orthogonal. Therefore, when light passes through a linear compensator, the phase of the electric vector that is parallel to the slow axis is retarded by δ_c and the amplitude attenuated by T_c with respect to the component parallel to the fast axis, as illustrated in Fig. 9.10c. The Jones matrix for a compensator can thus be written as

$$\mathbf{T} = K_c \begin{pmatrix} 1 & 0 \\ 0 & \rho_c \end{pmatrix} \qquad \rho_c = T_c e^{i\delta_c} \tag{9.39}$$

where K_c is a constant that accounts for the attenuation and phase shift along the fast and slow axes of the compensator.

A compensator, also called a retarder, then is an optical component that introduces a phase shift between orthogonally polarized components without affecting their relative amplitude. Retarders used in ellipsometry are linear retarders that have two light propagation directions: a fast axis and a slow axis. The component of incident light parallel to the slow axis is retarded in phase relative to the component along the fast axis. When the phase retardation is $\pi/2$, the retarder is called a quarter-wave retarder that is often used in ellipsometry. There are two types of retarders based on two different mechanisms: birefringent retarders and reflection retarders (the Fresnel rhomb). The more common birefringent retarder has two refractive indices n_o the ordinary and n_e the extraordinary refractive index. Light propagating through the component will have speeds of c/n_o and c/n_e, along the two directions, respectively. For a component of thickness L, this difference in propagation speed will result in a phase shift δ given as

$$\delta = \frac{2\pi L(n_o - n_e)}{\lambda} \tag{9.40}$$

Then for a material with known n_o and n_e, L can be varied to produce a specific desired retardation.

With the Jones matrices for these elements we can now formulate the propagation of light through an ellipsometer and keep track of the changing polarization state.

9.6.5 Detectors

Three types of detectors are commonly used for ellipsometry: photomultiplier tubes (PMTs), semiconductor photodiodes, and charge-coupled devices (CCD) in arrays. PMTs have been the primary detector for ellipsometry for many years. However, they exhibit significant polarization sensitivity and nonlinear intensity response. Semiconductor photodiode detectors are inexpensive, insensitive to polarization state, and linear over a broad range of beam intensity. Among semiconductor photodiode detectors, silicon photodiodes are most commonly used detectors in the ultraviolet–visible (UV–VIS) range, while InGaAs and HgCdTe detectors are often used for near-infrared (NIR) and infrared (IR) applications. A silicon diode array allows multiple wavelengths to be detected simultaneously rather than sequentially, which increase the data acquisition speed dramatically. The third type of detector is the CCD array, which is becoming popular for ellipsometry detection.

9.6.6 Null Ellipsometry

For a null ellipsometry measurement the optical elements are often arranged as shown in Fig. 9.8 in the PCSA configuration with a light source and photodetector. From the Jones matrix formulation above, the light exiting from a linear polarizer can be represented by a Jones vector as follows:

$$\mathbf{E}_{PO}^{te} = A_c \begin{pmatrix} 1 \\ 0 \end{pmatrix} \quad (9.41)$$

where A_c is the amplitude attenuation constant through the polarizer element. The subscript index PO refers to the polarizer (P) output (O), and the superscript te refers to the transmission–extinction reference system of the polarizer, as was shown in Fig. 9.10b. This output wave is the input for the compensator. However, the fast-slow (fs) reference system of the compensator is not in general aligned with that of the polarizer. Hence a rotation of coordinate system from te of polarizer to fs of the compensator is necessary, as shown in Fig. 9.10c. This can be accomplished by multiplying a coordinate rotation matrix $R(\alpha)$, where α is the angle of rotation from the old system to the new one as

$$R(\alpha) = \begin{pmatrix} \cos \alpha & \sin \alpha \\ -\sin \alpha & \cos \alpha \end{pmatrix} \quad (9.42)$$

9.6 MANUAL AND AUTOMATED ELLIPSOMETRY TECHNIQUES

The input wave for the compensator after rotation is then

$$\mathbf{E}_{CI}^{fs} = R(P - C)\mathbf{E}_{PO}^{te} = A_c \begin{bmatrix} \cos(P - C) \\ \sin(P - C) \end{bmatrix} \qquad (9.43)$$

where P and C are the azimuth angles of the polarizer and compensator, respectively, defined by rotation around the transmission axis of the polarizer or fast axis for the compensator rotating counterclockwise looking into the optical system from the detector. The output of the compensator is obtained by applying the Jones matrix, $\mathbf{T}_c^{fs}$, for the compensator to the input wave to yield:

$$\mathbf{E}_{CO}^{fs} = \mathbf{T}_c^{fs}\mathbf{E}_{CI}^{fs} = K_c A_c \begin{pmatrix} \cos(P - C) \\ \rho_c \sin(P - C) \end{pmatrix} \qquad (9.44)$$

It can be shown that the combination of a polarizer and compensator yields light exiting the compensator that is in general elliptically polarized. In fact, all possible elliptical polarization states can be obtained exiting a PC combination by setting the compensator (a quarter-wave retarder) with its fast axis at 45° and only rotating the polarizer. This is a typical configuration for null ellipsometry. The output of the compensator then becomes the input to the sample surface, which has an x-y reference system, so another rotation using Equation 9.42 from the fs system of compensator to the x-y system of sample is needed and is expressed as

$$\mathbf{E}_{SI}^{xy} = \mathbf{E}_{CO}^{xy} = R(-C)\mathbf{E}_{CO}^{fs} \qquad (9.45)$$

The light will then interact with the sample and the effect of the reflection from the sample surface can be expressed by another Jones matrix, $\mathbf{R}_S^{xy}$, as follows:

$$\mathbf{E}_{SO}^{xy} = R_S^{xy} S_{SI}^{xy} \quad \text{where} \quad R_S^{xy} = \begin{pmatrix} R_p & 0 \\ 0 & R_s \end{pmatrix} \qquad (9.46)$$

where R_p and R_s are as before the reflection coefficients for the p and s components of the light and can be derived from the Fresnel equations with known angle of incidence and refractive indices of the media involved (see Equations 9.6 to 9.14). The final optical element before reaching the photodetector is the analyzer, which is also a linear polarizer and thus has a t-e reference frame. Therefore, the light reflected from the sample surface is rotated to this t-e reference frame (from the x-y frame of the sample) as follows:

$$\mathbf{E}_{AI}^{te} = \mathbf{E}_{SO}^{te} = R(A)\mathbf{E}_{SO}^{xy} \qquad (9.47)$$

where A is the azimuth angle of the analyzer similarly defined in the same way as P for the polarizer. The output wave from the analyzer can thus be obtained by applying the Jones matrix that is characteristic of an analyzer (which has the same form as that for a polarizer):

$$\mathbf{E}_{AO}^{te} = \mathbf{T}_A^{te} \mathbf{E}_{AI}^{te} \quad \text{where} \quad \mathbf{T}_A^{te} = K_A \begin{pmatrix} 1 & 0 \\ 0 & 0 \end{pmatrix} \tag{9.48}$$

in which K_A is the attenuation constant.

In summary, the propagation of light from source to detector can be expressed as follows:

$$\mathbf{E}_{AO}^{te} = \mathbf{T}_A^{te} R(A) \mathbf{T}_S^{xy} R(-C) \mathbf{T}_C^{fs} R(P-C) A_c \begin{pmatrix} 1 \\ 0 \end{pmatrix} = K \begin{pmatrix} E_t \\ 0 \end{pmatrix} \tag{9.49}$$

where

$$K = K_A K_C K_P \tag{9.50}$$

and

$$E_t = R_p \cos A \left[\cos C \cos(P-C) - \rho_c \sin C \sin(P-C) \right]$$
$$+ R_s \sin A \left[\sin A \cos(P-C) + \rho_c \cos C \sin(P-C) \right] \tag{9.51}$$

Finally, the intensity of the light collected by the photodetector is

$$I = \left| E_{AO}^{te} \right|^2 \tag{9.52}$$

which is a function of P, C, A, ρ_c, R_p, and R_s. For a null ellipsometry system, usually a quarter-wave plate is chosen as the compensator, so ρ_c is known (because T_c and ρ_c are known from calibration). During the measurements, P, C, and A are arranged so that the light intensity detected by the photodetector becomes zero (null), which means that $E_t = 0$. With this condition, we finally obtain from Equation 9.51 the following relationship:

$$\rho = \frac{R_p}{R_s} = -\tan A \left[\frac{\tan C + \rho_c \tan(P-C)}{1 - \rho_c \tan C \tan(P-C)} \right] \tag{9.53}$$

where ρ was defined in Equation 9.9 as the complex reflection coefficient for the sample. Therefore, it is seen from Equations 9.53 with 9.11 to 9.14 that null ellipsometry measures Δ and Ψ, which are the relative changes in phase and amplitude of the two eigenpolarization states when light is reflected from the sample surface.

9.6.7 Rotating Element Ellipsometry

There are several schemes that are used to automate the ellipsometric measurement. Early attempts to automate the ellipsometry measurement simply used motors and

9.6 MANUAL AND AUTOMATED ELLIPSOMETRY TECHNIQUES

detectors to find the null condition for P and A and then employ the formulas in the previous section. Following this more sophisticated modulation techniques were developed. In principle, one can modulate any of the optical components and even add new ones to modulate. In practice, there is a literature on several successful schemes such as rotating analyzer ellipsometry (RAE), rotating polarizer (RPE), rotating compensator (RCE), and phase modulation ellipsometry (PME), with some schemes better than others for certain circumstances. To illustrate the general idea, RAE is described in more detail.

An RAE system, in principle, is the same as shown in Fig. 9.8 but where the analyzer is driven to continually rotate with a motor. In null ellipsometry the azimuth angles P, C, A are varied, in order to zero the light intensity exiting the analyzer with the information about the sample contained in the values of P, C, A, ρ_c, and ϕ. For RAE, the azimuth angles P and C (not usually used for spectroscopic RAE) are not changed during the rotation of A. At any time t the intensity of the light exiting from the analyzer rotating with frequency ω is a function of the instantaneous orientation of the analyzer from the zero azimuth position δ, in addition to the fixed values of P, C, A, ρ_c, and ϕ. Because the null condition is not required in rotating analyzer ellipsometry, the compensator is optional in the setup and often omitted. For a PSA arrangement of a RAE system, the electric field vector of the light impinging upon photodetector can be expressed similarly to the previous Jones matrix treatment as follows:

$$\mathbf{E}_{PMT} = E_0 \begin{pmatrix} 1 & 0 \\ 0 & 0 \end{pmatrix} \begin{pmatrix} \cos A & \sin A \\ -\sin A & \cos A \end{pmatrix} \begin{pmatrix} R_p & 0 \\ 0 & R_s \end{pmatrix}$$
$$\times \begin{pmatrix} \cos P & \sin P \\ -\sin P & \cos P \end{pmatrix} \begin{pmatrix} 1 \\ 0 \end{pmatrix} \quad (9.54)$$

where E_0 is a constant and now for RAE A is given as

$$A = \omega t + \delta \quad (9.55)$$

where δ is a phase constant offset as mentioned above. Equation 9.54 can be expanded as

$$\mathbf{E}_{PMT} = E_0 (R_p \cos A \cos P + R_s \sin A \sin P) \begin{pmatrix} \cos A \\ \sin A \end{pmatrix} \quad (9.56)$$

With the electric field at the detector given by Equation 9.56 and then using Equation 9.52 ($I = |E_{PMT}|^2$), the intensity is expressed as

$$I = I_0 \left[1 + \left(\frac{\tan^2 \Psi - \tan^2 P}{\tan^2 \Psi + \tan^2 P} \right) \cos 2A + \left(\frac{2 \tan P \cos \Delta \tan \Psi}{\tan^2 \Psi + \tan^2 P} \right) \sin 2A \right] \quad (9.57)$$

or

$$I = I_0 (1 + \alpha \cos 2A + \beta \sin 2A) \quad (9.58)$$

The constants I_0 (the dc signal component), α and β (2nd Fourier coefficients) can be obtained from a Fourier analysis of the detected signal, and thus the two ellipsometric parameters Δ and Ψ are obtained as

$$\tan \Psi = |\tan P|\sqrt{\frac{1+\alpha}{1-\alpha}}$$

$$\cos \Delta = \frac{\beta}{\sqrt{1-\alpha^2}} \qquad (9.59)$$

The RAE system is readily automated and usually in such a way that data acquisition and analysis are done simultaneously with a computer. This automation is especially important for spectroscopic *in situ* real-time monitoring. Typical automation hardware configuration is illustrated in Fig. 9.12a, in which an optical angular encoder synchronized with the rotating analyzer is used to digitize the signal after the analyzer and the start pulse at the beginning of each analyzer revolution indicates the period of the data signal for each analyzer revolution. In the case shown the optical encoder generates one start pulse and 360 equally spaced data pulses so that the analog-to-digital (A/D) converter digitizes the analog output signal from the detector (in this case a photomultiplier tube, PMT) into 360 data points at the trigger of the start pulse as shown in Fig. 9.12b. The digitized signal is then collected by a computer that also performs the Fourier analysis to obtain α and β (and thus Δ and Ψ), and then the modeling regression analysis as was outlined above. The overall speed of RAE measurements is appropriate for many processes interesting in microelectronics such as film growth or removal (etching) in real time using PC class computers. Below spectroscopic ellipsometry (SE) is discussed and SE often uses scanning monochromators. The usual scanning monochromators may take minutes to scan the visible spectrum that is used for most SE with shorter times for more limited spectral scans. Within the last 10 years detector arrays (both one-dimensional and two-dimensional) have been used to obtain complete spectra in seconds and less, thereby increasing the kinds of processes monitored using ellipsometry. The detector arrays are usually diodes or charged-coupled devices (CCDs). Phase modulation ellipsometry (PME), which usually provides a varying phase to the incident polarized light, can even be faster.

9.6.8 Spectroscopic Ellipsometry

For simple materials systems with one or two unknowns such as for a transparent film ($k = 0$) on a known substrate, single-wavelength ellipsometry (SWE) provides sufficient measurables Δ and Ψ at one known wavelength λ and angle of incidence ϕ_0 to yield n and L for the film. However, when the system becomes more complicated with multiple unknowns, as in a film–interface–substrate system, the number of parameters that can be obtained from SWE are no longer adequate to reliably quantify the physical properties of the system. For this example the film would have three

9.6 MANUAL AND AUTOMATED ELLIPSOMETRY TECHNIQUES

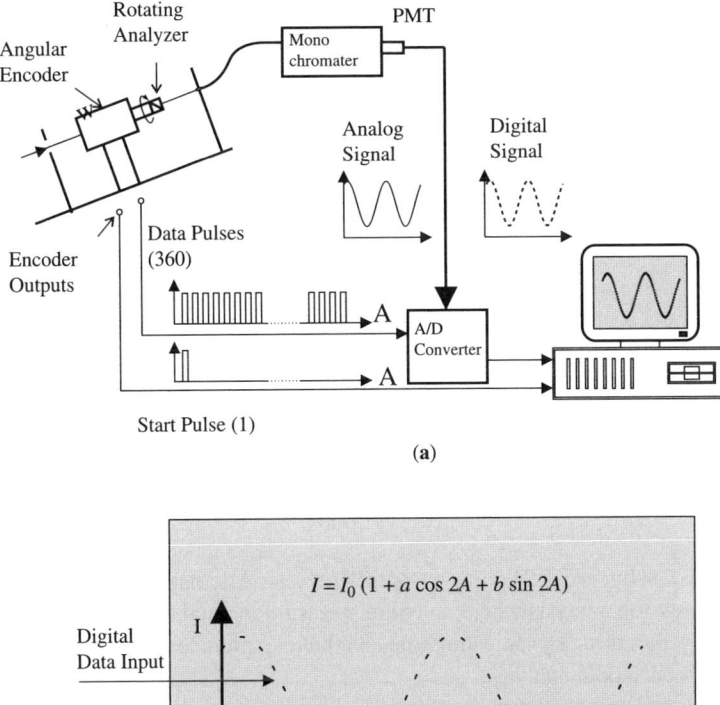

FIGURE 9.12 Rotating analyzer ellipsometry (RAE): (**a**) hardware and (**b**) data output. The hardware in (**a**) shows the analog output intensity variation as a result of the rotating analyzer, and the digital output shown in (**b**) is obtained as a periodic wave for Fourier analysis.

unknowns (n, k, and L), and likewise for the interface and the substrate has n and k unknown. This simple example has 8 unknowns with only the two measurables Δ and Ψ. So, in order to obtain more information from a multilayer or a thin-film–interface system, multiple-wavelength ellipsometry, that is, spectroscopic ellipsometry (SE), is employed. However, it should be noted that SE is nothing more than SWE that was described above but repeated at a number of selected wavelengths. The Δ and Ψ measurements at each wavelength can be evaluated and the results tabulated as a

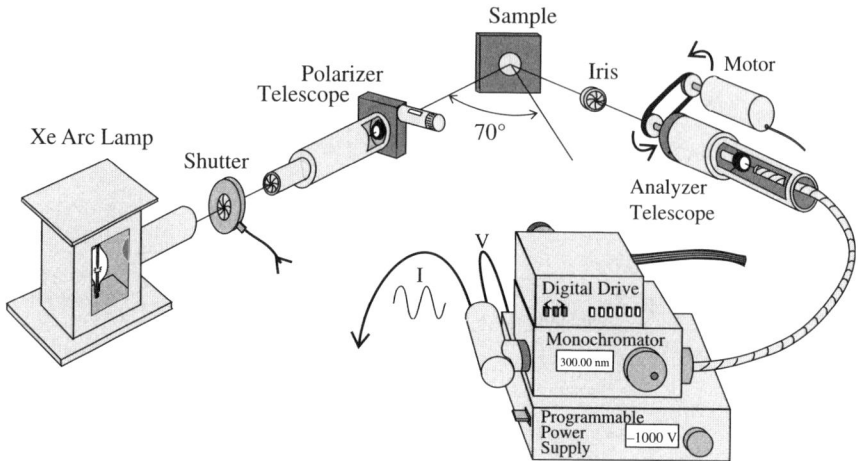

FIGURE 9.13 Components for a rotating analyzer ellipsometer.

function of λ. This is called point-by-point analysis. Alternatively, the results can be used in regression analysis where an entire spectrum is fitted to a known spectrum by minimizing fit errors by adjusting the available parameters as was discussed in Section 9.5 on model analysis.

Spectroscopic ellipsometry, as shown in Fig. 9.13, can be accomplished using a wide-band light source that can produce the desired spectrum (a Xe high pressure lamp for the UV–visible spectrum), a polarizer, sample stage, and a rotating analyzer (discussed above) (PCSA), and then an automated monochromator followed by a detector. The monochromator drive is controlled by a computer that also controls the rotating analyzer, and whose angular position is sensed by an optical angular encoder, to allow data to be acquired and analyzed at the same time at each wavelength. Alternatively, the light dispersed by the monochromator can be impinged upon a linear detector array (diode or CCD) and the entire spectrum accessed in one measurement. This hardware, the detector array and associated electronics to scan the array, is called an optical multichannel analyzer, or OMA, and is presently in wide use in commercial SE systems.

9.6.9 Ellipsometry Alignment and Calibration

In Section 9.4 the effect of some alignment errors on ellipsometry were discussed. Now we revisit this topic to further discuss principles of alignment for the optical axes, angle of incidence, and optical components, but without delving into specific alignment procedures for specific ellipsometers.

For any optical system an optical axis needs to be defined according to which light propagates through the optical system. Each of the two arms of the ellipsometer (to the left and right of the sample shown in Fig. 9.8) could have a different axis. For mechanical alignment each of the two axes need to be found, and then these axes need to be made to coincide through the mechanical center of the instrument. In this way ϕ_0 of 180° is found. From this value other values of ϕ_0 are obtained as the difference from 180° using precision engraved plates. This procedure is called autocollimation. One way to autocollimate is to send a well-defined and collimated light beam (such as from a low power HeNe laser) down each arm to impinge upon a transparent precision optical flat that has polished parallel faces and that is located on the sample stage and as close to the mechanical center of the instrument as possible. The light will reflect back toward its origin after reflecting from both polished surfaces of the flat. A Gauss eyepiece placed after the source enables the projection of crosshairs, which then reflect from both surfaces. When the two reflected crosshairs from one arm are exactly coincident, then the light beam traveling down the telescope is normal to the flat within the accuracy of the flat. Adjustments of the arm and centering the beam are made to the telescope to ensure this condition for both arms without moving the flat. This nontrivial maneuver ensures that the optical axes of both arms are precisely parallel but not necessarily coincident. With the use of apertures and with the ability to move the parallel beams on both arms horizontally and vertically, the parallel beams can be made coincident. Most ellipsometers are constructed with the ability to rotate each arm usually together so as to adjust to any ϕ_0 and with precision engravings to read the value to 0.01° or better. With the ellipsometer arms at 180° the engraving can be adjusted to exactly 180°, or, if that is not possible, then the offset noted for corrections to be made to scale readings (in software). The mechanical center of an instrument is the place for a sample and is at the center of rotation where the two arms of the ellipsometer pivot. After the two parallel axes are made coincident, as discussed above, the beams also intersect at the mechanical center of the ellipsometer optical bench.

Polarizer and analyzer can be aligned on the optical axis, and the offsets relative to the plane of incidence, the POI, are obtained by a variety of procedures. All the procedures use the facts that the components will totally extinguish light when the polarizing axes are at 90° and that at Brewster's angle the $\mathbf{E}_p$ wave is zero for reflected light. The compensator C is calibrated after P and A offsets are known and with the optical bench at 180°. With $P = 0°$ and $A = 90°$ and C mounted in between (PCA configuration), a minimum in the light at the detector is obtained at C (the fast axis) $= 0°$.

9.7 ELLIPSOMETRY MEASUREMENTS

In this section specific ellipsometry calculations and results are shown to illustrate important aspects of the ellipsometry measurement technique. More detailed ellipsometry studies and results are discussed in previous and later chapters, particularly Chapters 10, 11, and 12 dealing with topics that have benefited from the use of the ellipsometry technique such as roughness, surface cleaning, nucleation, and thin-film growth and deposition.

9.7.1 The Si Surface

In the visible region of the optical spectrum from the near infrared (NIR) at low photon energies to the near-ultraviolet (NUV) at higher energies that extends from about 1.5 to 5 eV where ellipsometry is often performed, semiconductors provide good examples of the variety of spectral features for surfaces, and Si being the often studied semiconductor is chosen as an example. Figure 9.14 shows the spectra in terms of ε_1 and ε_2 for both crystalline (c-Si) and amorphous (a-Si) Si. As was discussed above ε_1 can be associated with n and ε_2 with k using Equation 9.18.

The ε_2 spectrum for crystalline Si (c-Si) is characterized by several sharp features that have direct physical significance. Specifically, there are two peaks or strong absorptions in the displayed spectral region at 3.4 and 4.3 eV, which are associated with the Si electronic band structure that displays optical or direct (where the wave vector $k = 0$) transitions at these energies. At the energies of these transitions seen in the ε_2 spectrum of c-Si, the ε_1 spectrum displays a precipitous decrease with ε_1 actually dropping below 0 above 4.3 eV. The sharp drop in ε_1 due to the strong absorption is called a region of anomalous dispersion. In contrast to the c-Si spectrum the amorphous Si (a-Si) spectrum is seen to be both smooth and relatively featureless for both ε_1 and ε_2, although the broad features of the a-Si spectrum align with the sharper features in the c-Si spectrum. Since in principle there can be many amorphous and partially amorphous structures, the a-Si spectrum shown in Fig. 9.14 is only one example of an a-Si sample.

Figure 9.15 shows the spectra for both c-Si (solid lines) and amorphous SiO_2 (dashed lines) in terms of ε_1 and ε_2. SiO_2 is a dielectric with a bandgap near 9 eV. Therefore, there are no absorptions in the spectral region shown; k is approximately 0 as is ε_2, and ε_1 for $SiO_2 = n^2$. For SiO_2 n is about 1.47 at low photon energies and increases slightly with photon energy in the visible spectrum.

The spectrum for absorbing materials in terms of ε_2 or k can be informative about the identity and crystallinity of the material, as shown in Fig. 9.14, while for non-absorbing materials, as shown in Fig. 9.15, less definitive information is obtained from ε_1 and ε_2 (or n and k), although the real part of the refractive index, n is often used to assess the identity and state of materials.

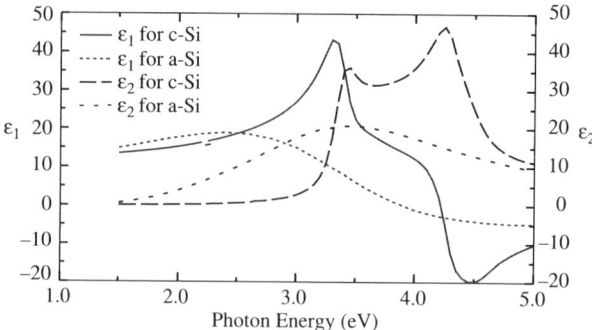

FIGURE 9.14 Dielectric functions for crystalline (c-Si) and amorphous Si (a-Si) in terms of the real ε_1 and imaginary ε_2 components in the visible light range.

9.7 ELLIPSOMETRY MEASUREMENTS

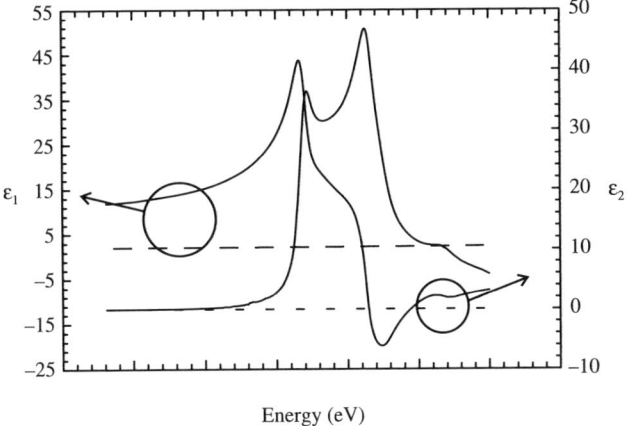

FIGURE 9.15 Dielectric functions for c-Si (solid lines) and SiO$_2$ (dashed lines) in terms of the real ε_1 and imaginary ε_2 components.

9.7.2 Surface with Overlayer

For a specified incident photon energy and a specified angle of incidence, a substrate can be represented as a single point in Δ, Ψ space. For example, for c-Si at $\phi_0 = 70°$ and $\lambda = 632.8$ nm, Δ and Ψ are calculated to be $179.257°$ and $10.448°$, respectively, and this point in $\Delta\Psi$ space for bare Si is shown in Fig. 9.16. If there is a film on this Si substrate, the initial Δ, Ψ point will shift, and as the film changes thickness the Δ,Ψ point will continue to shift position. As a film grows on a substrate, we anticipate that Δ, Ψ will trace a trajectory that is indicative of this growth. From real-time ellipsometry monitoring of this trajectory the film growth can be followed. Specifically, as the film thickness L_f increases, the optical film thickness is computed from the product of thickness and index as $L_f \cdot N_f$. This optical thickness causes a shift of phase Δ. Because some of the light is reflected from the top film surface and some from the film substrate interface, there will also be a change in intensity for the reflected light or a change in Ψ. Figure 9.16 shows the initial bare Si substrate Δ, Ψ point and two growth trajectories for different nonabsorbing films that have n values of 2 and 1.47 corresponding approximately to Si$_3$N$_4$ and SiO$_2$, respectively. Each symbol on the trajectories corresponds to 1 nm of film thickness. It is seen that near-zero film thickness, the change in Δ is large compared to that for Ψ while after about 30 nm the major change is in Ψ. Furthermore, the separation of the two trajectories is more than $0.02°$ in Δ for most of this thickness range and therefore readily discernable using ellipsometry. Figure 9.17 shows that if the Si$_3$N$_4$ and SiO$_2$ trajectories are continued further, another feature of ellipsometry is revealed, namely that the results are cyclic for materials where $k = 0$. The periodicity P is given as

$$P = \tfrac{1}{2}\lambda(\tilde{N}_1^2 - \tilde{N}_0^2 \sin^2 \phi_0)^{-1/2} \tag{9.60}$$

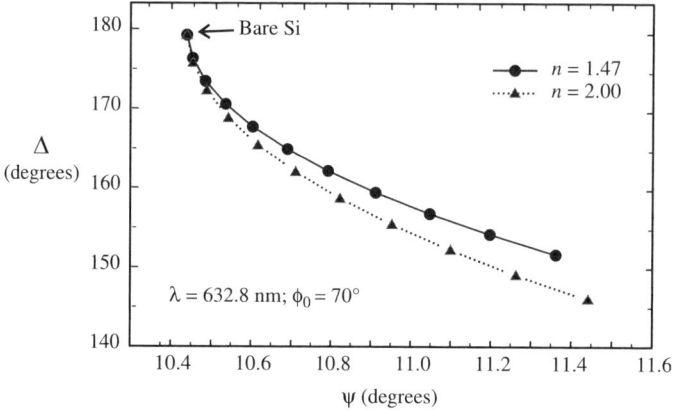

FIGURE 9.16 Δ and Ψ as a function of film thickness for two films on Si. The points are 1 nm apart. One film has $n = 1.47$, $k = 0$, which is close to SiO_2, and the other is $n = 2$ and $k = 0$, which is close to Si_3N_4.

and P depends on λ, ϕ_0, and the complex index for the film and ambient. It is observed from Fig. 9.17 that for the higher index the trajectory has more rapidly changing regions, indicating more sensitivity in Δ, Ψ space by virtue of the larger changes in Δ and Ψ with thickness. Also shown in Fig. 9.17 is the trajectory for an a-Si film ($k \neq 0$) on a c-Si surface. The small difference in $\tilde{N}$'s between a-Si and c-Si yields only small sensitivity to the change in the a-Si film thickness. In continuing the sensitivity discussion, the angle of incidence can also bear on this issue. Figure 9.18 shows a calculation of Δ, Ψ versus thickness space for an SiO_2 overlayer on a Si surface for 632.8 nm light at three values of ϕ_0. Recall that in the previous figures

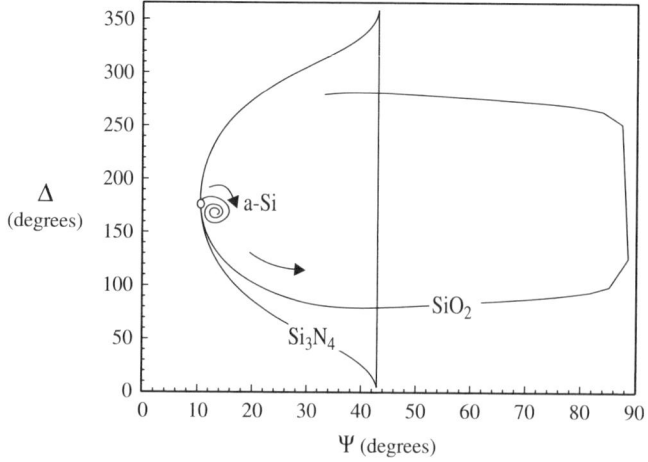

FIGURE 9.17 Δ and Ψ as a function of film thickness for three materials films on c-Si: a-Si, SiO_2, and Si_3N_4.

9.7 ELLIPSOMETRY MEASUREMENTS

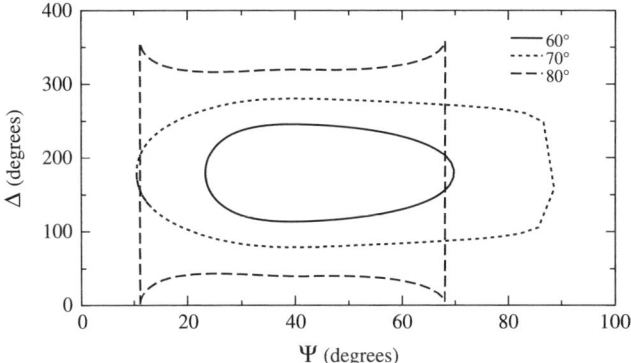

FIGURE 9.18 Δ and Ψ as a function of film thickness for SiO_2 on Si for light with $\lambda = 632.8$ nm at three angles of incidence: $\phi = 60°$, $70°$, and $80°$.

for SiO_2 on Si a value of $\phi_0 = 70°$ was used without justification. Figure 9.18 shows that for the same materials system there is very different trajectory size and therefore sensitivity of the ellipsometric measurement with changes of ϕ_0. Thus, the selection of the angle of incidence is important in determining the best parameters for an ellipsometric measurement. For this materials system the largest trajectory is seen at $\phi_0 = 70°$ with an obviously smaller trajectory at $\phi_0 = 60°$, and at $\phi_0 = 80°$ the changes in Δ and Ψ are also small except where the top and bottom parts of the trajectory are connected (at $0°$ and $360°$). Thus, a selection of the ϕ_0 that maximizes the size of the trajectory is important to optimize the sensitivity of the ellipsometric measurement. Current ellipsometric practice uses measurement at several ϕ_0 to improve data analysis. This practice is useful so long as the ϕ_0's that are used yield sensitive trajectories, and it should be understood this procedure can be counterproductive with the use of insensitive ϕ_0's.

The ellipsometry measurement of an overlayer on a substrate can be summarized using Figs. 9.19a and 9.19b that show the spectroscopic ellipsometry determined spectra in terms of ε_1 and ε_2 for thin SiO_2 films on a Si surface. Figure 9.19a displays the ε_1 and ε_2 spectra for the bare c-Si substrate as the solid line. The two dashed spectra are the ε_1 and ε_2 spectra for a 1- and 5-nm SiO_2 film on the c-Si substrate with the 1-nm SiO_2 film on Si being the spectrum closest to the c-Si spectrum. The most interesting region of these spectra is the 3- to 4.5-eV region of the ε_2 spectra. For bare c-Si this region contains the two sharp absorptions at 3.4 and 4.3 eV. The spectra for the 1 and 5 nm SiO_2 films on the c-Si substrate show the c-Si absorption features albeit with diminished intensity. The intensity decrease is larger at the Si absorption energies and larger at higher energies. Therefore, these spectral regions near substrate adsorption peaks are the most sensitive for film thickness and optical properties assessment in the thin-film regime to around 10 nm. Figure 9.19b displays spectra for 10 nm SiO_2 on c-Si (solid lines) and 50-nm SiO_2 on c-Si (broken lines). In the 10-nm spectra the c-Si spectral features at 3.4 and 4.3 eV in ε_2 are barely discernable and for the 50-nm SiO_2 film are not apparent.

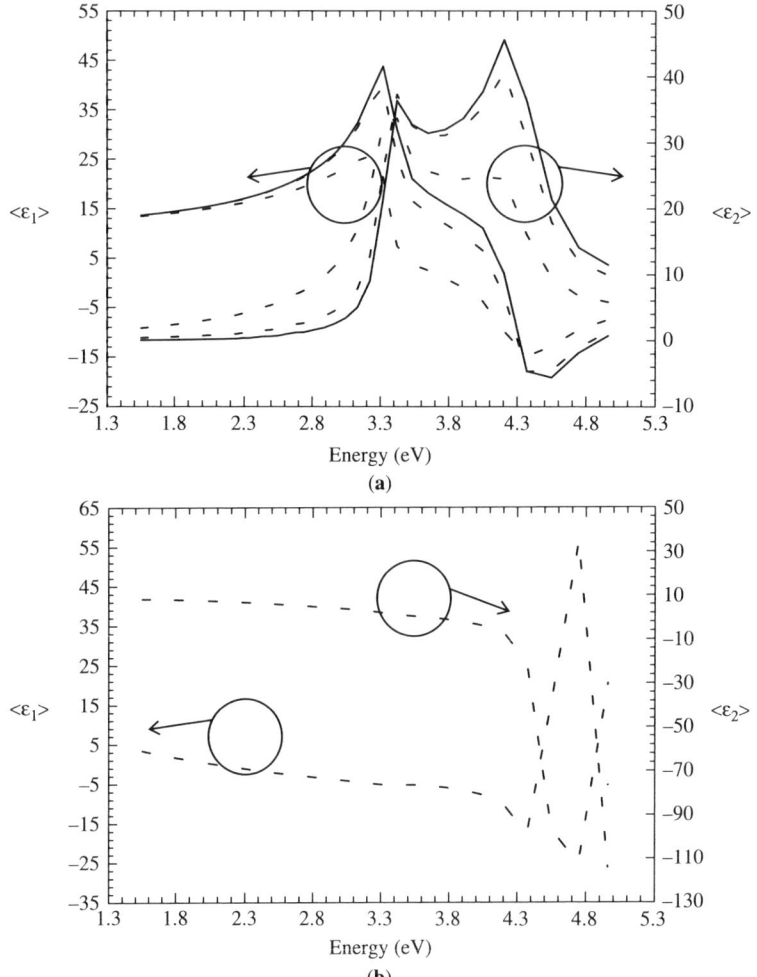

FIGURE 9.19 (a) Dielectric functions for crystalline (c-Si) (solid lines) in terms of the real ε_1 and imaginary ε_2 components in the visible light range along with the pseudodielectric functions (in brackets) for two SiO_2 films on Si (deshed lines). The top pseudodielectric functions are for 1-nm SiO_2 on Si and the bottom is for 5-nm SiO_2 on Si. (b) Pseudodielectric functions $<\varepsilon_1>$ and $<\varepsilon_2>$ for two SiO_2 films on Si. The solid lines correspond to 10-nm SiO_2 on Si and the dashed lines are for 50-nm SiO_2 on Si.

This is due to the fact that as the SiO_2 film thickness the optical interference between reflections from the top and bottom film interfaces dominates the reflection of light and therefore the dielectric function spectra.

Further detailed applications of ellipsometry alone and in combination with other techniques are discussed in Chapters 10 and 11.

REFERENCES

1. Drude P. Ann Phys Chem 1889;36:532,865.
2. Archer RJ. Ellipsometry in the measurement of surface and thin films. Washington, DC: US. Governement Printing Office; 1964.

SUGGESTED READING

R. M. A. Azzam and N. M. Bashara, 1979. *Ellipsometry and Polarized Light*, North Holland. A classic and authoritative text on ellipsometry. The applications are a bit out of date.

H. G. Tompkins, 1993. *A User's Guide to Ellipsometry*, Academic Press. A first book to look at when trying to learn about ellipsometry.

H. G. Tompkins and E. A. Irene (eds.), 2005. *Handbook of Ellipsometry*, William Andrews. A collection of authoritative chapters with theory, instrumentation, and updated applications.

H. G. Tompkins and W. A. McGahan, 1999. *Spectroscopic Ellipsometry and Reflectometry*, Wiley Interscience. A good introduction to spectroscopic ellipsometry and optical modeling.

PART II

MICROELECTRONICS APPLICATIONS

10

FILMS AND INTERFACES

10.1 INTRODUCTION

One main thrust of this book is surfaces. Chapters 1 to 9 dealt almost exclusively with the basic ideas about surfaces and interfaces and related issues are found in every chapter. Notwithstanding this emphasis on surfaces in the early chapters, it is important to realize that bare surfaces are rarely found in nature or in man-made applications. In fact, as was previously discussed, bare surfaces are reactive and therefore for most materials are difficult to prepare and to maintain. Rather most surfaces are found to be coated fully or partially with other materials. As such the surfaces act as substrates for films. Surface coatings if thin relative to the substrate are called films. It is observed that even very thin films display properties that are different from the substrate surface. The location where a film meets the substrate surface is called an interface. The extent of interfaces can be sharp with atomic dimensions indicative of virtually noninteracting or diffusing film and substrate materials or interfaces can extend beyond tens of nanometers indicative of considerable interaction and/or interdiffusion. Interfaces have compositions that are in general different from either the overlayer film or the substrate, and therefore interfacial properties are expected to be different from the film or substrate properties. Often surfaces are intentionally coated with films for the purpose of modifying the surface or producing desirable interface properties. In the field of microelectronics semiconductor surfaces are coated with various films for specific applications. In this chapter basic questions are addressed about how films and interfaces form with some

---

Surfaces, Interfaces, and Thin Films for Microelectronics. By Eugene A. Irene
Copyright © 2008 John Wiley & Sons, Inc.

applications. Chapters 11 and 12 will continue with several major applications of thin films in the fields of electronic materials and microelectronics, emphasizing the MOSFET as the vehicle to discuss film and interface issues.

Film and interface formation commences on a substrate surface. The first vestige of the film to form is called a nucleus, and the process by which nuclei of a new phase form is called nucleation. Thus, the subject of film nucleation is discussed first, and this is followed by film formation processes. A film can form by the reaction of the surface with another chemical to form a compound in film form, and this process is called film growth. Alternatively, chemicals can mix and react above a substrate surface and then deposit upon the substrate surface, and this film formation process is called deposition. These two film formation processes, growth and deposition, are discussed in this chapter.

10.2 NUCLEATION

The formation of nuclei of a new phase, for example, a solid particle of phase B in a gas phase A can be considered to form away from any surface, and the reaction is written as

$$A_g \rightarrow B_s \qquad (10.1)$$

where the solid phase nuclei of B composed of at least several atoms (the size of nuclei will be discussed later) form homogenously, that is, without a substrate to form upon. From the discussion of surface thermodynamics in Chapter 3, it was established that energy is required to create a new surface. Consequently, for a new surface to form in the process of nucleation the most often observed processes are those that minimize the required energy. Such pathways are the most probable pathways. Thus, when a B nucleus forms, a new surface of B atoms forms. By choosing the reactants and the conditions (concentrations, temperature, pressure, etc.), we could possibly render the reaction to form B_s in A_g to be thermodynamically favorable ($-\Delta G_{reaction}$), but even were we to adjust the conditions to achieve a favorable free energy for the formation of B_s, there is also an energy requirement per unit area ($+\gamma$) for the surface produced, called the surface energy. To minimize this surface energy per new area formed, γ, that is required for the formation of the new B_s phase, it would be energetically advantageous for the new phase to be spherical, since the spherical shape minimizes the surface area for a given volume of material in the B phase. In this way we maximize the $-\Delta G_{reaction}$ contribution and minimize the $+\gamma$ requirement so as to minimize the total energy requirement for the nucleation process.

The kind of nucleation considered above considers the formation of a spherical shaped phase away from any other surfaces. This is called homogenous nucleation. On the other hand we could also consider that B_s phase forms upon a substrate surface that is the same as B or different. This is called heterogeneous nucleation, and it will be shown below that heterogeneous nucleation requires less surface energy than homogenous nucleation and is therefore favored with all other variables equal.

10.2.1 Homogenous Nucleation

First, we consider the overall process where atoms (or molecules) of a particular material $A_{(g)}$ in the gas phase unite to form a volume of atoms in the solid phase $B_{(s)}$ that take a spherical shape as shown in Fig. 10.1. As mentioned above, there are two contributions to the free energy for this process. One component is the free energy for the formation a solid phase per mole from the gas phase, which is called ΔG_v and is related to $\Delta G_{\text{reaction}}$ where the subscript v is now used to indicate the thermodynamic free energy to form a volume of solid B from gas-phase A as given above by reaction 10.1. For this reaction involving atoms of A reacting to form B and then condensing to form a solid phase, the following is written for the free energy change:

$$\Delta G_v = G(B_s) - G(A_g) \tag{10.2}$$

The total free energy ΔG_{tot} can be obtained as the sum of the volume (ΔG_v) and surface energy (γ) terms. However, these terms ΔG_v and γ have different units and as such cannot be added directly. As mentioned above, ΔG_v has units of energy per mole, which is also energy per volume of material while γ has units of energy per area of surface produced. This dilemma can be obviated by using the good assumption above that the new phase is likely spherical in shape. With this assumption the two energy terms can be added when the volume and area components are removed by multiplying by the appropriate geometric factors for a spherical nucleus, namely the volume and area for a sphere as follows:

$$\Delta G_{\text{tot}} = \tfrac{4}{3}\pi r^3 \, \Delta G_v + 4\pi r^2 \gamma \tag{10.3}$$

where r is the radius of the spherical nucleus. As discussed above, the volume free energy term can be adjusted to be negative and therefore indicative of a spontaneous process by adjusting the various parameters (temperature, composition, etc.) of the nucleation reaction. However, the surface term is always positive, indicating that energy is required to drive the formation of a new surface. Whether ΔG_{tot} will be negative and therefore the nucleation process spontaneous will depend on whether the volume term or surface term dominates.

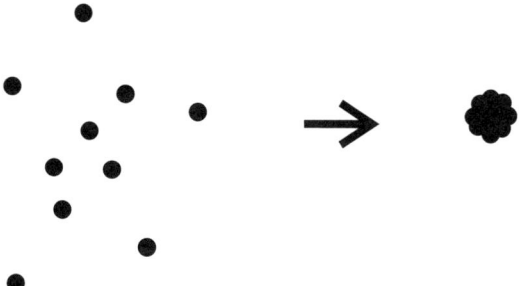

FIGURE 10.1 Spherical nucleus formed from atoms.

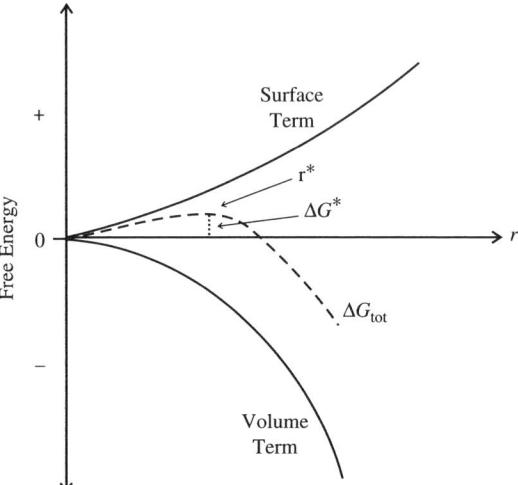

FIGURE 10.2 Free energy with surface and volume terms plotted versus the size r of a spherical nucleus.

The changes in the two energy terms on the right-hand side of Equation 10.3 as a function of nuclei size is shown in Fig. 10.2. As discussed above for the volume term, the conditions were adjusted so that this term is negative and it decreases (becomes more negative) with increasing r. Thus, with the surface term positive and growing more positive with increasing r as more surface is produced, it is inferred that the sum of these terms has an inflection point labeled as r^* in Fig. 10.2. For small nuclei sizes where $r < r^*$, the surface term is dominant because the square of small quantities is larger that the cube while for the case where $r > r^*$, the negative volume term dominates and pulls the transformation process given by reaction 10.1 to the right. Figure 10.2 teaches that a positive ΔG_{tot} results for small nuclei ($r < r^*$) and negative ΔG_{tot} for larger nuclei ($r > r^*$). This means that the small nuclei are thermodynamically unstable relative to larger nuclei. At r^* the barrier, ΔG^* is obtained and then r^* is calculated.

The inflection point r^* shown in Fig. 10.2 is obtained at $d\,\Delta G_{tot}/dr = 0$ as follows:

$$\frac{d\,\Delta G_{tot}}{dr} = \frac{d(4/3\pi r^3)\,\Delta G_v}{dr} + \frac{d(4\pi r^2 \gamma)}{dr} = 0 \qquad (10.4)$$

where $r = r^*$. Then solving for r^* and using the formula $\Delta G_v = \Delta E_v\,\Delta T/T_E$ obtained starting from the following relationship:

$$\Delta G_v = \Delta E_v - T_E\,\Delta S_v = 0 \quad \text{where } E_v \approx H_v \qquad (10.5)$$

10.2 NUCLEATION

The following equation is obtained:

$$r^* = \frac{-2\gamma}{\Delta G_v} = -2\gamma \left(\frac{T_E}{\Delta E_v \, \Delta T} \right) \qquad (10.6)$$

Using Equation 10.6 for r^* in Equation 10.3 yields an expression for the total free energy at the inflection point for homogenous nucleation:

$$\Delta G^* = \frac{16\pi \gamma^3}{3 \Delta G_v{}^3} \qquad (10.7)$$

The approximate equivalence of ΔE and ΔH is made for solids where under usual conditions of pressure P, $P\,dV$ is insignificant. From this we obtain for the entropy:

$$\Delta S_v = \frac{\Delta E_v}{T_E} \qquad (10.8)$$

Near equilibrium $T \approx T_E$ and with Equation 10.5 we obtain

$$\Delta G_v = \Delta E_v - T\left(\frac{\Delta E_v}{T_E}\right) = \frac{\Delta E_v(T_E - T)}{T_E} = \frac{\Delta E_v \, \Delta T}{T_E} \qquad (10.9)$$

with $\Delta T = T_E - T$; ΔT is called the driving force for the phase transformation and is also called supercooling. When $\Delta T \neq 0$, a transformation occurs.

If heat is evolved for the transformation given by the phase transformation summarized by reaction 10.1, the reaction is exothermic and ΔE_v is negative. Thus, at higher T and with $-\Delta E_v$, ΔT will be more negative ($T > T_E$), and then ΔG_v becomes more positive and the transformation as written ($A_g \rightarrow B_s$) is less favorable. If on the other hand the reaction is endothermic, ΔE_v is positive, the transformation is more favorable at higher T. The direction of the transformation is obtained from this thermodynamic treatment.

Another approach that yields further insight into the nature of nucleation can be obtained starting from an atomistic reaction scheme for the formation of nuclei of size i. If we consider a nucleus composed of B atoms to form starting from the interaction of 2B atoms to form a dimer and then the reaction of B dimers to form trimers, and so on, to finally form a nucleus consisting of iB atoms, B_i, as follows:

$$\begin{aligned} B_1 + B_1 &\rightarrow B_2 \\ B_2 + B_1 &\rightarrow B_3 \\ &\vdots \\ B_{i-1} + B_1 &\rightarrow B_i \end{aligned} \qquad (10.10)$$

The overall process is given as

$$iB_1 \rightarrow B_i \tag{10.11}$$

where i indicates the size of the cluster of B atoms. The free energy for the process is given as

$$\Delta G_i = \Delta G(B_i) - i\Delta G(B_1) \tag{10.12}$$

We recall the definition of the chemical potential for the molar quantities of the j component μ_j that is given by the partial derivative of the free energy with respect to the change in the moles of j, n_j:

$$\mu_j = \left(\frac{\partial G}{\partial n_j}\right)_{T,P,n_i} \tag{10.13}$$

Using Equation 10.12 and this definition but in the present case for i sized clusters rather than moles, the free energy is given as

$$\Delta G_i = \int_0^i \mu_i \, di - \int_0^i \mu_1 \, di \tag{10.14}$$

The chemical potential can also be written in terms of the standard state chemical potential $\mu°$, temperature T and pressure P as

$$\mu = \mu° + kT \ln P \tag{10.15}$$

This form can be used to express the chemical potential for individual atoms, μ_1, and for clusters of size i, μ_i, as follows:

$$\mu_1 = \mu° + kT \ln P \quad \text{and} \quad \mu_i = \mu_\infty + kT \ln P \tag{10.16}$$

where μ_∞ represents the bulk value. Using the Kelvin equation for P from Chapter 3:

$$\frac{2V_S\gamma}{r} = RT \ln \frac{P}{P_0} \tag{3.41}$$

Equation 10.16 for μ_i can be written as

$$\mu_i = \mu_\infty + \frac{V}{N_{AV}} \frac{2\gamma}{r} \tag{10.17}$$

10.2 NUCLEATION

where V is the molar volume, N_{AV} is Avogadro's number, and the Boltzmann constant $k = R/N_{AV}$ where R is the gas constant and is given as

$$N_{AV} = \frac{V}{\Omega} \quad (10.18)$$

where Ω is the volume per atom (or molecule). Also μ_∞ is given as

$$\mu_\infty = \mu^\circ + kT \ln P^\circ \quad (10.19)$$

where P° is the equilibrium vapor pressure. For a spherical nucleus i is the number of atoms in the cluster and is given as

$$i = \frac{\frac{4}{3}\pi r^3}{\Omega} \quad \text{and} \quad di = \frac{4\pi r^2}{\Omega} dr \quad (10.20)$$

Substituting for di into Equation 10.14 we obtain

$$\Delta G_i = \int_0^r \mu_i \frac{4\pi r^2}{\Omega} dr - \int_0^r \mu_1 \frac{4\pi r^2}{\Omega} dr \quad (10.21)$$

Then substitute for μ_i using Equations 10.17, 10.18, and 10.19 and μ_1 from Equation 10.16 into Equation (10.21) to obtain

$$\begin{aligned}\Delta G_i =& \int_0^i \mu^\circ \frac{4\pi r^2}{\Omega} dr + \int_0^i kT \ln P^\circ + \int_0^i \frac{\Omega 2\gamma}{r} \frac{4\pi r^2}{\Omega} dr \\ & - \int_0^i \mu^\circ \frac{4\pi r^2}{\Omega} dr - \int_0^i kT \ln P \frac{4\pi r^2}{\Omega} dr\end{aligned} \quad (10.22)$$

The first and fourth terms cancel, and the second and fifth terms can be combined to yield upon integration over dr:

$$\Delta G_i = \frac{4\pi r^3}{3\Omega} kT \ln \frac{P^\circ}{P} + 4\pi r^2 \gamma = \frac{4\pi r^3}{3} \Delta G_v + 4\pi r^2 \gamma \quad (10.23)$$

where

$$\Delta G_v = -\frac{kT}{\Omega} \ln \frac{P}{P^\circ}$$

and Equation 10.23 yields Equation 10.3 where $\Delta G_i = \Delta G_{tot}$.

10.2.1.1 Mixing

As discussed above using relationship 10.10, the nucleation processes commence with monomers that interact, and in time the system consists of dimers, trimers, and the like as well as the starting monomers. In essence the nucleation system evolves toward a mixture. Consequently, there exists a free energy of mixing that alters the energetics of the system. The change in the free energy of the system ΔG_{sys} is then given by a sum of the free energies of the components plus the free energy of mixing as follows:

$$\Delta G_{sys} = \sum_i n_i\, \Delta G_i + \Delta G_{mix} \qquad (10.24)$$

where as before n_i is the equilibrium number of nuclei of size i. From thermodynamics ΔG_{mix} in terms of the mole fraction X is given as

$$\Delta G_{mix} = nkT \sum_i X_i \ln X_i \qquad (10.25)$$

For the case above we have moles of i and 1 as n_i and n_1 and the total number of moles n is given by $n = n_i + n_1$ and for mole fractions $X_i = n_i/n$, $X_1 = n_1/n$. With these substitutions Equation 10.25 is given as

$$\Delta G_{mix} = nkT \left(\frac{n_i}{n} \ln \frac{n_i}{n} + \frac{n_1}{n} \ln \frac{n_1}{n} \right) = kT \left(n_i \ln \frac{n_i}{n_i + n_1} + n_1 \ln \frac{n_1}{n_i + n_1} \right) \qquad (10.26)$$

Equation 10.26 is inserted into Equation 10.24 and the following equilibrium condition is imposed:

$$\frac{\partial (\Delta G_{sys})}{\partial n_i} = 0 \qquad (10.27)$$

The result is obtained using the following relationship for the general function u:

$$\frac{\partial}{\partial x} \ln u = \frac{1}{u} \frac{du}{dx} \quad \text{and with } n_1 \gg n_i,\ n_i + n_1 \approx n_1 \qquad (10.28)$$

The derivative of the first term of Equation (10.24) with respect to n_i is given as

$$\frac{\partial (n_i\, \Delta G_i)}{\partial n_i} = \Delta G_i \qquad (10.29)$$

10.2 NUCLEATION

The derivative of the second term of Equation (10.24) with respect to n_i is given as

$$\frac{\partial(\Delta G_{\text{mixi}})}{\partial n_i} = kT \frac{\partial}{\partial n_i} \left\{ n_i \ln \frac{n_i}{n_1} + n_1 \ln \frac{n_1}{n_1} \right\} \qquad (10.30)$$

The second term in parentheses is zero and the derivative of the first term in parentheses yields two terms:

$$\frac{\partial(\Delta G_{\text{mixi}})}{\partial n_i} = kT \left\{ \ln \frac{n_i}{n_1} + n_i \frac{n_1}{n_i} \frac{1}{n_1} \right\} = kT \left\{ \ln \frac{n_i}{n_1} + 1 \right\} \qquad (10.31)$$

Now adding the terms according to Equation 10.24 and setting the sum equal to 0 according to Equation 10.27 yields

$$\Delta G_i + kT \ln \frac{n_i}{n_1} + kT = 0 \qquad (10.32)$$

Solving for n_i, we obtain the following:

$$\ln \frac{n_i}{n_1} = -\left(\frac{\Delta G_i + kT}{kT} \right) \qquad (10.33)$$

With $kT \ll \Delta G_i$ we obtain the final result:

$$n_i = n_1 e^{-\Delta G_i / kT} \qquad (10.34)$$

Combining Equations 10.23 with 10.32 where kT is small, we obtain

$$\Delta G_i = \frac{4\pi r^3}{3} \Delta G_v + 4\pi r^2 \gamma = -kT \ln \frac{n_i}{n_1} \qquad (10.35)$$

The size of a nucleus is related to both r the radius and n_1 the number of monomer units in the nucleus. Then from Equation 10.35 we learn that for $\Delta G_v > 0$ as r or i increase, the left side of Equation 10.35 decreases while for $\Delta G_v < 0$ as r or i increase, the left side increases. Also for $\Delta G_v < 0$ we have already seen there is a point of inflection at r^* or now i^*. Figure 10.3 displays a plot of the right side of Equation 10.35 (which is also ΔG_i) versus i for the $+$ and $-$ values of ΔG_v, which summarizes the facts above.

10.2.1.2 Volmer–Weber Theory The popular Volmer–Weber theory for nucleation provides a simple framework to explain nucleation and is the basis for more

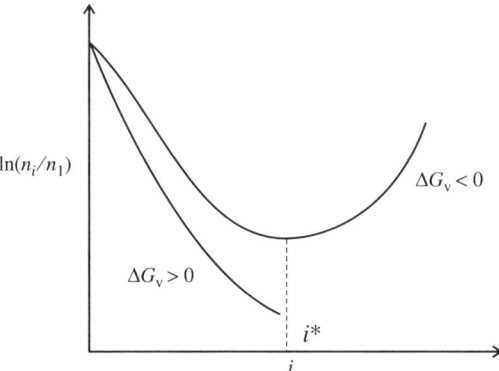

FIGURE 10.3 Plot of ΔG_i as the ratio of the number of nuclei of size i to the number of starting monomers versus the size i of nuclei from Equation 10.35.

complex theories. The basic idea of this theory is that a system of starting ingredients under a set of conditions maintains an equilibrium number of critical nuclei, n_i^*. In this system the critical nuclei are colliding with other species. When a productive collision occurs that increases the critical nucleus size, then the nucleus grows. The nucleus growth rate is then given by J_N, which is the product of number of critical nuclei, n_i^* multiplied by the number of fruitful collisions ω, multiplied by the cross-sectional area of the critical nucleus A^*. Recall Equation 10.34 for n_i that is modified to become n_i^*:

$$n_i^* = n_1 e^{-\Delta G_i^*/kT} \tag{10.36}$$

The number of collisions is given by the impingement rate of Chapter 8 (Equation 8.22) but now multiplied by the fraction of successful collisions α_c to yield the number of fruitful collisions ω:

$$\omega = \alpha_c \frac{P}{(2\pi MkT)^{1/2}} \tag{10.37}$$

The cross-sectional area of a spherical critical nucleus A^* is

$$A^* = 4\pi r^{*2} \tag{10.38}$$

The Volmer–Weber expression is the nucleus growth rate J_N:

$$J_N = A^* \omega n_i^* = \alpha_c (4\pi r^{*2}) \frac{P}{(2\pi MkT)^{1/2}} n_1 e^{-\Delta G_i^*/kT} \tag{10.39}$$

10.2 NUCLEATION

As the supercritical nuclei grow, ultimately they will encounter adjacent nuclei. A process of ripening occurs where nuclei will coalesce to form a smaller number of larger nuclei. Figure 10.4 shows this for a α phase in which spherical β phase nuclei are coalescing. The driving force for ripening is simply that the larger nuclei have less surface area relative to their volume (smaller surface/volume ratio) and thus is a lower energy configuration. If this ripening and coalescence occurs on a surface, the regions in between nuclei can offer sites for secondary nucleation, and ultimately the original surface will become completely covered with the new phase.

Before proceeding further with this development, it is useful to try to imagine what is occurring and why. It is straightforward to imagine atoms or molecules moving around and colliding with some of the collisions of α phase species converting to β phase, since we have already made this conversion probable by adjusting the conditions to yield a negative ΔG_v for the phase producing reaction. The initial result is that a small number of at first small nuclei of β form. If these β nuclei are smaller than the critical size ($r < r^*$), they are unstable, and there is a good chance that before they can grow larger via more of the fruitful collisions, they may disappear to reform α phase. Statistically, even some of the small nuclei may exist long enough to grow. Those lucky few of the $r < r^*$ nuclei that grow become more stable, and their chances of growing even larger steadily improve. We can imagine that these events are occurring with a huge number of atoms and/or molecules, so ultimately the number of statistical survivors is measurable, and we ultimately observe the formation of the β phase. This statistical notion is useful since it teaches that many events are occurring on a microscopic scale. Some of the events are favored and some are not, yet they occur because there are so many reactants and even low probability events can occur, however infrequently. Ultimately, the events that lead to observable stable phases are the favored events. Nature selects these via this statistical process and thermodynamics (with its laws) typically considers the initial and final states of a large number of reactants and products, to tally the overall energetics.

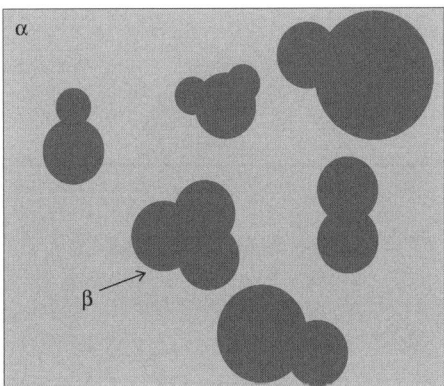

FIGURE 10.4 β-phase nuclei formed and ripening in α-phase medium.

10.2.2 Heterogeneous Nucleation

Thus far we have considered homogenous nucleation where a nucleus forms unfettered in three dimensions, and we specifically considered spherical nuclei. It is well known that heterogeneous nucleation, for example, nucleation on a surface, occurs more readily than homogenous nucleation. This is attributed to the fact that less γ is required in the presence of another surface. Recall that a spherical nucleus was considered for homogenous nucleation because the spherical shape minimized the surface/volume ratio and was therefore favored. However, in the presence of a surface, a hemisphere can form and further lower the surface area (from sphere to hemisphere), as shown in Fig. 10.5. Thus, it is often the case that heterogeneous nucleation is favored. One poignant example is that of cooling of water to well below the freezing temperature where ice does not form until a dust particle with its surface is present or the container surface is scratched exposing two-dimensional surface nucleation sites. In general, two-dimensional nucleation is complicated because there is the interaction of the surface with the nucleus, which may or may not be suitable for nucleus stability and which may dictate the nucleus shape and hence its energy. This is called wetting behavior and will be discussed below. This subject derives from the discussion of surface energies in Chapters 1 and 3.

First, we consider three important cases for nucleation on a surface. Figure 10.6a shows the simplest case to envision where a layer fully forms and then another layer fully forms and so on. In this case of so-called layer-by-layer growth, also called Franck–van der Merwe growth, the critical nucleus size $i^* = 1$. Or one atom or molecule comprises a critical nucleus. It is interesting to observe that when the first layer forms, the attraction between impinging and surface atoms is greater than the attraction between impinging atoms. Thus, the atoms form a monolayer rather than clumping together. Another way to express this is to state that the adhesion is greater than the cohesion for the atoms (molecules) that are nucleating upon the surface. Often for layer-by-layer growth there exists a structural relationship between the nucleating atoms that form the surface film and the preexisting surface, and the result is called epitaxial growth. In the field of microelectronics, epitaxial growth is often desirable in that the disorder at the film–substrate interface is minimized and carrier transport is

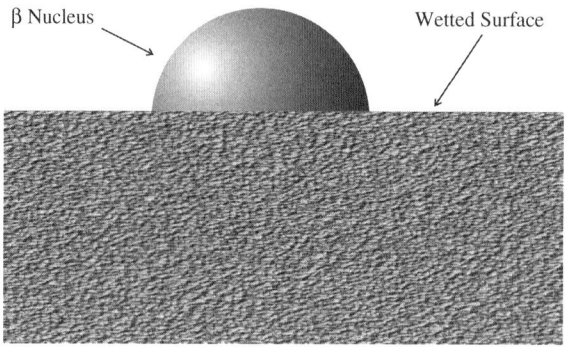

FIGURE 10.5 Hemispherical nucleus formed on a surface.

10.2 NUCLEATION

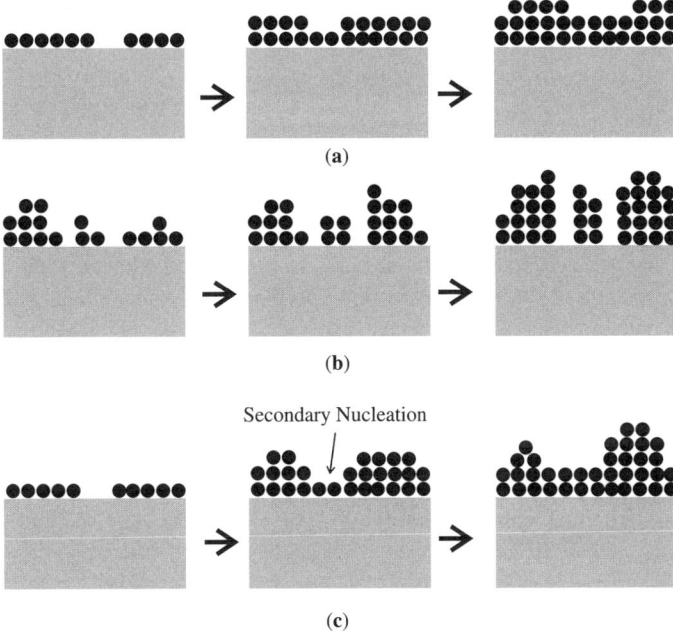

FIGURE 10.6 Three cases of heterogeneous nucleation: (a) layer-by-layer growth where $i^* = 1$; (b) cluster of Vollmer–Weber nucleation; and (c) combination of (a) and (b) or Stranski–Krastanov nucleation.

relatively unimpeded as compared to a disordered interface. The opposite to layer-by-layer growth is so-called Vollmer–Weber growth where the atoms (molecules) that nucleate have cohesive forces greater than adhesive forces and thus form clusters on the surface, as shown in Fig. 10.6b. These clusters grow and finally touch. This coalescence ultimately results in a continuous layer. Metal atoms on insulators often result in Vollmer–Weber nucleation. The resulting surface films are typically rougher than layer-by-layer films, since the nuclei are higher for Vollmer–Weber nucleation, that is, $i^* > 1$. The cases of layer-by-layer and Vollmer–Weber heterogeneous nucleation represent two extremes. Oftentimes surface sites for nucleation vary across a surface due to structural and/or stoichiometry variations and/or defects. Consequently, the dominance of either cohesive or adhesive interactions between impinging atoms and surface atoms can vary across an existing surface leading to incomplete layer-by-layer growth followed by Vollmer–Weber nucleation. This kind of mixed nucleation is called Stranski–Krastanov nucleation and is depicted in Fig. 10.6c, which shows initial but incomplete nucleation followed by three-dimensional nuclei formation. The spaces in between nuclei can provide nucleation sites, albeit smaller probability sites, for secondary nucleation at some point in the development of the primary nuclei. Stranski–Krastinov nucleation appears to be the transition from layer-by-layer to Vollmer–Weber nucleation. It is often observed in metal–metal and metal–semiconductor systems. In some cases

Stranski–Krastinov nucleation has been attributed to stress relief as a phase nucleates producing strain at the surface.

The role of the surface on heterogeneous nucleation can be understood by considering the surface energies. Figure 10.7 shows a spherical cap-shaped nucleus (cross-hatched) on a surface. The nucleus results from atoms (or molecules) from the vapor phase impinging upon a solid surface. Some of the atoms stick and some desorb. Those that adhere to the surface can move on the surface and interact with the surface atoms and with each other, ultimately forming a nucleus with a shape that minimizes the surface energy. The nucleus is a localized film on the surface and has a contact angle θ, and the interaction energies can be formulated according to the Young equation (see Chapter 3) as follows:

$$\gamma_{sv} = \gamma_{fs} + \gamma_{fv} \cos \theta \tag{10.40}$$

Also the total free energy for a nucleus of size i, ΔG_i can be obtained as before as the sum of volume and surface energy terms with each term multiplied by the appropriate geometrical term that converts the energy per volume or area to pure energy as follows:

$$\Delta G_i = c_v r^3 \Delta G_v + c_{a1} r^2 \gamma_{fv} + c_{a2} r^2 \gamma_{fs} - c_{a3} r^2 \gamma_{sv} \tag{10.41}$$

where for the spherical capped-shaped nucleus the geometrical constants are given as

$$\begin{aligned} c_v &= \frac{\pi}{3} (\cos^3 \theta - 3 \cos \theta + 2) \\ c_{a1} &= 2\pi (1 - \cos \theta) \\ c_{a2} &= \pi \sin^2 \theta \\ c_{a3} &= \pi \sin^2 \theta \end{aligned} \tag{10.42}$$

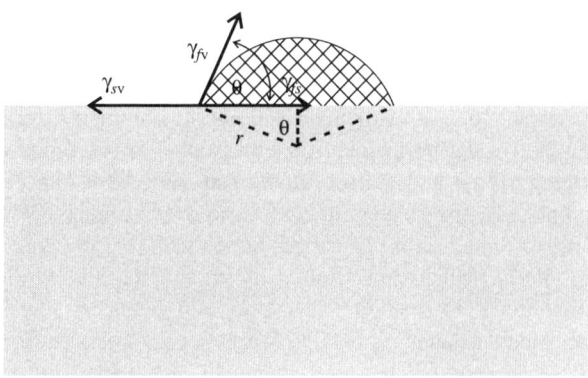

FIGURE 10.7 Cap-shaped nucleus (cross-hatched) on a surface with the three energy terms: γ_{sv}, γ_{fs}, γ_{fv} where s indicates surface, v vapor, and f is the formed nucleus.

10.2 NUCLEATION

These constants for the spherical cap are: $c_v r^3$ is the volume of the cap, $c_{a1} r^2$ is the curved surface area of the cap, and $c_{a2} r^2$ and $c_{a3} r^2$ are equal and represent the basal area of the cap. For layer-by-layer growth the contact angle $\theta = 0$, which indicates that there is total wetting of the surface, and that the interaction of atoms with surface is strong and the Young equation is as follows:

$$\gamma_{sv} = \gamma_{fs} + \gamma_{fv} \tag{10.43}$$

The other extreme is for weak interaction with the surface with $\theta > 0$ or far less wetting, and the Young equation is given as:

$$\gamma_{sv} < \gamma_{fs} + \gamma_{fv} \tag{10.44}$$

For Stranski–Krastinov nucleation the Young equation is

$$\gamma_{sv} > \gamma_{fs} + \gamma_{fv} \tag{10.45}$$

that could indicate that the strain energy is large relative to γ_{fv} and for homoepitaxy $\gamma_{fs} = 0$. Now an expression for ΔG_i^* can be obtained from Equation 10.41 by setting $d\,\Delta G_i/dr = 0$ and solving for r^* as follows:

$$\frac{d\,\Delta G_i}{dr} = 3c_v r^2 \Delta G_v + 2c_{a1} r \gamma_{fv} + 2c_{a2} r \gamma_{fs} - 2c_{a2} r \gamma_{sv} = 0 \tag{10.46}$$

Solving Equation 10.46 for r yields r^* of the form:

$$r^* = \frac{-2(c_{a1} \gamma_{fv} + c_{a2} \gamma_{fs} + c_{a2} \gamma_{sv})}{3c_v \Delta G_v} \tag{10.47}$$

Replacing r^* into Equation 10.41 and using Equation 10.42 we obtain the desired result for ΔG_i^*:

$$\Delta G_i^* = \left[\frac{16\pi \gamma_{fv}^3}{3\Delta G_v^2}\right] \left[\frac{\cos^3 \theta - 3\cos\theta + 2}{4}\right] \tag{10.48}$$

This expression for ΔG_i^* has two bracketed terms on the right. The first of these terms in brackets is identical with Equation 10.7, which is the critical free energy for homogenous nucleation. The second term in brackets must then be the term that defines the difference between homogenous and heterogeneous nucleation. This term is called the wetting factor, since it is a function of the contact angle θ. Using the wetting factor, we can now explore the effect of a surface on nucleation using Equation 10.36 that expresses the number of critical nuclei of size i, n_i^*, in terms of ΔG_i^* that is a function of the surface energy through θ.

$$n_i^* = n_1 e^{-\Delta G_i^*/kT} \qquad (10.36)$$

For perfect wetting θ = 0°, and the wetting factor is 0 and therefore $\Delta G_i^* = 0$. This means that in the presence of a surface where wetting is perfect, there is no barrier to nucleation. In this case $n_i^* = n_1$, which means that the critical nucleus size is one atom (molecule). In the absence of other surface heterogeneities this situation corresponds to layer-by-layer nucleation and growth. At the other extreme of no wetting θ = 180° and the wetting factor is 1 and ΔG_i^* is then given by the homogenous nucleation term. With reality yielding θ values somewhere in between these extremes, the presence of a surface decreases the barrier for the formation of critical nuclei relative to homogenous nucleation.

10.2.3 Nucleation Studies

The use of ellipsometry (see Chapter 9) as a nondestructive technique to study the nucleation and growth phenomena of Si films is well documented in the literature. In the application discussed below (1), the nucleation and growth of polycrystalline Si (poly-Si) films was studied using ellipsometry along with other techniques. Poly-Si that was heavily doped and therefore conductive is often used as the gate contact to SiO_2 in MOSFETs and modern bipolar devices as well. The compatibility of poly-Si with subsequent high temperature processing allows efficient integration of this material into advanced integrated circuit (IC) processing.

The initial nucleation stage of Si film formation was observed using *in situ* real-time ellipsometry (RTE) using an electron cyclotron resonance (ECR) plasma chemical vapor deposition (CVD) system under low pressures of about 10^{-2} Torr. The ECR plasma and CVD will be discussed later in the chapter. Also *ex situ* observations were made using AFM. The Si films were deposited onto <100> Si single-crystal wafers that were previously thermally oxidized to have about 24-nm-thick SiO_2. The temperatures and SiH_4 (as the gaseous source of Si via pyrolysis) partial pressures were varied, and the effects on nucleation and growth of Si nuclei were monitored. The samples were first heated to the desired temperature by a lamp heater behind the wafer. The deposition of poly-Si was performed at 400 W and 2.45 GHz microwave power in an Ar plasma under ECR conditions with +5 V substrate bias and was monitored using *in situ* ellipsometry. A mixture of 3% SiH_4 in Ar gas was used. A rotating analyzer spectroscopic ellipsometer was used with an angle of incidence of 70°. Spectroscopic ellipsometry (SE) data was obtained in the photon energy range from about 2.5 to 5.0 eV and used for optical modeling, and single-wavelength ellipsometry at a selected wavelength (365 nm, 3.4 eV) was used for real-time measurements.

10.2 NUCLEATION

For the ellipsometric study of Si nucleation, the measured Δ and Ψ data were compared to the calculated values as obtained from models to be discussed below. The nuclei were assumed (with further justification presented below) to grow as hemispherical islands upon the underlying SiO_2 film in a hexagonal network. The nuclei form a nonuniform film of thickness L that is composed of Si nuclei and voids, as shown in Fig. 10.8a. To model this layer of nuclei the Bruggeman effective medium approximation (BEMA) as discussed in Chapter 9 was used, and the film is considered to be uniform in thickness but with inhomogeneous composition (Si and voids), as shown in Fig. 10.8b. In this approach, a one-film model is used, with film thickness identical to the nuclei height and the film dielectric properties constructed from that of bulk Si and voids (air) with a BEMA of the form:

$$f_v \frac{\tilde{N}_{Si}^2 - \tilde{N}^2}{\tilde{N}_{Si}^2 + 2\tilde{N}^2} + (1 - f_v) \frac{1 - \tilde{N}^2}{1 + 2\tilde{N}^2} = 0 \quad (10.49)$$

where f_v is the volume fraction of Si in the film calculated based on the hexagonal geometric model assumed above, $\tilde{N}_{Si}$ and $\tilde{N}$ are the complex refractive indices for Si and the mixed composition film, respectively. The optical data used as input to the model for Si at high temperature was obtained from SE measurements of a-Si and c-Si at temperature. The substrate is SiO_2 film covered Si but with known (measured) optical properties, and thus the composite is considered as a substrate with known and unchanging optical properties for the model. To prevent crystallization of a-Si, lower temperatures (600°C) were used. For regression analysis Δ and Ψ were converted to the pseudodielectric function using the formula:

$$\langle \varepsilon \rangle = \sin^2 \phi_0 \left[1 + \left(\frac{1 - \rho}{1 + \rho} \right)^2 \tan^2 \phi_0 \right] \quad (10.50)$$

where $\rho = \tan \psi e^{i\Delta}$ and ϕ_0 is the angle of incidence. A nonlinear curve-fitting routine was used to fit the experimental data both in SE and real-time ellipsometry. Upon coalescence of the nuclei, voids are considered to be present under the film at the interface or on top of the film, in order to represent surface roughness.

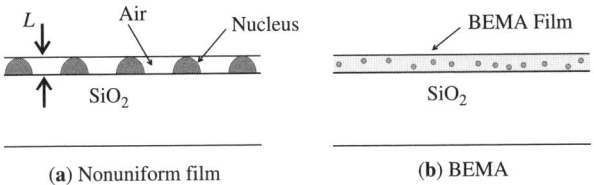

(a) Nonuniform film   (b) BEMA

FIGURE 10.8 (a) Nuclei on an SiO_2 film as a film composed of Si and voids; and (b) the nonuniform film in (a) is modeled as a mixed composition film using the BEMA.

As discussed in Chapter 9, prior to following the nucleation using real-time ellipsometry, it is necessary to construct a reasonable optical model that is based on the Si optical properties. To accomplish this characterization, a combination of SE and XTEM was performed and the results are summarized in Fig. 10.9. Figure 10.9a shows the imaginary part of the pseudodielectric function $<\varepsilon_2>$ for ECR-deposited Si films at sample temperatures of 605 to 700°C (solid lines).

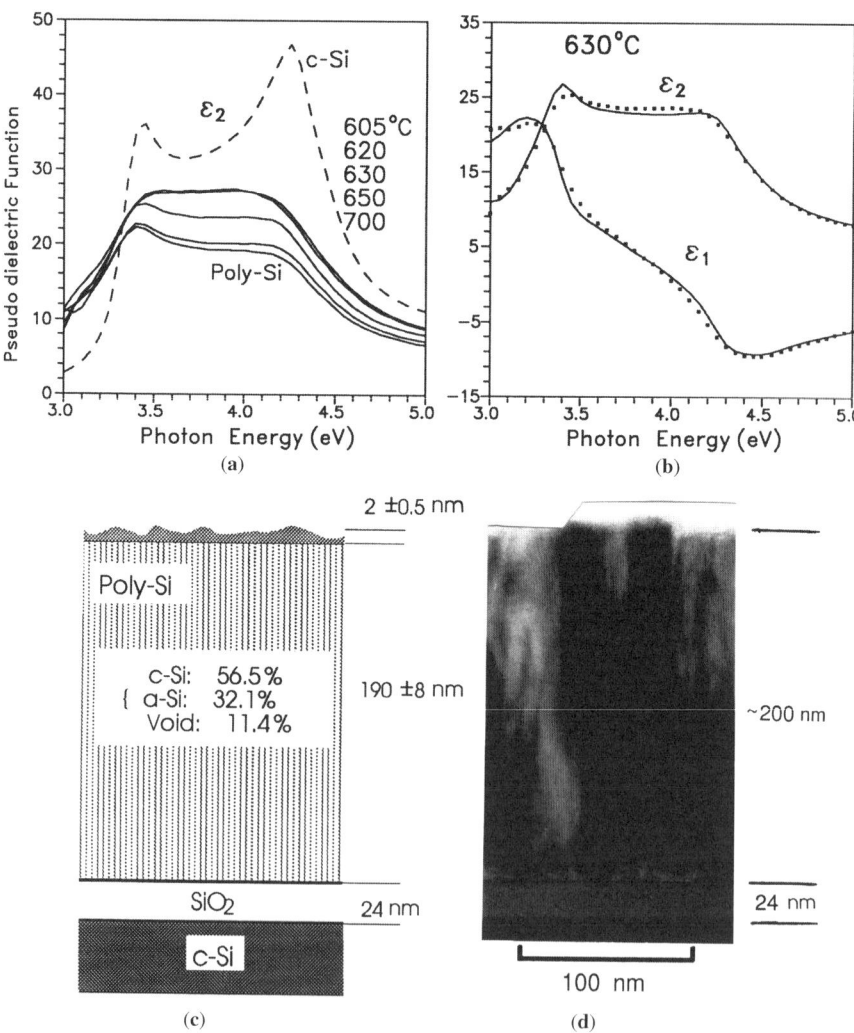

FIGURE 10.9 (a) Pseudodielectric function $<\varepsilon>$ vs. photon energy for ECR-deposited Si films at different temperatures along with reference c-Si spectrum (dashed line); (b) experimental (dotted lines) and calculated (solid lines) spectra using the optical model shown in (c) for Si film deposited at 630°C; (c) optical model; (d) cross-sectional TEM of the poly-Si film. [Adapted from Li et al. (1), Figure 2.]

10.2 NUCLEATION

TABLE 10.1 Parameters in Optical Model Obtained from Regression Analysis

Temperature (°C)	Roughness Layer Thickness (nm)	Si Film Composition (%)		
		c-Si	a-Si	Voids
605	1.5	50.4	41.7	7.9
620	1.7	53.1	39.9	7.0
630	2.2	56.5	32.1	11.4
652	3.0	56.2	30.6	13.2
700	3.3	58.5	27.3	14.2

All the spectra show features of c-Si, which has peaks located at photon energy of 3.4 and 4.3 eV (dashed spectrum), as was shown in Chapter 9, Fig. 9.14. Shown in Fig. 10.9b is the calculated spectra (solid lines) as obtained from the assumed (but reasonable) optical model shown in Fig. 10.9c, for a Si film grown at 630°C. Figure 10.9c displays the model parameters such as the various film thicknesses and composition as obtained from the regression analysis or provided as input for SiO_2 and c-Si. From the regression analysis the deposited poly-Si film contains both c-Si and a-Si. On top of the Si film is a roughness layer that is found to be necessary to include in order to improve the data fitting. The roughness layer is considered to be a mixture of a-Si, voids, and SiO_2 in a BEMA. Table 10.1 compares all samples in terms of the roughness layer thickness and Si film compositions for ECR-deposited Si films. The XTEM result in Fig. 10.9d reveals that the deposited Si film is poly-Si, which has columnar structure and has a thickness comparable to the calculated result from SE as shown in Fig. 10.9c. Thus, the optical model is consistent with XTEM, but the a-Si fraction obtained from the model must be due to the large amount of grain boundaries. Overall, the ECR-deposited films are uniform, with a thin surface roughness layer, and have the characteristics of c-Si. It is also noticed that the volume fraction of voids in the poly-Si film increased significantly from 7 to 11% with a deposition temperature rise from 620 to 630°C.

To represent nucleation, two geometric models with precedent in the nucleation literature were used. Both of the models consider nuclei with different shapes that coalesce: (1) hemispherical nucleation model and (2) cylindrical (or disklike) nucleation model. Figure 10.10 shows calculated Δ, Ψ trajectories for Si nucleation on SiO_2 for the two models and with the models depicted in the insets. For both models D is the distance between the nuclei assuming a hexagonal nuclei array, and $D = 0$ simulates layer-by-layer growth and is shown as the dashed line in Figs. 10.10a and 10.10b. Figure 10.10a for hemispherical nucleation shows that the radius of the nuclei increases continuously until the hemispheres come into contact. The positions in the Δ, Ψ plane where the nuclei come into contact (coalescence) are marked by squares. In Fig. 10.10b, which is for cylindrical nucleation, the nuclei grow to form columnar structures until contact, and then grow vertically. In this model, the growth ratio, κ, which is defined as the growth rate in the plane parallel to the surface to the out-of-plane or vertical growth direction is assumed to be one for

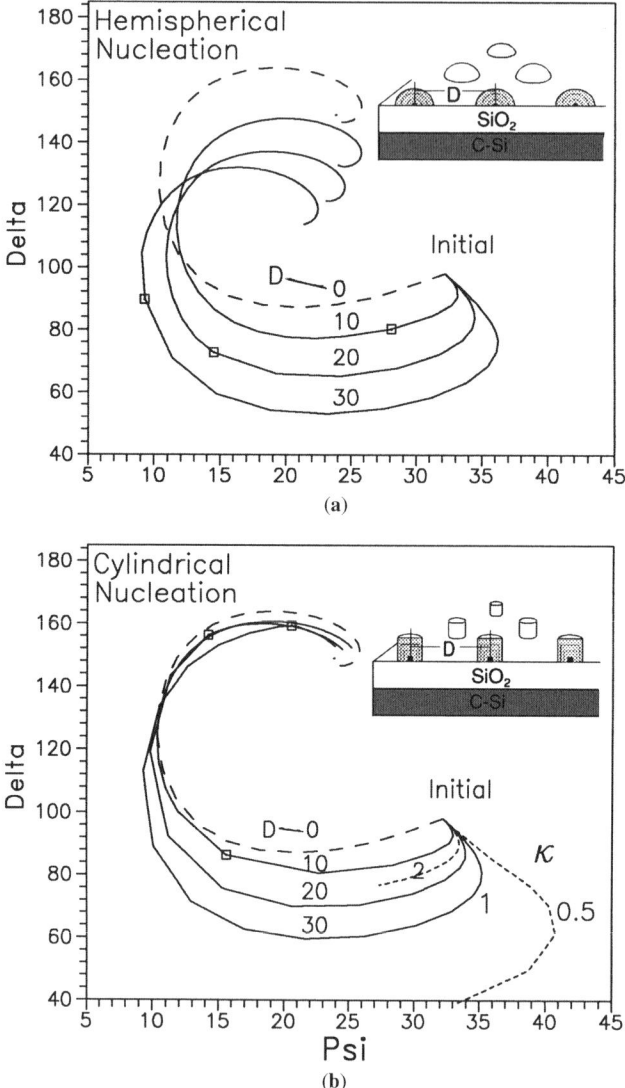

FIGURE 10.10 Δ, Ψ simulations assuming (**a**) a hemispherical model and (**b**) a cylindrical model. D is the distance between nuclei and is varied from 0 to 30 nm; the growth ratio κ is the ration of vertical to horizontal nuclei dimensions was 0.5, 1, and 2, and coalescence is indicated by squares. [Adapted from Li et al. (1), Figure 2.]

the solid lines with other values indicated for the dashed lines. Both simulations assume that surface roughness evolves from the initial nucleation mode and remains unchanged after coalescence. A constant growth rate and film refractive index ($\tilde{N} = 5.0 + 2.8i$) are assumed in both models and shown as the solid lines.

10.2 NUCLEATION

The other values of κ shown in Fig. 10.10b as dashed lines indicate that κ has a significant effect on nucleation. The treatment of coalescence for cylindrical nucleation turned out to be more complicated than for the hemispherical model in that a cone-type model was found to be necessary to interpret the data for microcrystalline Si or μc-Si growth on c-Si. From the simulation, island growth can be clearly distinguished from the layer-by-layer growth from the initial slope of the Δ,Ψ) trajectory (recall that layer by layer is for $D = 0$). Also, the ellipsometry trajectory provides information about the Si cluster density, as calculated from the initial nuclei distance and geometric distribution of the nucleus at the very early stage of film growth. The evolution of Δ,Ψ and the endpoint provide details about film coalescence and surface roughness.

As pointed out previously, it is important and useful to have complimentary techniques available to provide added credibility to any surface analytical technique. For this purpose atomic force microscopy (AFM) was used to image surfaces for which ellipsometry indicated nuclei were present, and thus AFM was used to establish and confirm the ellipsometry model. Figure 10.11a shows an AFM image of the deposited Si film for a measured Δ decrease of about 7°, as shown in Fig. 10.11b for the data point with the box. Si islands are observed with approximate height of 5 nm and a separation of approximately 30 nm. It was also found that for different temperatures that significantly influence Si film growth, the AFM images for the early stages of film nucleation show similar morphologies. From the image in Fig. 10.11a the asymmetry of the scales should be noted with a 1-μm horizontal scale and a 10-nm vertical scale the sharp peaks would really appear as small rolling hills in an image where the horizontal and vertical scales were equal.

Now using the optical models and information from SE, XTEM, and AFM presented above, the real-time evolution of Si nucleation on SiO_2 was followed using single-wavelength ellipsometry. The effects of substrate temperature and SiH_4 partial pressure on the nucleation and film growth are shown in Figs. 10.12a and 10.12b.

To explore the deposition temperature effect Fig. 10.12a shows real-time measured Δ,Ψ trajectories for five different temperatures with 0.06-min time interval between data points. The growth at lower temperatures (605 and 620°C) follow a similar lobe-shaped contour expected from a-Si, with Δ decreasing by about 7°, and Ψ increasing by about 0.5° in the initial region. The nucleation and growth at temperatures above 630°C is significantly different. At temperatures around 630°C, a larger lobe shape results from nucleation. After coalescence Δ,Ψ return to the opaque condition of $\Psi = 24°$, $\Delta = 145°$, which is the same endpoint for the two lower temperature data sets. The nucleation and growth at 650 and 700°C show similar initial nucleation but a different later stage of growth. Figure 10.13a shows Δ,Ψ data for 620°C. The solid line is a best-fit curve calculated from the cylindrical model using a refractive index $\tilde{N} = 4.4 + i3.0$. As mentioned earlier, the a-Si and c-Si data at high temperatures were measured by SE. However, both a-Si and c-Si have refractive indices that depend on temperature, and the a-Si index is dependent on the specific growth procedure. Also, it is never easy to decide whether the a-Si or the c-Si index is appropriate for modeling. For this reason, the refractive index chosen in the modeling is from a

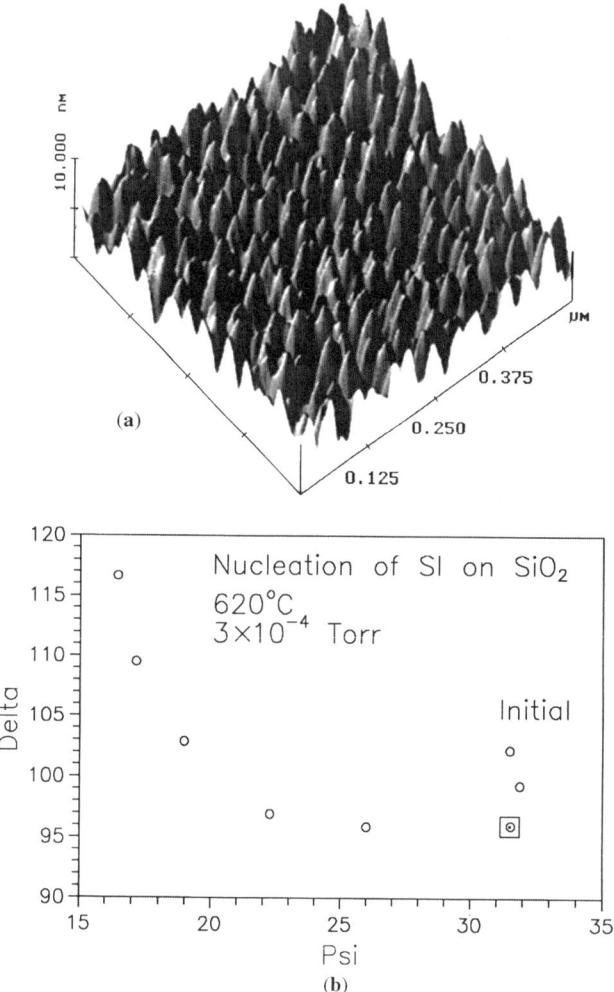

FIGURE 10.11 (a) AFM image of initial nucleation stage and (b) corresponding Δ, Ψ plot ν taken at square. [Adapted from Li, (1), Figure 3.]

measurement, although the final value was deduced from curve fitting. Also shown for comparison is the homogeneous layer-by-layer growth model, which is seen to be different from the experimental result. The cylindrical island nucleation model fits the data with an island distance of 30 nm as determined by AFM (see Fig. 10.11) as a fixed parameter, and with a growth ratio, $\kappa = 2$. A schematic of this model is shown in the insert of Fig. 10.13a. However, the hemispherical geometry model is also adequate for data fitting, but it yields an island distance D of about 5 nm, which does not agree with the AFM measurement. It was thought that in this temperature range, the Si precursors are more likely absorbed on the surface, so

10.2 NUCLEATION

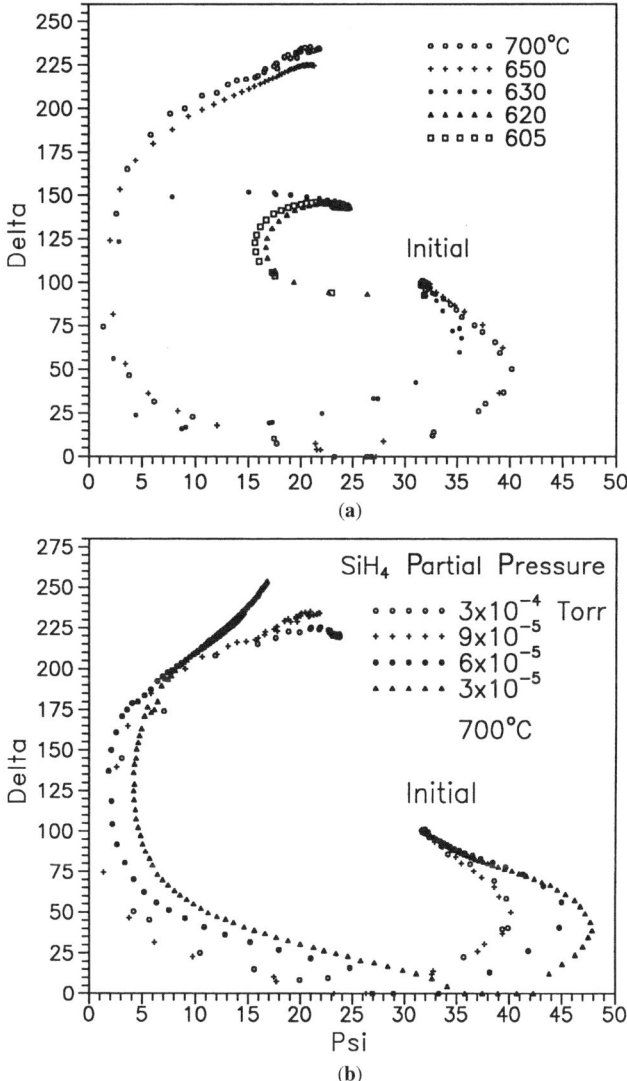

FIGURE 10.12 Real-time single-wavelength ellipsometry results in terms of Δ, Ψ for ECR CVD of poly-Si films: (**a**) as a function of deposition temperature at SiH$_4$ pressure of 9×10^{-5} Torr; and (**b**) as a function of SiH$_4$ partial pressure at total pressure of about 6×10^{-3} Torr at 700°C. [Adapted from Li et al. (1), Figure 4.]

that Si nuclei are able to contact to form a continuous film. This is the called the Kaschiev model by which the deposition is accomplished by the formation of two-dimensional critical nuclei, followed by lateral extension. Upon coalescence of the Si nuclei, a continuous film is formed that progresses in thickness with a thin roughness layer on top. As seen in Fig. 10.12a, the nucleation and growth behavior at

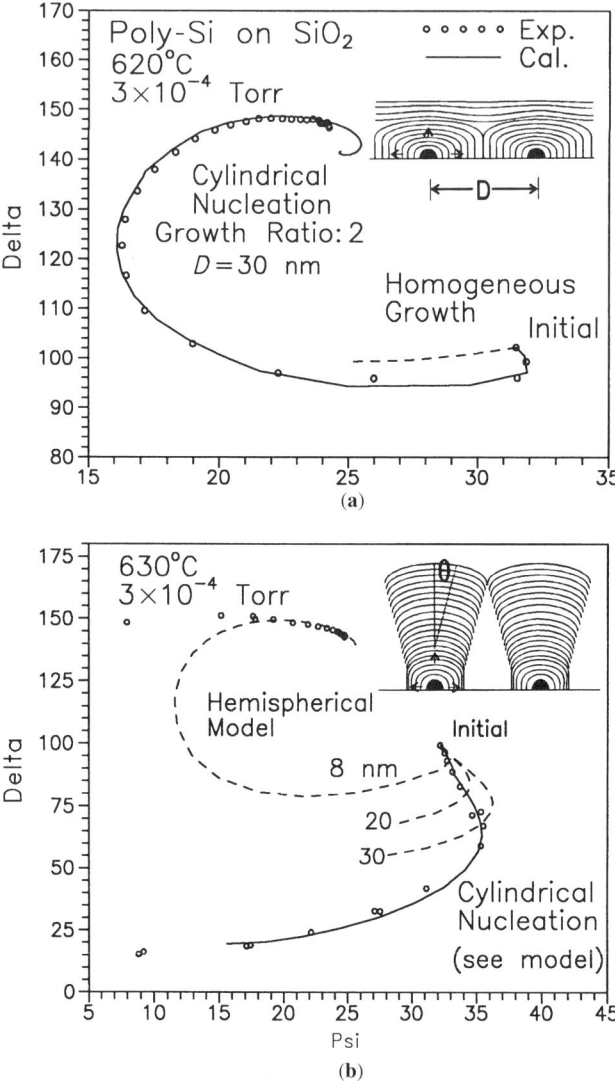

FIGURE 10.13 Nucleation data and fitting at (a) 620°C and (b) 630°C using the cylindrical model with $D = 30$ nm (solid line). The dashed line is for homogeneous or hemispherical growth. [Adapted from Li et al. (1), Figure 5.]

630°C is different. From the Δ,Ψ plot, Fig. 10.13b shows a large change of both Δ and Ψ in the initial region. This corresponds to nucleation with a large D or sparse nucleation. However, neither the hemispherical nor the cylindrical-type model fits the data very well over the range of growth. Three dashed lines in the figure are for the hemispherical model with $D = 8$, 20, and 30 nm, respectively. A conelike model, which is the solid line in Fig. 10.13b was found to be necessary to fit the initial part of the data. In this model, the Si nuclei grow with a growth ratio of 1.3. The nuclei grow faster out of plane prior to contact. An angle θ is introduced to simulate the conical shape of the nuclei, and this angle along with the contact factor S, the initial nuclei width divided by D, determine whether nuclei coalesce to form a continuous film. The physical meaning of $S < 1$ from the model is that the nuclei do not coalesce at the interface but rather above, leaving voids. For $S = 0.6$, this interface has about 36% of bulk material and 64% voids according to this model.

For higher temperatures of 650 and 700°C, the initial nucleation can be similarly fitted to the cone model. However, the later stage of growth did not fit well. From the SE results for these temperatures the surfaces are rougher and there are larger void fractions in the poly-Si film, and thus roughness likely dominates the optical measurement.

To explore SiH_4 partial pressure effects, Fig. 10.12b displays the Δ, Ψ data for different SiH_4 partial pressures at a constant deposition temperature of 700°C. The total pressure was in the range of 4 to 8×10^{-3} Torr. It is seen that clear features of nucleation were observed that varied with SiH_4 partial pressure. The initial data were fitted using the same model as in Fig. 10.13b. In the model, $\theta = 0$ indicates that the film barely coalesces to form a continuous film. When SiH_4 partial pressure is low, Si atoms are more likely to form isolated nuclei or crystal facets. Also, it is seen that increasing the SiH_4 partial pressure does not change this behavior very much.

The nucleation and growth behavior of poly-Si in ECR CVD have been studied using XTEM, AFM, SE, and single-wavelength real-time ellipsometry. The data for nucleation and growth of poly-Si from SiH_4 on oxidized Si substrates can be interpreted using an island nucleation model, where there are cylindrical-type islands with an initial spacing in agreement with AFM measurements. The initial spacing, D, does not depend strongly on temperature and is around 30 nm, which is larger than that (10 nm) for a-Si growth. Complete coalescence is observed for lower temperatures. At higher temperatures, a conelike model has to be used to fit the data. It was shown that the wafer temperature and gas pressure affect nucleation and growth. Higher temperatures seem to increase the surface roughness as well as increase the growth rate, while the low SiH_4 partial pressures can result in discontinuous films.

10.3 FILM FORMATION

As discussed in the previous sections, the nucleation process is the first process that takes place on a surface to produce the first vestige of a new phase. When nuclei continue to grow and ultimately intersect, a continuous film forms. Thus, nucleation can be considered as the first step in the film formation process, although nucleation may be fast and/or proceed with small critical nuclei and in some film studies is unobserved. Nevertheless, nucleation is an important part of the film formation process that must be considered to fully understand the nature and morphology of films.

In this section the formation process that occurs subsequent to nucleation is discussed. The present treatment of the vast area of film formation is by no means exhaustive. Rather, some fundamental principles are discussed. In Chapters 11 and 12 more examples of film growth and characterization studies that derive from the field of microelectronics are presented that use many of the principles from the present chapter. For microelectronics applications many thin-film materials are used that are prepared in different ways for different functions. In fact, the development of microelectronics since the 1960s has paralleled the development of thin-film understanding, processes, and hardware for thin-film studies. In microelectronics differing film morphologies are sometimes required. For some application amorphous films are desirable. This is typically the case for a gate dielectric where a polycrystalline material is not desirable because grain boundaries could lead to nonuniform electric fields, and a single-crystal film is difficult and costly to achieve. Also the enhanced conductivity afforded by single-crystal films is not necessary. Polycrystalline films are usually acceptable for gate contact metals because they provide higher conductivity than amorphous conductors and can be fabricated without the expense of single-crystal films. In some applications only single-crystal films will suffice, and the production of single-crystal films on a substrate is called epitaxy, and, because epitaxy implicates many of the surface and interface science principles discussed throughout this text, it is worthy of further discussion.

Epitaxy refers to the growth or deposition of a single-crystal film upon a single-crystal substrate. Processes that yield epitaxial films are used in microelectronics applications where the highest electrical quality of the film is required. For example, epitaxy could find applications in the gate region of a MOSFET or in bipolar transistor applications where PN junctions are fabricated. Typically, the carrier mobility is enhanced in an epitaxial film relative to the polycrystalline or amorphous structures of the same material. This can be understood by considering that carriers can scatter from grain boundaries, and they have little coherency in amorphous materials and thus lose mobility. Furthermore, the bonding and crystal structure defects associated with grain boundaries and amorphous materials could also lower the carrier lifetime and generation rate. The problem faced with implementing epitaxial processes is that it is not always easy and/or inexpensive to achieve epitaxy. The typical epitaxial processes (three commonly used epitaxy processes are discussed below) require high process cleanliness, accurate control of the process temperatures, pressures and compositions that are far greater than for conventional processes. Thus, while almost always desirable, epitaxial processes are used only when necessary.

There are two major kinds of epitaxy: where film and substrate structures are the same and where they are different. If the structure of the film and substrate are the same, then achieving epitaxy depends primarily on the degree of misfit between the film and substrate lattices and then on cleanliness and appropriate process conditions. Even more simply if the substrate and film material are the same, for example, an Si epitaxial film on Si (but perhaps differently doped to produce a vertical PN junction), then there is no crystallographic misfit and only cleanliness and appropriate process conditions are of concern. However, even with this simplification, it has been found that either high process temperatures or great cleanliness is required to achieve Si epitaxy on Si substrates. It is often the case that low process temperatures produce higher device yields and better quality for microelectronics processes. This is especially true for small devices where high temperature causes out diffusion or trans-interface diffusion of dopants and thus

10.3 FILM FORMATION

loss of junction sharpness and integrity. Thus, modern Si epitaxy processes have endeavored to achieve high surface cleanliness and use pure reactants and thereby lower the required epitaxial process temperatures. Typically, the processes are performed in clean ultra-high vacuum chambers.

If the film and substrate do not have the same structure, so-called dissimilar materials, then, as for the similar materials discussed above, structure and misfit remain important, but other factors such as chemical bonding and wetting (surface and interface thermodynamic properties; see Chapter 3) become important as well.

To achieve epitaxy for dissimilar materials, there must be some relationship between the crystallography of film and substrate. That means there must be some way for the film lattice to fit upon the substrate lattice. This may involve rotation if the film lattice with respect to the substrate lattice and may also require only certain substrate orientations to achieve the fit. It is more than likely that the fit will not be exact, and this is the case even for different materials with the same structure. The misfit will cause a stress to develop. As the film grows, the force due to the stress will increase. When a critical force value is attained, the film (or substrate) will mechanically fail possibly with delamination as the final state. A sufficiently large mismatch might totally prevent epitaxy.

With sufficiently small mismatch the next hurdle to overcome to achieve epitaxy is whether the chemical bonds that form at the epitaxial film–substrate interface are stable. For different materials it is possible if not likely that the interfacial bonding is incomplete, yielding excess interfacial charge. This charge build can preclude epitaxy. In some cases with compound structures such as GaAs, there is the possibility of having polar surfaces that could also affect the possibility to attain epitaxy. In one case for Ge films on GaAs it has been reported that Ge epitaxy can be achieved only on the nonpolar (110) GaAs surface. The reason is thought to be related to charge accumulation at the interface (2).

For epitaxy it is important to achieve layer-by-layer growth (discussed in Section 10.2 on nucleation). For this mode of film growth the critical nucleus size must be small, and this is achievable when the film completely wets the substrate, that is, when the contact angle (see Chapter 3) is near zero. This will occur when the surface energy for the film γ_f is less than that for the substrate, γ_{sub}, $\gamma_f < \gamma_{sub}$. If $\gamma_f > \gamma_{sub}$, then the critical nucleus size will be large and island formation will occur, which reduces the chances of achieving epitaxy. To reduce critical nucleus size, it is common to use high film growth rates and low process temperatures, and another way is to use surfactants to alter the surface energy. The use of surfactants has been discussed by Copel et al. (3) for the formation of Ge epitaxy on Si(100). In this study molecular beam epitaxy (MBE) to be discussed below was used to deposit Ge on Si and the reverse where As was used as a surfactant to alter the surface properties of the substrate. Arsenic was known to form a monolayer on both Si and Ge and to form a 2 × 1 overlayer on Si(100) and Ge(100) where it occupies the dangling bonds on these surfaces. In this study Si(100) single-crystal wafers were carefully cleaned and coated with a monolayer of As. The Ge was deposited. After Ge deposition the Ge surface was covered with As and Si was deposited. Epitaxial Ge and then Si layers were produced through the use of As, and the As was not found to alloy with either the Ge or the Si. There is a 4% lattice mismatch between Ge and Si, and thus island growth of Ge upon Si is expected. However, with the As monolayer no strain was observed, and the Ge was found to be epitaxial with no As alloying, and in fact the As was found to deplete during epitaxial growth. Next a Si capping layer was deposited epitaxially upon a thin (1.5 monolayers) Ge layer. The Ge was thin to eliminate strain due to the lattice mismatch. The procedure could be extended, producing 1.5 layers of Ge followed by 15 monolayers of Si for a total of 4.4 nm. Arsenic was used only prior to Ge deposition.

The following sections on film formation consider the two major ways to prepare films on surfaces: growth and deposition. As mentioned above, film growth refers to the process that uses the surface atoms/molecules as one of the reactants, that is, atoms from the substrate partly comprise the film composition. Deposition is a process in which atoms and/or molecules usually from the vapor phase are sometimes reacted and then condensed upon an existent surface to form a film. Figure 10.14 shows three common variations (I, II, and III) of deposition and growth processes. For all three cases the film that is formed is composed of the material labeled C. For case I, reactant A reacts with D, which is the solid surface yielding the product C with some of the surface D being consumed. This case is film growth and involves the solid surface as a reactant. Case II involves the reaction of gas-phase reactants A and B that react and form the gas-phase product C that then condenses to solid C on the surface D. Case III involves the reaction of A and B as for case II. However, in case III the reactants react at the surface. The surface material does not react but rather mediates the reaction and in effect may act as a catalyst for the reaction.

10.3.1 Film Growth

A straightforward example for case I is the oxidation of the single-crystal Si surfaces in oxygen [a review of Si oxidation is found in Irene (4)]. As discussed in Chapter 1, Si is diamond cubic and each major Si plane would expose a different surface area density of Si atoms to the O_2. When any Si crystal surface is in contact with sufficient O_2, the oxygen reacts with Si forming SiO_2. Thus, the Si surface supplies Si atoms to

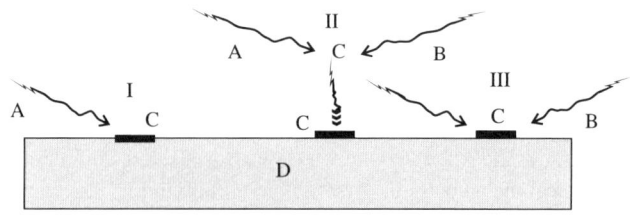

I Reaction with Surface
II Gas-Phase Reaction
III Reaction at Surface

FIGURE 10.14 Three (I, II, III) kinds of film formation mechanisms.

10.3 FILM FORMATION

the oxidation reaction and hence is an example of case I. Also, as discussed in Chapter 1, the different Si planes will oxidize with different rates of SiO_2 film formation. Here we consider more details for this important film growth system. Some established facts about the oxidation of Si is that the nucleation stage for this reaction is fast because the critical nucleus size is of the order of the molecular size of SiO_2 or $n_i^* = n_1$, and thus the nucleus formation is highly probable and fast. The growth is layer by layer, thereby forming uniform conformal and smooth amorphous SiO_2 films on a flat Si single-crystal surface. In a mechanistic scheme consisting of a series of reaction and/or mass transport steps, it is the slow steps that are observed. Therefore, the nucleation stage for Si oxidation is not kinetically significant in that it is fast. After the rapid formation of the first layer of SiO_2 on the Si surface, the reactants (Si and O_2) are separated by the already formed SiO_2 layer. Thus, for the formation of more than one layer of SiO_2, the Si atoms must migrate through the SiO_2 layer to react with O_2 or the O_2 molecules or O atoms must migrate through the SiO_2 to react with Si or both Si and oxygen must migrate. The established fact for this particular case is that the O_2 (most likely and predominantly the molecular O_2 species) migrates through the SiO_2. For many metal oxidation systems the metal atoms move and so the identity of the diffusing species must be experimentally verified for the case under study. With these facts about Si oxidation the mechanism for the growth of SiO_2 films on single-crystal Si can be envisioned as three processes in series, as shown in Fig. 10.15.

In this process a cleaned Si surface, typically a single-crystal surface, is exposed to oxygen at temperatures above 500°C. Almost instantly an SiO_2 film forms, and from this point in time forward the oxide grows. To follow the film formation kinetics, one observes the increase in the thickness L of SiO_2 as a function of time, dL/dt. As shown in Fig. 10.15, the first process in the series process scheme is the dissolution of O_2 in the SiO_2 at the gas–solid (O_2–SiO_2) interface. This process is typically fast relative to the other two processes in the series scheme and results in a concentration of O_2 in the SiO_2 of C_1 at the outer interface. The flux corresponding to this process is

$$J_1 = k_{sol} P_{ox} \qquad (10.51)$$

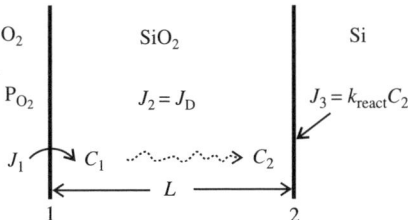

FIGURE 10.15 Film growth process: the oxidation of Si in O_2. Three fluxes in series J_1, J_2, J_3 across two interfaces 1 and 2 with oxidant concentration at the interfaces of C_1 and C_2 and SiO_2 film of thickness L.

This process is essentially the Henry law dissolution of O_2 in SiO_2, which is proportional to the pressure of O_2 with the Henry law constant k_{sol}. The second process in the series scheme is the diffusion of O_2 through the SiO_2 of thickness L to the SiO_2–Si interface, resulting in a smaller concentration of O_2, C_2. In the steady state the flux corresponding to this process is given as

$$J_2 = \frac{D_{ox} \, dC}{dL} = \frac{D_{ox}(C_1 - C_2)}{L} \tag{10.52}$$

The third process in the series scheme is the reaction of the O_2 at concentration C_2 with Si at the inner SiO_2–Si interface. This process continually removes O_2 and prevents accumulation at the interface. The flux corresponding to this process is expressed by a first-order reaction between O_2 with Si:

$$J_3 = k_{react} C_2 \tag{10.53}$$

A first-order rate expression with k_{react} the first-order rate constant has been chosen initially for simplicity to describe the reaction of the O_2 arriving at the Si surface to react with a constant concentration of Si at the crystal surface. Later research has confirmed the assumption of first order in O_2. This three-flux scheme can be simplified by considering that J_1, a gas-phase flux, is much faster than the other two fluxes that occur in the solid state. As mentioned above, with processes in series the slowest process determines the observed rate. Therefore, because J_1 is much faster than the other fluxes, we need not consider it further. Furthermore, fluxes J_2 and J_3 must be equal after an initial transient. That this is the case can be understood by considering the consequences using the formulas above for J_2 and J_3. If the fluxes were not equal, then either J_2 or J_3 would be the larger flux. If J_2 is larger, then the supply flux of oxidant to the SiO_2–Si interface (that depends on the difference $C_1 - C_2$) would increase, thereby supplying more oxidant to the interface, C_2, that would in turn increase J_3, which would then reduce C_2 and regulate the process. Likewise if J_3 were larger than J_2, then J_3 would reduce C_2 and in so doing increase the difference $C_1 - C_2$ and thereby increase J_2, again regulating the series steps. Thus, this self-regulating set of series fluxes must be equal, and hence a steady state obtains and the use of Fick's first law for the transport of O_2 is justified. Now with $J_2 = J_3$, the equivalence of Equations 10.52 and 10.53, C_2 can be obtained as follows:

$$k_{react} C_2 = \frac{D_{ox}(C_1 - C_2)}{L} \tag{10.54}$$

and

$$C_2 = \frac{D_{ox} C_1}{k_{react} L + D_{ox}} \tag{10.55}$$

10.3 FILM FORMATION

Using this value for C_2, a rate expression can be formed using the relationship:

$$J = J_2 = J_3 = \frac{\Omega \, dL}{dt} = k_{\text{react}} C_2 \tag{10.56}$$

With C_2 substituted from above and where Ω is the conversion from moles of oxygen gas in J_2 to moles of oxygen in solid SiO_2 J_3, and Ω has the value of about 2.2×10^{22} cm^{-3}. This value is obtained from the density of SiO_2 of 2.2 g/cm^3 divided by 60 g/mol for the molecular weight of SiO_2 all multiplied by Avogadro's number. A rate equation can be written with these results as follows:

$$\Omega \frac{dL}{dt} = \frac{D_{\text{ox}} C_1}{k_{\text{react}} L + D_{\text{ox}}} \tag{10.57}$$

Upon integration this yields a linear-parabolic dependence of the film thickness L on time of oxidation t as follows:

$$t + \text{const} = AL^2 + BL \tag{10.58}$$

where $A = 1/\Omega D_{\text{ox}} C_1$ and $B = 1/\Omega k_{\text{react}} C_1$; A and B are called the parabolic and linear rate constants, respectively. Diffusion is represented in the parabolic constant through D_{ox} while the linear constant depends upon the interface reaction through k_{react}. This linear-parabolic formula has been verified for the thermal oxidation of silicon. This real example illustrates how mass transport in terms of diffusion couples into the permeation problem, and it illustrates that often the overall permeation process is observed in which diffusion is only a part of the process.

From Equation 10.53 the first-order rate constant k_{react} was used. This rate constant assumes a constant Si atom concentration per area for a given orientation. However, the value for k_{react} must be different for different Si surface orientations as was given by the different area densities for Si atoms in Chapter 1, Table 1.1, where the Si atom area density for selected planes in single-crystal Si is given. Table 1.1 implies that in the early stage of Si oxidation where the SiO_2 film is thin, and thus where the oxidation kinetics are dominated by the interface reaction through the interface reaction rate constant k_{react}, rather than being dominated by the transport of O_2 as would be the case for thick SiO_2 films, k_{react} and hence the oxidation rate should be different for the various Si orientations. The order for the rate of oxidation should be that given in Table 1.1. Figure 1.6 shows experimental ellipsometry data for the Si oxidation with the three lowest Miller index planes for Si. The data was obtained using an ellipsometer attached to the oxidation furnace so that the experiment could be performed on the same sample without removing the sample during the oxidation, that is, *in situ*. It is seen that for that thinnest films of less than 20 nm (200 Å) the rate order is (110) > (111) > (100) as indicated by the order of the area density of Si atoms. Beyond this thickness there is a crossover with the (111) displaying a greater rate than the (110) orientation. This crossover has been attributed to SiO_2 film stress effects on the diffusion of O_2 in SiO_2 and that has been found to be smallest for the (111) Si orientation (this stress effect on Si oxidation is discussed further in Chapter 12). Figure 1.7 displays similarly obtained ellipsometry data at lower oxidation temperature where the order for the three lowest Miller index planes is the same as for the 800°C data in Fig. 1.6, but the crossover is barely visible at the thinnest SiO_2 film thicknesses. Interestingly, this data also shows oxidation data for the (311) and (511) orientations and once again shows the oxidation rate order dictated by the Si atom area density given in Table 1.1.

The example presented above of Si oxidation to produce SiO_2 films is a technologically important process that exemplifies growth as opposed to film deposition. Also this film growth system consists essentially of the processes of reaction and transport in series that can be modeled in a straightforward manner. The basic concepts about surface and interface structure can be used to understand the details of the surface reactions for film growth. Finally, the film growth process discussed above is a relatively simple process to implement. Figure 10.16 shows the usual method to perform Si oxidation in the microelectronics industry using a resistance-heated fused silica tube in which oxygen is flowed and the Si substrates are positioned. The fused silica tube end caps are designed to ensure that back flow is minimized. Of course, all aspects of this process can be automated. While for ultimate temperature control and process simplicity resistance heating is energy source of choice, other forms of energy input are described in the literature including the use of plasmas and microwaves. These forms of energy are more appropriate to chemical vapor deposition discussed below and typically are more costly, complicated and often do not yield as uniform a SiO_2 film as resistance heating.

It should be understood that while Si oxidation provides a powerful and well-studied example the details may not be applicable to other materials systems. The Si oxidation process is of great importance in microelectronics, and more about the silicon oxidation process was previously presented in Chapter 4 on roughness and more will be discussed in Chapters 11 and 12.

It is worth returning to the fact that Si oxidation in pure O_2 at oxidation temperatures below 1200°C result is amorphous SiO_2. This result is expected based on the fact that Si and SiO_2 do not have conformable crystal structures plus the large volume change that takes place upon oxidation disorders the interfacial region. To produce epitaxy the film–substrate interface must at least be closely matched in crystal structure and lattice size, and the process must produce a film that forms in

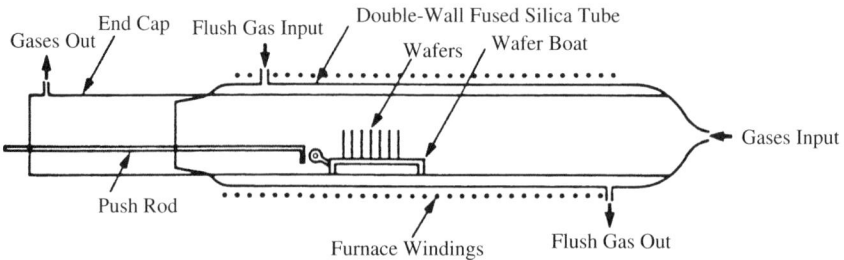

FIGURE 10.16 Fused silica double-walled Si oxidation system.

a nearly layer-by-layer mode. Most often polycrystalline and amorphous films are produced owing to the stringent requirements for epitaxy.

10.3.2 Film Deposition

Deposition processes commence with nucleation and after nucleation can be classified in broad terms as either chemical deposition or physical deposition, and the most important deposition processes involve the vapor phase and are called chemical vapor deposition (CVD) or physical vapor deposition (PVD). In reality, there may be both chemical and physical aspects to a given deposition process, and in some instances the appropriate acronym is not clear. Nevertheless, for the purpose of distinguishing the predominant basic principles of these deposition processes and in light of the common usage of the terms, CVD and PVD will be discussed separately with emphasis on the basic ideas of CVD and PVD. Some microelectronics examples of CVD and PVD will be discussed in the following chapters.

10.3.2.1 Chemical Vapor Deposition Chemical vapor deposition (CVD) can be defined as a method to produce a desired material usually in film form on a substrate material (that may be the same or different as the film) through the use of predominantly gas-phase chemical reactions. This kind of film formation process would most closely resemble mechanism II in Fig. 10.14 where a predominantly gas-phase reaction dominates the process. Figure 10.17 shows a generalized CVD system with a dashed line boundary in which a substrate is placed. One or more gaseous reactants are flowed into the system, mixed, and made to react, leading to the desired product that can nucleate and form a film on a substrate surface placed in the reactor. Often there is a source of energy to stimulate the gas-phase reaction. This energy input may be as simple as resistive heating using a tube furnace, as shown in Fig. 10.16 for Si oxidation, or as esoteric as a plasma. The source for the energy input is one

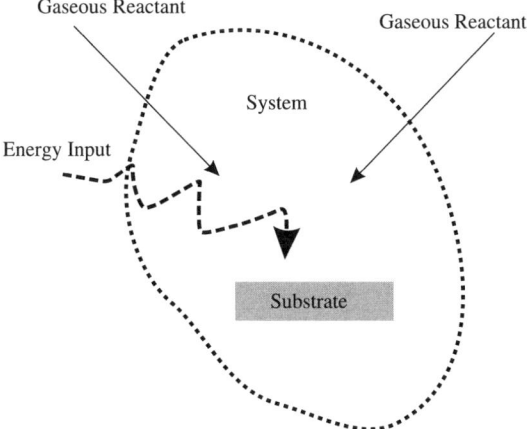

FIGURE 10.17 Generalized CVD system.

complication, and the reactant gas supply is another. Also some CVD reactors are operated at total pressures below 1 atm.

The characteristics and complications of CVD are illustrated using Fig. 10.18, which is a schematic of a radio frequency (RF) heated low pressure CVD system. As mentioned above a CVD apparatus will consist of several basic components:

- Gas Delivery System   For the supply of reactive precursors to the reactor chamber.
- Reactor Chamber   Chamber within which deposition upon substrates takes place.
- Energy Source   Provide the energy/heat that is required to facilitate the precursors to react/decompose.
- Vacuum System   A system to maintain a low operating pressure and also for removal of all other gaseous species other than those required for the reaction/deposition.

Along with the above-listed major subsystems a typical CVD system will also have an exhaust system for removal of volatile by-products from the reaction chamber, especially toxic by-products. When the exhaust gases are not suitable for release into the atmosphere or into an exhaust system, the exhaust products may require treatment or conversion to safe/harmless compounds. This can often be done by pyrolysis. Filters are also usually used to remove particulate that would harm mechanical pumps. CVD systems will have controls to monitor process parameters such as flow, pressure, temperature, and time. Many commercial CVD systems can automatically introduce and remove substrates and/or boats loaded with substrates. Some research-oriented

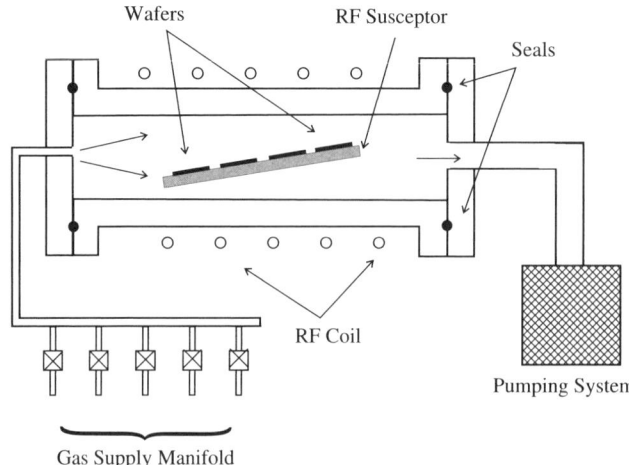

FIGURE 10.18   Schematic of a low pressure RF-heated CVD system showing major components.

CVD and PVD systems will also have various gas stream, surface, and film analytical techniques in order to perform *in situ* and real-time characterizations.

In Fig. 10.18 the samples in the form of circular wafers are placed on an RF susceptor, often called a boat, that is typically fabricated from graphite (a slab of machined graphite that can be coated with chemically and thermally inert BN, and sometimes slotted to prevent samples from shifting position). The graphite boat is placed within a fused silica enclosure, and the boat is heated using an RF generator that supplies RF energy to a metal coil (usually Cu tubing) around the fused silica tube. The graphite boat acts as a susceptor for the RF energy and heats the substrates that have been placed on the boat. In this form of energy input only a relatively compact region near the graphite susceptor is heated as compared with resistance furnace heating that heats the entire reactor volume within the furnace. The walls of the RF-heated enclosure are typically near room temperature. Among the advantages of this kind of heating is that the energy input is smaller and only needed for a short time before the process is started. Resistance heating is slow to change temperature and thus needs to be brought to the desired temperature literally hours before starting the process for a modestly large system. The fact that a smaller volume is heated also reduces impurities from the walls of the enclosure to be released during the process. In addition, with a large heated volume the reactants could become depleted before the desired product is near the substrates. Depletion of the reactants leads to low deposition rates and perhaps shifts in stoichiometry during deposition. Depletion effects can be partially compensated by tilting the wafer boat as shown in Fig. 10.18. Among the negative attributes of RF heating is that temperature uniformity is more difficult to achieve and the system is expensive and complex relative to resistance heating.

To supply reactants, a supply manifold is seen in Fig. 10.18 enabling a variety of reactant gases, dopants, and diluent gases to flow into the reaction vessel and thereby enable the system to perform many different CVD depositions. Automated systems have gas flow controllers to precisely meter the reactants into the reactor. In addition, the CVD reactor shown in Fig. 10.18 is operated at low pressure for several reasons that will be discussed below. For now, Fig. 10.18 displays a typical high gas capacity vacuum system that consists of a Roots pump backed up by a mechanical vane pump. The Roots pump (sometimes referred to as a Roots blower) uses rotating blades and is capable of high gas loads flowing through the reactor. This pump compresses the gas at the input of the vane pump enabling more efficient removal of the precompressed gases. Since there are usually particles of various products that have formed in the reactor, the effluent from the reactor needs to be filtered before reaching the pumps to reduce wear and damage to the pumps.

In a CVD reactor operated at atmospheric pressure (APCVD) the gases flow by convective transport through the reactor vessel. Because of this mode of transport, there are layers of gases that have different velocities due to frictional interaction with components in the vessel and the vessel walls, and there are layers with different temperatures and density. As the reactive gases enter, a heated zone reaction commences, and, if substrates are not placed in this region of the reactor, the resulting condensate is not accumulated on the substrates but rather depleted as mentioned above. The depletion is more noticeable in the dense gases at 1 atm. For all these reasons 1-atm reactors notoriously yield nonuniform depositions with respect to film thickness and stoichiometry across a substrate load in a CVD run. In addition, much of the reactants merely flow out of the reactor and are wasted. At lower pressures near 0.1 atm gaseous diffusion can become the dominant transport mechanism, and with the random motion of the gas atoms or molecules in the reactor both temperature and composition are more uniform. This is desirable for resulting CVD film uniformity as well as the economy associated with using smaller quantities of often expensive and toxic reactants. Therefore, for larger sample loading and for expensive reactants low pressure CVD (LPCVD) is the process of choice.

Plasma-assisted CVD (PACVD) or plasma-enhanced CVD (PECVD) has been developed to achieve low process temperatures. Hundreds of degrees lower process temperatures can be employed using plasmas to initiate CVD. Using microwaves or an electric field, a plasma can be created in a low pressure gas. A plasma is a dense gas of ions, electrons, and energetic neutrals all of which can promote chemical interactions. The energy for starting the chemical reactions generating the films comes from the collision of the plasma ions, neutrals, and electrons. More on plasmas will be introduced in the section on PVD below.

Besides resistance heating, RF heating, and plasmas, for promoting CVD reactions, photonic sources both intense incoherent and laser sources are also often used in research applications.

CVD precursor materials fall into a number of categories such as:

- *Halides* $TiCl_4$, $TaCl_5$, WF_6, and so on
- *Hydrides* SiH_4, GeH_4, $AlH_3(NMe_3)_2$, NH_3, and the like
- Metal organic compounds such as:
  - *Metal Alkyls* $AlMe_3$, $Ti(CH_2tBu)_4$, and so forth
  - *Metal Alkoxides* $Ti(OiPr)_4$, for example
  - *Metal Dialylamides* $Ti(NMe_2)_4$, for example
  - *Metal Diketonates* $Cu(acac)_2$ and the like
  - *Metal Carbonyls* $Ni(CO)_4$ and so forth
  - *Others*   Include a range of other metal organic compounds, complexes, and ligands

There are a variety of chemical reactions used in CVD, and among the most common used in microelectronics are the following:

- *Oxidation*: $SiH_4(g) + 2O_2(g) \Rightarrow SiO_2(s) + 2H_2O(g)$
- *Hydrolysis*: $AlCl_3(g) + 3H_2(g) + 3CO_2(g) \Rightarrow Al_2O_3(s) + 6HCl(g) + 3CO(g)$
- *Pyrolysis*: $SiH_4(g) \Rightarrow Si(s) + 2H_2(g)$; $Si(OC_2H_5)_4 \Rightarrow SiO_2 + \cdots$
- *Disproportionation*: $Si(s) + SiI_4(g) \Rightarrow 2SiI_2(g)$
- *Direct reaction*: $(CH_3)_3Ga(g) + AsH_3(g) \Rightarrow GaAs(s) + 3CH_4(g)$
- *Nitridation*: $3SiH_4(g) + 4NH_3(g) \Rightarrow Si_3N_4(s) + 12H_2(g)$

10.3 FILM FORMATION

There are many additional acronyms that have been widely used. For example, there has been a recent trend toward organic precursors for CVD owing to the diverse chemistry that can be obtained, and the process is called metal organic CVD, or MOCVD. When this process is practiced in conjunction with a vacuum system, it is often called metal organic vapor-phase deposition, MOVPD, and, if conditions can be adjusted to achieve epitaxy, MOPVD becomes MOVPE where the E is for epitaxy.

10.3.2.2 Physical Vapor Deposition The "physical" part of the acronym PVD refers to the binding energy that an atom or molecule has in a solid. PVD commences with a solid composed of the desired atoms or molecules. These atoms or molecules are then removed from the solid to the vapor phase by a physical process that overcomes the binding energy. Typically, the binding energies are of the order of about 10 kcal/mol (about 0.4 eV/atom) and often less while as discussed previously chemisorption binding energies can range upward of 100 kcal/mol. Even weak or nondissociative chemisorption, such as the binding of chemisorbed water vapor to vacuum chamber walls, has binding energies on the order of 20 kcal/mol. Then the key difference between PVD and CVD is the energy input to the process. For CVD, chemical reactions were sought and conditions adjusted to initiate gas-phase reactions. In PVD processes various energy sources are used to remove atoms/molecules from a solid to the vapor phase and permit the desired moieties to recondense upon a desired substrate elsewhere in the system. The general picture for PVD process is shown in Fig. 10.19 is different from CVD in that the reactant source is usually solid and the energy is directed at the source so as to eject the desired atoms/molecules from the source. In the following sections several common PVD methods are discussed: evaporation, molecular beam epitaxy, sputtering, and plasma.

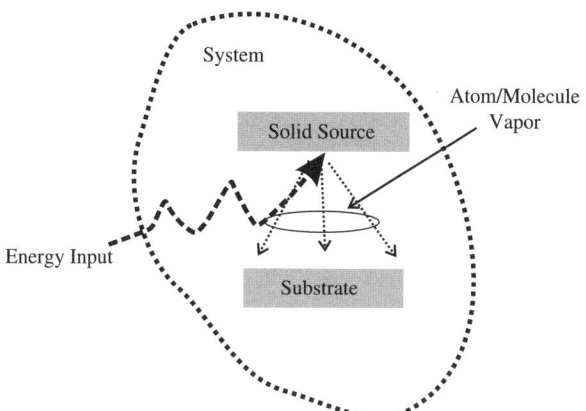

FIGURE 10.19 Generalized PVD system.

Evaporation One simple and important PVD process is evaporation where the energy input to dislodge atoms and/or molecules is derived from the heat added to the solid source. As the temperature is increased, the vapor pressure of the material increases, yielding more atoms/molecules to the vapor. Evaporation is the process whereby atoms or molecules in a condensed phase gain sufficient energy to enter the vapor state. Then with a substrate placed in the system in which evaporation is occurring but at a temperature below the condensation temperature, the atoms/molecules will condense upon the cooled surface, nucleate a condensed phase, and form a film as was discussed above.

Evaporation and condensation can be understood by first considering the flux $J(\#/At)$ of atoms/molecules striking a surface. Consider the shaded surface shown in Fig. 10.20 and atoms/molecules in the cylindrical volume element traveling in the x direction normal to the surface. Atoms/molecules in the volume with a velocity component in x, $\mathbf{v}_x$, will strike the area A in time t to yield a flux J. This flux can be obtained from the following integral:

$$J = \rho \int_0^\infty f(\mathbf{v}_x) \, d\mathbf{v}_x \tag{10.59}$$

In this integral ρ is the density in units of (# particles)/volume and $f(\mathbf{v}_x)$ is the Maxwell–Boltzmann distribution of velocities given as:

$$f(\mathbf{v}_x) = \left(\frac{M}{2\pi kT}\right)^{1/2} \exp\left(-\frac{m\mathbf{v}_x}{2kT}\right) \tag{10.60}$$

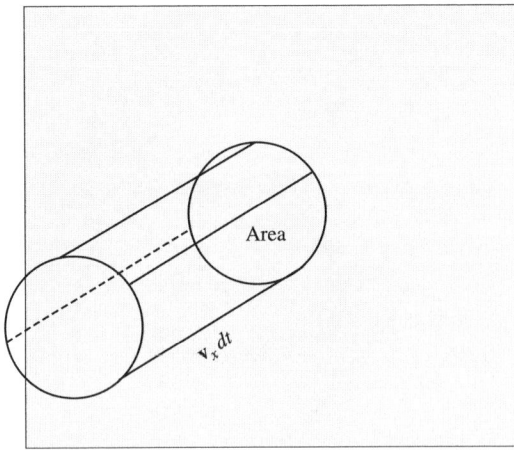

FIGURE 10.20 Cylindrical volume element for particles traveling in the x direction normal to the surface (shaded) with circular area and height $\mathbf{v}_x \, dt$.

10.3 FILM FORMATION

where M is the molecular weight of the atoms/molecules in the vapor. The integration to obtain J is obtained using the definite integral:

$$\int_0^\infty x e^{-ax^2} = \frac{1}{2a} \tag{10.61}$$

where $a = M/2kT$; $\frac{1}{2}a = kT/M$. The result for the impingement flux J is found to be

$$J = \rho \left(\frac{kt}{2\pi M}\right)^{1/2} \tag{10.62}$$

From this result and the ideal gas law:

$$PV = nRT = nN_A kT \tag{10.63}$$

an expression for density is obtained as follows:

$$\rho = \frac{nN_A}{V} = \frac{P}{kT} \tag{10.64}$$

This value for ρ when reinserted in Equation 10.62 for J yields the so-called Langmuir equation for the impingement flux as follows:

$$J = \frac{P}{(2\pi MkT)^{1/2}} = 2.89 \times 10^{22}(MT)^{1/2} \quad \text{for P in Torr} \tag{10.65}$$

Recall that this formula for flux was introduced as Equation 10.37 and in Chapter 8 as Equation 8.22. Some of the impinging atoms/molecules will stick and some will reevaporate. The number of impinging particles that stick can be expressed as

$$N_s = \alpha J = \alpha \frac{P}{(2\pi MkT)^{1/2}} \tag{10.66}$$

where α is called the sticking or accommodation coefficient. If the system comes to equilibrium, then the evaporation rate will equal the number that stick N_s as is given above. The maximum evaporation rate will be when P is the equilibrium pressure $P°$ and at $\alpha = 1$. If the vapor were removed so that evaporation takes place into vacuum and we are evaporating a solid (rather than a monolayer that had previously adsorbed) so that there is no shortage of evaporant, the maximum rate of evaporation will be given by the above formula rewritten for evaporation as

$$J_{\text{evap}} = \alpha \frac{P°}{(2\pi MkT)^{1/2}} \tag{10.67}$$

It is true that as evaporation proceeds the surface will cool because a larger fraction of the most energetic particles will leave the surface, and, therefore, the rate of evaporation will readjust to the surface temperature.

From Equation 10.67 the evaporation flux J_{evap} can be converted to the mass evaporated Γ_{evap} by multiplying by the factor M/N_A to yield for $\alpha = 1$:

$$\Gamma_{evap} = 5.834 \times 10^{-2} P° \sqrt{\frac{M}{T}} \, (g/cm^2 s) \tag{10.68}$$

This formula applies to a point source evaporating and can be integrated over time to yield the amount evaporated in a given time interval, M_{evap}. To calculate the amount striking a substrate M_{sub} from the amount evaporating M_{evap}, we can use the following ratio:

$$\frac{M_{sub}}{M_{evap}} = \frac{A_{sub}}{4\pi r^2} \tag{10.69}$$

where the ratio on the right is the area of the substrate relative to the area of a sphere into which evaporation from a point source occurs. If the substrate is not orthogonal to a radius from the point source but rather at some angle θ, then the formula is modified:

$$\frac{M_{sub}}{M_{evap}} = \frac{A_{sub} \cos \theta}{4\pi r^2} \tag{10.70}$$

The evaporation from a surface rather than from a point source is more likely and then involves the angle of the surface of the source ϕ as well:

$$\frac{M_{sub}}{M_{evap}} = \frac{A_{sub} \cos \theta \cos \phi}{4\pi r^2} \tag{10.71}$$

Thus, two factors determine the amount of evaporant striking a surface, the geometry (angles θ and ϕ) and the separation of source and substrate (as r^{-2}).

The evaporation PVD method is then simple to practice. Within a vacuum enclosure the evaporant charge, usually a solid, is raised in temperature. It is usual to melt the evaporant to increase the vapor pressure and hence the rate of evaporation. In this case the evaporant is placed in a container that is inert to the molten evaporant. This container is called a boat. It is usual to use a boat material that is both inert to the evaporant and is a good conductor of electricity. The boat can then be easily heated via the passage of a high electrical current. For example, to evaporate Au or Al a Ta or W boat can be used, although the reactivity of molten Al is an issue. However, to evaporate W with a melting point of over 3400°C is problematic because to achieve this temperature a large volume of the apparatus would also rise to high temperatures, and there is the issue of the boat material to withstand the temperatures and remain inert. On solution for the evaporation of low vapor pressure materials and alloys where each component has a different vapor pressure is to heat rapidly and locally to high temperatures using focused electron beams. This so-called e-beam evaporation uses a high voltage electron beam (typically 10 to 100 kW) that is focused tightly upon the evaporant that is contained in a water-cooled crucible. In this way only a small area of the evaporant is heated to a high temperature, which results in a small pool of the molten evaporant contained within the solid evaporant. Insulators and alloys as well as metals can be e-beam evaporated.

Similar to e-beam evaporation is laser ablation. Ablation refers to evaporation directly from the solid phase caused by the intense local heating. Laser ablation then is the use of a laser, typically a pulsed laser, impinging on the evaporant to cause evaporation. The rapid heating gives rise to an expanding plume of the ablated material and some of the evaporated material is ionized.

Molecular Beam Epitaxy There are many applications particularly in the field of electronic materials that require epitaxial films; often the desired epitaxial material has complex chemical stoichiometry. For these applications the chemical constituents must be supplied with great control to achieve a high level of crystal perfection as well as the desired precise chemical stoichiometry. For this purpose molecular beam epitaxy (MBE) systems that utilize Knudsen cells are in current use. A Knudsen cell is a sealed container with one small diameter orifice in which an evaporant is placed, as shown in Fig. 10.21a. The cell containing evaporant is heated to produce a vapor of the evaporant with a vapor pressure inside the Knudsen cell of from about 10^{-3} to 10^{-7} atm. In this range kinetic theory predicts that the number of vapor particles that strike the cell orifice of area A in a unit time and escape from the cell is a function only of the vapor pressure within the cell that can be closely controlled by the temperature. Thus, the amount of material in the beam of effluent from the Knudsen cell can be well controlled. By using a Knudsen cell to supply each constituent for a film, the stoichiometry can be controlled, and, by controlling the rate of film formation, the crystallinity of the deposit can also be controlled. In a sense MBE is nothing more than tightly controlled co-evaporation. Figure 10.21b shows a partial MBE system within a vacuum chamber where several cells are shown all with effluent beams directed at a sample on a rotatable substrate holder for uniformity. The substrate holder can also usually be heated to enable highly crystalline deposits to form, and the Knudsen cells can be independently heated and directed at the sample. The vacuum systems for MBE are usually capable of ultra-high vacuum so as to achieve high levels of purity in the deposits, and often the MBE systems will house characterization tools to determine crystallinity (e.g., LEED) and composition (e.g., XPS) and film formation dynamics (e.g., SE).

Formula 10.68 above for evaporation from a surface can be used to determine the amount of material arriving at a substrate from a Knudsen cell. The amount of material emanating from a Knudsen cell can be obtained from the pressure within the cell that is obtained from the temperature and the area of the cell opening.

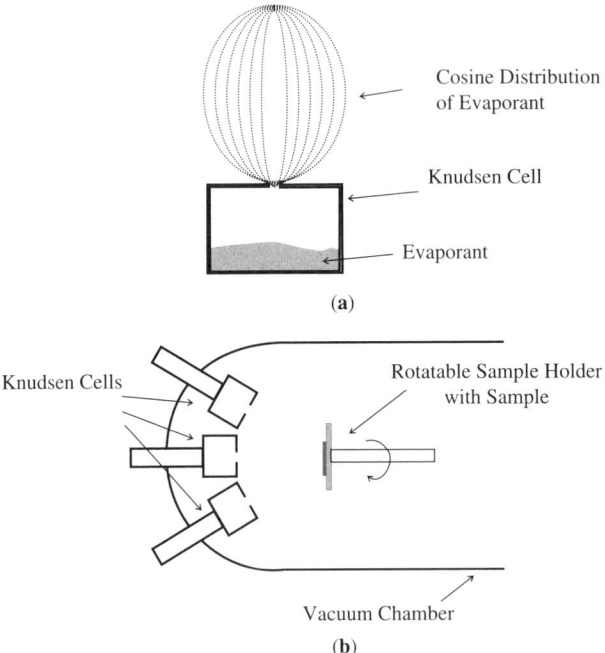

FIGURE 10.21 (a) Knudsen cell showing cosine distribution of effluent and (b) MBE system with several Knudsen cells and a rotatable sample holder.

If the opening or orifice is cylindrical rather than knife edge in thickness, then appropriate factors (Clausing factors) are necessary to correct for a smaller effective orifice area. Also many Knudsen cells are operated at sufficiently high pressures that the material flowing out of the orifice is not simply related to the pressure within the cell, and in those cases the amount of material in the beam is not easily calculated.

Sputtering The basic sputtering process is shown in Fig. 10.22 where atoms/molecules are physically knocked off one surface, called a target, by impinging high energy ions (in Fig. 10.22 Ar^+ ions), and the ejected target ions condense on another surface, called a substrate, that is separated from the target. The entire apparatus is enclosed within a vacuum chamber. The incident ions are typically produced using electrons from an electron gun (a filament with accelerating grids) or from a plasma interacting with a gas such as Ar. There are many ways to generate the plasmas and hence many kinds of sputtering sources. The source of the impinging ions is the main factor that distinguishes one method of sputtering from another, and all sputtering processes include a target and a substrate with varying geometries that suit the specific application. Below the most common sputter deposition methods are briefly discussed in terms of the essential differences.

10.3 FILM FORMATION

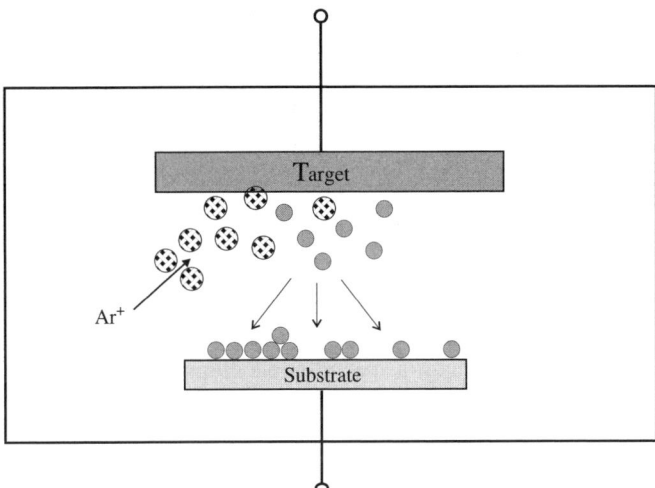

FIGURE 10.22 Schematic of the sputter process with incoming Ar ions sputtering target atoms that condense on the substrate.

The direct current (dc) diode sputtering method applies static potentials in the kilovolt range to a conducting target and substrate that are in the form of metal plates, and the vacuum chamber has pressures in the low milliTorr range. The negative plate (cathode) is the target and is bombarded by ions from the plasma set up between these two plates, thereby sputtering target atoms. These atoms now vaporized could then deposit on the anode surface or on other surfaces inside the vacuum system and form films. The dc diode sputtering systems usually display slow deposition rates and require high plate voltages. This system is not usually useful for the deposition of dielectric films because of charging.

For RF diode sputtering RF power that excites the plasma is coupled to the powered electrode through a capacitor. The discharge plasma acts like a rectifier, and the capacitor is charged to a negative voltage known as the dc bias potential. This dc bias potential is generally a measure of the kinetic energy of the bombarding ions that actually perform the sputtering. The substrate is also bombarded by energetic electrons and that can cause some damage, but RF sputtering has many practical applications and avoids most of the problems of charging.

Direct current and RF magnetron sputtering use crossed electric and magnetic fields in which dense plasma is generated. This method reduces impedance and permits operation at relatively low pressure and also substantially reduces substrate bombardment. The high deposition rates result from the high power that can be applied. RF magnetron sputter sources operate in much the same manner as DC magnetron sputter sources. There are, however, some important complicating factors, and the operating potential for the RF magnetron source is generally higher than for a dc magnetron source.

Ion beam sputtering makes use of broad beam ion sources where the ion source, the sputter target, and the substrate are electrically decoupled. This allows a high degree of independent control and is often used in research deposition systems where high deposition rates are not required. Figure 6.23 in Chapter 6 displays a research film deposition and analysis system in which an ion gun (Kaufman sputter ion gun) that is supplied with Ar gas is used to create a beam of sputter ions (Ar^+ ions) that are made to impinge upon a selected target. In fact this system can have many targets on a rotatable carousel. Then a particular target material can be selected and complex film materials and structures can be fabricated. The sputtered atoms from the target can impinge and condense in film form upon the substrate that can be rotated for uniformity and heated or cooled if needed. This system also has an analytical capability consisting of a spectroscopic ellipsometer (polarizer and detector) and a low energy ion gun that is used to analyze the deposited film either by analyzing recoiled ions and neutrals emitted from the substrate surface by means of the impinging ions from the telefocus ion source using the mass spectrometry recoil ions (MSRI) detector and/or by analyzing the backscattered primary ion beam using the ion scattering spectrometry (ISS) detector. A number of turbomolecular pumps are used to achieve vacuum suitable to the process under study, and a load lock chamber separated from the main process chamber is used to insert samples without significantly disturbing the vacuum and cleanliness of the main process chamber. Applications using this sputter deposition and analysis system are discussed in Chapter 11.

Plasma Deposition   As indicated in the previous section on sputtering, there are many ways to form a plasma. Once generated either by dc or ac (RF) means the plasma being composed of energetic species, electrons, ions, and neutrals can be used to initiate both gas-phase and surface reactions and thereby enhance the chemical vapor deposition. Plasma-enhanced chemical vapor deposition (PECVD) has emerged as an important method for film formation mainly because the process temperature can be significantly reduced using PECVD. Higher process temperatures can cause unwanted reactions to occur such as oxidation reactions and interdiffusion, and delicate structures can be damaged by thermal expansion and contraction.

Starting again from Fig. 10.19, PECVD mostly impacts the energy input to the chemical reaction. There are two general methods to excite a plasma, via dc or ac. Figure 10.23a illustrates a dc plasma where an anode and cathode are increased in potential to the point where atoms in the gas between the electrodes become ionized. If deexcitation is avoided by having sufficiently low gas pressures (but not so low so that there are too few species to ionize), the ionization causes the production of electrons that gain energy in the electric field and can in turn ionize more neutrals. With sufficient applied potential a cascade effect can occur, creating a gas of electrons, ion, and neutrals—a plasma. The density of the plasma will depend on specific conditions (dc field, pressure, and on the ionized gas). A dc plasma requires high applied potentials that are often difficult to control and maintain, and therefore ac plasmas generated at high frequencies (radio or microwave frequencies), so-called RF plasmas, are preferred. There are a variety of RF plasma methods. Figure 10.23b

10.3 FILM FORMATION

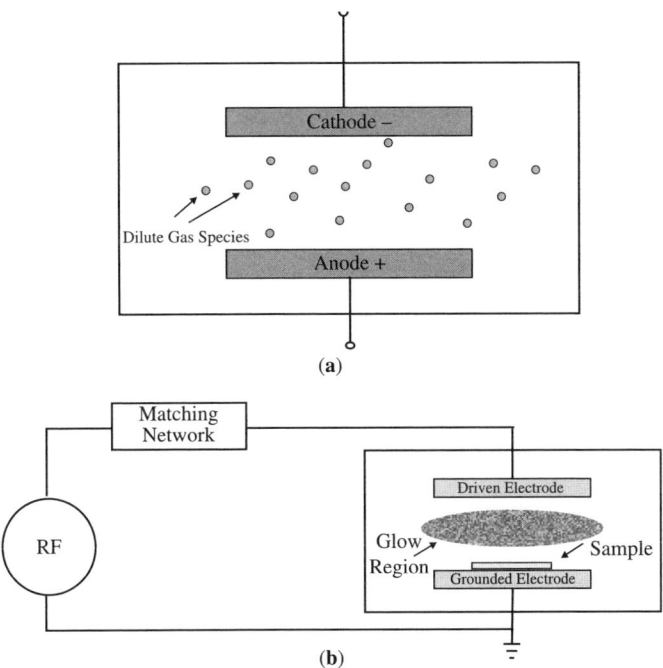

FIGURE 10.23 (a) Schematic of a dc discharge plasma system and (b) schematic of an RF plasma system.

shows a general setup for an RF plasma system. With electrode spacing of centimeters, at pressures below about 5 Torr and with a grounded and driven electrode, a plasma is formed in the same manner as for a dc plasma. However, in the RF case a plasma is formed near the center of the electrode spacing rather than close to one electrode (the anode) as for dc systems. Reactive gas mixtures are admitted near the plasma where the energetic plasma species initiate and/or accelerate the gas-phase reactions. Also energetic ions can bombard the sample surface and further activate surface reactions. Damage can also occur as a result of this surface bombardment, and in some systems attempts are made to minimize damage by simultaneous substrate annealing.

More complex plasma systems are used to generate higher density plasmas that are more efficient at promoting gas and surface reactions. One of the high density plasma methods previously mentioned is the electron cyclotron resonance plasma, or ECR plasma. A typical ECR geometry is shown in Fig. 10.24a and displays a microwave energy input of 2.45 GHz into a region that has a magnetic field of around 875 Gauss that provides resonance conditions. The 2.45 GHz microwaves are generated by a magnetron source and are guided through the microwave circuit to a coupler. The choice of the microwave frequency (2.45 GHz) is made based on the available magnetron sources that are widely used for domestic microwave applications. The microwaves propagate through a quartz window to the vacuum plasma source chamber,

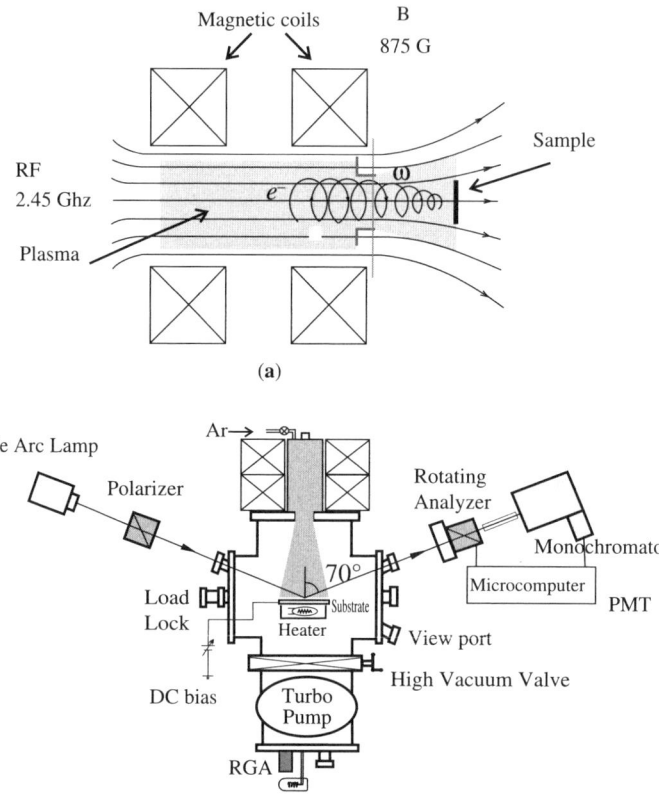

FIGURE 10.24 (a) Schematic of an ECR plasma system and (b) ECR plasma system with a spectroscopic ellipsometer and vacuum system.

which has a selected gas at low pressure (typically Ar in the 10^{-3}-Torr range). The electromagnets around the plasma source chamber generate a nonuniform static magnetic field inside the chamber. Electrons created in the chamber will move cycloidally in the magnetic field inside the plasma and will gyrate under the influence of the Lorentz force ($\mathbf{F} = -e\mathbf{v} \times \mathbf{B}$) with a frequency ω given as

$$\omega = \frac{e\mathbf{B}}{m_e} \tag{10.72}$$

where m_e is the mass of electron. In a magnetic field of $\mathbf{B} = 875$ Gauss, the cyclotron frequency of electrons f is given as

$$f = \frac{\omega}{2\pi} = 2.45 \text{ GHz} \tag{10.73}$$

10.3 FILM FORMATION

and this resonance enables efficient energy transfer and a long electron path. With **B** higher than the 875 Gauss but with a gradually decreasing magnetic field, the microwave propagates into the substrate region. The long cycloidal electron paths through the system increase the probability for more fruitful ionizations. The result is a high density plasma with an ion density of more than $10^{11} cm^{-3}$, which is more than 3 orders greater density than a simple RF plasma and thus with relatively high efficiency for promoting chemical reactions. Figure 10.24b displays an ECR plasma system with an associated spectroscopic ellipsometer. This plasma and SE system was used for the poly-Si nucleation application discussed above.

Many types of plasma sources have been developed. Based on the plasma density (i.e., the density of electrons or positive ions) these sources can be generally divided into two groups, that is, low density sources such as dc and RF sources, and high density sources that include ECR plasma, helicon plasma, and inductively coupled plasma or ICP sources.

Self-Assembly, Langmuir–Blodgett and Spin Casting of Organic Films While most surface science past and present deals with highly crystalline surfaces of inorganic materials, in recent years research emphasis has shifted toward organic materials. The original dominance of inorganic materials has been the impetus from the metals and semiconductor industries to understand and process metals and semiconductors and related materials. Furthermore from a basic surface science perspective, crystalline surfaces, in particular single crystalline surface, simplify the studies of surfaces and reactions at surfaces by fixing the geometry and stoichiometry of the constituent atoms. More recently organic materials for structural, electronic, and medicinal applications have prompted interest in organic surfaces, interfaces, and films and consequently methods to prepare and characterize the potentially technologically important materials and structures made from organic starting constituents. Organic molecules are typically more thermally fragile than inorganics, and many of the methods to prepare organic films use room temperatures, although there are instances where mild thermal evaporation has been used. In the following paragraphs a brief introduction to a vast literature on organic films is attempted. Three methods are chosen and treated separately as methods to produce thin films of organic materials: self-assembly, Langmuir–Blodgett, and spin casting.

Self-assembly (SA) refers to the fact that under certain conditions atoms, molecules, or molecular assemblies can arrange themselves in an ordered array without external manipulation. Thus, self-assembly is a process in which there are specific local interactions and constraints between a set of components with the result that components will autonomously assemble into a final structure without external assistance during this process. SA can be intermolecular or intramolecular. Intermolecular SA is a vast area in biology and biochemistry research where workers are attempting to find conditions where biologically active molecules form from other constituents. Here the focus is on how atoms or already formed molecules can assemble into films. In this limited context our previous discussion about film nucleation is useful to recall. A classic example of SA is illustrated with Fig. 10.25. Figure 10.25a shows a graphic of an asymmetric molecule of some organic substance where one end of

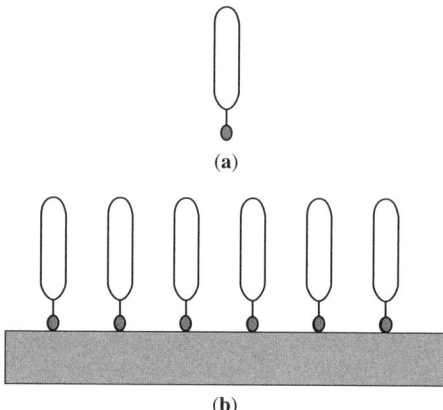

FIGURE 10.25 (a) Depiction of an asymmetric molecule and (b) the molecules in (a) have a group attracted to a surface.

the molecule (shaded) has a functional group that has a chemical and/or physical affinity for a particular solid substrate and/or the other end of the molecule is repelled by the substrate. Figure 10.25b shows that when the solid substrate is exposed to a mobile concentration of the molecules in Fig. 10.25a, the molecules self-assemble in such a way as to maximize attractive interactions and minimize repulsive interactions. As shown in Fig. 10.25b, an orderly array of molecules is produced, namely the molecules self-assembled into a crystalline film. While many large organic molecules and polymer chains are anisotropic in many ways, many different kinds of SA can be envisioned. Furthermore, with different regions having different interactions, the substrate can act as a template and produce SA films with a desired ordering. If there is insufficient material to cover a surface uniformly, then SA could result in isolated or connected islands rather than continuous films. Multiple layers are also possible, but the interactions that occur after the first full layer are not with the substrate and may therefore be stronger or weaker than interactions with a layer of molecules. In this case for weaker interactions and even for strong substrate interactions, the packing of molecules is likely to have numerous packing defects with voids, irregular coverage, and clumping. Consequently, there are only a few ideal systems where nearly perfect multimolecular layered films can be produced.

The problem of defects with SA films can be largely overcome using the Langmuir–Blodgett (LB) technique. The LB technique also is predicated on anisotropic interactions such as for SA. However, the LB technique applies lateral pressure to the forming layer to compress the molecules and thereby provide more substantial ordering. This is shown in Fig. 10.26a. With the molecules compressed and ordered on the surface of a liquid such as water, a solid substrate is lifted vertically through the compressed layer, and the film on the liquid then adheres to the substrate forming a more dense film with greater perfection than usually possible with SA alone. Figure 10.26b shows that as the area per molecule is reduced by the application of the compressive force, the film undergoes phase transformations.

10.3 FILM FORMATION

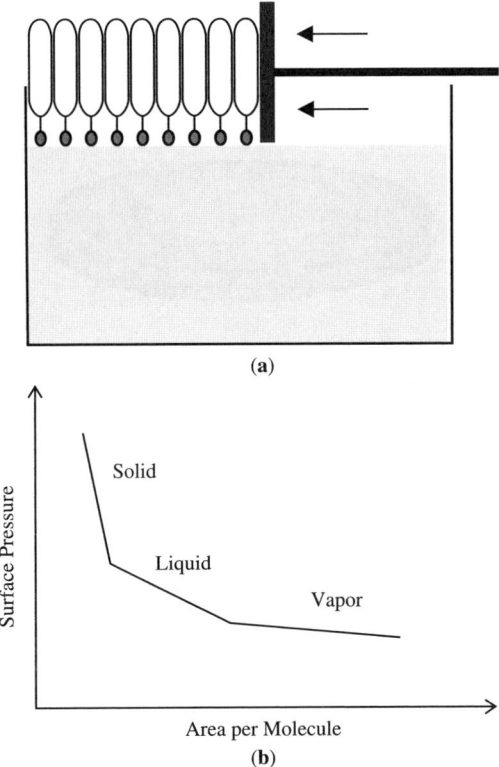

FIGURE 10.26 (a) Langmuir–Blodgett method to form a film through the application of a force. (b) Plot of the surface pressure applied versus the area per molecule with the surface phases appearing as a function of pressure.

Finally, if a high degree of ordering is not required and in the case where sufficient surface molecule interactions are not found, the molecules can be deposited from a volatile solution using the spin-casting method shown in Fig. 10.27a. A drop of a solution of desired molecules is made to fall onto a spinning substrate. The solution spreads and the solvent evaporates leaving a film behind. The main parameters that determine the thickness of the film are spin speed and solution concentration, and spin time is also important. This technique is predicated on finding a suitable volatile solvent with appropriate viscosity for the organic material to be deposited. Solvent can become occluded in the film, and so annealing is usually needed to achieve a stable material. Figure 10.27b shows a commercially available spin-casting tool that can be covered to exclude air and spin-cast in an inert and particle free ambient.

Atomic Layer Deposition/Epitaxy Atomic layer deposition (ALD) or atomic layer epitaxy (ALE) comprises a modern method that is useful for the preparation of very thin films, and it has been saved for last for several reasons. One is that relative to all

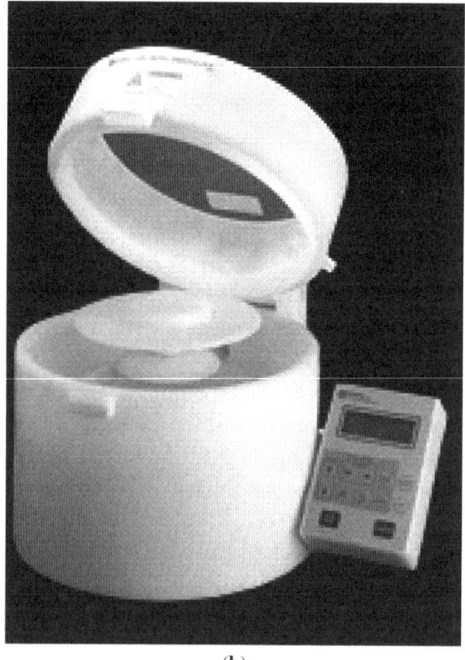

FIGURE 10.27 (a) Illustration of the spin-cast method to prepare organic films and (b) a commercial spin-cast apparatus from Laurel Technologies Inc.

10.3 FILM FORMATION

the other techniques for film making discussed above ALD or ALE is the newest. However, more importantly these related techniques are best categorized somewhere in between the CVD and PVD categories because they are made of both chemical and physical aspects of film formation.

Atomic layer deposition/epitaxy was introduced as a method to prepare epitaxial thin films by Suntola (5). ALD/E can be thought of as a four-step process in which precursor vapors are sequentially introduced into the reaction chamber, usually a vacuum chamber, and the substrate exposed to the vapors adsorb the first precursor. The steps are as follows:

1. First precursor is introduced to substrate and adsorbed.
2. Chamber is evacuated to remove the vapor precursor.
3. Second precursor is introduced and reacts with the first to produce the desired product at the substrate surface.
4. Chamber is evacuated to remove the vapor precursor.

This sequence will produce one layer of product. However, the cycle can be repeated as many times as necessary to produce the desired thickness of product. Figure 10.28 graphically illustrates this process. Figure 10.28a shows part of the first step in ALD where a surface is exposed to a molecule (the first precursor) in the gas phase that contains at least some desired film atoms. Usually, one end of this molecule strongly chemisorbs to the substrate. This exposure is typically performed in a clean vacuum system, and the substrate is cleaned and prepared prior to the introduction of the first precursor. Figure 10.28b shows that the first precursor has adsorbed. The adsorption shown is epitaxial, although this is not necessary. In the second step (not shown in the figure) the ALD/E system is purged of the vapor of the first precursor, leaving a monolayer of the first precursor. The second precursor is then introduced and reacts with the adsorbed monolayer and forms the desired film shown in Fig. 10.28c. This second precursor and the products formed are purged or pumped away. The result is the desired product film on the surface. If another layer or layers are desired, then the process is repeated. For this it is necessary that the first precursor also strongly adsorbs to the film produced in the previous iteration of the process or that the previous layer can be chemically modified to adsorb the first precursor.

There are many published ALD/E reaction schemes that have been found to operate well. Some schemes yield epitaxy and others simply yield well-controlled films with a desired thickness. One example of an ALE process is for ZnO films epitaxially deposited upon GaN (6) and example of an ALD process (nonepitaxial) and is more appropriate to MOSFETs is SiO_2 deposited nonepitaxially upon Si (7).

In the above-cited ZnO ALE study (6) several ZnO deposition processes were used, but here only one process is discussed in which the Zn precursor, $ZnCl_2$, was vapor deposited upon a GaN substrate. The lattices for ZnO and GaN are hexagonal with a mismatch of only 1.8%. Thus, epitaxy is likely. After purging of this first precursor, H_2O vapor, the second precursor was introduced and was used to react

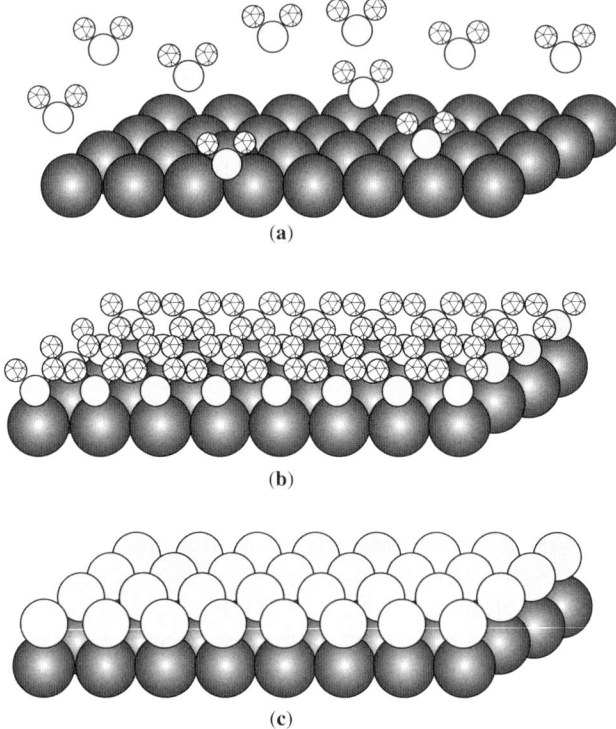

FIGURE 10.28 Possible ALD/E mechanism: (**a**) shows a substrate layer exposed to a reactant molecule; (**b**) displays the reactant adsorbed epitaxially upon the surface and; (**c**) displays the final desired monolayer film.

with the adsorbed $ZnCl_2$ to form the desired epitaxial ZnO film on the GaN substrate and release HCl to the vapor phase. The overall reaction is as follows:

$$ZnCl_{2(ads)} + H_2O_{(v)} \Rightarrow ZnO_{(ads)} + 2HCl_{(v)}$$

where the subscript ads refers to adsorbed surface species and v to vapor-phase species.

In the above-cited SiO_2 study (7), SiO_2 was formed on Si(100) surfaces using an ALD process. The result was not epitaxy and in fact was an amorphous SiO_2 film. As discussed many times previously, this film is at the heart of MOSFET technology, and, as the technology advances, thinner and thinner dielectric films are required with good film thickness control that can be achieved using ALD. The first step was to expose a cleaned Si(100) surface to an H_2O plasma to hydroxylate the surface forming SiOH groups. Then the first part of the ALD reaction was effected using $SiCl_4$ as the precursor via the following reaction:

$$SiOH_{(ads)} + SiCl_{4(v)} \Rightarrow SiOSiCl_{3(ads)} + HCl_{(v)}$$

After purging the reactant H_2O vapor is introduced and the reaction is as follows:

$$SiCl_{(ads)} + H_2O_{(v)} \Rightarrow SiOH_{(ads)} + HCl_{(v)}$$

Now the SiOH groups regenerate the process for the next cycle.

REFERENCES

1. Li M, Hu YZ, Wall J, Conrad K, Irene EA. J Vac Sci Technol 1993;11:1686.
2. Harrison WA, Kraut EA, Waldrop JR, Grant RW. Phys Rev B 1978;18:4402.
3. Copel M, Reuter MC, Horn von Hoegen M, Tromp RM. Phys Rev B 1990;42:11682.
4. Irene EA. Crit Rev Solid State Mater Sci 1988;14(2):175–223.
5. Suntola T. US patent 4,058,430. 1972.
6. Godlewski M, Szczerbakow A, Yu Ivanov V, Ghali M, Langer R, Barski A. Electron Technol 2000;33:416.
7. Klaus JW, Ott AW, Johnson JM, George SM. Appl Phys Lett 1997;70:1092.

SUGGESTED READING

M. Ohring, 1992. *The Materials Science of Thin Films*, Academic Press. A good first book to read about thin-film science and preparation.

L. I. Maissel and R. Glang, 1970. *Handbook of Thin Film Technology*, McGraw-Hill. A good book for reference and to read about applications if a bit out of date.

11

ELECTRONIC PASSIVATION OF SEMICONDUCTOR–DIELECTRIC FILM INTERFACES

11.1 INTRODUCTION

This chapter deals with applications that bear on a major scientific and technological issue in the area of surface and interface science within the scope of microelectronics, namely electronic passivation. In several previous chapters the fundamentals of surface states and surface sites were discussed, and now the focus is on how the surface states that have been shown to affect the operation of electronic devices can be controlled through interface formation. An understanding of electronic passivation is tantamount to an understanding of surface and interface electronic states that was begun in Chapters 5 and 7. In this chapter several more detailed studies about surface and interface states are presented that extend the discussions in Chapter 5 and 7, and electronic passivation is further elucidated using several published studies that were aimed at achieving electronic passivation for important semiconductor surfaces. Each of the application studies includes the use of a variety of techniques that were presented in earlier chapters for materials and electronics characterization.

11.2 INTERFACE ELECTRONIC STATES

The notion of surface and interface electronic states is a central issue in microelectronics, and the origin of surface states was discussed in Chapter 5 along with the use of photoemission for the measurement of the surface and interface states. For many electronics

---

Surfaces, Interfaces, and Thin Films for Microelectronics. By Eugene A. Irene
Copyright © 2008 John Wiley & Sons, Inc.

applications semiconductors surfaces are purposely covered with dielectric films where surface states become interface states. The most notable example in microelectronics is the MOSFET device, which provides the heart of integrated circuits (ICs). In Chapter 7 some of the electronic characteristics of MOSFET devices were discussed, and an application was discussed that tied together with the structure of Si surfaces, namely the effect of Si surface orientation on the number of interface electronic states at the Si–SiO_2 interface. In the present chapter several additional interface states and microelectronic issues are addressed using studies from the scientific literature.

It was discussed in Chapter 7 that the surface or interface electronic states between a semiconductor such as Si and a dielectric film on the semiconductor such as SiO_2 can determine many of the operational characteristics of MOSFETs. Pertinent to the Chapter 5 subject matter on the fundamentals of surface and interface states and modern microelectronics is a seminal study by Allen and Gobeli at Bell Laboratories that dates to the beginning of the microelectronics revolution in which surface states on Si were measured using photoemission techniques (1). In this study Si crystals were cleaved under ultra-high vacuum (10^{-10} Torr), and the work functions were measured using the Kelvin probe technique, and the photoemission

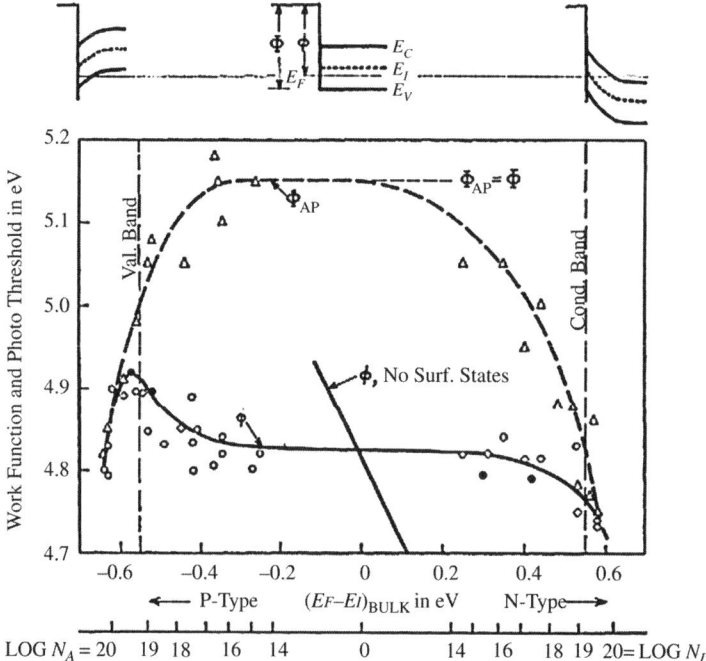

FIGURE 11.1 Dashed top curve is the photoemission results Φ, the straight line is the work function ϕ for the case of no surface states, and the solid line is ϕ measured with surface states. The data were taken from various doped Si(111) samples as is indicated on the horizontal axes. The definitions of Φ and ϕ are in the inset above the data figure. [Adapted from Allen and Gobeli (1), Figure 3.]

11.2 INTERFACE ELECTRONIC STATES

of electrons was also measured. The Si samples used included a range of doping from intrinsic Si (undoped) to highly P- and N-type doped samples. Figure 11.1 displays the data from the Allen and Gobeli study where the photoemission results Φ are represented by the dashed top curve and, the experimental work function results ϕ are shown by the solid curve. The solid straight line in Fig. 11.1 represents the anticipated values for ϕ if there were no surface electronic states. Clearly, the experimental ϕ results differ greatly from the no surface state calculation of ϕ and therefore show that surface states are present on the freshly cleaved Si(111) surface. From the difference in Φ and ϕ and the knowledge of the Si doping levels, the amount of charge in interface electronic states was calculated. Figure 11.2 displays the calculated surface state trapped charge Q_{ss} versus surface potential for the Si(111) surface as the solid curve. (Note that in the older literature Q_{ss} was commonly used for surface state charge, but later the scientific community agreed to use Q_{ot} for the charge trapped in the surface or interface states, as in Chapter 7.) The shape of the results was used to fit to various surface state distribution models as is seen by the dashed S-shaped curve that fits well and the straight line that does not. From these experimental and modeling results, the authors deduced that a two-discrete-state model fits best, and later results from other independent studies have given credence to this early study and the interface state model. In addition Allen and Gobeli were able to deduce that the number of surface electronic states on Si was approximately the same as the number of Si atoms on the surface. Of crucial significance to microelectronics technology was a somewhat passing remark made by these early workers that when O_2 was admitted to the vacuum system ϕ rose to nearly the Φ value, which is indicative of no surface states. This remarkable result means that as the Si surface is oxidized the surface states on Si are removed. Surface reactions would be expected to tie up the dangling bonds and thereby reduce the surface/interface electronic states, and this early study made the experimental observation that provides the groundwork for the microelectronics revolution using Si.

Following the pioneering work of Allen and Gobeli (1), there were several high quality studies using spectroscopic techniques that not only verified the early work but when taken in combination with the other available studies led to a better understanding of the results. Chiarotti and co-workers (2) using visible–ultraviolet (Vis-UV) optical reflection spectroscopy examined the optical transitions associated with surface states. Figure 11.3 shows the optical spectra for a clean Ge(111) surface along with the Ge surface after some oxidation. The increase in the plot of $\ln(I_0/I)$ where I_0 is the incident light intensity and I the reflected light is essentially a light absorption spectrum. Thus, this representation shows that there is more optical transmission as the Ge surface is oxidized, and a similar result was obtained (not shown) for Si(111). The absorption α_s is given as

$$\alpha_s = \frac{1}{5}\left[\ln\left(\frac{I_0}{I}\right)_{\text{clean}} - \ln\left(\frac{I_0}{I}\right)_{\text{ox}}\right] \quad (11.1)$$

where the factor $\frac{1}{5}$ is from the multiple reflection for the specific apparatus used. Figure 11.4 shows the adsorption for both Ge (left) and Si (right), and the peak

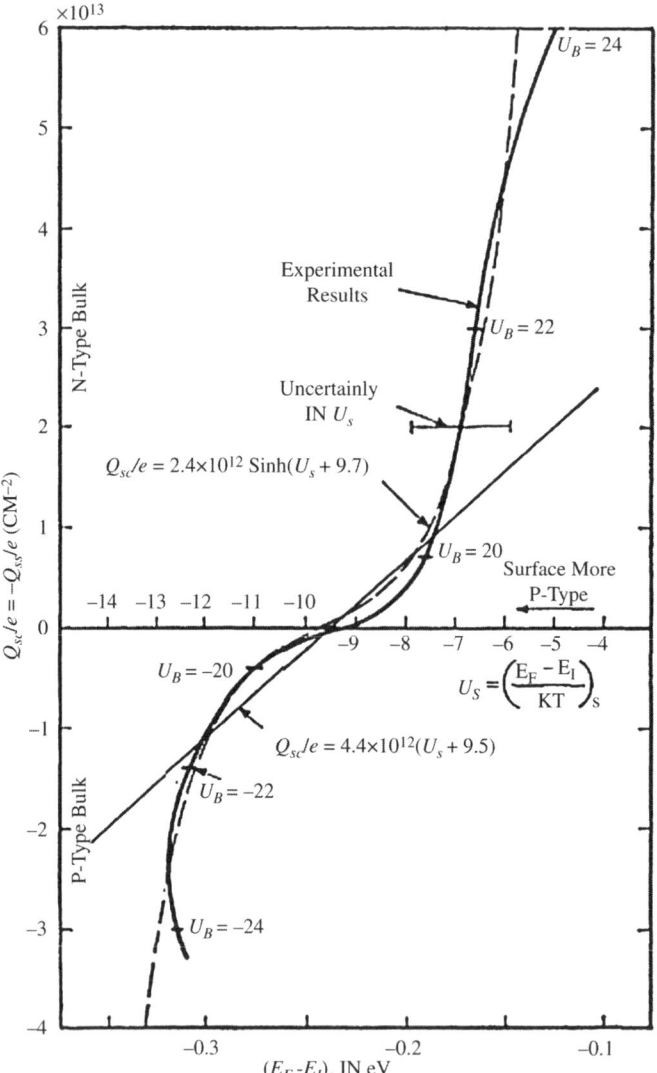

FIGURE 11.2 Surface state charge Q_{ss}/e versus surfaces potential u_s for Si(111). The solid curve is experimental results and the dashed curve and solid line are different surface state distribution models. [Adapted from Allen and Gobeli (1), Figure 4.]

position shows the energetic location of the optical transitions between the states that have given rise to the adsorption. Also the spectra are consistent with the two-state model introduced earlier by Allen and Gobeli (1), and they deduce that the number of surface states is of the order of 10^{14}, which is also consistent with one state per surface atom proposed by Allen and Gobeli.

In Chapter 5 a review by Eastman and Nathan on photoemission to access surface electronic states was summarized using Fig. 5.17. The photoemission results for

11.2 INTERFACE ELECTRONIC STATES

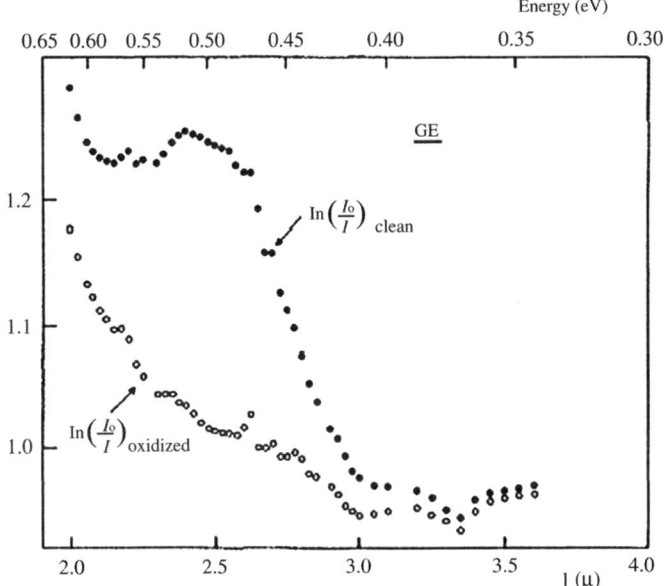

FIGURE 11.3 Logarithm of the ratio of the incident light I_0 to reflected light I as a function of the incident photon energy for cleaved Ge(111) surface before (freshly cleaved in 10^{-10} Torr) and after oxidation (10^{-6} Torr). [Adapted from Chiarotti et al. (2), Figure 2.]

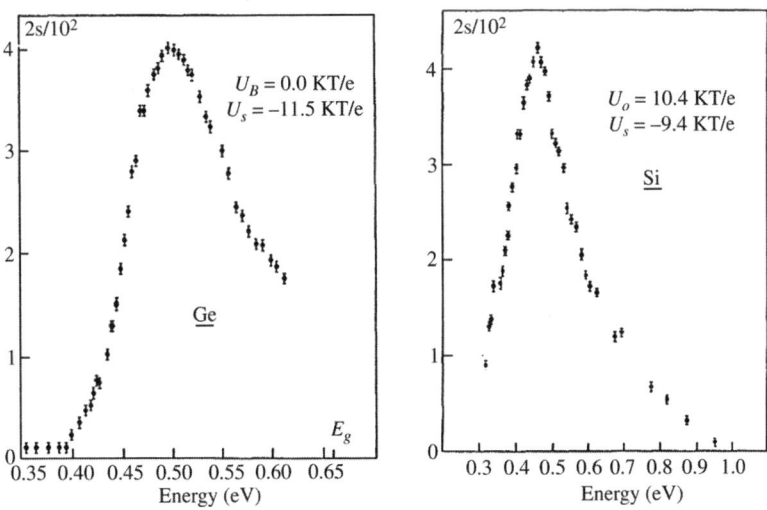

FIGURE 11.4 Absorption constant versus incident photon energy for cleaved Ge and Si in UHV. [Adapted from Chiarotti et al. (2), Figures 3 and 4.]

Si and GaAs in Fig. 5.17 are similar to the results cited above for filled surface states and photoemission yield spectra for empty states. More complete photoemission results are shown in Fig. 11.5 where UV photoemission (UPS) was obtained at three incident photon energies (20, 12, and 10 eV) in ultra-high vacuum (UHV) at 10^{-10} Torr on cleaved and oxidized Si(111). The surface state emission is shown at several incident photon energies as the difference before and after oxidation of the surface since after oxidation the states decreased. Figure 11.6 shows similar results obtained at one incident energy for both Ge and GaAs. In all cases the intrinsic surface states decreases after reaction of the surface. With this study the earlier results using different techniques were extended to other semiconductors, thereby generalizing both the results and understanding. Empty states were accessed using inverse photoemission (IPES) where electron transitions into empty surface states yield photon energies characteristic of the empty states (see Chapter 5). In Figs. 11.7 and 11.8 empty state spectra for Ge and GaAs, respectively, were called yield spectra by the authors.

The reaction and adsorption at semiconductor surfaces was also studied by monitoring surface electronic states. That this is the case should be intuitive since both chemical reaction and adsorption (chemisorption and physisorption) involve some measure of charge exchange with the solid surface where surface electronic states reside. The results of an early study by Garner et al. (5) show important findings about the interaction of the Si surface with oxygen. Figure 11.9 shows the result of exposing a clean UHV-cleaved Si surface (Fig. 11.9a) to O_2 (Fig. 11.9b) namely the surface electronic states disappear as was discussed in several studies above and in Chapters 5 and 7. The interesting fact uncovered in this study (5) is that subtly different O_2 exposures of the Si surface yielded different results. Exposure procedure 1 was a milder exposure where the freshly cleaved surface was exposed to 1 L oxygen (1 Langmuir = 10^{-6} Torr s and the exposure was 10^{-8} Torr for 10^2 s) followed by consecutive exposures of 9, 90, and 900 L O_2. Exposure procedure 2 was one exposure of 10^3 L (10^{-6} Torr for 10^3 s). The results of these two different O_2 exposures on the photoemission are shown in Fig. 11.10. First, it should be mentioned that both exposures removed the filled surface electronic states, as shown in Fig. 11.9. Figure 11.10 displays the Si $2p$ core levels obtained using 130 eV incident photons. Curve a (bottom right) is the unexposed clean Si surface in UHV. Curve b is the clean surface exposed using procedure 1, the milder exposure procedure. The surface states disappear but the Si $2p$ core levels are unaffected, that is, there is no energy shift in the core levels. This means that no appreciable amount of charge is displaced in the interaction of the O_2 with the Si surface and may indicate a covalent-like interaction. However, after exposure procedure 2, as shown in curve c in Fig. 11.10, there is more than 2 eV shift in the Si $2p$ core level, indicating a more ionic interaction. An Si $2p$ core level shift of 3.8 eV is expected for SiO_2. These findings indicate that a number of different interactions are simultaneously possible on an O_2-exposed Si surface. The mix of resultant states is a function of the exposure conditions.

In summary, the sampling of studies about surface and interface electronic states presented in Chapters 5, 7, and 11 selected from a myriad of such studies in the

11.2 INTERFACE ELECTRONIC STATES

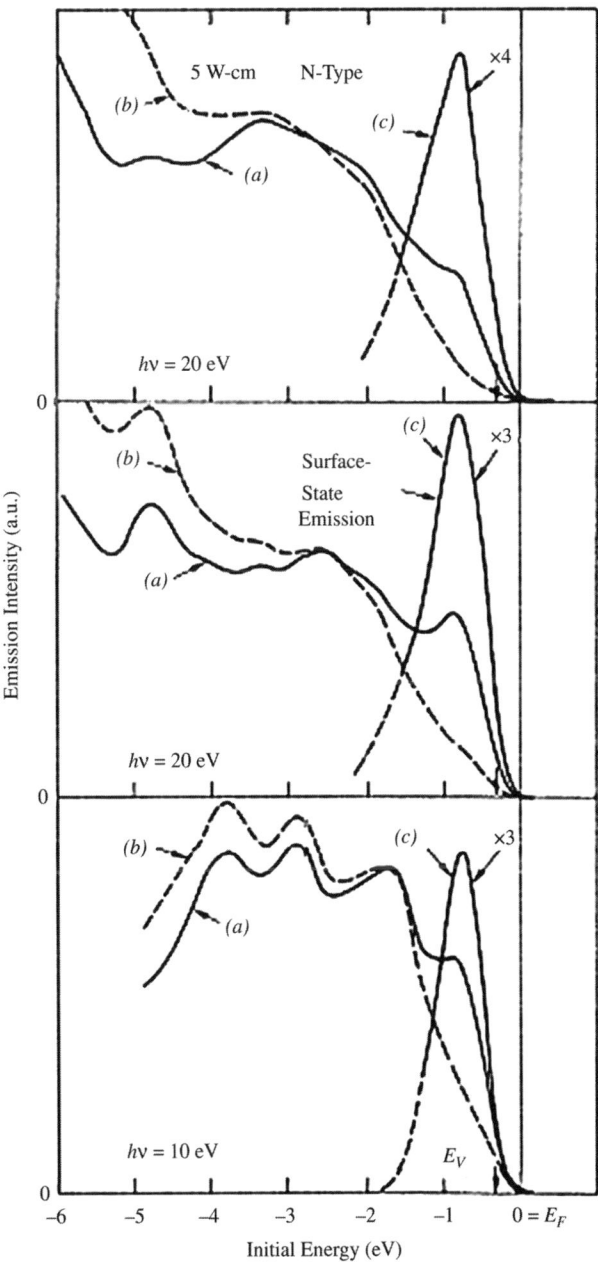

FIGURE 11.5 Photoemission results for freshly cleaved Si(111) in UHV (within 30 min at 10^{-10} Torr) as the solid lines (curves labeled *a*) and 7 h later shown as dashed curves (curves labeled *b*). The difference is given as the peaked solid curve that is the difference between *a* and *b* curves. [Adapted from Eastman and Grobman (3), Figure 1.]

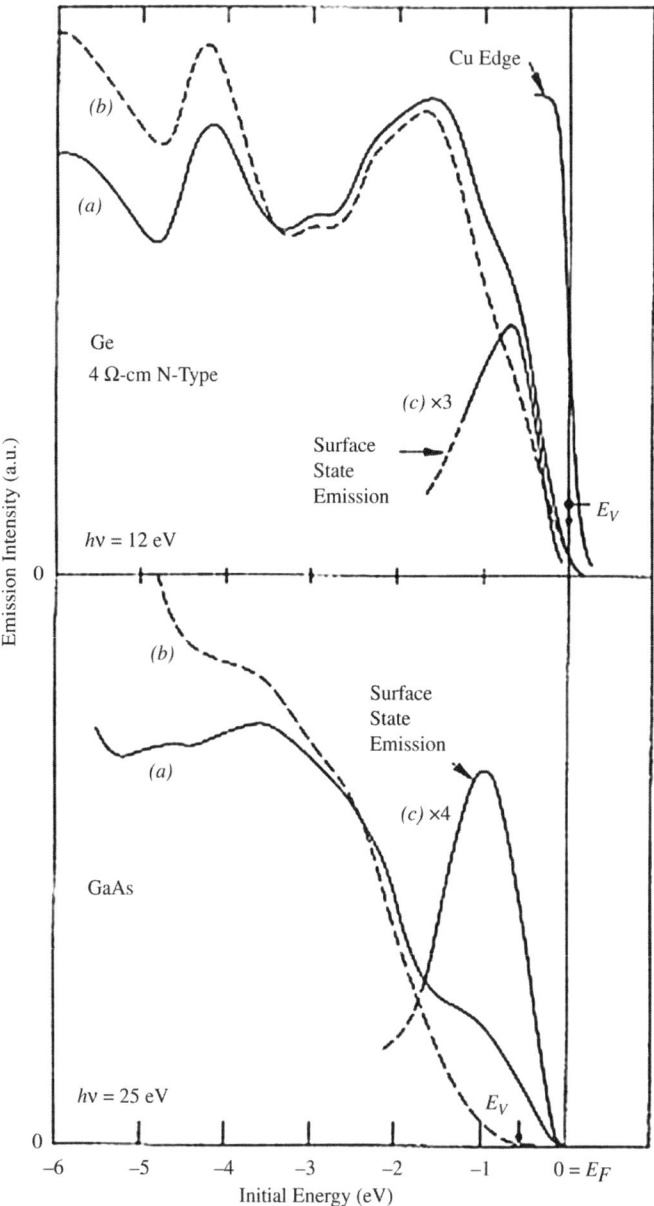

FIGURE 11.6 Photoemission results for freshly cleaved Ge and GaAs in UHV (at 10^{-10} Torr) and for Ge about 1 h after cleavage and about 15 min for GaAs. Curve (*a*) was clean surface (solid curves) and (*b*) after some measure of exposure to air (dashed curves). The difference is given as the peaked solid curve (surface state emission) that is the difference between (*a*) and (*b*) curves. [Adapted from Eastman and Grobman (3), Figure 2.]

11.2 INTERFACE ELECTRONIC STATES

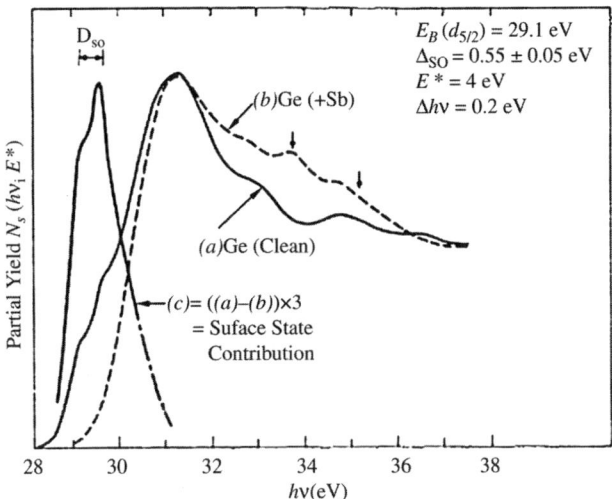

FIGURE 11.7 Photoemission partial yield for Ge(111) with the clean surface (solid line) and Sb monolayer-coated surface (dashed line) and empty states at the left (solid peaked curve). [Adapted from Eastman and Freeouf (4), Figure 1.]

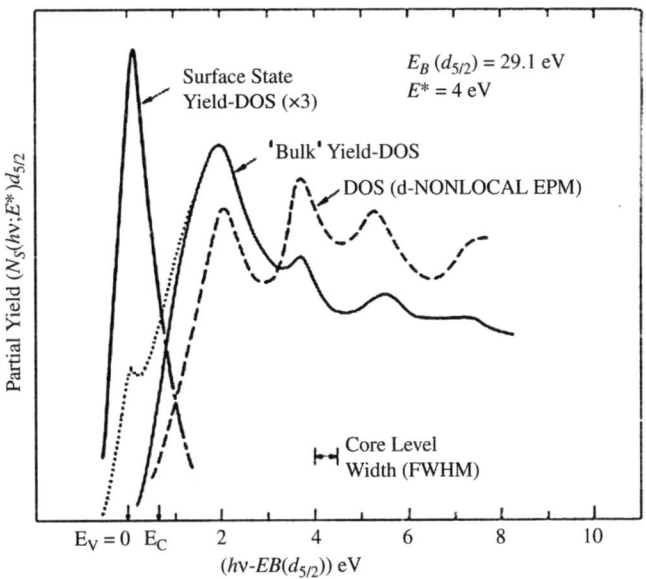

FIGURE 11.8 Photoemission partial yield for GaAs(110) (solid line) with adsorption edge spectrum (dotted line) and conduction band density of states (dashed line) and empty surface states at the left (solid peaked curve). [Adapted from Eastman and Freeouf (4), Figure 3.]

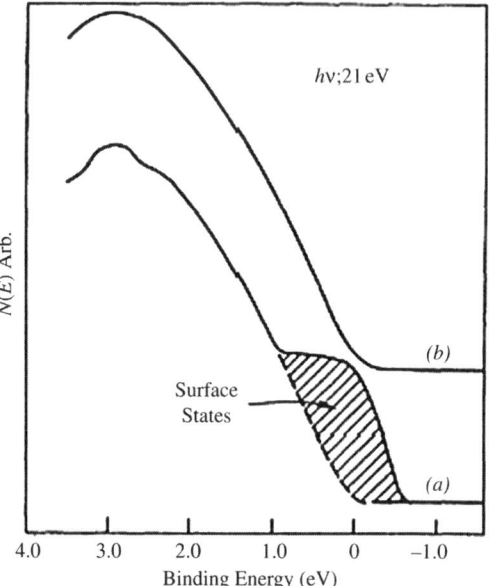

FIGURE 11.9 Photoemission from clean (*a*) and O_2 exposed (*b*) reconstructed S(111) surfaces. [Adapted from Garner et al. (5), Figure 1.]

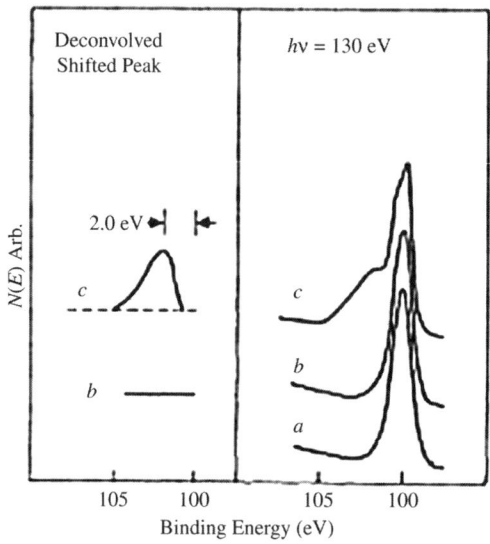

FIGURE 11.10 Photoemission from clean (*a*) and O_2 exposed (*b*) (milder exposure by procedure 1) and (*c*) (exposure by procedure 2) reconstructed S(111) surfaces. [Adapted from Garner et al. (5), Figure 2.]

11.3 ELECTRONIC PASSIVATION

In the previous section several important studies about surface and interface electronic states were discussed adding to those in Chapters 5 and 7. In Chapter 7 the deleterious electronic implications of the existence of interface electronic states at the Si–SiO$_2$ interface was introduced. In this and the following sections of this chapter the discussion and cited applications turn to methods for eliminating detrimental interface electronic states, and this elimination is called electronic passivation. Passivation is a term frequently used in the fields of electrochemistry and corrosion and in those contexts refers to rendering a surface stable or unreactive. Often this kind of passivation is accomplished by coating the surface to be passivated with a film that is relatively unreactive. Electronic passivation refers to rendering a surface electronically stable and reproducible. In most cases electronic stability is achieved only with sufficiently low surface electronic states so as to enable electronic device operation with the electronically passivated surface. Keeping the MOSFET discussion in Chapter 7 in mind with the MOSFET operation dependent upon the ability to change the Si surface potential from accumulation (off) to inversion (on), then it is essential to have a stable reproducible semiconductor surface where the Fermi level is not pinned. With this definition of electronic passivation as the rendering of the semiconductor surface free of surface states (or at least at a low enough level to unpin the Fermi level), it is straightforward to understand that electronic passivation is among the most important issues in microelectronics as presently practiced.

In Chapter 5 the occurrence and spectroscopic measurement of surface electronic states were discussed. In Chapter 7 the electronics consequences in terms of Fermi level pinning (Section 7.4) and electronic measurement (Section 7.51) of the charges trapped in the surface or interface electronic states were discussed. Fermi level pinning as was discussed in Chapter 7 was shown to occur when there are a large number of available interface states that dominate the exchange of charge across the interface. Consequently, for a pinned Si Fermi level, the surface potential of Si in a MOSFET does not change appreciably with an external bias. Thus, the MOSFET device with a pinned Fermi level would not respond to the external bias applied to the device and not turn on or off, that is, the device would not operate. Essentially, electronic passivation refers to the unpinning of the Fermi level by reducing the interface electronic states, and thus a passivated interface enables device operation. It then follows that electronic passivation of semiconductors is key to achieving useful interface-dominated electronic devices such as MOSFETs.

Throughout this book the MOSFET has played a dominant role due to the ubiquitous presence of MOSFETs in microelectronics. As discussed in earlier chapters, MOSFET operation is dominated by the electronics properties of the interface of a semiconductor that is typically Si and a dielectric that is typically SiO$_2$, and the crucial interface referred to is the Si–SiO$_2$ interface that is adjacent to the channel region of a MOSFET (see Fig. 7.5a). Si is the most used semiconductor in modern microelectronics and SiO_2 the corresponding most used dielectric in MOSFETs and therefore in microelectronics. The reasons for this choice of materials for MOSFET fabrication lies with the ability to electronically passivate the Si surface using SiO_2 films. In the context of microelectronics and the Si-based MOSFET, electronic passivation refers to the lowering of the number of surface electronic states or perhaps better interface electronic states at the Si–SiO_2 interface below the level at which the Si MOSFET device operates. Roughly speaking, this is about the level at which the Fermi level is unpinned, and for Si the surface/interface states need to be below about 10^{12} cm^{-2} eV^{-1}. However, Si MOSFET devices using SiO_2 as the gate dielectric can achieve levels below 10^{10} cm^{-2} eV^{-1} (see Chapters 7 and 10), and levels much above 10^{10} cm^{-2} eV^{-1} deleteriously affect device operation and reliability.

If one cannot electronically passivate a particular semiconductor material for a particular device such as a MOSFET, then that semiconductor cannot be used in the surface/interface-dominated MOSFET. Thus, in one sense electronic passivation separates those semiconductors that can and cannot be used in various microelectronics applications. In fact Si is the predominant semiconductor in microelectronics not because of its semiconductor properties that are mediocre but rather because of the ability to electronically passivate Si. Since MOSFETs operate based on the conduction of electrons (N-channel) or holes (P-channel) in the channel, high carrier mobility can dictate device speed with other factors equal. Table 11.1 shows a comparison of electron and hole mobilities for several common semiconductors. It is clear from this table that Si has the lowest electron mobility and the best hole mobility but only by a small amount over GaAs. Overall GaAs has the best mobilities, yet GaAs is not used for MOSFETs.

As pointed out in Chapter 5, surface states arise from a number of sources, and the common progenitors were discussed using Fig. 5.6. It is seen that all the features shown in Fig. 5.6 are defects that give rise to an altered interatomic potential, which in turn yields new states that are different from the electronic states obtained from the bulk crystal potential. The mere termination of the crystal lattice at a surface gives rise to dangling bonds, which result in intrinsic surface electronic states. Other electronic states that arise due to some kind of external effect such as surface defects or impurities are termed extrinsic surface states. Thus, to

TABLE 11.1 Electron and Hole Mobilities for Common Semiconductors

Semiconductor	Electron mobility (cm^2 V^{-1} s^{-1})	Hole Mobility (cm^2 V^{-1} s^{-1})
Si	1.4×10^3	4.8×10^2
Ge	3.9×10^3	1.9×10^2
GaAs	8.5×10^3	4×10^2
InP	4.6×10^3	1.5×10^2

electronically passivate a semiconductor surface, the intrinsic and extrinsic electronic states must be reduced to tolerable levels. Now several important semiconductors listed in Table 11.1 are discussed relative to electronically passivating the surface.

11.4 SEMICONDUCTOR PASSIVATION STUDIES

As discussed above, success was achieved in the early 1960s in electronically passivating Si surfaces using SiO_2 that resulted from the thermal oxidation of Si, and this achievement led to modern microelectronics. However, Si was not the first semiconductor to be tried in transistors, MOSFETs, and other devices. To this day only modest success has been achieved in MOSFETs with semiconductors other than Si despite significant effort with other semiconductors. In the following sections studies are selected that are focused on the electronic passivation of Si, Ge, InP, and GaAs. From these experimental studies some of the important issues are uncovered that relate to the criteria to achieve electronic passivation of semiconductors. Furthermore, these selected studies illustrate some of the techniques and methodology that has been used in the development of modern microelectronics.

11.4.1 Si Thermal and Plasma Oxidation

From numerous experimental and theoretical studies of the Si surface, such as the studies discussed above and in previous chapters, most notably Chapters 5, 7, and 10, and from similar studies on other semiconductor surfaces, the number of intrinsic electronic surface states are found to be of the same order as the number of surface atoms, and for Si nearly 10^{15} cm^{-2} (see Table 1.1). Thus, the density of intrinsic states is expected to vary with the crystal orientation. A study of the effect of Si orientation on the density of interface states D_{it} was discussed in Chapter 7 where Si orientation was observed to be an important factor. Considering that the intrinsic states originate from the unsatisfied bonds at the surface, it then seems intuitive that the formation of a stable chemical reaction product on the Si surface such as SiO_2 would tie up the dangling bonds and reduce the number of intrinsic states. How effective for electronic passivation one reaction product may be relative to another is a complicated problem and at the minimum depends on reactivity and structure. Also the reaction product itself must be electronically innocuous. This means that the electronically passivating film cannot obvert the operation of the semiconductor in a particular device. For example, a metal film on Si may reduce intrinsic states, but a metal cannot support the electric field needed for MOSFET operation and indeed could provide a short circuit to device operation (a source-to-drain short in a MOSFET). The fact that the Si Fermi level was unpinned as a result of SiO_2 formation on Si was known empirically to workers in the early 1960s, which enabled the advent of MOSFET technology with Si as the semiconductor of choice for that technology and the thermal oxidation of Si a cornerstone process for MOSFET technology. The studies previously discussed and cited and many other studies confirmed the intuition and led to a substantial understanding of the process of electronic passivation. However, while the ability to electronically passivate Si was reality since the early days of microelectronics, the precise nature of the Si oxidation reaction on Si and the reactions on other semiconductors are not now fully understood even some 50 years after the advent of microelectronics. For example, reactions that are performed under energetic ion beam and/or plasma stimulation enable low thermal process temperatures and thus reduce interface intermixing and interdiffusion problems such as dopant migration and the resulting loss of device junction tolerances. However, these low thermal temperature processes provide considerable excess energy that may induce vacancies and/or interstitials or even other unwanted reactions producing different products than desired and may ultimately produce extrinsic surface states. Furthermore, while it is clear that the Si–SiO$_2$ interface is an electronically passivated interface that allows MOSFETs to operate, there are debates about the chemical and physical nature of this interface, and thus the subject of electronic passivation remains as an issue for research. Chapter 12 will address what is known and unknown about the Si–SiO$_2$ interface while in the sections to follow the processes that lead and do not lead to electronic passivation will be discussed.

The thermal oxidation and electronic passivation of the Si surface has been studied extensively during the past 40 years [see, e.g., the review by Irene (6) and references therein]. In Chapters 1, 5, 7, and 10 various aspects of the thermal oxidation of single-crystal Si was presented with the linear-parabolic oxidation model for SiO$_2$ growth presented in Chapter 10. Figure 10.15 indicates that the time evolution of the SiO$_2$ thickness during Si thermal oxidation has linear and parabolic (squared) SiO$_2$ film thickness–time dependence, and there are essentially two processes represented by two fluxes in a steady state. One flux represents the inward diffusion of oxidant through the growing oxide film, J_2, while the other, J_3, represents the flux of oxide being produced by the interfacial reaction between Si and oxidant. The fact that we consider only the inward flux of O is the result of O_2^{18} tracer and other studies (see the 1988 review (6) references to the original studies). In Chapter 1 Fig. 1.6 and 1.7 display some Si thermal oxidation results that were obtained using *in situ* real-time ellipsometry. Even though these thermal oxidation data are for the very initial oxidation regime, the parabolic nature is seen in the data for the thickest films and linear behavior in the initial regime. Thus, a physically consistent model that represents the Si oxidation process is available as was presented in Chapter 10. What is in considerable doubt and not included in the linear-parabolic model is the nature of the Si–SiO$_2$ interface. As a result of clean thermal oxidation of high quality single-crystal Si wafers, the surface electronic states are reduced from about 10^{15} cm^{-2}, the number of atoms on an Si surface, to less than about 10^{10} cm^{-2} or more than 5 orders of magnitude. In Chapter 7, Table 7.1 shows D_{it} levels in the low 10^{11} range from the university study cited. In the microelectronics industry under the most rigorous processing conditions an order magnitude lower D_{it} is routinely obtained. So for the case of Si there is an easily and controllably grown film of SiO$_2$ that electronically passivates the semiconductor Si surface so as to enable MOSFET operation.

Typically, Si thermal oxidation is carried out at temperature greater than 800°C to obtain the highest quality Si–SiO$_2$ interface. However, as mentioned above, there are

11.4 SEMICONDUCTOR PASSIVATION STUDIES

decided MOSFET processing advantages to perform Si oxidation without using high process temperatures. One way to achieve oxidation at low process temperatures is by means of an electron cyclotron resonance (ECR) plasma (see Chapter 10) to provide excitation for the oxidation mechanism. For semiconductor oxidation the use of an ECR plasma appears promising because of the high ionization ratio associated with ECR, yet the low energy ionic species enables low processing temperatures with little damage. In one study (7) of the damage associated with ECR oxidation of Si, it was found that the damage was due to the oxidation reaction performed at low process temperatures and not typical of ion bombardment damage, and there was a dc bias effect on the oxidation kinetics. Since the ECR technique enables the production of thin oxides at low temperatures, the study was aimed at understanding the mechanism for oxidation and includes ECR plasma oxidation of Si at temperatures between about 80 and 400°C using *in situ* ellipsometry for the film growth characterization. To obtain a reliable description of the growth of the oxide layer, two complementary *in situ* ellipsometric measurements were performed: static spectroscopic and dynamic real-time ellipsometry measurements at a fixed incident photon wavelength. The former SE technique enables a more precise determination of the correct optical model and the determination of model parameters such as damage layer thickness, and film and damage layer composition, and the latter technique provides dynamic information on film growth as a function of oxidation time. It was found that the sample bias and temperature are the main parameters for controlling the oxidation kinetics. The apparatus for this study shown in Fig. 11.11 consists of a high precision rotating analyzer spectroscopic ellipsometer system with an independent vacuum process chamber equipped with a custom ECR plasma source. The plasma conditions used were: 300 W input power at a frequency of 2.45 GHz, with

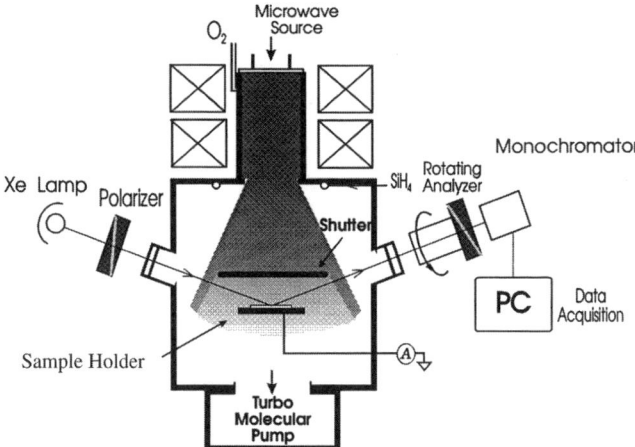

FIGURE 11.11 Schematic illustration of microwave electron cyclotron resonance (ECR) plasma system with vacuum process chamber and spectroscopic ellipsometer system. [Adapted from Joseph et al. (8), Figure 1.]

oxygen pressure of 5×10^{-4} Torr. The distance from the microwave cavity to the Si sample wafer was 20 cm. The shutter depicted in Fig. 11.11 will be discussed later and it enables a reduction in damage.

The spectroscopic ellipsometer measurements were taken at 41 photon energies between 2.5 and 4.5 eV. This spectral region was chosen because it contains the main features of the optical spectrum of Si, namely the two interband transitions at 3.4 and 4.3 eV (see Fig. 9.14). The real-time single-wavelength measurements were carried out at 340 nm (3.65 eV). This energy was chosen for several reasons. First, the values of Δ and Ψ for the bare silicon and thin oxides are in a region where the accuracy of the rotating analyzer ellipsometer is good. Second, the dependence of the dielectric function of Si on sample temperature is the smallest. This discovery, shown in Fig. 11.12 where it is seen that near 3.65 eV the dielectric functions for Si at different temperatures converge, enables the attainment of highly accurate ellipsometric data despite small temperature changes that occur during processing. For real-time ellipsometry measurements at elevated temperatures, either the variation of the dielectric functions must be known and accounted for or spectral regions of insensitivity to temperature must be determined. For reported sample temperatures above ambient the heating is regulated; but for measurement without applied heating (referred to here as room temperature), the temperature floats to about 80°C due to plasma exposure depending on the duration of the oxidation experiments. Third, there is high sensitivity to the thickness variation of the oxide layer. Figure 11.13 shows a typical variation of the pseudodielectric function for different

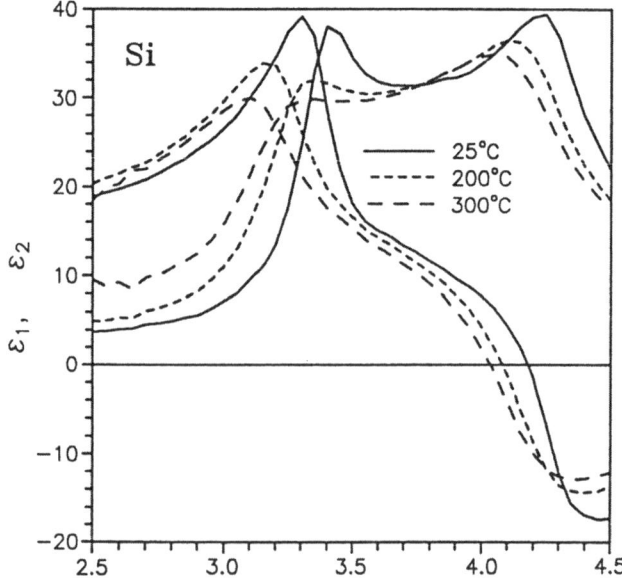

FIGURE 11.12 Dielectric function for c-Si as a function of temperature. ε_1 are single peaked and ε_2 are double peaked curves. [Adapted from Joseph et al. (8), Figure 2.]

11.4 SEMICONDUCTOR PASSIVATION STUDIES

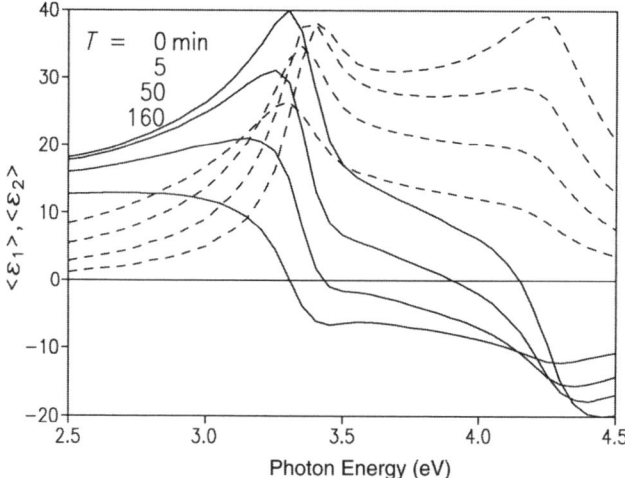

FIGURE 11.13 Dielectric function for the ECR plasma oxidation of Si(100) in O_2 as a function of oxidation time T. ε_1 are the solid curves and ε_2 are the dashed curves. [Adapted from Hu et al. (7), Figure 1a.]

oxidation times, and at 3.65 eV a proportional change is seen with increasing thickness. Lastly, at this energy the transparency of the oxide is high and Si is low, rendering the two major components optically dissimilar, hence improving the measurement contrast. ECR oxidations were performed using about 10 Ω-cm P-type Si(100) oriented single-crystal and freshly cleaned Si wafers that were placed in the load lock of the vacuum chamber. For all samples an initial spectrum was obtained prior to ECR processing to ensure the same starting surface. The initial spectra have shown that the cleaning procedure is reproducible and that the cleaned samples could be precisely described by a one-film model using the Bruggeman effective medium approximation (BEMA) and with the film composed of equal percentages of SiO_2 and amorphous silicon and with a thickness close to 1 nm on a crystalline Si substrate. It was found to be necessary to model the native oxide film on the Si substrates as a mixture of oxide and a-Si is an indication of some roughness and/or contamination.

For the analysis of the ellipsometric measurements different optical models were used for the spectroscopic data and the single-wavelength data (8). As discussed in Chapter 9, spectroscopic ellipsometry has been shown to yield reliable information for complex films with the use of stratified layer models. Each of the layers is either a medium for which the optical properties are known, such as SiO_2, or where the properties can be described by a mixture of components. For this latter case, the optical properties are determined from the dielectric functions of the components and the volume fractions using the BEMA. To limit the number of parameters, it was assumed that the entire oxide overlayer was described by two discrete layers, with the top layer being pure SiO_2 and an interface layer composed of no more than two known components. Therefore, four unknown parameters remain: two thicknesses and two volume fractions. The layers were assumed to be composed of one or two of following components with well-known dielectric functions: c-Si, a-Si, SiO_2, and voids. Four sensible models shown in Fig. 11.14a were evaluated and the results are shown as well as the unbiased estimator δ that measures the fit of the data to the model. The results indicate that model 4, made up of a pure SiO_2 top layer and an interface layer composed of SiO_2 and a-Si, always yielded the best fit in terms of the lowest unbiased estimator. Figure 11.14b shows a typical fit of

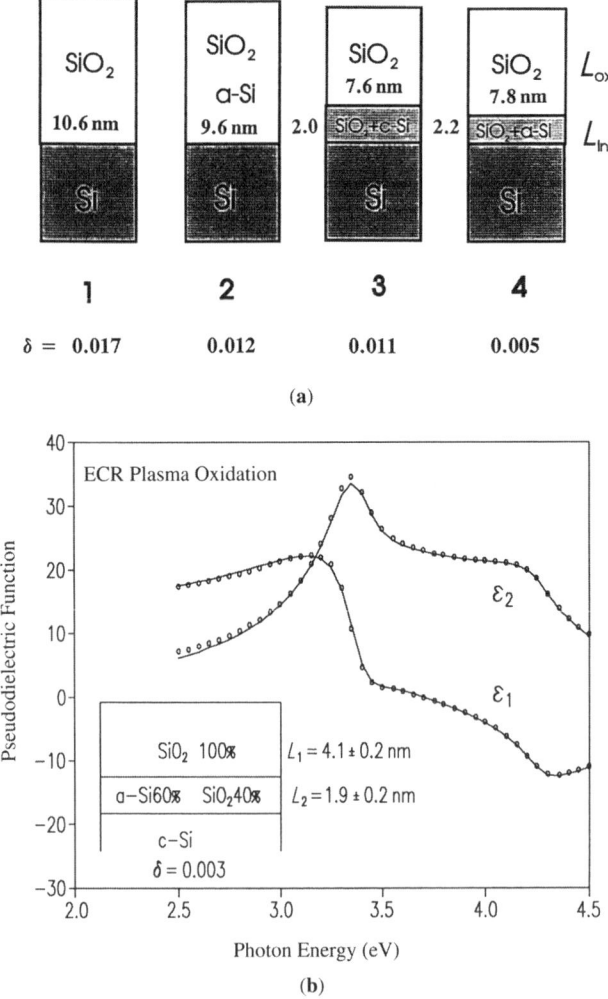

FIGURE 11.14 (a) Various optical models considered for ECR plasma oxidation of Si with typical values for the unbiased estimator δ. (b) Shows a typical data fit to model 4 that consistently displayed the best fit to the model (inset). [Adapted from Hu et al. (7), Figure 1b.]

11.4 SEMICONDUCTOR PASSIVATION STUDIES

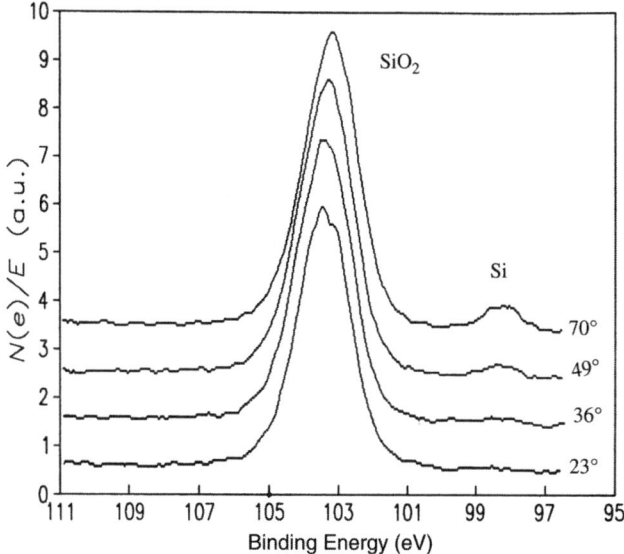

FIGURE 11.15 XPS spectra for ECR plasma grown SiO_2 film on c-Si at different take-off angles.

the data to the model with δ and the model parameters obtained from regression analysis in the inset. To interpret the single-wavelength real-time ellipsometry measurements in terms of oxidation kinetics (SiO_2 thickness versus oxidation time), a trajectory method was used. In this procedure a measured (pseudo) refractive index for the starting cleaned Si surface was used, and the complex multilayer system is simulated by one layer with an adjusted composite refractive index. For this calculation the index of the substrate is calculated from the first experimental point, and using this value a trajectory in the Δ,Ψ plane corresponding to the growth of the layer with an arbitrary index was compared to the experimental points. For this comparison an error function was defined as the sum of the distance between the experimental points and the calculated trajectory in the Δ,Ψ plane. With the use of a minimization procedure, the index of the layer is varied to obtain the minimum of the error function. The thickness of the layer corresponding to each experimental point is assumed to be that of the closest point to the trajectory. From this procedure a calculated value for the index of the layer is also obtained. This value could give some information about changes in the layer during oxidation, but it is not as precise as the spectroscopic results and thus should only be looked at as a relative value. It was verified that there is no measurable difference for the growth of a thin oxide between this one-layer model using an effective index for the substrate, and the more exact two-layer model, where the substrate is described by a native oxide layer over bare c-Si. It is worth noting that the thickness of the layer in the one layer model is only the thickness above the initial layer, but in the two-layer model the thicknesses of the layers are measured from the bare Si substrate.

This best-fit model (model 4) is considerably different from that found for the case of ion beam damage where the interface layer was composed of c-Si, a-Si, and voids. This suggests a different damage level and extent associated with ECR plasma processing. The BEMA fits also enable the extraction of SiO_2 film thicknesses, which were independently checked by angle-resolved x-ray photoelectron spectroscopy (XPS) shown in Fig. 11.15, from which the Si $2p$ spectra showed the presence of an SiO_2 film with a peak near 103.5 eV and the unoxidized Si peak near 99 eV. SiO_2 film thickness values obtained from the appearance of the Si substrate peak at various angles were compared with the SE BEMA value on the same sample. In one typical case XPS gave an average thickness of 6.7 ± 0.5 nm and ellipsometry yielded a value of 7.2 nm.

To measure film thickness using XPS [see, e.g., Mitchell et al. (9)], the electron escape depth (see Chapters 5 and 6) or attenuation length (AL) for the photoelectron through the film is required. This is typically measured in a separate experiment or obtained from the literature. The measurement of AL requires knowledge of the film density, which is usually not known accurately since it can be different from the bulk density. The XPS film thickness is obtained from a formula of the form

$$L_{ox} = (AL)\sin\theta \ln\left(1 + \frac{1}{Q}\right) \quad (11.2)$$

In this formula θ is the XPS detector take-off angle and Q is a product of two ratios of intensities measured in separate experiments. Q is given as

$$Q = \left(\frac{I_{sub}}{I_{film}}\right)_{Si\ 2p} \left(\frac{I_{film,\infty}}{I_{sub,0}}\right)_{Si\ 2p} \quad (11.3)$$

The first term on the right in brackets is the ratio of the XPS intensity of the Si $2p$ peak from the substrate to that of the oxide film, and the second term in brackets is the ratio of the XPS intensity of the Si $2p$ peak for an infinitely thick film to that of a bare Si substrate.

Figure 11.16 shows the BEMA-analyzed thickness values for the plasma oxide and the interface damage layer as a function of oxidation time for three substrate biases. For the longer oxidation times a self-limiting oxidation is seen without a sample bias. That the oxidation becomes self-limiting implies that about a 3-nm oxide provides a barrier to further oxidation, which slows to produce less than 5 nm in 2 h. With the addition of sample bias the situation is altered considerably in that the positive sample bias enhances the oxidation rate by fivefold beyond a film thickness of 3 nm, producing oxide at about 2.5 nm/h as compared to 0.5 nm/h for zero bias in the same thickness regime. Alternatively, the negative bias significantly reduces the rate, if not completely stops growth, in the same time interval. It was concluded that negative oxygen-related plasma species are primarily responsible for the oxide growth in the transport-limited regime beyond 3 nm in

11.4 SEMICONDUCTOR PASSIVATION STUDIES

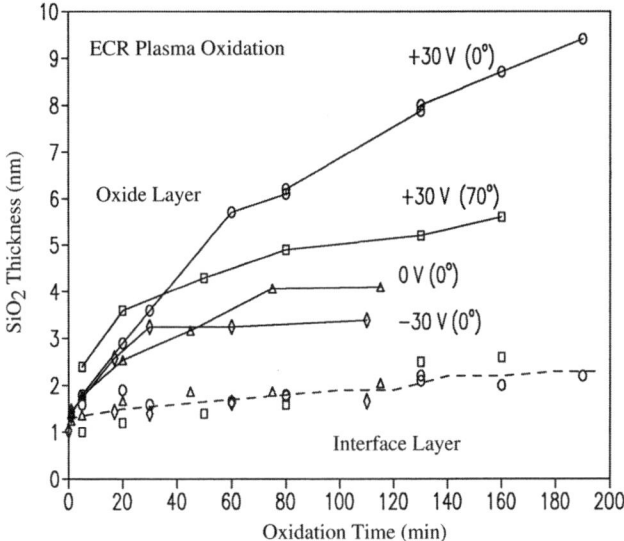

FIGURE 11.16 ECR plasma oxidation results from BEMA fits to the real-time SE data yielding the oxide thickness (solid curves) and damage layer thickness (dashed curve) for various applied voltage (V) biases and sample to plasma tilt angles (0° and 70°) to the plasma. [Adapted from Hu et al. (7), Figure 2.]

accord with previous studies of plasma oxidation. In the initial growth regime all the applied sample bias conditions yield about the same oxidation rate.

For an explanation consistent with the results, it was first considered that the high density of electrons in the plasma promotes electron attachment to O_2 via a favored interaction to produce a molecular ion, O_2^- (see a discussion of this in Chapters 8 and 10), that is less stable than O_2 and more readily decomposes to atomic species according to the following:

$$O_2 + e^- \rightarrow O_2^- \rightarrow O^- + O \tag{11.4}$$

The O^- can readily migrate through the oxide with positive sample bias. For negative sample bias, however, the molecular ion O_2^+ predominates at the outer oxide surface, and this larger molecular species (compared to O^-) is less likely to migrate rapidly through an oxide at low temperatures or to decompose to atomic species. Prior to oxide barrier formation, when the oxide is too thin to present a diffusion barrier, the oxidant species react readily and similarly with the Si surface, and bias presents only second-order effects. In this case both bias polarities simply attract ionic species in addition to neutrals, thereby slightly increasing the oxidation rate over the unbiased case in the earliest oxidation regime.

As mentioned above, the damage layer observed as a result of ECR plasma oxidation shows no crystallographic damage in the Si surface, but rather an inhomogeneous oxide interlayer in between the c-Si substrate and stoichiometric SiO_2. It is seen in Fig. 11.16 that for all bias conditions most of the damage layer forms within seconds during the oxidation process and afterwards changes little with further oxidation. The fact that the damage layer reaches a steady-state thickness is not easily rationalized because as oxidation proceeds the Si surface with damage is consumed. The fact that a decrease in the damage layer is not observed during oxidation strongly suggests that the oxidation reaction itself, with the attendant large change in molar volume at the $Si-SiO_2$ interface during the conversion of Si to SiO_2 can contribute to the damage layer in the form of a-Si, as is required by the BEMA to obtain the best fit. The low process temperature for ECR plasma oxidation was incapable of annealing away any of the damage.

It was also found that the activation energy for transport of oxidant during ECR oxidation of Si is about $5\times$ smaller than thermal oxidation, and plasma oxidation displays a pronounced substrate bias voltage effect indicating the dominance of charged atomic species, likely O^- moieties. From optical measurements the same phases are found to be present at the interface for ECR plasma oxidation as for thermal oxidation. Finally, many literature studies indicate that under careful processing it is possible to obtain interfaces nearly as good electrically as those from thermal oxidation. Therefore, ECR plasma oxidation can provide a low temperature process for SiO_2 growth on Si and provide good quality electronic passivation of Si.

11.4.2 Ge Passivation via Thermal and ECR Plasma Oxidation and SiO_2 Deposition

For the case of the thermal oxidation of Ge, the resulting GeO_2 is thermodynamically stable at room temperature but decomposes at the GeO_2-Ge interface by a disproportionation reaction at temperatures near 500°C and higher. Si behaves similarly but the disproportionation reaction is not significant until above 900°C, thereby permitting a wider thermal processing window. Also, GeO_2 is soluble in H_2O and thus degrades when exposed to atmospheric conditions. For these reasons Ge has never achieved the notoriety of Si despite the fact that Table 11.1 shows that the electron mobility in Ge is considerably higher than for Si.

A comparison has been made between the thermal and ECR plasma oxidation of single-crystal Ge (10,11). The ECR plasma system was the same as discussed above (see Fig. 11.11), and thermal oxidations were also performed in the same apparatus using 1 atm O_2. Ge(100) N-type commercially available single-crystal wafers were used. The oxidation was monitored using real-time single-wavelength (SWE) and spectroscopic ellipsometry (SE) as was done for Si as discussed above, and the ellipsometric analysis of the data was also done as for the Si ECR plasma oxidation study discussed above. The single wavelength used for Ge real-time studies was 335 nm, and this was chosen for the same reasons as 3.65 eV was chosen for Si. SE was done in the 2.0- to 4.5-eV range that includes the two Ge interband transitions at 2.3 and 4.4 eV. Experimental ellipsometry results are shown in Fig. 11.17 where Figs. 11.17a and 11.17b are for thermal oxidation and 11.17c and 11.17d are for ECR plasma oxidation of Ge(100). Except for the fact that lower temperatures can

11.4 SEMICONDUCTOR PASSIVATION STUDIES

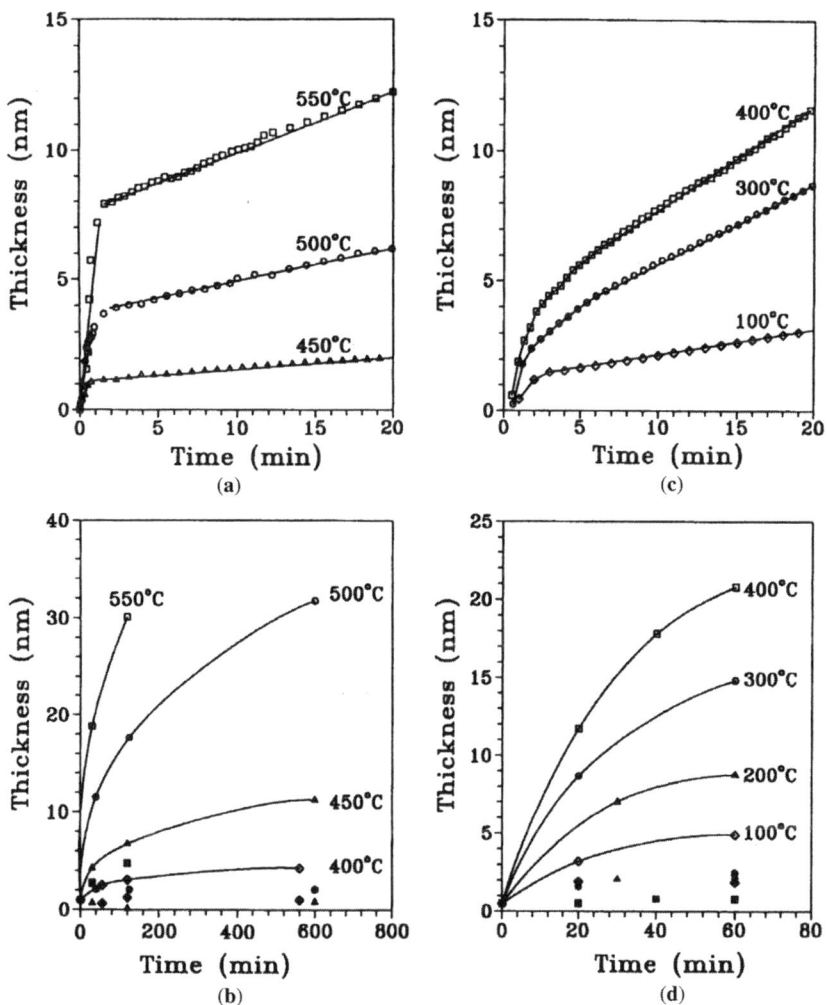

FIGURE 11.17 Ge oxidation results in terms of oxide thickness versus oxidation time for different temperatures. (a) and (b) are for thermal oxidation and (c) and (d) are for ECR plasma oxidation at +60 V sample bias. (a) and (c) were obtained from single-wavelength real-time ellipsometry using a single-layer model and (b) and (d) were obtained from SE using a two-layer model. The oxide thicknesses are given by the symbols with lines drawn through and for (b) and (d) below the thickness data the interface layer thickness is given by solid symbols. [Adapted from Wang et al. (11), Figure 1.]

be used to achieve the same oxide thicknesses for the ECR plasma oxidation, the overall oxidation kinetics was found to be similar for thermal and ECR plasma oxidation and also similar to the comparison above for Si thermal and ECR oxidation. It should be noticed in Fig. 11.17 that the SE results (Figs. 11.17b and 11.17d) require a

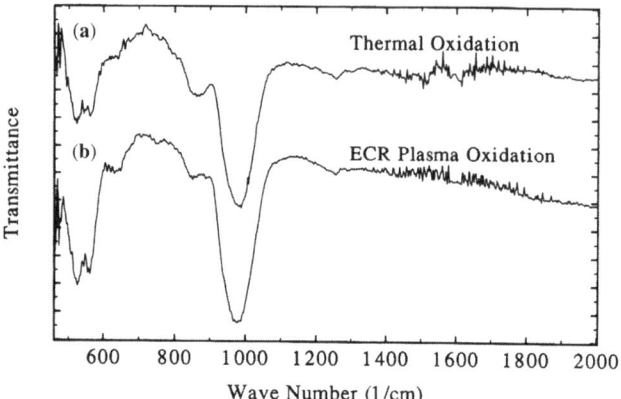

FIGURE 11.18 FTIR spectra of (**a**) thermal oxide grown on c-Ge at 550°C oxidation temperature and (**b**) an ECR plasma-grown oxide on c-Ge at +60 V bias and 400°C. [Adapted from Wang et al. (11), Figure 3.]

two-layer model where an interface layer of about 2 nm exists between the GeO_2 overlayer film and the Ge substrate. The interface layer was modeled as for the Si case in Fig. 11.14a, model 4, where the interface layer is composed of GeO_2 and a-Ge analogous to Si.

Figures 11.18 and 11.19 each show a comparison of thermal oxidation and ECR plasma oxidation results for Fourier transform infrared (FTIR) spectra and XPS spectra, respectively. These results demonstrate that the two oxidation methods yield the same products, namely GeO_2.

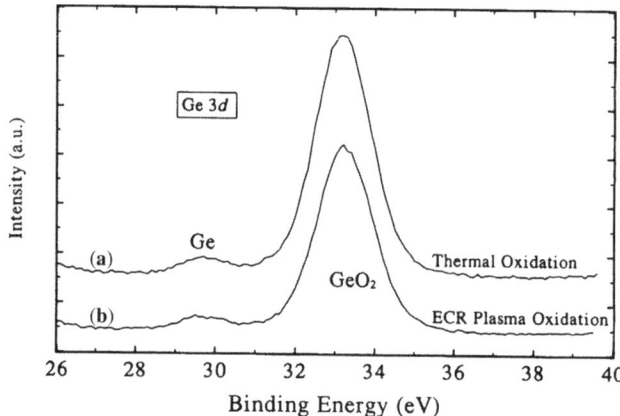

FIGURE 11.19 XPS spectra of Ge $3d$ electron for typical (**a**) thermal oxide grown on c-Ge at 550°C oxidation temperature and (**b**) an ECR plasma-grown oxide on c-Ge at +60 V bias and 400°C. [Adapted from Wang et al. (11), Figure 4.]

11.4 SEMICONDUCTOR PASSIVATION STUDIES

Figure 11.20 shows a thermal and ECR plasma oxidation comparison of high frequency $C-V$ measurements for various GeO_2 thicknesses on Ge(100) single-crystal surfaces. In terms of electronic passivation, the high frequency $C-V$ measurements provide clear evidence for carrier inversion (formation of depletion region), which is a good indication of low interface state levels ($<10^{12}$ cm^{-2}) and reasonably low levels of interface charge ($<10^{11}$ cm^{-2}) without the use of cleanroom processing or high quality materials. Therefore, there is substantial reason to believe that Ge behaves much like Si, and under ideal conditions (low T and humidity) thermal oxidation would electronically passivate the Ge surface. However, in this study both the

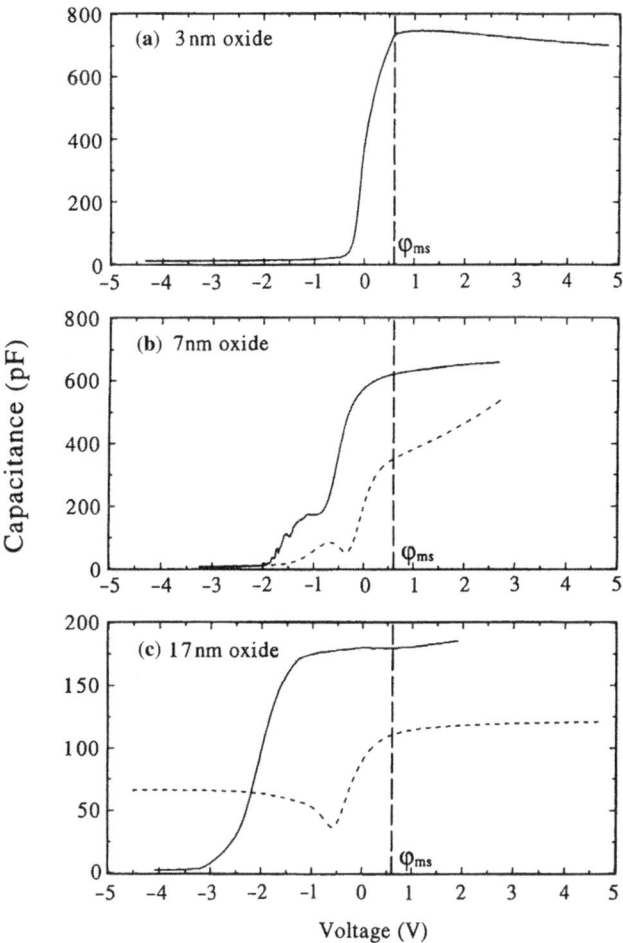

FIGURE 11.20 High frequency (1 MHz) capacitance versus voltage results for thermal (solid line curves) and ECR (dashed line curves) for oxides grown to different thicknesess on c-Ge at 550°C for thermal oxidation and +60 V bias and 400°C for ECR oxidation. [Adapted from Wang et al. (11), Figure 5.]

thermal and ECR plasma oxides displayed considerable leakage current that prevented further electronic characterization.

For the electronic passivation of Ge reasonable success has also been achieved using PECVD SiO_2. In one study (Wang et al. (12)) used *in situ* real-time SWE and SE to monitor the ECR PECVD of SiO_2 on both Si and Ge substrate temperatures from floating temperature ($<50°C$) to $400°C$. The resulting dielectric layers were evaluated by capacitance–voltage measurements and XPS. XPS results have shown that oxidation occurred along with SiO_2 deposition at the initial stage of PECVD. This interface oxidation occurred under the deposited film and is therefore called subcutaneous oxidation.

Figure 11.21a shows the Δ,Ψ trajectory, obtained by SWE, up to 4 min of SiO_2 deposition on Ge at floating temperature (FT is about $40°C$), 200 and $400°C$. Figure 11.21b shows oxide thickness versus deposition time calculated from data shown in Fig. 11.21a using the trajectory method. An SiO_2 single layer on Ge was used for the optical model. Figure 11.22 shows similar results for Si substrates. It is seen in Figs. 11.21b and 11.22b that prior to the characteristic linear thickness increase expected for SiO_2 or any film deposition, the oxide thickness increases at a rate significantly higher than linear for the initial 10 s. In the previous ECR plasma Si and Ge oxidation studies discussed above, it was reported that the initial oxidation was fast, the rate increased with temperature, and was virtually bias independent. Therefore, it seemed likely that the initial fast oxide thickness increase is due to Ge (or Si) oxidation instead of or in addition to SiO_2 deposition. To verify this, the SiO_2 deposition was stopped at points A and B, shown in Fig. 11.21b, for the FT (about $40°C$) and $400°C$ depositions. SE modeling indicated that a two-layer model gave the best fit. The top layer is composed of pure SiO_2, and the interface layer is made up of GeO_2 and amorphous Ge (a-Ge) between the SiO_2 film and the c-Ge substrate. Figures 11.23a and 11.23b show XPS results corresponding to points A and B in Fig. 11.21b, respectively. The binding energy for elemental Ge $3d$ is 28.7 eV. For both samples, a chemical shift of 3.2 eV relative to elemental Ge peak was observed, which indicated the formation of GeO_2. The XPS data were modeled to determine the thickness of SiO_2 and GeO_2 layer upon Ge, by assuming there is a GeO_2 layer between SiO_2 and substrate elemental Ge, and the results agree qualitatively with the SE modeling results. Given that the initial GeO_2 thickness on a cleaned Ge substrate was about 0.3 nm, both points A and B show measurable increases in GeO_2 thickness upon ECR plasma deposition of SiO_2. Thus, all the results in this study led to the conclusion that the initial oxide thickness increase is mainly due to Ge oxidation instead of SiO_2 deposition. The initial Ge oxidation rate (i.e., without the SiO_2 overlayer) increases with increasing temperature, and it is faster than the linear subcutaneous oxidation rate discussed below.

The properties of MOS capacitors made from PECVD SiO_2 films on Si and Ge substrates were measured using the high (HF) and low frequency (quasi-static QS) C–V method as was discussed in Chapter 7. Analysis of C–V data yielded flat band voltage, V_{fb}, and interface trap density, D_{it}. The interface trap density D_{it} was calculated using high–low frequency C–V data shown in Figs. 11.24 and 11.25

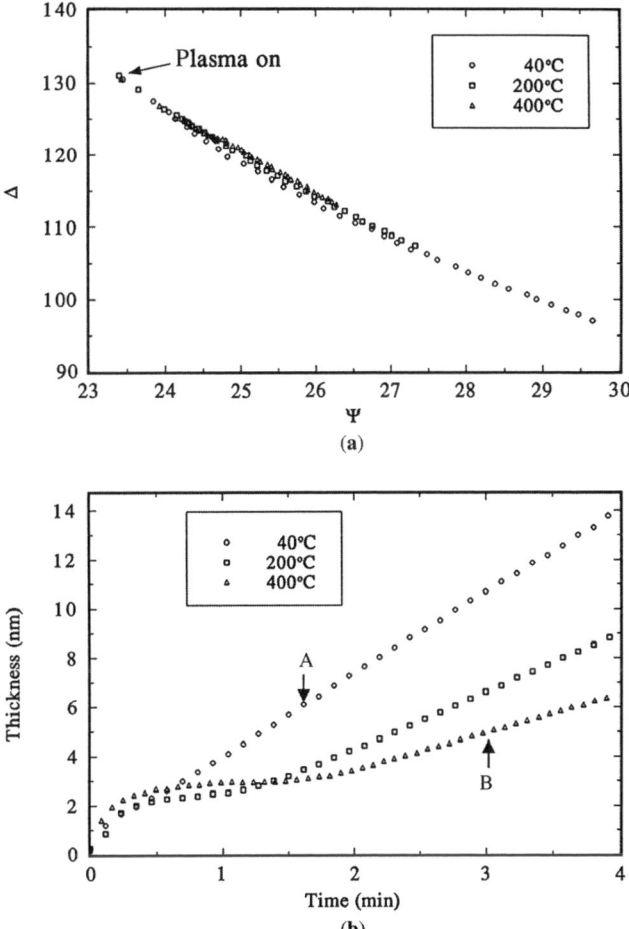

FIGURE 11.21 (a) Δ,Ψ trajectory from SWE, in the initial period (4 min) of SiO₂ deposition on Ge at floating temperature (FT), 200 and 400°C. (b) Oxide thickness versus deposition time calculated using the trajectory method. [Adapted from Wang et al. (12), Figure 2.]

on Si and Ge MOS structures, respectively. For ECR PECVD of SiO_2 on Si Fig. 11.24 displays both high frequency (HF) and quasi-static (QS) or low frequency representative $C-V$ data for PECVD SiO_2 on N-type Si. Figure 11.24a is for PECVD SiO_2 at FT. Midgap D_{it} was 2×10^{11} cm^{-2} eV^{-1}, V_{fb} was $+0.21$ V. A hysteresis of 0.64 V was shown by the retrace upon sweeping from accumulation to depletion, indicating a trapped charge density of 2.7×10^{11} cm^{-2}. A predeposition oxidation step was performed at temperatures from FT up to 400°C, and the $C-V$ results did not improve. Postdeposition oxidation was also attempted at different temperatures, but also without significant improvement. If the SWE analysis in Fig. 11.21 is examined carefully, it can be seen that oxidation occurs at the initial stage (about 10 s) of PECVD.

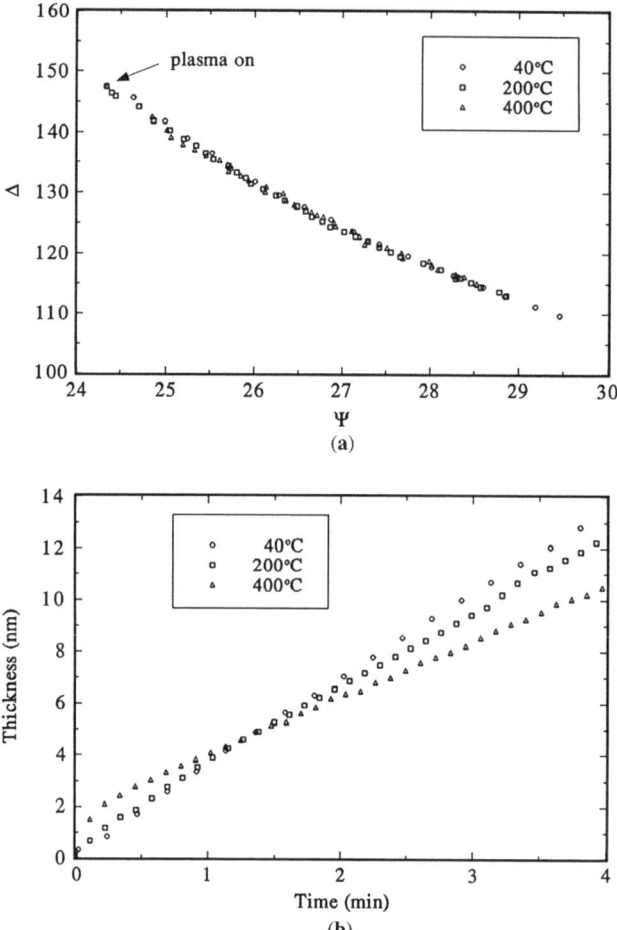

FIGURE 11.22 (a) Δ, Ψ trajectory from SWE, in the initial period (4 min) of SiO_2 deposition on Si at floating temperature (FT), 200 and 400°C. (b) Oxide thickness versus deposition time calculated using the trajectory method. [Adapted from Wang et al. (12), Figure 3.]

It was argued that if the plasma-assisted oxidation step before deposition provides surface cleaning and oxide passivation, a separate postoxidation step will not make further improvement of the electrical properties. It is also reported that if the temperature is too low (<150°C), OH groups can be incorporated into the films and results in a degradation of the oxide bulk properties and electron trapping. Therefore, postdeposition annealing was attempted. Figure 11.24b shows representative $C-V$ data for the same sample as in 11.24a but received a rapid thermal anneal (RTA) in vacuum at 800°C for 1 min. Heating was accomplished in vacuum using banks of quartz iodine lamps at 5×10^{-7} Torr during RTA. After RTA D_{it} was reduced to 5×10^{10} $cm^{-2} eV^{-1}$, and V_{fb} was +0.16 eV. A hysteresis in the $C-V$ trace of

11.4 SEMICONDUCTOR PASSIVATION STUDIES

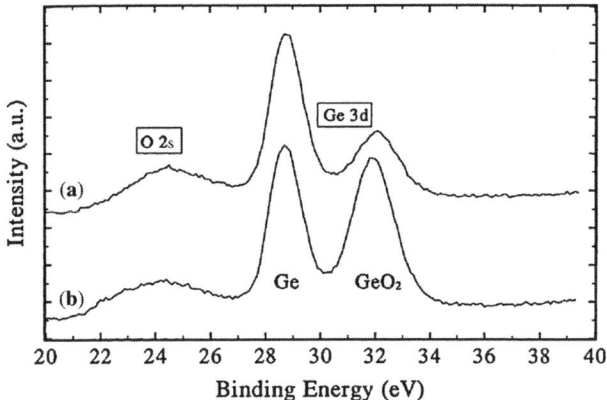

FIGURE 11.23 (a) and (b) are XPS results corresponding to points **A** and **B** in Fig. 11.21b, respectively. [Adapted from Wang et al. (12), Figure 4.]

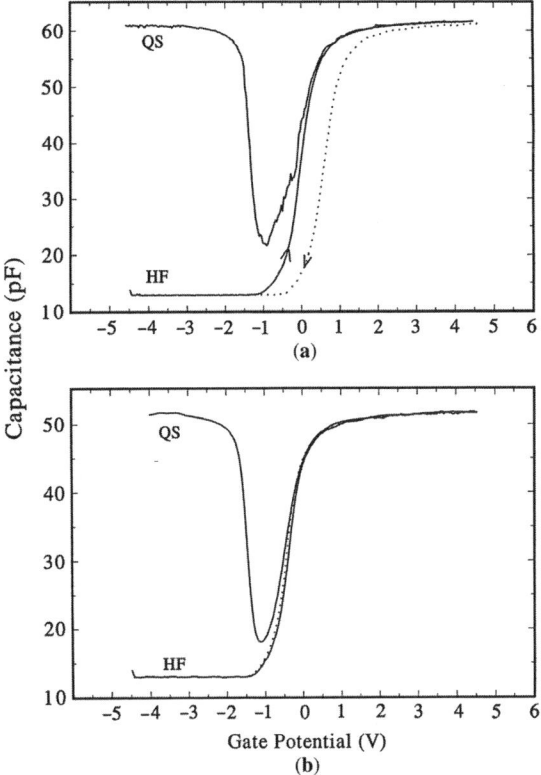

FIGURE 11.24 High frequency (HF) and quasi-static (QS) C–V results for ECR plasma SiO_2 deposited on N-type Si: (a) deposited at FT without rapid thermal anneal (RTA) and (b) deposited at FT but with RTA at 800°C in vacuum for 1 min. [Adapted from Wang et al. (12), Figure 11.]

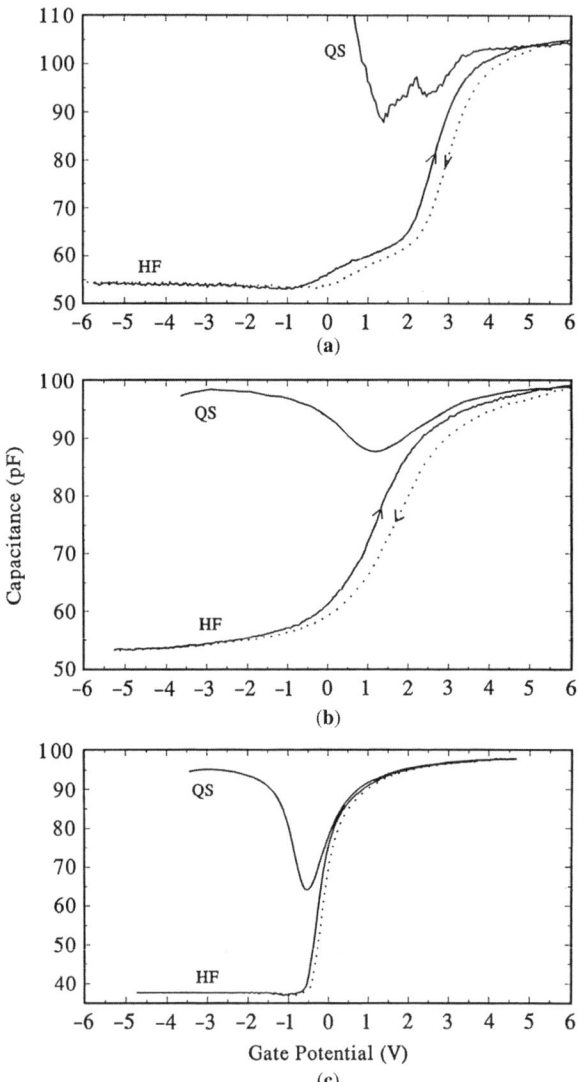

FIGURE 11.25 High frequency (HF) and quasi-static (QS) $C-V$ results for ECR plasma SiO_2 deposited on N-type Ge: (**a**) deposited at 400°C, (**b**) deposited at 200°C and with a 2-nm Si underlayer, and (**c**) deposited at 200°C and with a 2-nm Si underlayer and followed by ECR plasma oxidation to consume the Si. [Adapted from Wang et al. (12), Figure 12.]

0.05 V indicated a trapped charge density of 1.8×10^{10} cm^{-2}. Figure 11.26 shows FTIR spectra of (1) oxide deposited at FT and (2) oxide with the same deposition condition with (1) but with RTA. OH groups at about 3550 cm^{-1} were greatly reduced after RTA. Therefore, postdeposition RTA of oxide in high vacuum was shown to

11.4 SEMICONDUCTOR PASSIVATION STUDIES

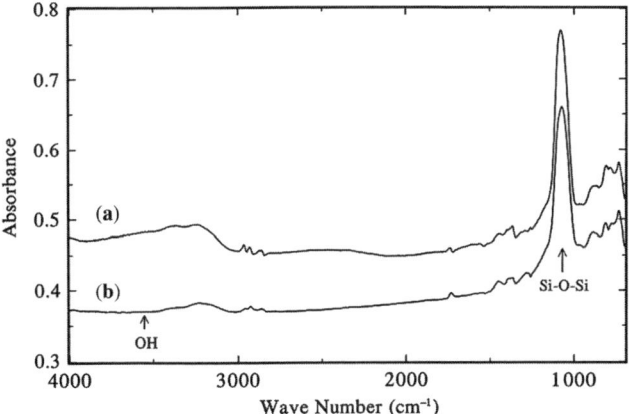

FIGURE 11.26 FTIR spectra for ECR SiO$_2$ deposited at FT: (a) without RTA and (b) with RTA at 800°C in vacuum for 1 min. [Adapted from Wang et al. (12), Figure 13.]

effectively reduce the OH group incorporation and thus reduces D_{it} and charge trapping states.

For ECR PECVD of SiO$_2$ on Ge SE showed that Ge-rich GeO$_2$ was formed at the beginning of the deposition processes. A thin Si interlayer is necessary to prevent the undesirable Ge-rich GeO$_2$ formation. Figure 11.25a is from a c-Ge substrate with PECVD SiO$_2$ deposited at 400°C. The high frequency $C-V$ curve shows evidence for leakage in the accumulation region. The low frequency $C-V$ curve is noisy, and the capacitance is large in the inversion region. The sample exhibited a midgap D_{it} of 3×10^{12} eV^{-1} cm^{-2} and V_{fb} of $+2.6$ V. The retrace of the high frequency curve shows 0.38 V of hysteresis, indicating a trapped charge density of 2.4×10^{11} cm^{-2}. Figure 11.25b is for a c-Ge sample where a 2-nm Si interlayer was deposited at room temperature before SiO$_2$ deposition at 200°C. The SE model shows 20.0-nm SiO$_2$ and 1-nm Si remained at the interface. Midgap D_{it} was 2×10^{12} eV^{-1} cm^{-2}, V_{fb} was $+1.3$ V, and trapped charge density was 3.8×10^{11} cm^{-2}. The data in Fig. 11.25c is from a c-Ge substrate with a 2-nm Si interlayer deposited at room temperature and SiO$_2$ deposition at 200°C, and this was followed by ECR plasma oxidation at 400°C and $+20$ V dc bias for 5 min. The SE model shows 25.6-nm SiO$_2$ single layer on Ge. Midgap D_{it} was 3×10^{11} eV^{-1}cm^{-2}, which is one order of magnitude smaller than the above two cases. V_{fb} was smaller than $+0.1$ V with 0.11 V of hysteresis, indicating a trapped charge density of 7.4×10^{10} cm^{-2}, which is three times lower than the case without Si. It was reported that if the deposition temperature is too low ($<150°$C), OH groups can be incorporated into the films and results in a degradation of the oxide bulk properties and electron trapping. But if the deposition temperature is too high, subcutaneous oxidation also degrades the electronic properties. Therefore, low temperature deposition plus moderate temperature ($\sim 400°$C) oxidation provides a method to improve the electrical properties.

The real-time oxidation analysis shows that the subcutaneous oxidation follows parabolic growth kinetics during SiO_2 deposition. The parabolic rate coefficient is proportional to the voltage drop across the subcutaneous oxide layer and yields temperature-activated transport. The subcutaneous oxidation of Ge yields a Ge-rich oxide that degrades the electrical properties of metal–oxide capacitors. A thin Si layer deposited at room temperature before SiO_2 deposition protects the Ge surface from undesirable oxidation during SiO_2 deposition. The required thickness of Si layer was predicted in this study. ECR plasma oxidation at 400°C after SiO_2 deposition reduces the charge trapping states. An $Al/SiO_2/Si/Ge$ structure showed improved electrical properties of metal–oxide Ge capacitors. For c–Si substrates, a rapid thermal anneal of the deposited SiO_2 in high vacuum was shown to effectively eliminate the OH group incorporation, and thus reduces both interface trapped charge and charge trapping states.

11.4.3 InP Passivation via Thermal and ECR Plasma Oxidation and SiO_2 Deposition

The thermal oxidation of InP has been studied using SE and a variety of other characterization techniques (13). The thermal oxidation of InP has been found to be slow at 340°C but increased rapidly with temperature with the initial rate leveling and finally displaying self-limiting growth. In Liu et al. (13) a variety of oxidations of InP in O_2 up to 500°C were performed and analyzed. Figure 11.27 shows thermal oxidation results that indicate self-limiting growth, hence a diffusional rate-limiting mechanism was proposed that is in agreement with the findings of previous workers. From ellipsometry and etch rate studies discussed below, the resulting oxide was found to have two distinct layers. The first formed layer is a stoichiometric oxide with the formula $InPO_4$. After this layer forms, further oxidation yielded a mixed outer layer composed of In_2O_3 and $InPO_4$ that coats the $InPO_4$ layer and the transport mechanism results in

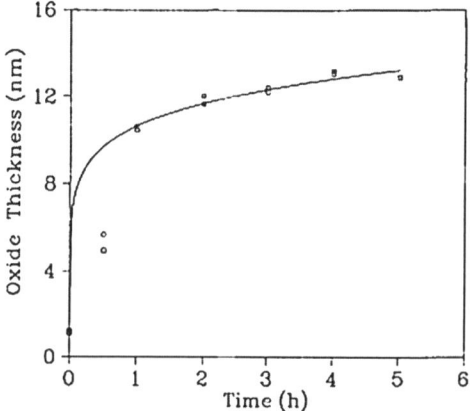

FIGURE 11.27 Thermal oxidation of InP(100) in O_2 at 360°C. [Adapted from Liu et al. (13), Figure 1.]

excess P in the InP substrate. To elucidate the thermal oxidation mechanism an oxidation study was performed using O_2^{18}, as the oxidant. For these experiments an InP thermal oxidation was first carried out in O_2^{16}, and after some time the ambient was switched to O_2^{18}, and then secondary ion mass spectrometry (SIMS) analysis was done where the surface was sequentially ion etched and the composition determined in order to obtain the cross-sectional composition. SIMS is similar to MSRI discussed in Chapter 6 and yields the mass spectrum of surface species as does MSRI, but by a secondary ion collision–ejection process rather than by direct recoil.

Consequently, for SIMS there is a larger matrix effect than for MSRI, which can lead to inaccuracies in composition. A matrix effect is essentially the sensitivity of an element to analysis depending on its chemical and/or physical environment in a material (see Chapter 6). Nevertheless, SIMS is a commonly used surface and interface analysis technique. The results of the tracer experiments are displayed in Fig. 11.28. Figure 11.28a, top, shows the SIMS depth profiles obtained for an oxide grown in O_2^{16} for 2 h at 450°C. The ion counts for In are nearly constant through the oxide layer and increase slightly near the oxide–InP interface. This interface was approximately located near 9 min sputtering time based on where the ion counts reach half of maximum (recall that the sputtering time could be linearly converted to thickness with the proper calibration). The oxide interface is also indicated by the O^{16} and O^{18} profiles, as the counts drop more than 3 orders from the oxide layer to the substrate. Since no O^{18} was introduced during the oxidation of the sample, the ratio of the O^{18} to O^{16} counts shown as the bottom ratio curve in Fig. 11.28b should be representative of the natural abundance of about 500:1, but was found to be closer to 300:1 and attributed to the ion yield differences. The steady increase of the phosphorus counts in the oxide layer was consistent with the reports in the literature that showed a P-rich inner layer and an In-rich outer layer. However, the SIMS data do not show an abrupt interface between the two oxide layers, and this was expected and attributed to the energetic bombardment used in the SIMS analysis that is known to cause ion mixing. The SIMS depth profiles shown in Fig. 11.28a, bottom, are for oxide samples grown in the same batch as those used for Fig. 11.28a, top, profiles but then received an additional O^{18} oxidation for 2 h. Although the oxide is slightly thicker than that of the O^{16} oxidation only (about 475 Å rather than 450 Å), the O^{16} and In profiles were similar to those in Fig. 11.28a, top, profiles except that the P profile shows a small plateau between a short initial rise and a long steady rise in the oxide layer. The initial rise was attributed to either a surface measurement artifact or a real deficit of P near the surface, and the plateau suggests that a uniform oxide with an abrupt boundary exists in the oxide layer near the surface. The difference in P profiles in Fig. 11.28a, top and bottom, indicates that the chemical composition of the oxide varies with the oxidation time, which was consistent with other reported studies. However, the SIMS profile of P did not show a pile up of P at the oxide–InP interface, as was reported earlier in the literature. This was thought to be due to the fact that SIMS detects the secondary ions emitted from a sample by ion bombardment, and the detected ions include contributions from all the oxidation states of the element. SIMS cannot distinguish elemental phosphorus from a phosphorus oxide nor from InP. The P ion counts in

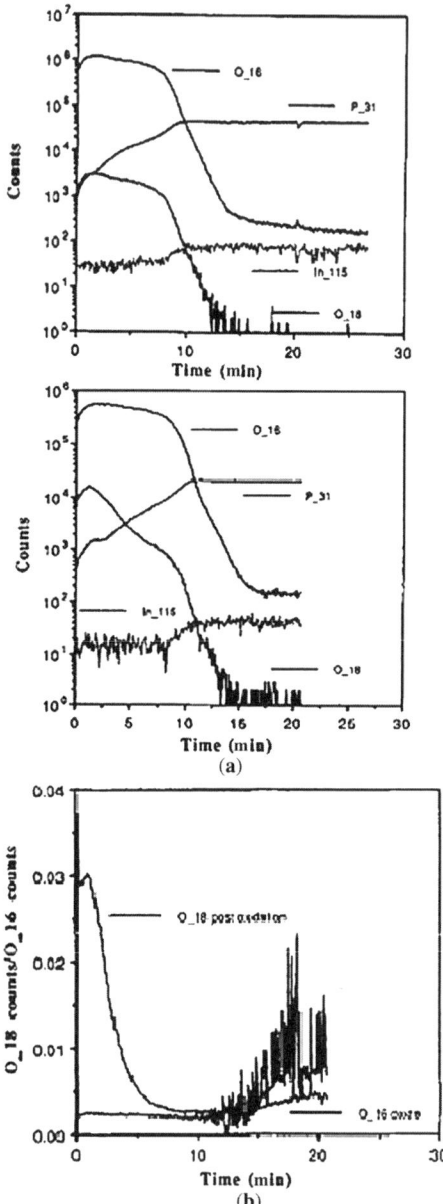

FIGURE 11.28 (a) SIMS depth profiles (as counts for an atom versus sputter time) for the thermal oxidation of InP: top panel, 450°C 2 h in O_2^{16}; bottom panel, same as top and then 2 h in O_2^{18}. (b) Ratio O^{18}/O^{16} for the oxidations in (a) labeled as O^{18} postoxidation. The bottom curve is the ratio with all growth in O_2^{16}. [Adapted from Liu et al. (13), Figures 2 and 3.]

the InP matrix (2×10^4) were sufficiently high to conceal the response of a possibly more modest P excess at the interface. The additional oxidation in O^{18} of the InP oxide results in a localized O^{18} distribution. The O^{18} profile in Fig. 11.28a, bottom, shows a peak near the oxide surface. The location of the excess O^{18} can best be represented by calculating a ratio of the relevant O^{18} profile relative to a matrix profile. This ratio of O^{18} to O^{16} counts for the oxides with and without O^{18} postoxidation is shown in Fig. 11.28b, where the top curve in Fig. 11.28b shows that the O^{18} is localized near the oxide surface indicating that the oxidation occurs at this surface. The slight rise of the ratio after about 12 min of sputtering was attributed to the fact that the detection limit for O^{18} has been reached. The O^{18} marker experiments confirm that oxygen is not the diffusing species during oxidation and that the oxidation reaction takes place at the outer interface via the diffusion of In and P from the InP substrate. The mechanism proposed was that at the outset of InP oxidation a stoichiometric $InPO_4$ is formed at the bare or thin oxide covered InP. As the oxide thickens, the InP and oxidant are separated, and transport must occur in order to continue the reaction. In the InP case the In and P must migrate outward (in contrast to the Si and Ge cases where O migrates inward). If the In^{3+} (radius = 0.08 nm) and P^{3-} (radius = 0.21 nm) come from the substrate, then the smaller In^{3+} ion would likely migrate faster leaving excess P behind. This model is summarized by Fig. 11.29.

The ECR plasma oxidation of InP was also studied (14) in an attempt to determine if excess P is also found at the InP–oxide interface. Prior to these studies the growth of thermal oxide on InP was found not to yield an electronically passivated interface, and many workers believed that this was due to the impurity states resulting from the excess P at the interface.

The hardware for the ECR plasma oxidation was the same as shown in Fig. 11.11. As was the case for ECR oxidation of Si discussed above, to obtain a reliable description of the growth of the oxide layer, two complementary *in situ* ellipsometric

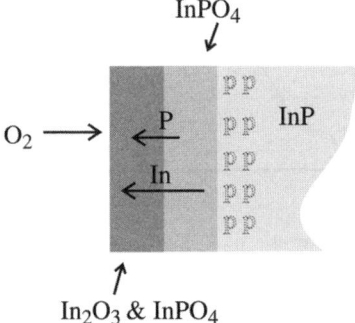

FIGURE 11.29 InP thermal oxidation model showing P and In diffusing outward after an initial formation of $InPO_4$. P diffuses more slowly yielding an In-rich outer layer and excess P near the InP–$InPO_4$ interface. [Adapted from Liu et al. (13), Figure 4.]

measurements were performed: static spectroscopic (SE) and dynamic real-time measurements at a single fixed wavelength (SWE). SWE was performed at 3.5 eV or 354 nm for several reasons, as was the case for Si oxidation above. One reason is that both Δ and Ψ are accurately measured for both bare and film-covered InP at that energy; and another is that ε is relatively insensitive to substrate temperature in the temperature range investigated. Figure 11.30 shows measured ε data for InP at several temperatures of interest in this study where in the range of 3.5 eV the convergence of ε_2 is seen, and while ε_1 is not as insensitive to temperature its excursion is small. For many important semiconductors (Si, Ge, InP, InSb) particular wavelength regions were observed that are appropriate for temperature-sensitive and temperature-insensitive measurements, and the authors humorously described these particularly useful wavelengths chosen for specific measurements as "magic wavelengths." To interpret the single-wavelength measurements in terms of oxidation kinetics (oxide thickness versus oxidation time) a trajectory method was again used and was described above.

As discussed above for the case of thermal oxidation of InP, there is significant evidence indicating the existence of two distinct layers with the outermost layer being predominantly In_2O_3 and the inner layer a mixture of $InPO_4$ and P and perhaps some other oxides, and these chemically different layers were differentiated by different chemical etch rates. Figure 11.31 shows SE results analyzed using a two-film InP substrate model (discussed below) on the etching of the ECR-grown oxide in two different HF/H_2O solutions. The sharp change in film etch rate strongly indicates a two-layer system, as found previously with the HF etching of InP thermal oxides

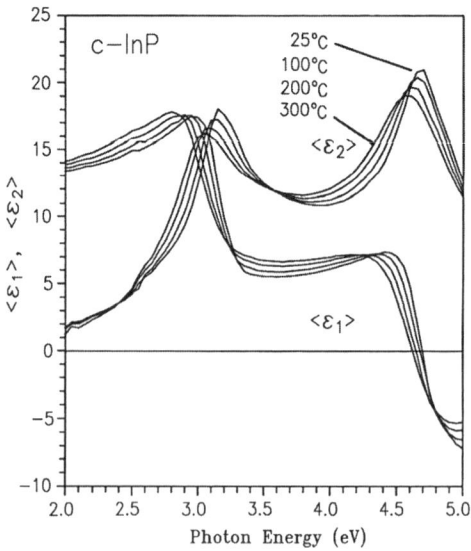

FIGURE 11.30 Pseudodielectric functions of InP at different temperatures. [Adapted from Hu et al. (14), Figure 2.]

11.4 SEMICONDUCTOR PASSIVATION STUDIES

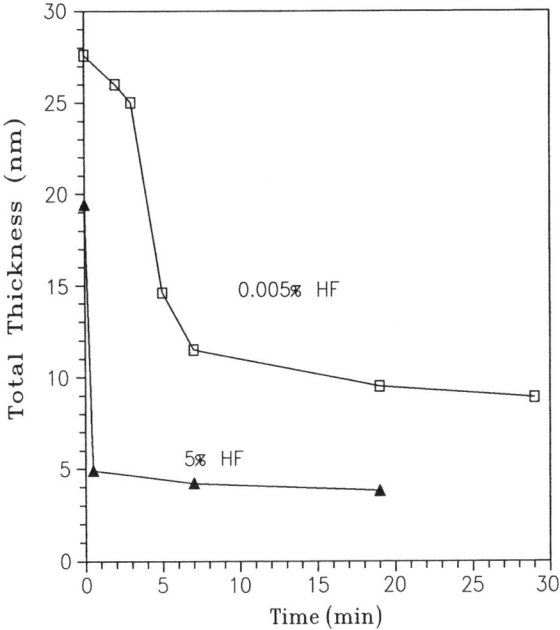

FIGURE 11.31 Etch results of ECR plasma oxide grown on InP. [Adapted from Hu et al. (14), Figure 3.]

with the outermost oxide etching relatively fast. Thus, there is strong evidence that two distinct layers exist.

Figure 11.32a shows an XPS survey spectrum after ECR plasma oxidation of InP, but before etching, and shows the outer layer to be essentially In_2O_3. After etching, Fig. 11.32b shows that once the fast etching outer layer is removed, a slow etching predominantly P layer is seen on the InP surface with only small amounts of In and O. Figure 11.32c is a typical XPS spectrum of the P spectral region. At 134 eV the small peak was attributed to oxidized P and, at 129 eV, is also a small peak for P in InP. In between the peaks at around 131 eV are the free P peaks. These results are concordant with literature findings for the thermal oxidation of InP, and have led to the adoption of a two-layer optical model where the outer layer is assumed to be pure In_2O_3 for simplicity, and the inner layer is a mixture of amorphous InP (a-InP) and $In(PO_3)_3$. For the inner layer, the dielectric function for a-InP was used in place of that for P because no dielectric function data was available for P, and likewise the dielectric function for $In(PO_3)_3$ was used as the inner layer oxidized species based on previous work.

The SE data was fit to the two-layer model based on the XPS and etching results and several additional reasonable models shown in Fig. 11.33a along with the fitted parameters. The figure of merit of the fit, δ, is always lowest for two-film models, and this lends further credibility to the model. The two-film models have the same outer layer of In_2O_3 since that is in agreement with the XPS data and

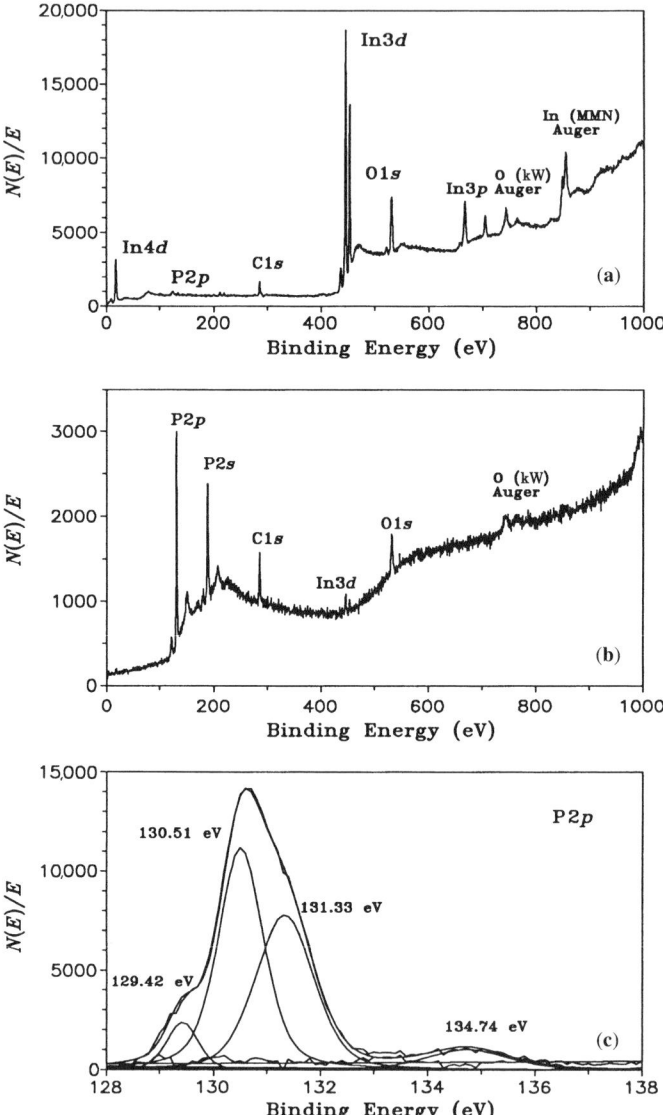

FIGURE 11.32 XPS spectra of ECR plasma-oxidized InP samples: (a) survey spectrum after oxidation; (b) survey spectrum after oxidation and etching; (c) same sample as in (b) but spectrum of P 2p region. [Adapted from Hu et al. (14), Figure 4.]

models D and E compare inner layers of a-InP with $In(PO_3)_3$ or In_2O_3. There were cited reports that the innermost layer is either $In(PO_3)_3$ from electrochemical oxidation or $InPO_4$ from thermal oxidation. However, since there were no available dielectric functions for $InPO_4$, this possibility has not been tested, but a good fit

11.4 SEMICONDUCTOR PASSIVATION STUDIES

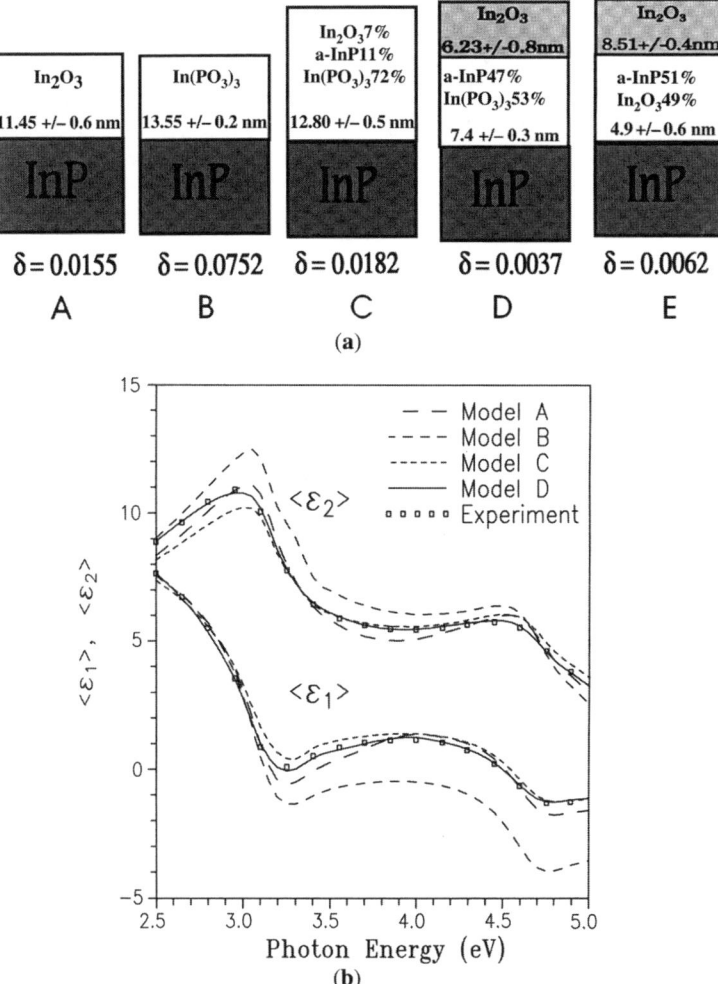

FIGURE 11.33 (a) Possible models for InP ECR oxidation and (b) SE fitting results to the models A–D. [Adapted from Hu et al. (14), Figures 5 and 6.]

has been obtained with D, and so this model was used for the analyses that follow. Figure 11.33b shows the excellent fit of model D (solid line) the experimental data (open circles).

Figure 11.34 shows the total thickness versus $t^{1/2}$ as obtained from SE and model D. It was seen that the oxide growth is enhanced by both higher substrate temperature and positive bias. Positive bias affects the rate of oxidation after an initial regime of about 10 nm indicating the onset of diffusion control. Diffusion control was evidenced by linearity of the plot after an initial regime that was positive bias

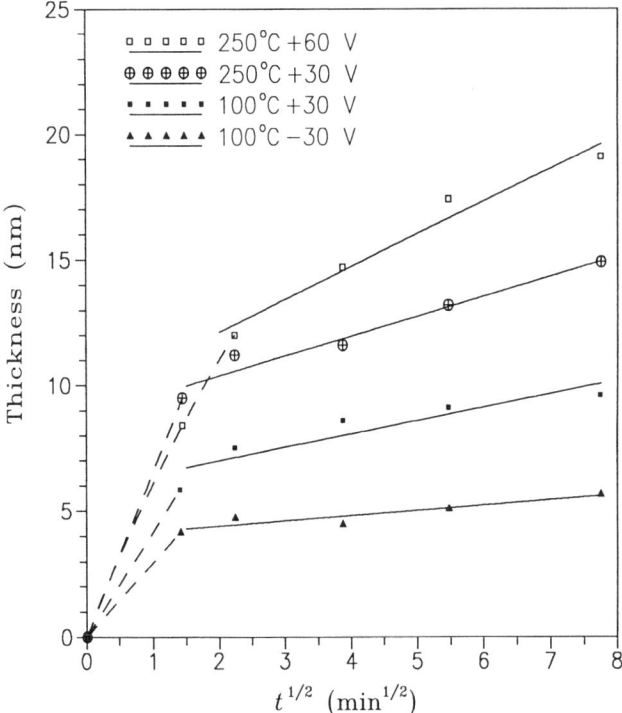

FIGURE 11.34 ECR plasma-oxidized InP in terms of thickness versus time as $t^{1/2}$ as obtained from SE and model D. [Adapted from Hu et al. (14), Figure 10.]

independent but temperature dependent. The negative bias condition shows limited growth indicative of growth first but then growth balanced by etching.

Considering the thermal and ECR plasma studies above on Si, Ge, and InP, there is a strong resemblance of ECR plasma oxidation to thermal oxidation kinetics. This is likely the result of the low electric field across the oxide for ECR oxidation. The larger part of the potential drop is across the plasma sheath rather than across the growing oxide.

Beyond a very thin initial oxide that is stoichiometric $InPO_4$, both the thermal and ECR plasma oxidation of InP lead to an excess of P at the InP–oxide interface. This fact renders both oxidation techniques useless for the production of InP MOS devices where a low number of surface electronic states is important. To obvert this problem, it was thought that to electronically passivate InP one could first thermally oxidize to produce less than 5 nm $InPO_4$ and before the buildup of excess P that is then followed by the deposition of a capping layer of SiO_2 to the required gate insulator thickness (15). The SiO_2 is deposited using ECR plasma deposition at low temperatures to prevent intermixing or interface reaction. The ECR plasma hardware for this process was the same as shown in Fig. 11.11. Notice that there is a shutter in the plasma that reduces the plasma near the sample. This was found to be a key feature in the study since the shutter reduced the ion and photon damage caused by sample exposure to the dense ECR plasma. With the shutter in place both oxidation and deposition processes were possible albeit at reduced film formation rates. The initial oxidation of InP was performed without the shutter at 2×10^{-3} Torr O_2 to yield about 9 nm oxide. This oxide was etched back in dilute HF–H_2O to about 3 to 4 nm at which time an XPS spectrum was obtained and is shown in Fig. 11.35a. This spectrum shows the free P peak at 130.9 eV (doublet). To avoid the free P, the shutter was used and for the same time and conditions 4 nm of oxide was produced, and the XPS spectrum is shown in Fig. 11.35b and displays the 134 eV peak for oxidized P, the 130-eV peak for P in InP, and the broad In peak at about 125 eV. Thus, with the shutter a stoichiometric oxide is detected at the InP surface. This result was also confirmed using SE data where the best fit with shutter does not include a layer with excess P at the interface. Using the stoichiometric $InPO_4$ grown by ECR plasma oxidation with the shutter in place, an ECR plasma-deposited SiO_2 film about 70 nm thick was deposited to cap the thin $InPO_4$ interface layer. This SiO_2 film was deposited from a stream of 3% SiH_4 in Ar + O_2 in a 1/7 ratio at about 150°C and a total pressure of 3×10^{-3} Torr. After metallization high frequency C–V measurements were done with the results summarized in Fig. 11.36. The solid line is the forward trace for the capacitor and the dot–dash line close to the solid line is a theoretical C–V curve for a perfect structure with no interface states or fixed charge. The small stretch out indicated by the difference in these two curves is indicative of low interface electronic states. The dashed curve is

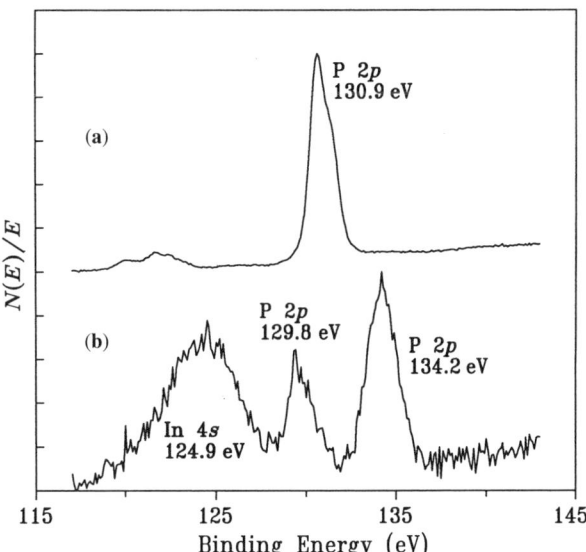

FIGURE 11.35 XPS spectra for ECR plasma-oxidized InP. (a) Direct ECR plasma oxidation to about 9 nm and etched back to about 4 nm and (b) ECR plasma oxidation to 4 nm using a shutter. [Adapted from Hu et al. (15), Figure 2.]

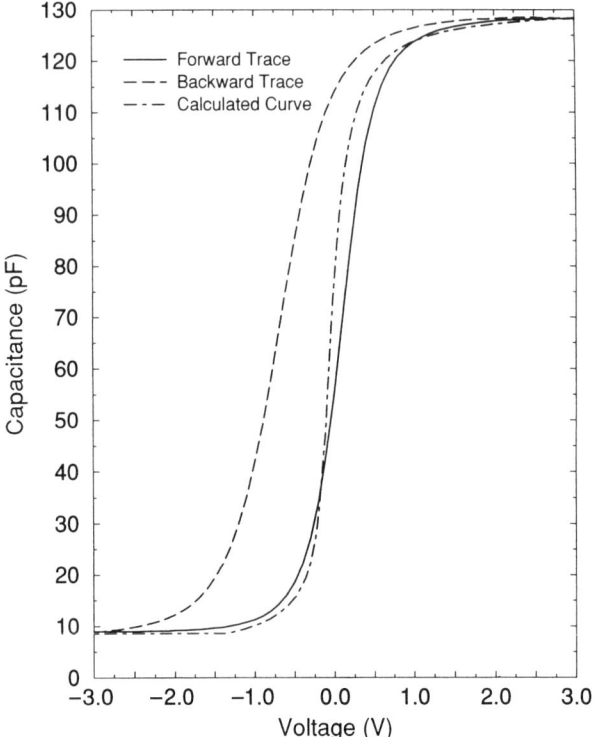

FIGURE 11.36 High frequency (1 MHz) $C-V$ traces for ECR plasma-deposited SiO_2 (7 nm) on ECR plasma-oxidized InP yielding 3 nm oxide and Au–Cr metal top contact. Solid line is forward trace (inversion to accumulation), dot–dash curve is a calculated trace for 10 nm SiO_2 on InP and the dashed line is the reverse trace. [Adapted from Hu et al. (15), Figure 4.]

the reverse $C-V$ trace that shows hysteresis that indicates interface trapped charge. The hysteresis was reduced to less than 1 V by a rapid thermal anneal at 500°C in O_2 for 1 min to a level less than 3×10^{11} cm^{-2}. The InP surface could be inverted, and thus the interface state level is not high enough to cause Fermi level pinning. While the interface with InP is not perfect, this study demonstrated that careful processing and analyses can yield electronically passivated InP.

11.4.4 GaAs Passivation via Thermal and ECR Plasma Oxidation and SiO_2 Deposition

Compound semiconductors, such as InP and GaAs, offer materials properties advantages over Si, such as increased carrier mobility (see Table 11.1) and direct band gap. Devices made from these semiconductors are useful for microwave and/or optoelectronic applications. However, neither InP nor GaAs MISFET technology (where I is for insulator as opposed to O for oxide for MOSFET) has developed to any significant

11.4 SEMICONDUCTOR PASSIVATION STUDIES

degree, and this fact stems from the high density of interface states at the InP and GaAs–insulator interface as well as other poor electrical properties of the GaAs and InP oxides. GaAs has received the most attention as an alternative semiconductor to Si because overall it potentially offers the most performance enhancements.

One study (16) compares the thermal and plasma oxidation of GaAs as was done above for Si, Ge, and InP. In this study the emphasis was on the interfaces obtained after the two kinds of oxidation processes, and also there was a comparison made with Si. The structural differences between the oxides obtained thermally and with ECR plasma as well as the effect of the plasma conditions on the resulting oxides, and orientation effects in the kinetics of oxidation are included. Thermal oxidations were included in this study mainly for the purposes of comparison with ECR plasma results and were performed on different GaAs orientations, at 420°C, and on Si(100) at 500 and 800°C. ECR plasma oxides were grown on various GaAs and Si substrates. Substrate bias and temperature effects were observed. The oxidations were monitored *in situ* and in real-time with single-wavelength ellipsometry (SWE), *in situ* with SE, and *ex situ* with XPS. Optical models of the oxides were established from SE data and checked with XPS.

Figure 11.37 shows thermal oxidation results for GaAs at two temperatures and three orientations. A single-layer model shown in the inset to Fig. 11.37a with two components, GaAs oxide plus a-As, was used to determine the total oxide film thicknesses from spectroscopic ellipsometry measurements. Figure 11.37a shows that the rate of oxidation increases with temperature as would be expected and that the decreasing oxidation rate with time indicates that there is transport control as was the case for Si, Ge, and InP. Figure 11.37b shows that the (100) and (110) GaAs surfaces exhibit essentially the same thermal oxidation behavior while the (111)A samples oxidize more slowly. Figure 11.38 shows the (100), (110), and (111) GaAs crystallographic faces. A (111)A surface has Ga surface atoms attached to 3 As atoms below the surface plane, and the valency of Ga is satisfied. This situation is not favorable to oxidation (recall that oxidation is the transfer of electrons) because electrons are not readily available. On the other hand, the (110) orientation, with both As and Ga atoms at the surface, oxidizes faster than the (111)A orientation. The (100) orientation also oxidizes faster, and this is possibly attributable to the fact that on this surface each atom is attached to 2 atoms in the plane below, yielding uniform but weaker bonding than for the (111) surface. These results are also consistent with etch rate studies on various GaAs orientations, where it was found that the (111)A etch rate was the slowest for reaction–rate limited etching processes. Thus, for GaAs thermal oxidation the rates of oxide formation appear to be dominated by the chemical environment at the GaAs surface rather than simply by the atomic density, as was seen for Si monoatomic surface.

A careful SE modeling study revealed that the thermal oxidation of GaAs is best fit using a two-film model, as shown in Fig. 11.39. The top oxide film was found to be a single-phase material that has excess Ga and the interfacial layer has two components: a GaAs oxide similar to the top layer with excess Ga plus excess As segregated near or at the GaAs–GaAs oxide interface. The top oxide could be removed by chemical etching in water, and thus the compositions of both oxides could be separately

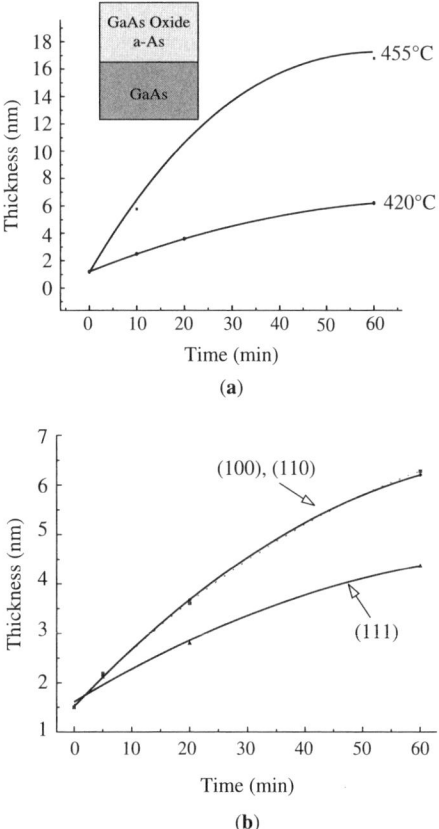

FIGURE 11.37 (a) Effect of temperature (420°C versus 455°C) on the thermal oxidation of (100) GaAs and the model used to analyze that data; (b) GaAs substrate orientation effect on thermal oxidation at 420°C. The lines are polynomial fits to the data. [Adapted from Lefebrve and Irene (16), Figures 2 and 3.]

examined using XPS. Representative XPS spectra for the top layer is shown in Fig. 11.40, whereupon this sample was etched and similar spectra were taken. The before and after XPS spectra were each deconvoluted yielding the interesting result that the Ga/As rations changed substantially from the outer to inner oxides. The outer oxide has a Ga/As ratio of about 7 and so is Ga rich, while for the interfacial oxide the ratio is 14 or double that of the outer oxide. This indicates that the inner oxide is far more Ga rich, which is likely due to the segregation of As at the GaAs–oxide interface.

Figure 11.41 shows ECR plasma oxidation kinetics data for three orientations of both Si (Fig. 11.41a) and GaAs (Fig. 11.41b) as obtained from both single-wavelength (SWE) and spectroscopic ellipsometry (SE) measurements. SE data were also taken at different stages of the growth to confirm the SWE results. It is seen that the Si and GaAs orientation has no significant effect on the kinetics of plasma

11.4 SEMICONDUCTOR PASSIVATION STUDIES

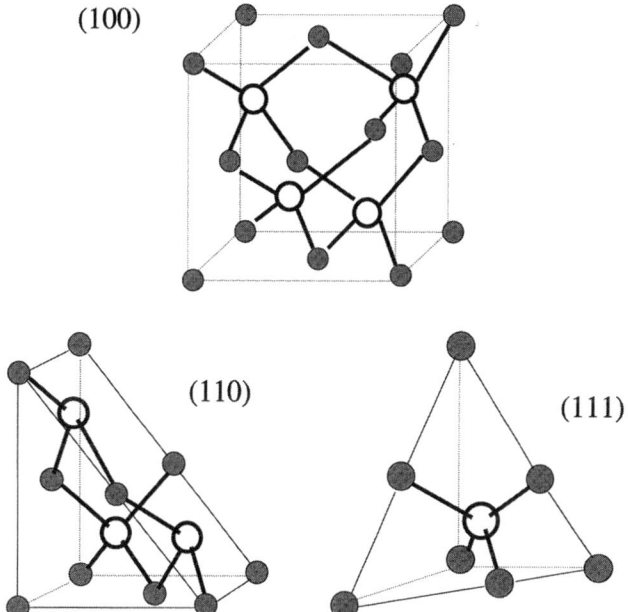

FIGURE 11.38 Three major GaAs orientations where open circles are Ga and shaded circles are As.

oxidation, which is contrary to the thermal results. In fact, if there is any effect on Si, it is interesting to note that it is the (110) surface that oxidizes the slowest and that for thermal oxidation is the fastest oxidizing orientation. The effect of substrate bias and temperature on the kinetics of ECR plasma oxidation has been studied for Si and GaAs where it was shown that in both cases a positive substrate bias, as well as higher temperature, enhances the oxidation rate, and also that the initial regime of oxidation for about the first 5 min is independent of substrate bias.

Figure 11.42 shows representative ECR plasma-oxidized GaAs $3d$ Ga and As spectra. Comparing this set of spectra with the thermally oxidized spectra shows clear differences. One remarkable difference is that the Ga/As ratio for the ECR

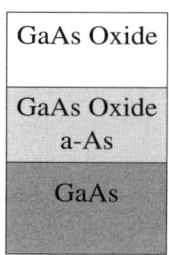

FIGURE 11.39 GaAs optical model used for the analysis of thermal oxidation ellipsometric data. [Adapted from Lefebrve and Irene (16), Figure 2.]

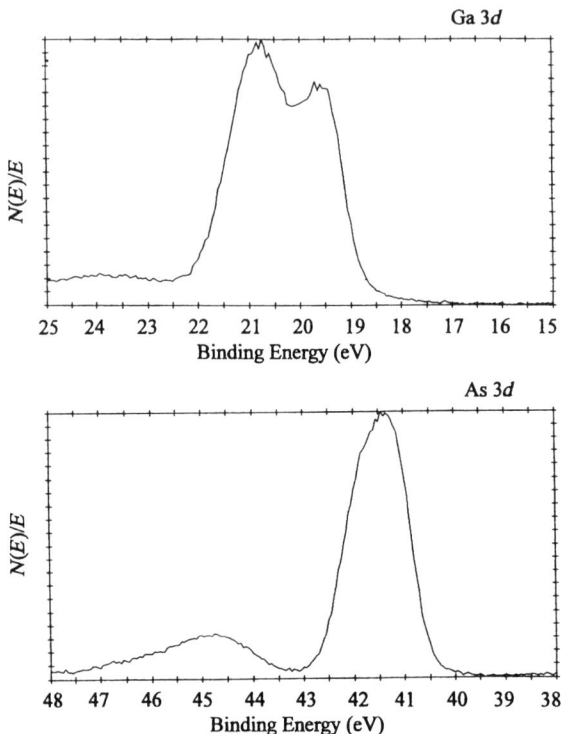

FIGURE 11.40 Ga and As 3d XPS spectra for thermally oxidized GaAs (420°C, 20 min), [Adapted from Lefebrve and Irene (16), Figure 5.]

plasma-oxidized spectra is close to 1 (about 1.1). A similar comparison of thermally and ECR-oxidized Si 2p XPS spectra was made, and the differences are far more subtle indicating that the ECR spectrum showed an Si^{2+} species that was not seen for Si thermal oxidation.

In a later related study (17) XPS was used to probe the valence bands of both the thermal and ECR plasma oxides and compare the resulting spectra for thermal and ECR plasma-grown oxides to published valence band spectra for different reference binary and ternary Ga and As oxides. In addition, electrical properties for the various interfaces that can be formed from the different oxides were probed. This study was motivated by the previous study (16) that indicated that ECR plasma oxides on GaAs are closer to the stoichiometric $GaAsO_4$, and if stoichiometry is important, it may lead to superior electronic interfaces with GaAS. However, in the previous study it was also reported that too long of an ECR plasma oxidation leads to a deviation from stoichiometry for the oxide. Thus, a similar methodology to that used previously utilized for the electronic passivation of InP, namely to first use a mild plasma oxidation to produce thin stoichiometric passivating oxide films and thereby reduce the number of intrinsic interface states, followed by ECR plasma SiO_2 deposition to cap and preserve the stoichiometry of the interface.

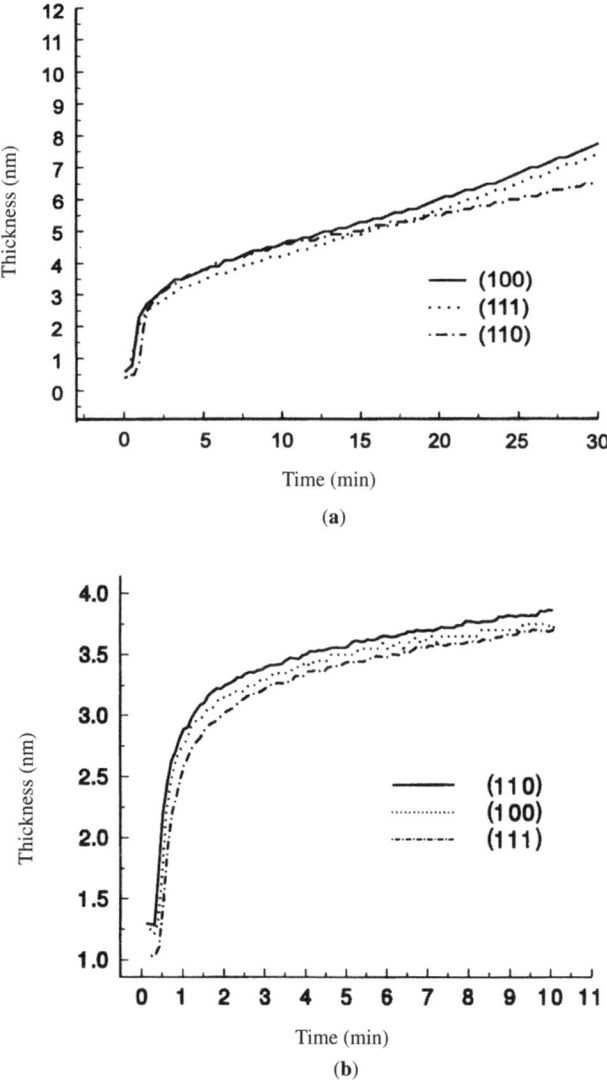

FIGURE 11.41 Comparison of the ECR plasma oxidation of three orientations of (a) Si (+30 V bias) and (b) GaAs (+10 V bias) with both at floating temperature and 300 W input power. [Adapted from Lefebrve and Irene (16), Figure 7.]

The ECR plasma oxidations were done at +10 V substrate bias, under O_2 pressure of 8×10^{-4} Torr, at a floating temperature that was less than 85°C. The plasma-enhanced CVD (PECVD) SiO_2 was also deposited on Si to provide a reference sample as well as GaAs as an insulator film on top of the grown GaAs oxides to form capacitors for electronics testing. In all cases, the SiO_2 was deposited for

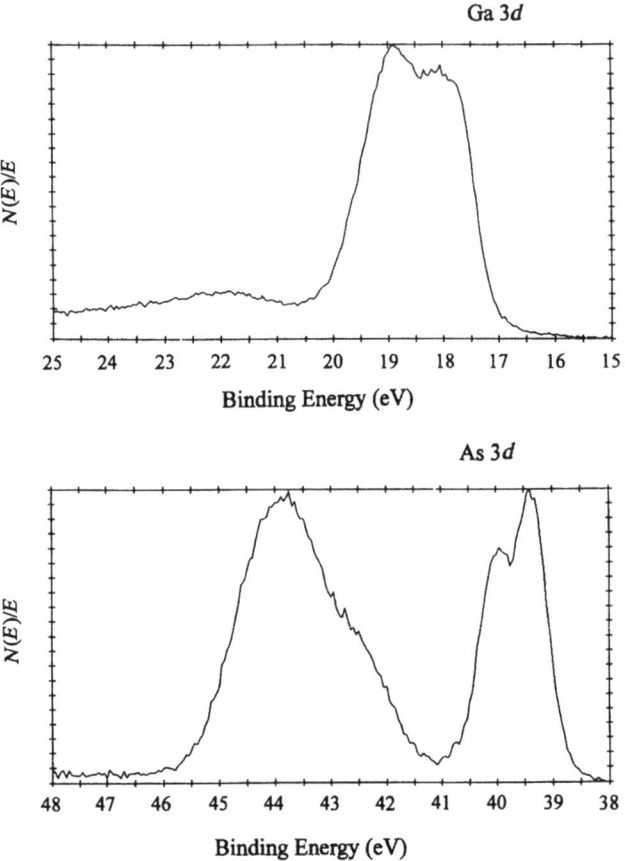

FIGURE 11.42 Ga and As 3d XPS spectra for ECR plasma-oxidized GaAs (300 W, +10 V bias, 10 min). [Adapted from Lefebrve and Irene (16), Figure 1.]

about 7 min at a substrate temperature of 100 to 120°C, which yields roughly 25 nm of SiO_2, as determined from SE. The plasma power was 300 W, and no substrate bias was applied during deposition. The flow rate of SiH_4 (3% in Ar) was kept around 4.8 sccm, and that of O_2 at around 26 sccm. During deposition, the ratio of O_2 to pure SiH_4 was about 15. The total pressure was kept below 1 mTorr because higher pressures result in rapid Si deposition on the walls of the chamber, which in turn can cause SiH_4 depletion and possibly an inhomogeneous film. All samples were located about 20 cm downstream of the plasma to reduce damage induced by active plasma-generated species. Low temperature (below 300°C) remote PECVD of SiO_2 on Si has been reported to yield high quality SiO_2 films with refractive indexes between 1.45 and 1.47. From prior studies it was concluded that pure SiO_2 with properties close to those of thermal oxide and low H concentration was obtained for a ratio of O_2 to pure SiH_4 greater than 9.

11.4 SEMICONDUCTOR PASSIVATION STUDIES

Figure 11.43, left, shows reference spectra for various binary and ternary Ga and As oxides, including As_2O_3, As_2O_5, $Ga(AsO_3)_3$, $GaAsO_4$, and Ga_2O_3. These spectra have been compared to tight binding calculations, and the experimental peaks can be explained in terms of atomic states. Figure 11.43, right side, shows the XPS valence band spectra for thermal oxide grown on GaAs at 420°C for 20 min, and for an ECR plasma oxide grown on GaAs for 10 min at $+10$ V substrate bias. Both oxides have thicknesses around 4 nm. As was reported in the previous study (16), the thermal oxide is Ga rich and the Ga_{ox}/As_{ox} ratio decreases to the free surface, indicating that there is more of the As appearing as As oxide toward the oxide free surface, even though the oxide remains Ga rich. On the other hand, the ECR plasma oxide is nearly stoichiometric and uniform and contains a larger amount of As^{5+} than the thermal oxide. The origins of the As^{3+} component is subject to question, and one possible origin reported was due to room temperature photo reduction of As oxides, which occurs during XPS measurements. In this work increasing amounts of As^{3+} were found, respectively, for 1, 7, and 18 h XPS acquisition times on a GaAs sample cleaned with $H_2SO_4 : H_2O_2 : H_2O$ (3:1:1) solution.

The features seen for the thermal oxide compare well with those of Ga_2O_3, plus another smaller contribution between 12 and 14 eV, possibly from an As oxide and/or the substrate since XPS probes deeper than UPS. The ECR plasma oxide's features closely resemble those of $GaAsO_4$, the stoichiometric GaAs oxide, modulated by the substrate and/or some other As oxides. This was expected since it is known that the thermal oxidation is driven by thermodynamics, that is, the products of thermal oxidation are those predicted by the Ga–As–O ternary condensed phase

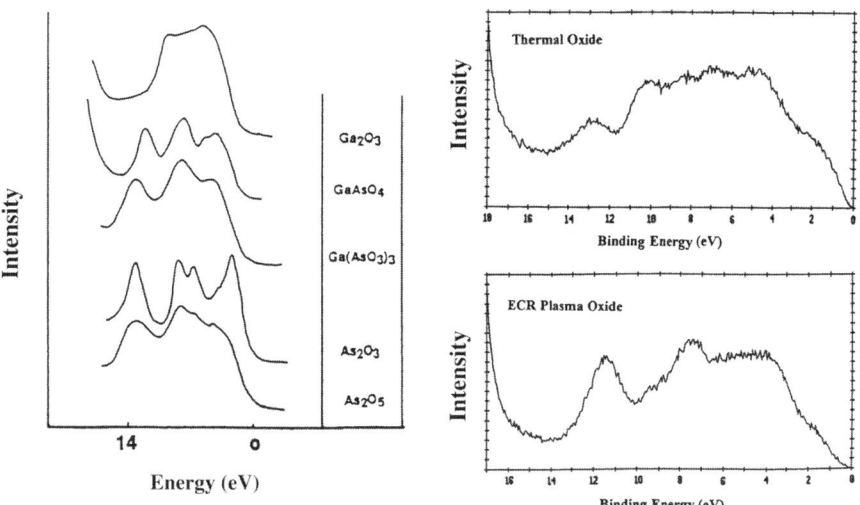

FIGURE 11.43 Left, XPS valence band spectra for different GaAs binary and ternary reference oxides. [Adapted from Hollinger et al. (18) right, XPS spectra for thermal and ECR plasma oxides grown on GaAs. [Adapted from Lefebrve et al. (17), Figures 2 and 3.]

diagram, taking into account the evaporation of the As oxides. On the other hand, the plasma oxidation is driven by kinetics.

As a reference sample, 25 nm of SiO_2 was deposited on Si and gave the following film composition: SiO_2 at 99.35 ± 0.28%, a-Si at 0.65 ± 0.28%. Al was evaporated onto the sample, and a postmetallization anneal was performed. $C-V$ measurements were then performed on the capacitor to determine the electrical quality of the interface and of the SiO_2. Typical results of the quasi-static and the high frequency $C-V$ measurements are shown in Fig. 11.44. The resistivity of the sample was of 1 to 2 Ω-cm, which leads to a C_{min} value of about 25 pF. The agreement between the theoretical values of C_{max} and C_{min}, and Fig. 11.44 is good, which means that accumulation and inversion were attained with the ECR PECVD SiO_2 on Si. The quasi-static $C-V$ curve reaches a C_{max} in accumulation that is slightly higher than that reached in the high frequency case and is possibly due to a leaky oxide. Analysis of the $C-V$ curves with the high and low frequency method gives the interface state density at midgap, D_{it}, with a value in the high 10^{11} cm^{-2} eV^{-1} range at midgap. This value is relatively high from a device standpoint and possibly indicative of the laboratory environment that was not a cleanroom. These results on Si were used for comparison with GaAs.

The previous studies have been done on SiO_2–GaAs structures, with and without an Si interfacial layer, using ellipsometry and $C-V$ measurements. The authors did not show quasi-static data and report Fermi level motion through at least one half of the band gap. The authors concluded that unoxidized Si at the interface, at least in the amount required to form an Si–GaAs heterojunction, did not improve the electrical quality of the GaAs–SiO_2 interface, as it had previously been assumed.

The electrical properties of GaAs–native oxide–SiO_2 structures were tested both before and after postmetallization annealing, and the $C-V$ characteristics were worse

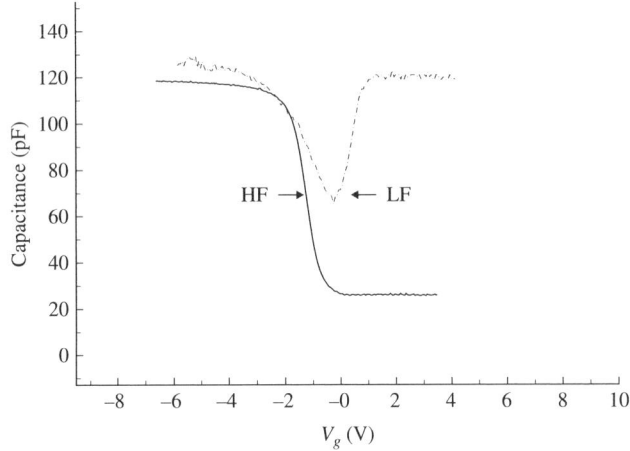

FIGURE 11.44 High and low frequency $C-V$ curves for 25-nm PECVD SiO_2 deposited on P-type Si at about 110°C. [Adapted from Lefebrve et al. (17), Figure 4.]

11.4 SEMICONDUCTOR PASSIVATION STUDIES

than those observed with a GaAs thermal or ECR plasma oxide interface to be shown below and were not shown.

For the SiO_2–thermal oxide–GaAs samples N-type GaAs samples were cleaned and thermally oxidized at 420°C for 15 min and then transferred to the ECR plasma processing chamber, where about 25 nm of SiO_2 was deposited. The SE results were best fit to a two-film model (GaAs oxide film and SiO_2 film) on a GaAs substrate. The $As^{(0)}$ content of the GaAs oxide was high, which is expected for a thermal oxide. Figure 11.45 shows low and high frequency data obtained on these capacitors. One problem in interpreting the results was that the static dielectric constant for the GaAs oxides was unknown and was assumed to be between 3 and 6. C_{max} was calculated to be 105 pF. The resistivities of the samples were of 1 to 2 Ω-cm, which leads to a C_{min} value of around 54 pF. When comparing the calculated values C_{max} to the data shown in Fig. 11.45, it was concluded that accumulation was not attained. On the other hand, the experimental and calculated values for C_{min} are close, and inversion or depletion may have occurred. Another uncertainty in the calculation of C_{min}, is the knowledge of the impurity concentration at the GaAs–oxide interface, which may have changed during deposition. The Fermi level at the semiconductor surface is probably pinned since it could not be swept across the entire band gap. The high frequency $C-V$ curves were also "stretched-out," which indicates the presence of a high density of interface states. Annealing does not have a significant effect on the high frequency $C-V$ curves, but the dip in the low frequency $C-V$ curve is larger in the case of the annealed sample. The low frequency data resemble what is usually seen in the literature for low frequency $C-V$ data on GaAs, that is, the dip in the quasi-static $C-V$ curve is very shallow, indicating a very small change in surface potential, and therefore very high surface state density with Fermi level pinning.

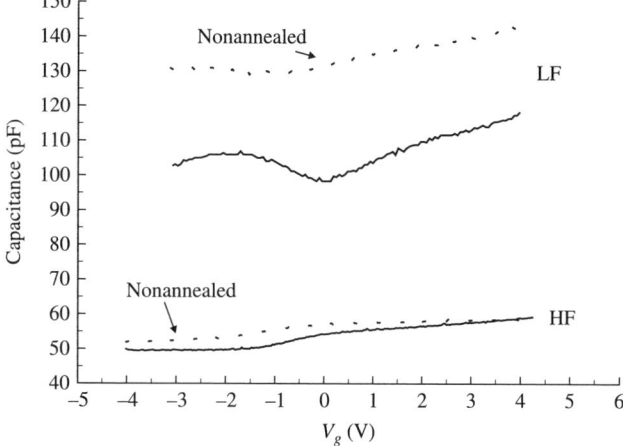

FIGURE 11.45 High and low frequency $C-V$ curves for 25-nm PECVD SiO_2 deposited on thermally oxidized GaAs at about 110°C. Solid lines are for annealed samples and dotted line are for unannealed samples. [Adapted from Lefebrve et al. (17), Figure 5.]

For SiO$_2$–ECR plasma oxide–GaAs samples cleaned N-type GaAs substrates were ECR plasma oxidized for 2 min at a substrate bias of +10 V. The SE modeling results indicated that the interfacial layer of the GaAs plasma oxide after PECVD was found to be abnormally As$^{(0)}$ rich, when compared to that found when modeling the grown ECR plasma GaAs oxide without subsequent SiO$_2$ deposition, which is probably indicative of a structural change of the oxide. The heating combined with UV light from the plasma, plus the effect of the SiO$_2$ deposition process, may have changed the GaAs ECR plasma oxide. Subcutaneous substrate oxidation has already been observed during PECVD growth of SiO$_2$ on Si and on Ge and may also have occurred here. Figure 11.46 shows representative quasi-static and high frequency C–V data for the Al–SiO$_2$–ECR plasma oxide–GaAs structure before and after postmetallization anneal. C_{min} was calculated to be 50 pF and C_{max} of 90 pF, which indicates accumulation is not reached and that inversion might be reached for the nonannealed samples. It is possible that annealing may have a detrimental effect on the interface by changing the stoichiometry of the GaAs plasma oxide. Annealing shifted both low and high C–V measurements toward lower C values. The dip in the quasi-static C–V curve is very shallow, again indicating a very small change in surface potential.

In this (17) and the previous GaAs study (16), a combination of SE and XPS indicated that the interfaces produced by thermal and ECR plasma oxidation of GaAs were significantly different. XPS valence band results showed that the thermal oxide is close to Ga$_2$O$_3$, whereas the ECR plasma oxide was close to the stoichiometric GaAsO$_4$. This result was anticipated, considering that the thermal oxidation is driven by thermodynamics and the plasma oxidation by kinetics. The stoichiometric plasma oxide was expected to be more desirable than the thermal oxide in

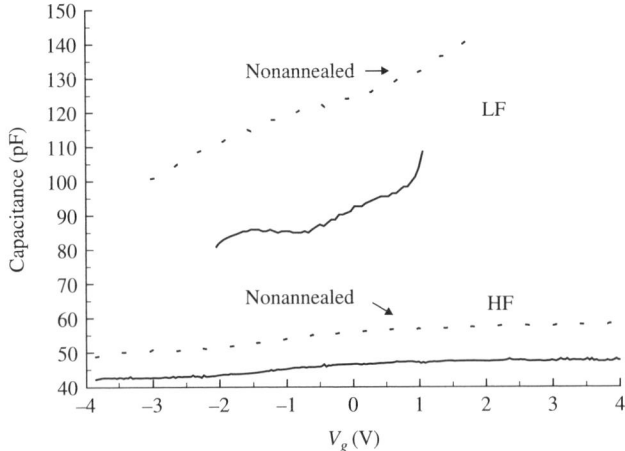

FIGURE 11.46 High and low frequency C–V curves for 25-nm PECVD SiO$_2$ deposited on ECR plasma-oxidized GaAs at about 110°C. Solid lines are for annealed samples and dotted line are for unannealed samples. [Adapted from Lefebrve et al. (17), Figure 6.]

terms of electrical applications. However, the C–V results show that the interfaces produced by thermal and ECR plasma oxidations of GaAs have similarly bad electrical properties in terms of interface charges. A possible explanation for the similarity of the electrical properties obtained with various interfaces is that the quantity of $As^{(0)}$ at the GaAs–oxide interface, which has previously been held responsible for the poor electrical properties of GaAs MIS-type devices, is similar in all cases. Another possible explanation is the deterioration of the ECR plasma oxide stoichiometry observed with *in situ* SE that occurs during the SiO_2 PECVD to form the overlayer. Valence band XPS from oxide films grown on GaAs surfaces by thermal oxidation and ECR plasma oxidation were compared. The studies clearly show that the ECR-grown oxides are nearly stoichiometric and close to $GaAsO_4$ while thermally grown oxides are closer to Ga_2O_3 having significant amounts of As^{3+} oxidation states. However, despite the significant difference in the GaAs oxides structure, most notably the $GaAsO_4$-like ECR plasma oxides, the interfaces prepared by both thermal and plasma oxidation were found to be equally unpassivated in terms of high levels of interface electronic states and Fermi level pinning.

11.5 HIGH STATIC DIELECTRIC CONSTANT GATE OXIDES

For present high performance MOSFETs, which are the mainstay of the microelectronics industry, ultra-thin SiO_2 films less than 2 nm in thickness are required and soon to be demanded by the relentless decrease in device size and commensurate increase in device density on an integrated circuit (IC) chip. As shown in Fig. 7.5 a MOSFET device can be made smaller by reducing the areas for the source, drain, and gate. To operate the device a gate potential is required that charges the gate that has the capacitance of the SiO_2 film C_{ox} given by Equation 7.47 rewritten as

$$C_{ox} = \frac{K\varepsilon_0 A}{L_{ox}} \qquad (11.5)$$

where K is the static dielectric constant of SiO_2 ($K = 3.9$), A is the area of the capacitor or the gate area, and L_{ox} is the SiO_2 film thickness. From this formula it is seen that as the area A decreases, C_{ox} also decreases. Thus, to maintain the capacitance of the SiO_2-based MOSFET, and therefore many of the device operating characteristics, the SiO_2 film thickness L_{ox} must also decrease. With the availability of virtually perfect Si single-crystal wafers with nearly perfect surfaces and that the Si can be processed in ultra-clean environments, there appears to be no reason why 1-nm SiO_2 films cannot be manufactured. However, the problem associated with ultra-thin films goes beyond having a method or process. Specifically, it is known that quantum mechanical tunneling dominates the conduction mechanism for SiO_2 at thicknesses below 2 nm. Quantum mechanical tunneling was introduced in Chapter 6 in reference to STM and will be discussed again in relation to MOSFETs in Chapter 12. For now it is an established fact that electron tunneling in ultra-thin SiO_2 films with thicknesses less than 2 nm results in intolerably large leakage currents in MOSFET devices. Consequently, if IC devices and designs are to advance in terms of size and speed, a substitute for SiO_2 as a gate dielectric is required. This conclusion is gleaned from Equation 11.5, which shows that another solution to reducing A (the size of the MOSFET) and in turn L (to maintain C) is to increase K. Of course, the dielectric constant is a property of the material (in this case SiO_2) and thus to change K, the gate dielectric material, SiO_2, must be changed. Early in the quest for a high K material that could substitute for SiO_2, the most exciting new materials considered were complex oxides such as $Ba_{0.5}Sr_{0.5}TiO_3$ (BST) that have K's at the minimum several times to more than a hundred times the value of K for SiO_2 (3.9) and are therefore attractive since usable film thicknesses for these high K materials would be at least tens of nanometers and therefore beyond the tunneling regime. At physical film thicknesses above 5 nm, the low electric field tunneling currents are near zero. The reason for the range of K's that appears in the literature for BST is the fact that for thin films K is dependent on the film thickness and preparation conditions. Considerable attention has been paid to the thermodynamic stability of the materials relative to SiO_2 as is prudent. Many if not most of the candidate materials are reactive toward Si and produce intermixed layers with K values intermediate between that of SiO_2 (3.9) and the value for the pure high K overlayer. Even the candidate materials such as Zr and Hf oxides that appear thermodynamically stable adjacent to Si based on equilibrium thermodynamics could be problematic at the atomic scale when forming an interface with Si, once again leading to intermixing at the interface. An intermixed interface layer(s) has several profound effects. The first effect is that such a layer provides a series capacitor that lowers the effective K of the film stack, and the second is that such a film may give rise to considerable densities of interface electronic states at the Si–film interface. Also, many of the high K candidate oxides enable the permeation of oxygen, which can lead to subcutaneous oxidation of the Si substrate in the oxygen-rich deposition ambient. Subcutaneous oxidation of Ge, InP, and GaAs was discussed above and leads to another low K film (SiO_2 or a silicate) and another series capacitor. In terms of interface electronic states the formation of subcutaneous SiO_2 on the Si is actually desirable because the SiO_2 lowers the interface electronic states, but in terms of effective K another dielectric is added to the series capacitance that further lowers the effective K of the dielectric film stack. The fact that several dielectric layers may be present and that some are homogenous and some intermixed and each film with a different K and thickness, presents a formidable thin film, interface, and surface metrology challenge. In the following discussions powerful *in situ* real-time characterization tools such as MSRI and ellipsometry are used in combination with TEM, AFM, and electronic measurements to identify and probe the nature of the films and interfaces. The electronic characterizations yield values for K's and interface states. However, it will be pointed out that great care must be exercised when determining K values when more than one film is present. In one study to be discussed below, it was shown that the relationship among the many possible film K's and L's in a gate stack is hyperbolic. It was shown that for certain values of K and L for the films that are present, even small errors in the values of K and L

for the films can change the values for the K by more than several hundred percent for the high K overlayer that is typically obtained from capacitance versus voltage ($C-V$) measurements.

11.5.1 Barium Strontium Titanate

In one study of the $Ba_{0.5}Sr_{0.5}TiO_3$ (BST) interface with Si (19), the BST films were deposited using ion sputtering on both bare and thermally oxidized Si, and the BST deposition was observed in real-time using *in situ* spectroscopic ellipsometry and time-of-flight ion scattering and recoil spectrometry techniques. At the outset of BST film deposition on Si, an approximately 30-Å interface layer formed on bare silicon, and this film has an intermediate static dielectric constant $K \sim 12$ and refractive index $n \sim 2.6$ at photon energies of 1.5 to 3.25 eV. The interface layer growth rate was greatly reduced on an oxidized Si substrate. The results were shown to have profound implications on the static dielectric constant of BST.

The BST films were prepared by reactive ion beam sputtering of a stoichiometric, $Ba_{0.5}Sr_{0.5}TiO_3$ target in a background of 1×10^{-4} Torr O_2 using both Ar^+ and Kr^+ sputter ion beams. The instrumental setup of the deposition and analysis system were previously described in Chapter 6 using Fig. 6.23. Deposition was monitored using *in situ*, real-time spectroscopic ellipsometry (for SE see Chapter 9), to measure growth dynamics and optical properties, and the mass spectrometry of recoiled ions (for MSRI see Chapter 6) was used to determine the film's surface composition. This technique's energy selection of ions with kinetic energies resulting from direct recoil ejection results in a depth resolution of approximately one monolayer for 10-keV incident ions. The films were deposited at a substrate temperature of either 450 or 50°C, on nominally bare, and thermally oxidized [in a conventional thermal oxidation furnace (see Chapter 10) at 950°C in dry O_2 to yield 80 Å SiO_2] single-crystal Si wafers that were cleaned and HF dipped prior to processing. The nominally bare Si was inserted into the vacuum system within a minute of the HF dip but prior to deposition were heated in 10^{-4} Torr of O_2, which resulted in a suboxide layer. SE revealed that this layer was optically equivalent to an SiO_2 layer approximately 15 Å thick. Hence, this was used as the starting substrate for BST deposition onto bare Si. Sputter deposition times varied from about 16 to 20 min for films deposited on bare and oxidized Si. MSRI spectra were acquired on separate samples upon halting the deposition at 2, 4, 6, and 10 min due to the alignment constraints and the nature of the charged environment during sputtering. Subsequent sample annealing was done on both kinds of samples by quickly (~ 1 min) ramping the substrate temperature to 650°C, in 10^{-3} Torr O_2 and holding for 15 min, while acquiring MSRI spectra at 5 and 10 min and after cooling to below 100°C. SE was acquired in real time during the depositions, as well as pre- and postanneal.

The optical properties of BST were determined using thicker 20-nm BST films and a two-film optical model that included the substrate, an interface layer, and a pure BST overlayer (see the inset in Fig. 11.47). The optical properties of the interface layers were obtained from thin films that were all intermixed interface layers (verified using MSRI with discussion below) at about 3 nm thickness. Using the initial, before

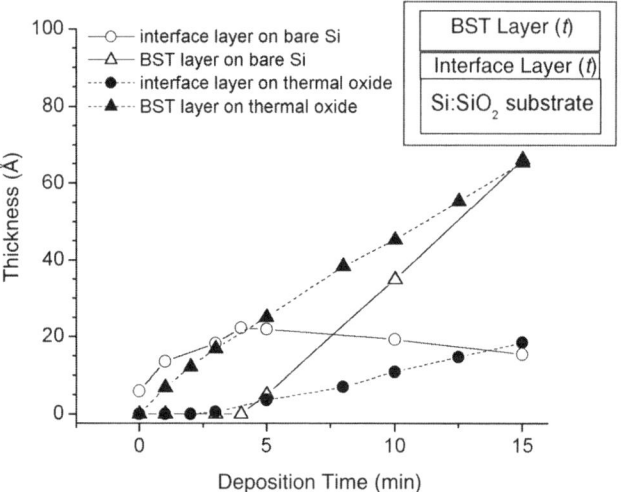

FIGURE 11.47 Real-time SE results for BST deposited on bare Si(solid line) and SiO$_2$ (dashed line) with BST layers and interface layers from the two-film model shown in the inset. [Adapted from Mueller et al. (19), Figure 1.]

deposition spectra, the substrate optical properties for the nominally bare and thermally oxidized Si substrates were obtained and used as the substrate database from which to model the growth of both the interface and BST layers in subsequent depositions. This two-film model shown in the inset of Fig. 11.47 yielded good fits [mean square error (MSE) <1.5]. Figure 11.47 shows that for the nominally bare substrate (solid lines) the initial film growth is entirely the mixed interface layer (○), and upon reaching a critical thickness of about 22 Å, the growth of the interface layer subsides in favor of the pure BST phase (Δ). These SE results are verified by the MSRI results in Fig. 11.48 that show the surface components of the sample at different times during the deposition. It can be seen that during the initial interface layer growth, up to 4 min of deposition, Si is present as a surface component, while after 4 min the BST layer begins to grow, the Si signal is all but gone from the spectrum. Figure 11.47 also shows the deposition of BST upon the thermal oxide with a very different behavior for interface formation. In this case the initial film growth is in the form of BST (▲), which then slowly changes after around 20 Å growth, apparently via reaction with the substrate, to form an intermixed interface layer (●) that continues to grow slowly. The existence of an incubation time before the interface layer forms is likely due to the time required for a sufficient amount of atomic Si to diffuse to the interface for reaction. Further evidence of this is obtained by comparing the 50 and 450°C depositions and subsequent annealing.

Figure 11.49 shows that for 50°C deposition bare Si displays the same sequential growth behavior as for 450°C albeit with different interface layer thicknesses. Figure 11.49 shows that for the low temperature deposition (dashed line, ●) the interface layer grows to only half the thickness of the 450°C interface layer. Additionally,

11.5 HIGH STATIC DIELECTRIC CONSTANT GATE OXIDES

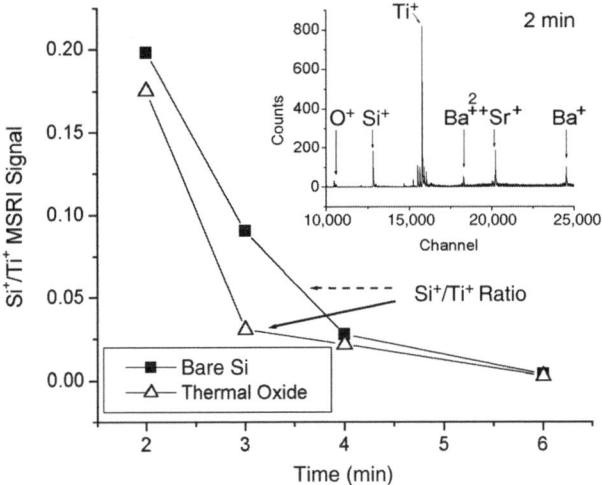

FIGURE 11.48 MSRI obtained Si^+/Ti^+ ratio versus BST deposition time on bare Si and SiO_2 with the MSRI spectrum in the inset. [Adapted from Mueller et al. (19), Figure 2.]

after approximately 10 min of deposition the interface layer for higher temperature deposition (solid line, ○) begins to increase in thickness, while it decreases for the low temperature deposition. These observations are consistent with the reduced supply (activity) of Si at the interface layer–BST boundary at lower temperatures.

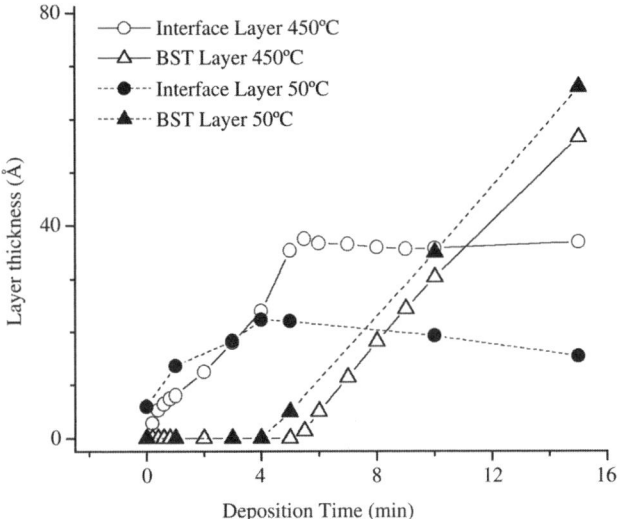

FIGURE 11.49 Real-time SE results for BST deposited on bare Si at 450 and 50°C [Adapted from Mueller et al. (19), Figure 3.]

Figure 11.50 displays MSRI spectra before (Fig. 11.50a) and after (Fig. 11.50b) annealing sputter-deposited BST. SE spectra acquired before and after annealing show small increases in both the interface thickness and in the underlying SiO_2 thickness. The appearance of Si signal in the MSRI spectrum after annealing is due to the more rapid diffusion of Si along the grain boundaries, while the slower expansion of the interface layer is due to the slower bulk diffusion of Si, through the interface and the subsequent reaction with BST. The uniform growth on the underlying SiO_2 layer is likely to occur by reaction of the Si substrate with unbound O migrating from the interface and BST layers. These observations point to two diffusion mechanisms occurring simultaneously.

This study shows that BST reacts strongly when the Si chemical activity is high, forming an interface layer that degrades the composite films overall properties. A critical thickness for this interface layer is revealed in which the Si activity decreases during growth. The SiO_2-covered Si shows that there is no immediate interface layer formation, but rather the interaction between the BST and the oxide is slow to form the intermixed layer and the thinner layer. The results of this study confirm the intuitive notion that a more reactive substrate will more rapidly form a thicker interface layer.

The MOS capacitors prepared on samples resulting from the deposition of BST on bare Si and on thermal SiO_2 were compared to assess the role of the different interfaces on electronics properties. The BST depositions were done at 450°C in a background of 2×10^{-4} Torr O_2 for 5 and 10 min on bare Si and 3.6-nm thermal SiO_2, respectively. MSRI spectra of the film prepared on the bare Si substrate showed Si

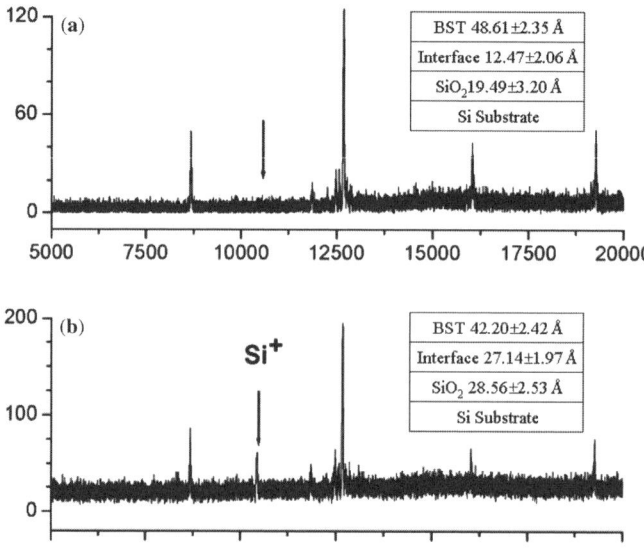

FIGURE 11.50 MSRI spectra of BST on (**a**) bare Si before and after (**b**) annealing at 650°C for nearly 5 min. The Si spectral region is shown with arrows and the insets show the SE model results. [Adapted from Mueller et al. (19), Figure 4.]

to be present at the surface, and SE modeling indicated 0.9 nm of subcutaneous oxide covered by a 5.8 nm intermixed layer, which would have been covered by pure BST had the deposition been allowed to continue. MSRI of the BST film deposited on SiO_2 showed no Si surface signal, and SE modeling yielded 6.8 nm of BST deposited on 3.6 nm of thermal oxide. Representative normalized $C-V$ scans are shown in Fig. 11.51a for both the bare Si (solid line) and thermal SiO_2 (dashed line) samples with the layer thickness obtained by SE also given in Fig. 11.51a. It can be seen that the $C-V$ of the bare Si sample exhibits a larger flat band shift of 0.23 V when compared to the thermal oxide sample, indicative of fixed oxide charge (Q_f), as well as a greater stretch out of the capacitance curve, which is attributed to interface trapped charge (Q_{it} or D_{it}). Both curves show hysteresis between a forward and reverse voltage sweep, with the bare Si sample exhibiting over 2 times the voltage shift of the thermal oxide sample (0.33 versus 0.14 V, respectively). In short, the sample with the thicker SiO_2 film has a higher quality Si–dielectric interface.

In a related study (21), it was found that for multiple layers of dielectric there is a parabolic relationship between the static dielectric constants (K's) and film thicknesses (L's) in a dielectric film stack. It was shown using experimental data that this relationship renders K values that are extracted from capacitance–voltage measurements very sensitive to the other input K and L values. Errors as large as

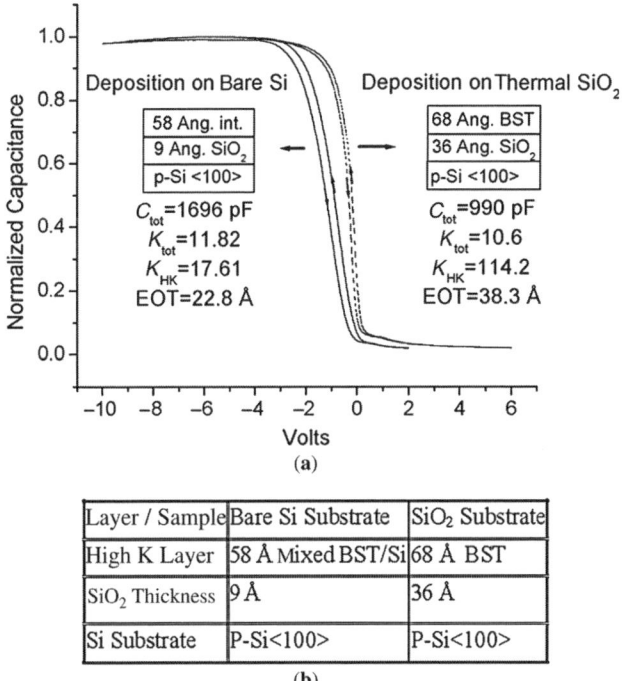

FIGURE 11.51 (a) Shows normalized high frequency (1 MHz) $C-V$ results for BST films grown on Si and SiO_2. (b) shows SE model results for the films in (a). [Adapted from Mueller et al. (20), Figure 10.]

100% and larger can be made in K from small changes in the input parameters. Thus, there is a requirement for accurate film thicknesses and a specific knowledge of the nature of interface film(s), and dielectric properties are necessary for accurate K determinations.

The use of high K dielectrics such as strontium titanate $SrTiO_3$ (STO) and barium strontium titanate $(Ba,Sr)TiO_3$ (BST) among many others with high bulk permittivities (K can be greater than 200) are attractive candidates to substitute for SiO_2. It would be ideal to deposit these materials directly onto Si and thereby preclude the lowering of K afforded by any intervening film such as SiO_2 with $K = 3.9$ in the series stack. The total capacitance C_{tot} of the stack is given as

$$\frac{1}{C_{tot}} = \sum_n \frac{1}{C_n} \qquad (11.6)$$

where n identifies a film in the stack. Thus, the existence of an intervening lower K film would reduce the K of the stack. However, most of the high K candidates are reactive materials and form intermediate mixed-oxide phases with Si, as was shown above for BST films deposited by RF sputtering. Also as shown above, to achieve low leakage and low interface state density (D_{it}), it would be desirable to have a thin layer of SiO_2 in contact with Si. Even without significant reaction the possibility of forming subcutaneous SiO_2 beneath the high K film that is deposited in an oxidizing ambient has also been reported. In all these cases the end result would be either a two- or three-film stack. With $C = K\varepsilon_0 A/L$ where A is the area of capacitor, L is the dielectric film thickness, and ε_0 is a permittivity of free space ($\varepsilon_0 = 8.85 \times 10^{-12}$ F/m), Equation 11.6 is solved for the dielectric constant of the high K film for the case of 2 or 3 films to yield, respectively:

$$K_{HK} = \frac{L_{HK}}{\frac{\varepsilon_0 A}{C_{tot}} - \frac{L_{int}}{K_{int}}} \quad \text{or} \quad K_{HK} = \frac{L_{HK}}{\frac{\varepsilon_0 A}{C_{tot}} - \frac{L_{int}}{K_{int}} - \frac{L_{SiO_2}}{K_{SiO_2}}} \qquad (11.7)$$

where the subscripts HK, int, and SiO_2 refer to high dielectric constant, interface, and SiO_2 films, respectively, and C_{tot} is the total capacitance of dielectric stack.

As shown in Fig. 11.52 for the general case [for an equation of the form $y = (a_1 x + b_1)/(a_2 x + b_2)$] the dependence K_{HK} on K_{int} is graphically represented as a hyperbola with center coordinates given as:

$$x_0 = \frac{L_{int}}{\frac{\varepsilon_0 A}{C_{tot}}} \qquad y_0 = \frac{L_{HK}}{\frac{\varepsilon_0 A}{C_{tot}}} \qquad (11.8)$$

$$x_0 = \frac{L_{int}}{\frac{\varepsilon_0 A}{C_{tot}} - \frac{L_{SiO_2}}{K_{SiO_2}}} \qquad y_0 = \frac{L_{HK}}{\frac{\varepsilon_0 A}{C_{tot}} - \frac{L_{SiO_2}}{K_{SiO_2}}} \qquad (11.9)$$

for the two and three-film models, respectively. The center coordinates of the hyperbola define a minimum for the dielectric constant of both interface and high K layers

11.5 HIGH STATIC DIELECTRIC CONSTANT GATE OXIDES

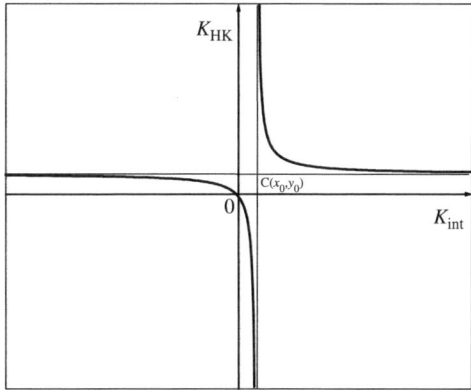

FIGURE 11.52 Hyperbolic dependence $K_{HK}(K_{int})$ with asymptotes and center $C(x_0, y_0)$. [Adapted from Mueller et al. (21), Figure 1.]

below which the values are nonphysical (negative). From here on we consider only the physically tenable positive values for the dielectric constants, namely the positive branch of the hyperbola.

The total capacitance for a two- or three-film stack is typically measured (C_{tot}) using high frequency C–V techniques and then appropriate values for the other(s) K's in the stack are inserted into Equations 11.7 and K_{HK} is extracted. The hyperbolic functional form for K_{HK} indicates that there is high sensitivity to the specific values for dielectric constant and thickness for the SiO_2 and/or the mixed layer, which are typically input and that could result in errors larger than 100% in the values for K_{HK} extracted from the analysis, if small errors are made in the input values (K's and L's) for the SiO_2 and/or the mixed interface layer. It is clear that the most sensitive region for the changes in the K_{HK} values with K_{int} values is near the center of hyperbola, and, actually, in many real situations this region is often the interesting region where the underlying SiO_2 film is thin, resulting from subcutaneous oxidation discussed above.

This is illustrated below where BST thin films were deposited on Si substrates by ion beam sputter deposition, and results and data from the literature where strontium titanate (STO) films are deposited on Si using molecular beam epitaxy (MBE). The parameters used in the calculations of $K_{BST}(K_{int})$ as well as $K_{STO}(K_{int})$ are included in Table 11.2.

Figure 11.53a shows a plot of the functions for K_{BST} (the high K material was BST in this study) versus K_{int} is presented for 3 films. The thicknesses of the film were

TABLE 11.2 Parameters Used for Calculations of Dependences of K_{HK} on K_{int} Shown in Fig. 11.53

#	Dielectric Film	L_{int} (K varies) (nm)	L_{SiO_2} ($K = 3.9$) (nm)	L_{HK}, (K varies) (nm)	C_M/A, (F/m^2)
A	BST (Fig. 11.53a)	1.531	1.032	4.437	0.0136
B	STO (Fig. 11.53b)	0.7	0	11.0	0.0316

obtained using SE. The capacitor structure Ir/BST/Si was prepared completely *in vacuo* and was characterized by capacitance–voltage measurements made at 1 MHz. For the calculations of $K_{BST}(K_{int})$ in Fig. 11.53a the parameters used are listed in row A of Table 11.2 and in Equation 11.7 for three-film stack, which included a BST layer and an interface layer with SiO_2 underneath. If it is assumed that the interface layer has the dielectric constant $K = 4$, then the value of

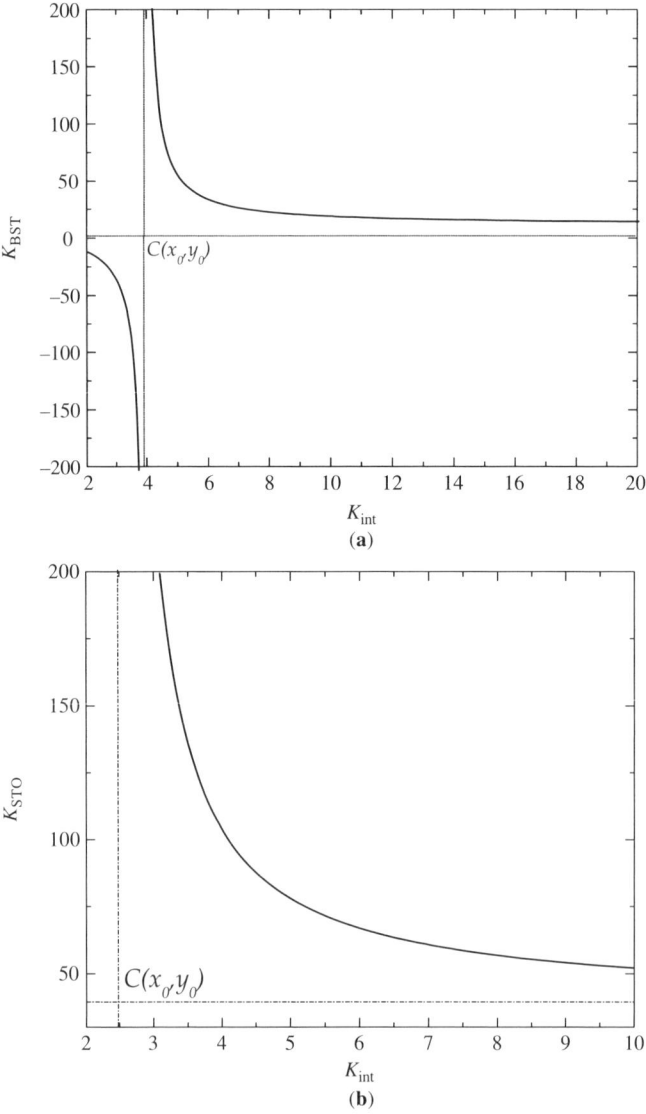

FIGURE 11.53 $K_{HK}(K_{int})$ for (**a**) BST and (**b**) STO thin films. [Adapted from Mueller et al. (21), Figure 2.]

11.5 HIGH STATIC DIELECTRIC CONSTANT GATE OXIDES

$K_{BST} = 1207$ is extracted, which is unreasonably large due to the asymptote near $x_0 = 3.95$. Interestingly, if we assume at $K_{int} = 3.9$ (dielectric constant for pure SiO_2) the value of K_{BST} is negative and therefore nonphysical. On the other hand if a mixed layer were present with a higher K, $K_{int} = 12$, then a K_{HK} of 35 is calculated. The following three-film model example illustrates the sensitivity of input values for the film thicknesses of both interface and SiO_2 layers in the region of the asymptote. At fixed value of $K_{int} = 4$ for interface layer and $K_{SiO_2} = 3.9$, the K_{HK} for the BST layer decreases by a factor of 3 as a result of the rounding off to an integer either L_{int} or L_{SiO_2} from Table 11.2.

Calculations of the dependence of K_{STO} on K_{int} of two-film models using the literature data in row B of Table 11.2 are shown in Fig. 11.53b. In this study a 11.0-nm STO film with 0.7-nm interface layer was prepared, which would yield a high sensitivity region with K_{int} near 2.49. Therefore, all values of K_{int} at or above 3.9 yield positive values for K_{HK} for STO. The assumption that the interface layer is SiO_2 ($K = 3.9$) yields the $K_{STO} = 108$. If instead the interface layer is a suboxide, then the dielectric constant of SiO_x is expected to be larger in value then 3.9, and a K_{int} for SiO_x of 10 will give $K_{STO} = 52$. An error of 0.1 nm will result in increasing the K_{BST} to 145 at $K_{int} = 3.9$. Thus, the accurate determination of the interface layer properties is crucial. Thus, it was shown that even small errors in the K's and L's for the films in a gate stack can lead to large errors in the K_{HK}'s obtained from the usual $C-V$ measurements.

Using great care with the evaluation of film thicknesses the extraction of dielectric constants from the absolute values of capacitance obtained indicates that the resultant high K films differ significantly in quality. The overall dielectric layers have similar values for their dielectric constant (K_{tot}), 11.8 and 10.6 for the bare Si and thermal oxide samples, respectively. Extraction of the high K layer dielectric constant value (K_{HK}) for each sample results in the value of $K_{HK} = 17.6$ for the mixed layer on bare Si, and $K_{HK} = 114.2$ for the BST layer on SiO_2. The supplied values for the thickness and K of SiO_2 necessary to extract the thermal oxide sample's high K layer dielectric constant fall near the asymptote region. Thus, a 1% variation in the thickness of the SiO_2 layer results in an 18% change in the extracted K_{HK}. The bare Si sample's K_{HK} dependence falls into a less sensitive region, with 1% errors in supplied thickness values of either SiO_2 or interface layer resulting in errors of 1% or less. It should be noted that while the errors from the fitting measurements induce an uncertainty of roughly ± 0.8 in the overall K measurement, the calculated dielectric constant of the BST layer on SiO_2, which lies near the asymptote region, will vary from values of 5125 to 55 when the uncertainties of the fitted thickness values are included. Regardless, deposition of nearly 7.0 nm of BST on a starting Si substrate with 3.6 nm of SiO_2 resulted in a capacitor with an equivalent oxide thickness (EOT) of 3.83 nm, while depositing nearly half that material on bare Si resulted in a mixed layer of 5.8 nm with an EOT of 2.2 nm as well as a low electrical quality subcutaneous oxide layer. While the last example indicated the mixed layer to have a $K = 17.6$, experiments have shown that this value is dependent upon deposition conditions and that the K_{tot} (~ 12) measured for the bare Si sample is a more accurate representation of the average K_{int}, which has been found to range from 8 to 18.

In another study a comparison was made of the interfaces for (Ba, Sr)TiO$_3$ films deposited on Si and SiO$_2$/Si substrates (22). (Ba, Sr)TiO$_3$ thin films were again deposited by ion sputtering on both bare and oxidized Si. SE results have shown that a SiO$_2$ underlayer of nearly the same thickness (2.6 nm on average) was found at the Si interface for BST sputter depositions onto nominally bare Si, 1 nm SiO$_2$ on Si or 3.5 nm SiO$_2$ on Si. This result was confirmed by high resolution TEM (HRTEM) analysis of the films, and it is believed to be due to simultaneous subcutaneous oxidation of Si and reaction of the BST layer with SiO$_2$. Using the conductance method, a decrease was observed in the interface trap density D_{it} of an order of magnitude for oxidized Si substrates with a thicker SiO$_2$ underlayer. Further reduction of D_{it} was achieved for the capacitors grown on oxidized Si and annealed in forming gas after metallization.

As discussed above, many of the candidate materials being considered for high static dielectric constant (high K) oxide dielectrics react with SiO$_2$ to form undesirable interface layers that have the potential to lower the overall K, add interface electronic states, and degrade the overall electrical reliability. One way to maintain the high quality interface and still provide a high overall capacitance for the gate dielectric film stack is to use an ultra-thin SiO$_2$ layer as an underlayer of a high K film. The study by Suvorova et al. (22) was aimed at identifying the factors that optimize the SiO$_2$ underlayer in terms of minimizing its effect on K yet maintaining interface quality. *In situ* characterization of barium strontium titanate (Ba,Sr)TiO$_3$ (BST) films grown on bare and thermally oxidized Si substrates was accomplished using SE. Electrical characterization has been used for *ex situ* studies of capacitors prepared completely *in vacuo* as well as for differently annealed samples. Composition and structural properties have been correlated with TEM and Rutherford backscattering spectroscopy (RBS).

The BST films were prepared by reactive ion beam sputtering of a Ba$_{0.5}$Sr$_{0.5}$TiO$_3$ target in an O$_2$ partial pressure of 2×10^{-4} Torr using a Kr$^+$ ion beam. The hardware was the same as in Fig. 6.23. Prior to deposition P–Si<100> wafers were cleaned and HF dipped and thermally oxidized in a conventional furnace in dry 100% O$_2$ at 1000°C and in 10 to 15% O$_2$ in Ar at 700°C to yield about 3.5 and 1.0 nm SiO$_2$, respectively. BST deposition was monitored using *in situ* SE from which optical constants and thicknesses of the grown films were obtained. As was done in previously cited studies, the films were deposited at a substrate temperature of 50 or 450°C and followed by annealing at 450 or 650°C in an oxygen background of 2×10^{-4} Torr. Iridium or platinum electrodes were deposited *in vacuo* by RF sputtering at room temperature through a shadow mask. Selected samples were annealed in forming gas (10% H$_2$ in N$_2$) at 450°C after high temperature oxidation and prior to BST deposition. Finally, an additional postmetallization anneal in forming gas was done prior to capacitance–voltage (C–V) and conductance–voltage (G–V) measurements.

Rutherford backscattering spectroscopy was utilized to characterize the composition of the BST films using a 1.3-MeV He$^+$ beam scattered at 170° and incident both at <100> and random directions to the Si substrate. The compositional and structural variations in the BST thin films were studied using a combination of

HREM and energy-filtered TEM (EFTEM). It should be noted that RBS is a high energy manifestation of ISS discussed in Chapter 6, and the high energy primary ion beam yields bulk rather than surface information. The cross-sectional TEM samples were prepared using standard methods discussed in Chapter 2 (mechanical thinning and Ar ion milling). The stoichiometry of the BST films as obtained from RBS analysis was $Ba_{0.43}Sr_{0.57}Ti_{1.57}O_x$ ($x < 3$), indicating that Ti-rich films were formed, with the likely presence of oxygen vacancies in the film. Overlapping of oxygen peaks from the BST film and SiO_2 underlayer complicated the quantitative determination of each constituent and yielded an error for O of 25%. The estimated error for Ba, Sr, and Ti was about 15%.

The thicknesses of the films in the stack were determined using the optical properties of BST, and the intermixed layer obtained from previous work cited above and the optical properties used for the ultra-thin interfacial SiO_2 layers were from an earlier study on ultra-thin SiO_2 films. SE model results along with electron microscopy results for film thicknesses (t) of selected samples are shown in Table 11.3. The BST studies discussed above indicated that during processing the BST reacts with the Si substrate and, to a lesser extent, with the SiO_2 resulting in formation of an intermixed layer and the SiO_2 underlayer grown subcutaneously.

The SE results in Table 11.3 show the formation of an interface mixed layer (t_{mix}) after deposition and anneal for the various substrates. For the cases of bare Si and 1-nm thermal SiO_2, the SiO_2 has grown to thicknesses of 2.6 and 2.9 nm, respectively, during film deposition and annealing. SE results for thicker SiO_2 substrates indicate a reduction of the underlying SiO_2 film from about 3.5 to 2.1 and 2.6 nm. Although the resultant thickness of SiO_2 is approximately of the same order, the real-time observations of interface formation discussed above showed that the growth rate of the intermixed layer is reduced on the thermally oxidized substrate. The BST deposited on bare Si had a thicker BST film that yields an increased thickness of the mixed layer compared to the other samples.

The fact that the different initial thickness of the SiO_2 underlayer after processing of the samples with different initial substrates, bare and oxidized Si, yield similar results was explained by considering that the subcutaneous oxidation and reaction occurring at two interfaces Si/SiO_2 and SiO_2/BST are taking place simultaneously. The BST studies discussed above indicate that the interface formation observed in real time using SE and MSRI techniques display dynamic interface growth as displayed in Fig. 11.54.

TABLE 11.3 Summary of Layer Thicknesses [t(nm)] for BST Samples on Bare and Oxidized Si Substrates Determined by SE, HREM, and EFTEM[a]

Technique	Bare Si	1 nm SiO_2	3.5 nm SiO_2	3.5 nm SiO_2
SE	11.1/7.2/2.6	3.7/1.7/2.9	4.9/1.8/2.1	4.6/2.2/2.6
HREM	20.3/2.2	6.0/2.0	6.3/2.7	5.6/2.6
EFTEM (Ti, O maps)	17.0/4.0	5.1/3.2	6.2/2.5	5.5/3.0

[a]For SE the thicknesses of the film stack are $t_{BST}/t_{mix}/t_{SiO_2}$, for HREM and EFTEM, t_{BST}/t_{SiO_2}.

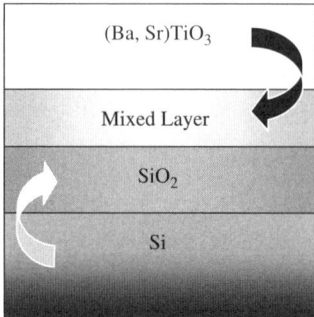

FIGURE 11.54 Pictorial representation of the interface reactions occurring during the deposition of BST on Si in the presence of O_2. The arrows represents the two interface reactions. [Adapted from Suvorova et al. (22), Figure 1.]

High resolution electron microscopy (HREM) images of two BST films deposited on thermally oxidized Si substrates, with initial SiO_2 thicknesses of 1.0 nm (Fig. 11.55a) and 3.5 nm (Fig. 11.55b) respectively, are shown in Fig. 11.55. Also these TEM images show that the deposited BST films are amorphous. The HREM measurements of the SiO_2 underlayer formed after BST deposition exhibit less difference from the starting value than was obtained from the SE analysis above. For the 1-nm initial SiO_2 thicknesses, HREM showed an increase to 2 nm after BST deposition as compared to about 2.9 nm for SE. For the 3.5 nm SiO_2 film, a reduction to 2.7 nm was observed by HREM rather than the 2.1 nm result obtained from SE.

The same trends are seen for both the SE and HREM measurements, though the absolute film thicknesses obtained from the two techniques are different. Despite the differences in the individual film thicknesses determined from the different techniques, the total thickness of the film stack was found to be the same for each sample. The differences between the SE and HREM analyses may result from

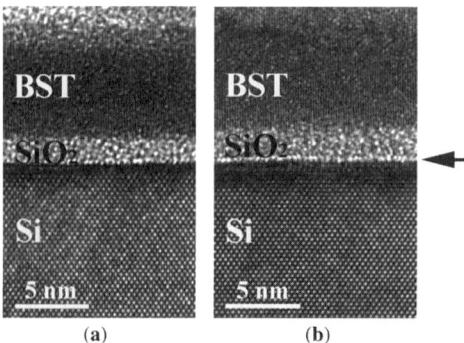

FIGURE 11.55 High resolution TEM cross-sectional images of Si(100)/SiO_2 BST structure where (**a**) had 1 nm and (**b**) had 3.5 nm initial SiO_2 film thickness. The arrow indicates the Si surface. [Adapted from Suvorova et al. (22), Figure 2.]

11.5 HIGH STATIC DIELECTRIC CONSTANT GATE OXIDES

difficulties in clearly identifying the boundary between the two amorphous layers from the HREM images, and the presence of any intermixing. The dark region along the Si/SiO_2 interface (indicated by the arrow in Fig. 11.55) is attributed to overlapping of the amorphous SiO_2 and crystalline Si regions, demonstrating the roughness of the interface. The observed difference between the two images suggests that the BST/SiO_2 interfacial layer for the 3.5-nm initial SiO_2 thickness, shown in Fig. 11.55b, is less uniform than in the sample with the 1.0-nm initial SiO_2 thickness (Fig. 11.55a). For a BST layer grown on bare Si substrates, the SiO_2 underlayer has a thickness of 2.2 nm and the BST/SiO_2 interface is similar to that observed for the thermal 1.0-nm SiO_2 grown on an Si substrate.

Chemical information was presented from energy-filtered TEM (EFTEM) images of constituent elements, as shown in Fig. 11.56. Figure 11.56 shows the zero-loss image and Ti elemental map of an $Si(100)/SiO_2/BST$ structure. The Ti map demonstrates continuous coverage within the BST layer (indicated between the arrows), as shown in Fig. 11.56b. The thickness of the oxide layers was estimated from a line profile across the layers using the full width half-maximum of the profile peak and was correlated with the HREM measurements (Table 11.3). The EFTEM data was found to be consistent with the presence of intermixing between the Ti- and Si-rich layers predicted from previous ellipsometry results.

The conductance method discussed in Chapter 7 was used to compare the interface trap density D_{it} for two types of structures: BST on Si and SiO_2-coated Si. The interface trap density for these samples were estimated in terms of maximum equivalent parallel conductance G_p per unit area using the expression for continuously distributed interface traps (see the discussion in Chapter 7 following Equation 7.58):

$$D_{it} = \frac{2.5}{q} \left(\frac{G_p}{\omega} \right)_{max} \tag{11.10}$$

where q is the electronic charge (1.6×10^{-19}C) and ω is a frequency. The D_{it} estimated for samples grown on initially 3.5-nm SiO_2 substrates were about an order of magnitude lower than for the samples grown on bare Si and 1-nm SiO_2 substrates

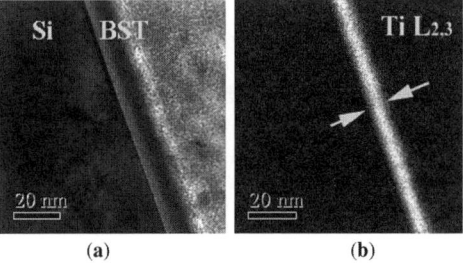

FIGURE 11.56 EFTEM images of $BST/SiO_2/Si$ where (**a**) is the zero loss image and (**b**) chemical map showing Ti distribution as the bright line indicated by arrows. [Adapted from Suvorova et al. (22), Figure 3.]

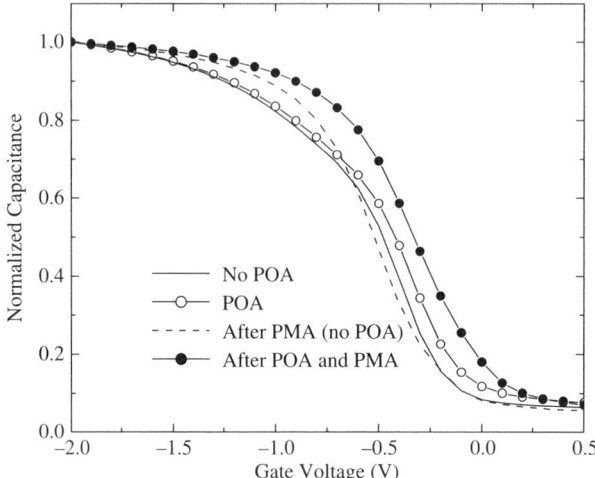

FIGURE 11.57 Normalized 1 MHz $C-V$ curves for the Pt/BST/Si structures on 3.5 nm SiO$_2$ coated Si with and without various treatments: postoxidation anneal (POA) and postmetal anneal (PMA). [Adapted from Suvorova et al. (22), Figure 4.]

with an average 3.0×10^{13} eV^{-1} cm^{-2}. While the value for D_{it} is high compared to typical production quality of SiO$_2$ gate dielectrics, the comparison with thinner SiO$_2$ layers suggests that a thicker SiO$_2$ underlayer can withstand the processing conditions without as much intermixing with BST overlayer and conserves its high quality interface. Conversely, thinner SiO$_2$ intermixes completely due to interface reaction with BST film and is replaced by a lower quality SiO$_2$ layer due to subcutaneous reaction of oxygen with Si substrate. To further improve the interface, annealing in forming gas (10% H$_2$ in Ar) was performed after oxidation of the Si substrate (postoxidation anneal, POA) and prior to a BST deposition as well as postmetallization anneal (PMA) were performed on BST structures grown on oxidized Si (t_{SiO_2} = 3.5 nm). The conductance–voltage method was used to determine D_{it} values throughout the depletion region (from flat band to midgap). Figure 11.57 shows the high frequency (1 MHz) $C-V$ curves of MIS capacitors grown on SiO$_2$/Si substrate. A decrease in "stretch out" is clearly seen of the samples after PMA anneal in forming gas, while no visible change is observed between the samples with and without POA anneal in the same ambient. The values of D_{it} (at midgap) of the samples are presented in Table 11.4. The D_{it} values were reduced by an order of magnitude after PMA treatment.

In this study it was demonstrated that initially for Si substrates with various SiO$_2$ thicknesses BST deposition produces similar SiO$_2$ thicknesses, but the resultant electrical properties of SiO$_2$/Si interfaces differ. BST films sputter deposited onto oxidized Si substrates yield significantly better interfacial electrical properties as compared to films deposited onto nominally bare Si and thin oxide (1 nm). In all cases a complex film stack is produced consisting not only of the BST and SiO$_2$ films but also an intermixed layer. The thicknesses of the layers change due to

TABLE 11.4 D_{it} Values of the Samples with Thermal SiO$_2$ (3.5 nm) Substrate Obtained from the Conductance Method

Anneal	Before PMA	After PMA
POA	2.6×10^{12} eV^{-1} cm^{-2}	1.7×10^{11} eV^{-1} cm^{-2}
No POA	3.0×10^{12} eV^{-1} cm^{-2}	2.3×10^{11} eV^{-1} cm^{-2}

reaction and subcutaneous oxidation. The thicker the starting SiO$_2$ layer the more control of the final stack capacitance and the better the electrical properties. Significant improvement in lowering D_{it} values was achieved after PMA in the presence of the forming gas compared to analogous postoxidation anneal prior to BST film deposition. Despite improvements in the interfacial electrical properties of the stack with an SiO$_2$ film present, the electrical properties are not yet comparable to the capacitors incorporating only SiO$_2$ as a gate dielectric. The message learned from extensive studies of BST and related promising high K candidate oxide dielectrics is that these materials are reactive with Si and to a lesser extent with SiO$_2$. The interfacial reactions were studied and reveal complex film formation with multiple films of varying composition. In terms of electrical properties all have been shown to be significantly inferior to SiO$_2$ and typically not usable as a gate dielectric in MOSFETs. In view of these findings other simpler and less reactive dielectrics have been studied with a few examples given in the next section.

11.5.2 ZrO$_2$, HfO$_2$, and MgO as Potential High K Dielectrics

In a recent study (23) of ZrO$_2$ as a potential high K gate dielectric, the interface formed by the thermal oxidation of sputter-deposited Zr metal onto Si(100) and SiO$_2$-coated Si(100) wafers was studied *in situ* and in real time using spectroscopic ellipsometry (SE) and mass spectrometry of recoiled ions (MSRI). SE yielded optical properties for the Zr and ZrO$_2$ films and interface and MSRI yielded film and interface compositions. An optical model was developed and verified using transmission electron microscopy. Interfacial reaction of the ZrO$_2$ was observed for both substrates with more interaction for Si substrates. Equivalent oxide thicknesses and interface trap levels were determined on capacitors with lower trap levels found on samples with a thicker SiO$_2$ underlayer as was found above for BST.

The apparatus and procedures were the same as used for the above discussed BST experiments. ZrO$_2$ thin films were grown on single-crystal MgO(100), Si(100), and thermally grown SiO$_2$/Si substrates by ion beam sputtering of Zr metal with subsequent thermal oxidation. The MgO substrates were selected in addition to Si and SiO$_2$ because MgO is relatively unreactive and provided pure ZrO$_2$ for characterization.

In situ real-time SE was performed at an incident angle of 70° in the spectral range of 1.5 to 4.5 eV. A 10 keV Ar$^+$ primary ion beam was used for MSRI analysis. Metal–ZrO$_2$–interface–substrate capacitor test structures were prepared *in vacuo* by film deposition immediately followed by sputter depositing platinum through a shadow mask. High frequency (1 MHz, 20 mV oscillator level) C–V measurements were performed, and the density of interface traps, D_{it}, was determined for fabricated samples using the conductance method.

The deposition of Zr onto MgO substrates was monitored by *in situ*, real-time SE until no further change in the pseudocomplex index of refraction was seen. After 30 min of depositing Zr it is seen that pseudorefractive index in terms of $<n>$ and $<k>$ stabilize, indicating that the thickness of the metal film has surpassed the penetration depth of the light. It should be recalled that pseudovalues indicated by the brackets $<>$ are values obtained for composite structures of film(s) and interfaces. The optical properties of Zr were then determined based on an optical model comprised of a pure infinite Zr substrate. The extracted optical properties were then used to determine the thicknesses of thin (30 nm or less) Zr metal films. The SE thickness values obtained for all the Zr metal films studied were within 1 nm of the thicknesses obtained from a quartz crystal oscillator film thickness monitor.

A Kramers–Kronig consistent parametric oscillator model was used to describe the dispersion of the optical properties of Zr metal. This model was comprised of a Drude and Lorentz oscillator. The Drude oscillator describes the collective oscillation of free electrons typical in metals. Interband transitions are described by a Lorentz damped oscillator. The complex dielectric function as a function of photon energy, E, for Zr then has the form

$$\tilde{\varepsilon}(E) = \varepsilon_1(\infty) - \frac{A_1 \mathrm{Br}_1}{E^2 + i\mathrm{Br}_1 E} + \frac{A_2 \mathrm{Br}_2 E_2}{E_2^2 - E^2 - i\mathrm{Br}_2 E} \qquad (11.11)$$

where $\varepsilon_1(\infty)$ is the offset for the real part of the dielectric function. The second term is the Drude oscillator with amplitude, A_1, and broadening, Br_1. The third term is the Lorentz oscillator with amplitude, A_2, broadening, Br_2, and center energy, E_2. The offset term and the Lorentz amplitude are dimensionless quantities. The other terms all have units of electron volts and the values for each of the parameters were given in the paper.

In order to determine the optical properties of ZrO_2, a Zr film of 15.3 nm (determined by SE) was deposited onto single crystal MgO (100). The Zr film was subsequently oxidized *in situ* at 250°C in 5×10^{-4} Torr O_2. The optical properties of the ZrO_2 film were modeled using the following Cauchy equation:

$$n = A + \frac{B}{\lambda^2} \qquad (11.12)$$

It should be recalled that this dispersion model is often used for non-absorbing films and ZrO_2 is reported to be non-absorbing in the measured spectral range.

The index of refraction of ZrO_2 for the measured spectral range is shown in Fig. 11.58. In this optical range ZrO_2 was found to be transparent and $k = 0$. The properties measured here for thin films are slightly different than literature values for ZrO_2 indicating an expected variability from film and bulk and even with film

11.5 HIGH STATIC DIELECTRIC CONSTANT GATE OXIDES

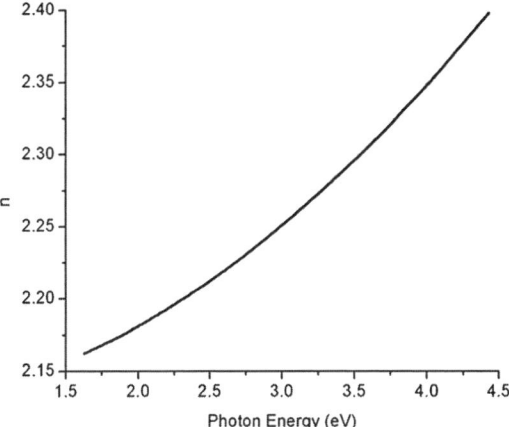

FIGURE 11.58 Index of refraction n for a 25-nm film of ZrO_2 ($k = 0$). [Adapted from Lopez et al. (23), Figure 3.]

thickness and the film deposition method. Therefore, it is important to determine the optical properties for the specific thickness range and deposition process.

Using the determined optical properties, the thickness from SE analysis of one ZrO_2/MgO sample agreed to within 0.8-nm of the thickness determined from cross-sectional TEM performed on the same sample. From the film thickness measurements a Pillings-Bedworth ratio of 1.6 was obtained for the transformation of Zr to ZrO_2 on MgO. The Pillings-Bedworth (P-B) ratio is a common materials parameter that estimates the volume expansion of a material upon oxidation. The P-B ratio assuming bulk densities (Zr-6.511g/cm^3, ZrO_2-5.680g/cm^3) is 1.548. The P-B ratio was used (below) to estimate ZrO_2 resulting film thicknesses.

In order to ensure complete oxidation of Zr to form ZrO_2 films, a ZrO_2 film was re-oxidized in the deposition chamber at the same O_2 pressure but at elevated temperatures, specifically 250°C and 300°C–700°C in 100°C increments. After each temperature increment the sample was brought to room temperature and optical properties were determined. The sample was then placed in a conventional resistance furnace and exposed to 100% flowing O_2 at a temperature of 600°C. The values for the refractive index as a function of the probe energy n(E) was determined after each heat treatment and showed a trend a lower n(E) as the re-oxidation temperature increases. The lowest n(E) was seen after the furnace re-oxidation yielding a difference in n of 0.02 when compared to the 250°C oxidation. Surface roughness was simulated using a Bruggemann effective medium approximation (BEMA) discussed in Chapter 9 that was composed of a 50-50 mixture of ZrO_2 and voids. As roughness increased (1 nm to 7 nm) the apparent index shifted to a lower value, especially for the higher energy region. The magnitude of the shift from simulations was smaller than that seen in the experiments. For a 4-nm roughness layer the index difference at 1.75 eV was 0.002 and at 4.0 eV the difference was 0.02. This increase in roughness can only account for some of the differences. Another possible explanation for the shift to lower index is the incorporation of oxygen at vacancy sites within the film. This incorporation would decrease the density of the film thereby lowering the index based on the Classius-Mossoti formula (see Chapter 9 equation 9.24). The fact that the lowest index was seen for furnace re-oxidation in 100% oxygen supports this hypothesis. It was concluded that a combination of surface roughness and oxygen incorporation would cause the observed shift. From the minimal change in index following various oxidation treatments, it is concluded that the ZrO_2 films are completely oxidized at temperatures as low as 250°C.

MSRI was used to determine the extent of intermixing between Zr/Si and Zr/SiO_2. Approximately 0.5 nm increments of Zr were deposited on the substrate (Si or SiO_2) and probed *in situ* after each increment. Figures 11.59a and b show the MSRI spectra for a bare Si surface and after a deposition of 1.3 nm Zr, respectively. It is seen that Si is present at the surface after a 1.3 nm deposition of Zr.

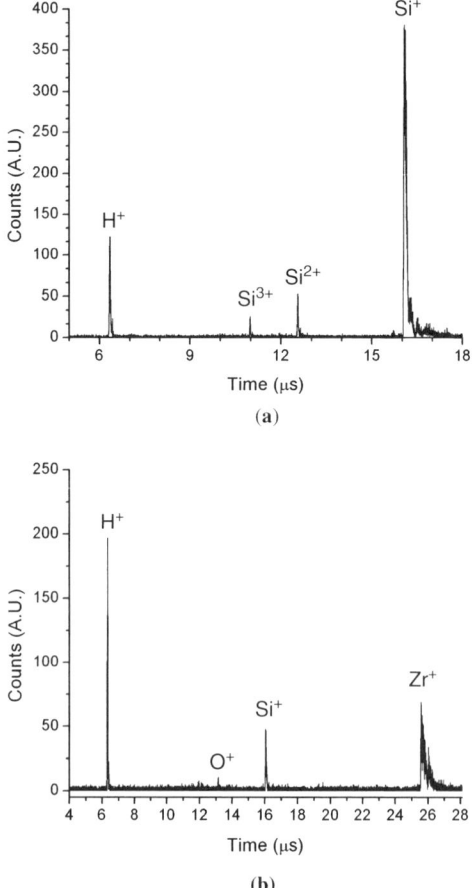

FIGURE 11.59 MSRI spectra for (**a**) bare Si and (**b**) 1.3-nm Zr on Si. [Adapted from Lopez et al. (23), Figure 5.]

11.5 HIGH STATIC DIELECTRIC CONSTANT GATE OXIDES

Features near 26 μs are the peaks for the naturally occurring isotopes of Zr. The Si^+ peak is completely attenuated after depositing 2.0 nm Zr for both the Si and SiO_2 substrates. After oxidation of the 2.0 nm Zr film for both Si and SiO_2 substrates, the Si peak was still not detected. Though, when the samples had an initial layer of 1.3 nm Zr, the Si^+ peak intensity did increase after oxidation relative to the unoxidized sample. To eliminate the possibility the MSRI Si signals were due to Zr film discontinuities.

The AFM measurements were conducted to obtain information about the roughness of the films as an indicator of possible discontinuities in the film and hence the cause for the Si MSRI signal for thin films of Zr. No significant differences were found when comparing the surface profile of 1- to 2-nm Zr films and rms roughness values for the substrates (Si and SiO_2) to that after deposition and oxidation. SEM analysis was also conducted on these samples and showed no evidence of island formation. Thus, it was concluded that the Si MSRI signal from thin Zr films was from intermixing and not discontinuities.

The interface width can be estimated by taking into account the volume expansion upon oxidation. If it is assumed that the PB ratio for this intermixed region to be that of Zr to ZrO_2, then the final thickness of this layer after oxidation would be 3.2 nm, ignoring intermixing of Si with Zr. The PB ratio for $ZrSi_2$ to $ZrSiO_4$ is 1.33 based on bulk densities ($ZrSi_2$—4.88 g/cm^3, $ZrSiO_4$—4.56g/cm^3). Using this PB ratio, the thickness of this intermixed layer after oxidation then becomes 2.6 nm. From this analysis the extent of intermixing on both Si and SiO_2 was estimated to be less than 3 nm.

Cross-sectional TEM was performed on a sample that was initially 19.6 nm Zr on bare Si and then oxidized at 250°C at 5×10^{-4} Torr O_2. The TEM result in Fig. 11.60 shows that the bulk ZrO_2 film was polycrystalline with a monoclinic structure and

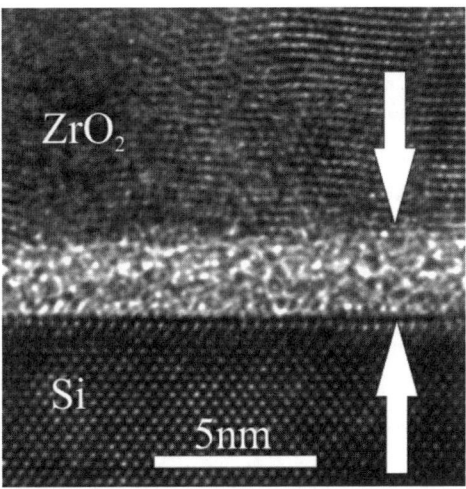

FIGURE 11.60 HREM cross-sectional image of ZrO_2/Si interface. [Adapted from Lopez et al. (23), Figure 6.]

28.7 nm thick and also shows an amorphous interface region of approximately 2.3 nm in thickness, and therefore concordant with the MSRI results above.

The intention was to deposit thin Zr oxide films on both Si and amorphous SiO_2 that are all interface (intermixed with Si) and then to characterize the optical properties of these interface layers. To that end, 4 experiments were conducted: 1 and 2-nm Zr films were deposited on both Si and amorphous SiO_2 and all four samples were oxidized. After oxidation the optical properties of the intermixed layers were extracted using upper and lower input thickness estimates derived from P-B ratios discussed above. The upper thickness estimate was obtained using a P-B ratio of 1.6 (assuming a Zr to ZrO_2 transformation) and lower estimate was based on a P-B ratio of 1.3 (assuming a $ZrSi_2$ to $ZrSiO_4$ transformation). Both estimates gave similar results.

$C-V$ measurements were done on $Pt/ZrO_2/Si$ and $Pt/ZrO_2/SiO_2/Si$ samples, in order to determine the dielectric constant for the film stack as well as the equivalent oxide thickness, EOT. The measurements were corrected for series resistance and quantum effects were not taken into consideration for the analysis but would be small due to the film thicknesses in this study. Figure 11.61 shows typical $C-V$ data for a $Pt/ZrO_2/Si$ sample. Based on the ellipsometric thickness of the ZrO_2/interface stack and the accumulation capacitance, the dielectric constant of this stack was calculated to be 20 with an equivalent oxide thickness (EOT) of 1.2 nm. $C-V$ data for a $Pt/ZrO_2/SiO_2/Si$ sample showed that although the starting thickness of the SiO_2 layer was 2.5 nm the E_{ot} value determined was 2.1 nm. Since the EOT was found to be less than the starting thickness of SiO_2, reactivity at the interface is confirmed.

The conductance method based on varying frequency $C-V$ and $G(\omega)$-V measurements was used to compare the interface trap density, D_{it}, for ZrO_2/Si and

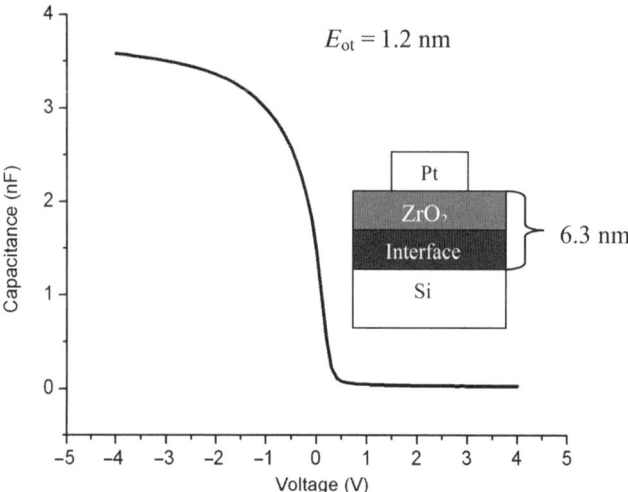

FIGURE 11.61 1 MHz $C-V$ of $Pt/ZrO_2/Si$ sample. [Adapted from Lopez et al. (23), Figure 9.]

11.5 HIGH STATIC DIELECTRIC CONSTANT GATE OXIDES

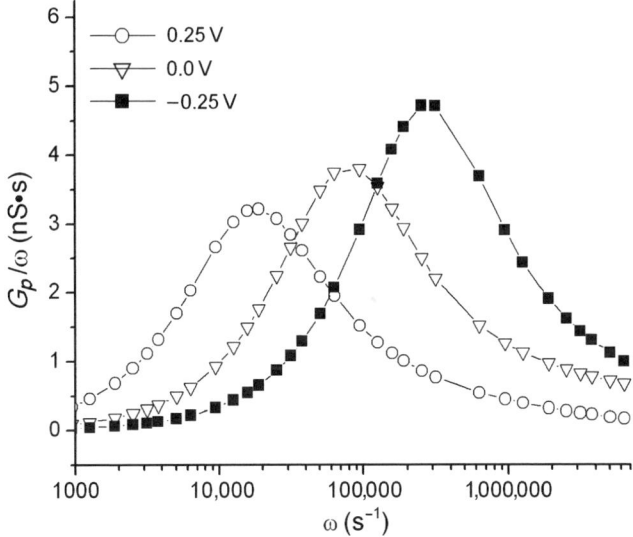

FIGURE 11.62 G_p/ω versus ω at various bias values. [Adapted from Lopez et al. (23), Figure 10.]

$ZrO_2/SiO_2/Si$ samples. The interface trap density for these samples was estimated in terms of maximum equivalent parallel conductance, G_p, per unit area using Equation 11.10 for continuously distributed interface traps. Figure 11.62 shows the G_p/ω versus ω data where the various bias values represent specific energy positions within the Si bandgap. The conductance method probes a portion of the Si bandgap depending on substrate type and for P-type Si the values of D_{it} are limited to the energies from the valence band edge to near mid-gap. From the maximum in the conductance curves the D_{it} values trace half of the traditional U-shaped dependence from band edge to midgap to band edge. The D_{it} values near mid-gap were used to construct Table 11.5 and have been normalized by the D_{it} value obtained with a 9-nm SiO_2 film (1×10^{12} cm^{-2} eV^{-1}) underneath the ZrO_2 film in order to compare results under similar processing conditions. Post-metallization anneals in forming gas were not conducted on these samples.

The results in Table 11.5 show that a thin SiO_2 underlayer (2.5 nm) is not sufficient to reduce interface traps and passivate the interface, and indeed equivalent to no initial SiO_2 at the interface. This is likely due to the reactivity of ZrO_2 with SiO_2.

TABLE 11.5 Interface quality comparison for ZrO_2 on Si and SiO_2

Sample	Normalized D_{it}
ZrO_2/Si	8
ZrO_2/SiO_2(2.5-nm)	9
ZrO_2/SiO_2 (9.0-nm)	1

The value of incorporating an ultra-thin underlayer of SiO_2 has been discussed above to maintain a SiO_2/Si interface quality while minimizing the dielectric offset of adding a material with a lower K to the system. The results indicate that for the case of ZrO_2 formed from oxidation of Zr metal there is no advantage to incorporating this ultra-thin SiO_2 underlayer.

A similar study to that on ZrO_2 cited above was performed on HfO_2 (24). The interfaces studied were formed by the thermal oxidation of sputter-deposited Hf metal onto Si(100) and SiO_2-covered Si(100) wafers and were also analyzed *in situ* and in real- time using SE and with MSRI. SE yielded optical properties and MSRI yielded film and interface composition. Reactivity between HfO_2 and both substrates was found to be similar based upon the optical properties of the interface layer. Equivalent oxide thicknesses and interface trap levels were determined, and unlike the ZrO_2 case significant reduction in interface traps was noticed for samples with a 2-nm SiO_2 film on Si.

The optical properties of MgO, the substrate used for Hf and HfO_2 optical property determination, were determined as before to deconvolute film properties for films subsequently grown on this substrate. Since MgO is transparent in the spectral range used, backside reflections were minimized by having only one side polished. The optical properties of the MgO substrate were modeled using the Cauchy dispersion model (see Equation 11.12) that is often used for nonabsorbing films.

The deposition of Hf onto MgO substrates was monitored using *in situ*, real-time SE until no further change in the pseudocomplex index of refraction was seen as was the procedure for Zr deposition onto MgO in the ZrO_2 study above. After 35 min of depositing Hf, it was found that $<n>$ and $<k>$ stabilized, indicating that the thickness of the metal film has surpassed the penetration depth of the light. The optical properties of Hf were then determined based on an optical model comprised of a Hf film on a pure infinite substrate. The Hf film was modeled using a Kramers–Kronig consistent parametric oscillator model. This model was comprised of a Drude and Lorentz oscillator as for Zr films. Using the Hf optical properties Hf thicknesses of thin (30-nm or less), Hf metal films were determined by SE. The SE thickness values obtained for the Hf metal films studied were within 1 nm of the thicknesses obtained from a quartz crystal rate monitor.

To determine the optical properties of a thin film of HfO_2, an Hf film of 16.7 nm was deposited onto MgO. From previous work on the ZrO_2 system cited above, complete oxidation of Zr to ZrO_2 occurred at temperatures as low as 250°C in 5×10^{-4} Torr O_2. Given the similar chemistries between Zr and Hf, these same oxidation conditions were tried. However, ellipsometry results unexpectedly revealed that the Hf film was not completely oxidized under these conditions. A two-film model comprised of an Hf metal underlayer with a transparent (Cauchy) overlayer adequately described the SE experimental data. These experiments indicated that the minimum temperature for the complete oxidation of Hf metal films was 500°C at an O_2 pressure of 5×10^{-4} Torr. The index of refraction of HfO_2 for the measured spectral range is shown in Fig. 11.63. A Pillings–Bedworth (PB) ratio of 1.7 was obtained for the transformation of Hf to HfO_2 on MgO. The PB ratio assuming bulk densities (Hf–13.31 g/cm^3, HfO_2—9.68 g/cm^3) is 1.62.

11.5 HIGH STATIC DIELECTRIC CONSTANT GATE OXIDES

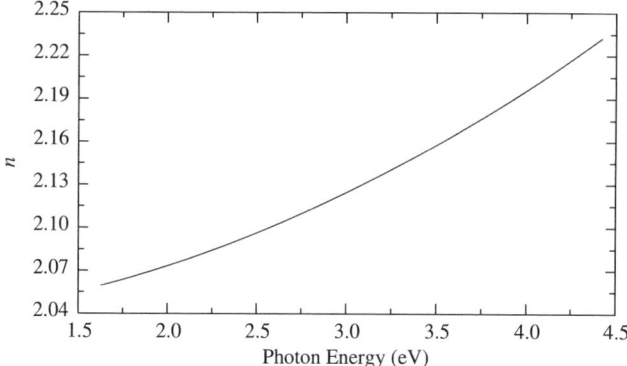

FIGURE 11.63 Index of refraction, n, for thin film (27.8 nm) HfO_2 on a MgO substrate. [Adapted from Lopez and Irene (24), Figure 5.]

The MSRI was used to determine the extent of intermixing between Hf/Si and Hf/SiO_2. For the MSRI experiments approximately 0.5-nm increments of Hf were deposited on the substrate (Si or SiO_2) and probed *in situ* after each increment. The surface composition after 0, 1, and 2 nm Hf on Si is shown in Figs 11.64a, 11.64b, and 11.64c, respectively. The Si^+ signal is located near 16 μs. The Hf^+ signal with isotopes is located near 34 μs. It can be seen from Fig. 11.64c that Si is no longer present at the surface after approximately 2 nm of Hf has been deposited. After oxidation treatment, the Si peak is barely discernable above background. Similar results were obtained when the deposition was carried out on the SiO_2 substrate. To eliminate the possibility that the MSRI Si signals were due to Hf film discontinuities, AFM measurements were done that yielded RMS roughness values comparable to the undeposited substrates and thus lends credence to the assertion that the thin Hf films were continuous below 0.5 nm.

The interface width was estimated as for the Zr study above by taking into account the volume expansion upon oxidation and the above MSRI results. If the P-B ratio for this intermixed region is assumed to be that of Hf to HfO_2 (1.7) then the final thickness of the HfO_2 layer after Hf oxidation would be approximately 3.4 nm ignoring any intermixing of Si with Hf. The P-B ratio for $HfSi_2$ to $HfSiO_4$ was calculated to be 1.32 based on bulk densities ($HfSi_2$-8.03 g/cm^3, $HfSiO_4$-6.97g/cm^3). Using this P-B ratio the thickness of this intermixed layer after oxidation was approximated to be 2.6 nm. When all results are taken together it was estimated that the extent of intermixed interface on both Si and SiO_2 is less than 3.4-nm and likely between 2 and 3 nm thick.

$C-V$ measurements were performed on $Pt/HfO_2/Si$ and $Pt/HfO_2/SiO_2/Si$ samples, in order to determine the dielectric constant for the film stack as well as the equivalent oxide thickness, E_{ot}. As for the Zr studies the measurements were corrected for series resistance and quantum effects were not taken into consideration for this analysis but would be small due to the film thicknesses in this study. Figure 11.65 shows typical $C-V$ data for a $Pt/HfO_2/Si$ sample. Based on the ellipsometric

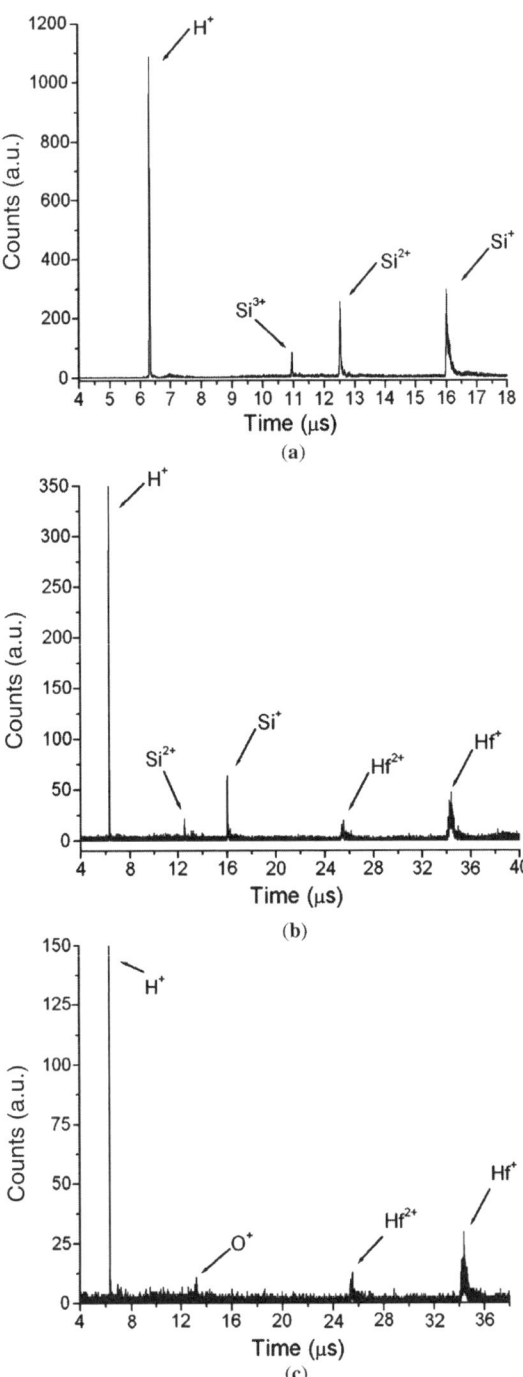

FIGURE 11.64 Time-of-flight MSRI spectra (a) for bare Si, (b) ~1 nm Hf on Si, and (c) ~2 nm Hf on Si. [Adapted from Lopez and Irene (24), Figure 6.]

11.5 HIGH STATIC DIELECTRIC CONSTANT GATE OXIDES

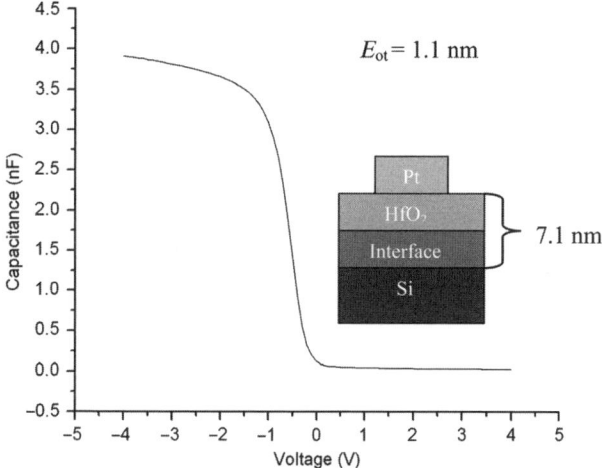

FIGURE 11.65 1 MHz C–V of Pt/HfO$_2$/Si sample. [Adapted from Lopez and Irene (24), Figure 9.]

thickness of the HfO$_2$/interface stack and the accumulation capacitance, the dielectric constant of this stack was calculated to be 25 with an E_{ot} of 1.1 nm. C–V data for a Pt/HfO$_2$/SiO$_2$/Si sample showed an E_{ot} value of 2.6 nm with a stack dielectric constant of 12. The underlying SiO$_2$ thickness was 2.0 nm. The results from the electrical characterization of similar ZrO$_2$ capacitors showed the E_{ot} to be less than the starting thickness of the SiO$_2$ underlayer thereby implying reactivity at the interface.

The conductance method was used to compare the interface trap density, D_{it}, for HfO$_2$/Si and HfO$_2$/SiO$_2$/Si samples. D_{it} was estimated in terms of maximum equivalent parallel conductance, G_p, per unit area using the expression for continuously distributed interface traps using Equation 11.10. Figure 11.66 shows the G_p/ω versus ω data where the various bias values represent specific energy positions within the Si band gap. The D_{it} values near midgap in Table 11.6 have been normalized by the D_{it} value obtained with an 8.7-nm SiO$_2$ film (1×10^{12} cm^{-2}eV^{-1}) underneath the HfO$_2$ film to compare results under similar processing conditions. Post-metallization anneal by forming gas was not conducted on these samples. The results in Table 11.6 show that HfO$_2$ on bare Si gives a low quality electronic interface as compared to SiO$_2$. By incorporating a thin SiO$_2$ underlayer (2 nm) the interface quality is enhanced by a factor of ~13. The value of incorporating an ultra-thin underlayer of SiO$_2$ has been proposed to maintain an SiO$_2$/Si interface quality while minimizing the dielectric offset of adding a material with a lower K to the system. The results indicate that for the case of HfO$_2$ formed from oxidation of Hf metal, the addition of a 2-nm SiO$_2$ underlayer greatly reduces the number of interface traps while maintaining a sufficiently high static dielectric constant (12) as compared to amorphous SiO$_2$ (3.9).

A recent study considered MgO as a potential high K dielectric (25). Recall that in several previously cited studies on BST ZrO$_2$ and HfO$_2$, MgO was used as an inert

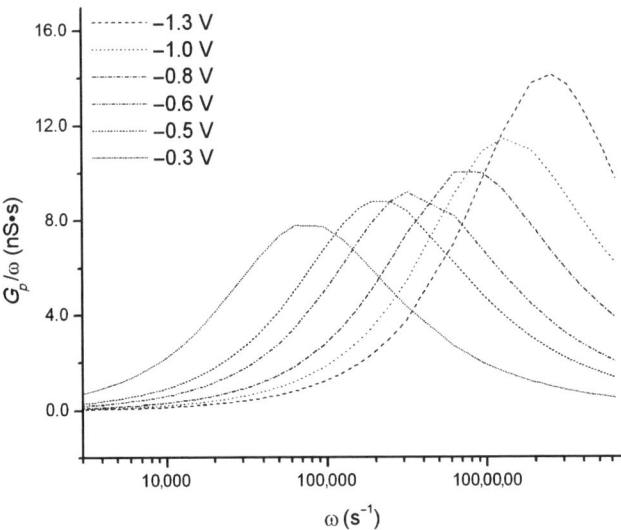

FIGURE 11.66 G_p/ω versus ω at various bias values. [Adapted from Lopez and Irene (24), Figure 10.]

substrate from which the optical properties of deposited films could be obtained. Thus, in the present study MgO was chosen as a candidate gate dielectric for several reasons. It is chemically inert and therefore should result in sharp interfaces. MgO has a wide band gap (7.3 eV), ensuring large band offsets with Si to minimize leakage. MgO has a K value of about 9.8 for bulk material; that would enable more than a 2× increase in gate dielectric thickness.

MgO thin films with sharp interfaces were deposited by sputtering of a Mg target on Si. The film stack was characterized using SE and TEM, and the film static dielectric constant (K) and interface traps were determined. An amorphous SiO_2 layer was found at the MgO/Si interface as a result of subcutaneous Si oxidation. K for the MgO films was found to be about twice that of SiO_2, and the interface trap densities of MgO/Si were found to be comparable with SiO_2/Si, rendering MgO competitive with all presently considered high K dielectrics.

Cleaned Si substrates (P-type, <100> orientation, $\rho = 1$ to 2 Ω-cm) were transferred in less than a minute into the load-lock chamber of the custom-built sputtering deposition system shown in Fig. 6.23. Once loaded, an Si substrate was annealed at

TABLE 11.6 Interface Quality Comparison for HfO_2 on Si and SiO_2

Sample	Normalized D_{it}
HfO_2/Si	38
HfO_2/SiO_2 (2 nm)	3
HfO_2/SiO_2 (8.7 nm)	1

11.5 HIGH STATIC DIELECTRIC CONSTANT GATE OXIDES

600°C for 10 to 15 min, first to remove the hydrogen and yield a bare Si surface for subsequent film deposition. The deposition of MgO thin films was carried out in high purity oxygen at a partial pressure of 1×10^{-4} Torr. An Mg target with a purity of 99.98% was used as the source material. All the MgO depositions were carried out at room temperature under an O_2 partial pressure $\sim 1 \times 10^{-4}$ Torr, with the Kr^+ ions accelerated to a beam voltage of 650 V and beam current of 30 mA.

The XTEM shown in Fig. 11.67 shows an amorphous interface region of approximately 0.6 nm in thickness between 14 nm of polycrystalline MgO and single-crystal Si. Energy-filtered TEM (EFTEM) shown in Fig. 11.68 revealed constituent elements, and sharp interfaces (the online version of this paper shows color: with the color red for Si and green for O, and yellow for the interface, resulting from color mixing and is indicative of a Si-O-rich layer, presumably SiO_2).

The SE was used to determine the MgO film thickness as well as optical properties using a 70° angle of incidence and a spectral range of 1.5 to 4.5 eV (276 to 763 nm). The optical properties of the transparent MgO film were modeled using the Cauchy equation as given above (Equation 11.12). The as-determined indices of refraction (n) for three MgO films of increasing thickness ($\sim$14, 26, and 37 nm) are shown in

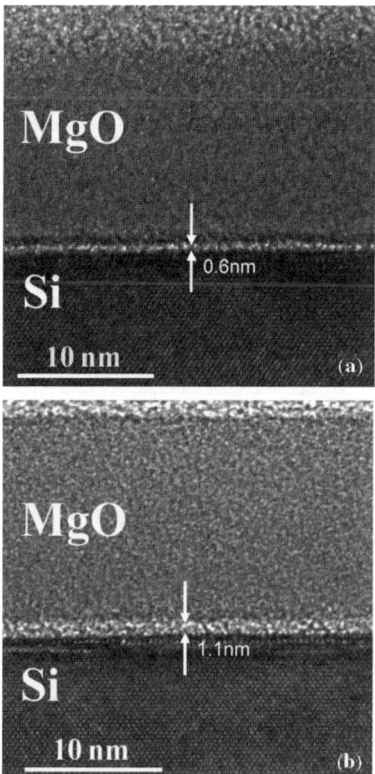

FIGURE 11.67 Cross-sectional TEM of 14-nm MgO film on Si: (**a**) as deposited and (**b**) after postdepositional anneal. [Adapted from Yan et al. (25), Figure 1.]

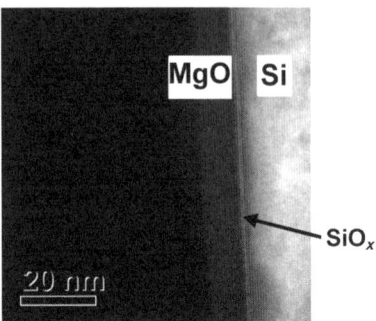

FIGURE 11.68 Energy-filtered cross-sectional TEM image of the 14-nm postdeposition annealed MgO film on Si. [Adapted from Yan et al. (25), Figure 2.]

Fig. 11.69 (solid curves) over the measured spectral range along with single-crystal bulk MgO values. The inset shows the two-film optical model with an SiO_2 interface. The n values for the MgO films in this work are lower than the bulk single-crystal values, as expected. Also shown in Fig. 11.69 are n values for halves of the same MgO samples but postdeposition oxygen annealed (PDOA, in 100% dry O_2 at 600 to 650°C for ~15 min) (dashed curves). The PDOA increased n with a larger effect for the thinner MgO films and virtually no change for the 37-nm film. SE spectra acquired before and after PDOA indicated a progressively smaller thickness decrease; namely 1.25 nm (8.4%), 1.09 nm (4.2%), and 0.47 nm (1.3%) for the 14-, 26-, and 37-nm MgO films, respectively. This, coupled with an increase in n, implies film densification through O_2 annealing. The TEM image of the 14-nm annealed MgO film (Fig. 11.67b) indicated a small interface layer growth (from 0.6 nm for the as-deposited to 1.1 nm) due to subcutaneous Si oxidation. The thickness of this SiO_2 interface was expected to be less than 1.1 nm for the other two

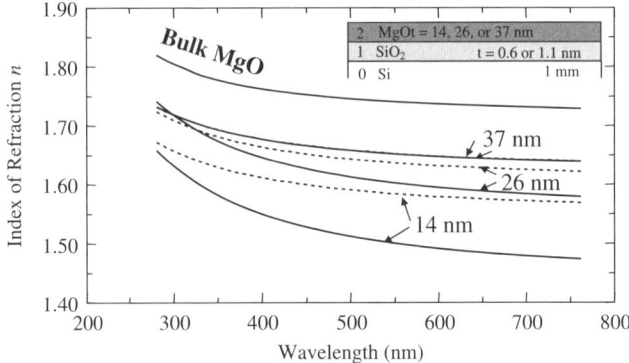

FIGURE 11.69 Refractive indices n for 14, 26, and 37 nm MgO films with solid lines for as-deposited films and dashed lines for postdeposition annealed films. The inset shows the model used, and n for bulk MgO is given for reference. [Adapted from Yan et al. (25), Figure 3.]

11.5 HIGH STATIC DIELECTRIC CONSTANT GATE OXIDES

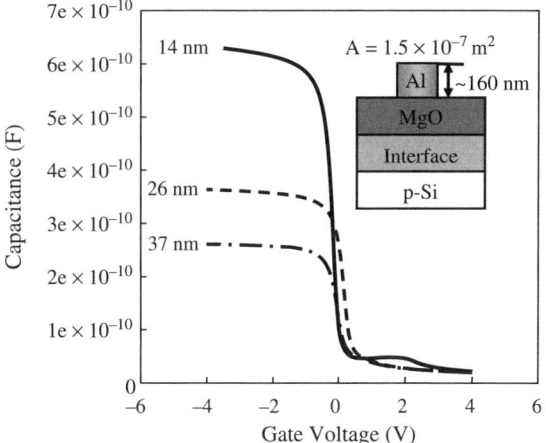

FIGURE 11.70 C–V results for three thickness MgO films (14, 26, and 37 nm) MOS capacitors with the structure shown in the inset. [Adapted from Yan et al. (25), Figure 4.]

MgO thicknesses (26 and 37 nm) owing to the longer diffusion path to the Si interface. A 1.1-nm SiO$_2$ interface was assumed for all three PDOA MgO films in the ellipsometric data analyses. The observed thickness dependence of the PDOA on both index increase (Δn) and thickness decrease (Δd) indicates that the density of the as-deposited MgO film increases with thickness and reaches a plateau value after reaching a critical thickness of $\sim$40 nm. The PDOA treatment is then required to ensure the film densification is saturated.

High frequency (1 MHz, 20 mV oscillator level) capacitance–voltage (C–V) and conductance–frequency (G_p-ω), were performed on fabricated MgO (PDOA only) and SiO$_2$ (as a reference) based MOS capacitors to determine K for MgO along with the equivalent oxide thickness, EOT, and the density of interface traps, D_{it}. Postmetallization anneal in forming gas (90% N$_2$ + 10% H$_2$, at about 450°C for 10 to 15 min) was conducted prior to the measurement.

Figure 11.70 shows C–V data for the three Al/MgO (14, 26, or 37 nm)/Si MOS capacitors, shown in the inset. The shape of the C–V curves indicates few interface traps; and the near-zero flat band voltage is indicative of a small number of negative charges at the interface. Table 11.7 shows that based on the ellipsometric film

TABLE 11.7 Summary of Electronic Results on MgO Films

	14 nm	26 nm	37 nm
(a) K	8.2	8.6	8.1
(b) EOT,[a] nm	7.5	12.5	18.9
(c) Normalized D_{it}	1	1.2	2.1

[a] $EOT = \dfrac{\kappa_{SiO_2} \varepsilon_0 A}{C_{acc}}$.

thickness, TEM oxide thickness, and total accumulation capacitance, the K value for the MgO films was calculated to be above 8. This value is lower than the bulk value from the literature (9.8) but double that of SiO_2 (3.9). EOT values of approximately 1/2 MgO film thicknesses are thus obtained, as shown in Table 11.7.

The interface trap density, D_{it}, was estimated in terms of maximum equivalent parallel conductance, G_p, per unit area using Equation 11.10 as was done in previous studies above. The midgap interface-state density D_{it} values for MgO/Si were normalized by the D_{it} value obtained with a 30-nm thermally grown SiO_2 film on Si to compare results under similar processing conditions. The D_{it} results in Table 11.7 indicate that the interface of our MgO/Si is comparable to the lab-grown quality of SiO_2 gate dielectric.

In summary, MgO not only provides a high dielectric constant alternative to SiO_2 but also interface quality with Si almost indistinguishable from thermally prepared SiO_2 and a wide band gap. These qualities render MgO superior to all presently considered alternative high K dielectrics.

REFERENCES

1. Allen FG, Gobel GW. Phys Rev 1962;127:150.
2. Chiarotti G, Nannarone S, Pastore R, Chiaradia P. Phys Rev B 1971;4:3398.
3. Eastman DE, Grobman WD. Phys Rev Lett 1972;28:1378.
4. Eastman DE, Freeouf JL. Phys Rev Lett 1974;33:1601.
5. Garner CM, Landau I, Su CY, Pianetta P, Miller JN, Spicer WE. Phys Rev Lett 1978;40:403.
6. Irene EA. CRC Crit Rev Solid State Mater Sci 1988;14(2):175–223.
7. Hu YZ, Joseph J, Irene EA. Appl Phys Lett 1991;59:1353.
8. Joseph J, Hu YZ, Irene EA. J Vac Sci Technol B 1992;10:611.
9. Mitchell DF, et al. Surf Interface Anal 1994;21:44.
10. Hu YZ, Wang Y, Li M, Irene EA. J Vac Sci Technol A 1993;11:900.
11. Wang Y, Hu YZ, Irene EA. J Vac Sci Technol A 1994;12:309.
12. Wang Y, Hu YZ, Irene EA. J Vac Sci Technol B 1996;14:1687.
13. Liu X, Denker MS, Irene EA. J Electrochem Soc 1992;139:799.
14. Hu YZ, Joseph J, Irene EA. J Vac Sci Technol B 1994;12:540.
15. Hu YZ, et al. Appl Phys Lett 1993;63:1113.
16. Lefebrve PR, Irene EA. J Vac Sci Technol B 1997;15:1173.
17. Lefebrve PR, Lai L, Irene EA. J Vac Sci Technol B 1998;16:996.
18. Hollinger G, et al. Phys Rev B 1993;49:1159.
19. Mueller AH, Suvorova NA, Irene EA, Auciello O, Schultz JA. Appl Phys Lett 2002;80:3796.
20. Mueller AH, Suvorova NA, Irene EA, Auciello O, Schultz JA. Appl Phys Lett 2003;93:3866.
21. Mueller AH, Suvorova NA, Irene EA. Appl Phys Lett 2002;80:3596.

22. Suvorova NA, Lopez CM, Irene EA, Suvorova AA, Saunders M. J Appl Phys 2004;95:2672.
23. Lopez CM, Suvorova NA, Irene EA. J Appl Phys 2005;98:033506.
24. Lopez CM, Irene EA. J Appl Phys 2006;99:024101.
25. Yan L, Lopez CM, Shrestha R, Irene EA, Suvorova AA, Saunders M. Appl Phys Lett 2006;88:142901.

SUGGESTED READING

V. Swaminathan and A. T. Macrander, 1991. *Materials Aspects of GaAs and InP Based Structures*, Prentice Hall. Materials preparation, processes, and basic physics of III–V semiconductors.

L. G. Meiners and H. H. Wieder, Semiconductor Surface Passivation, *Mat Sci. Repts.* 3, p. 139 (1988). An excellent place to start to understand passivation of semiconductor surfaces with many references to the early papers in the field.

C. W. Wilmsen (Ed.), 1985. *Physics and Chemistry of III–V Semiconductor Interfaces*, Seven excellent chapters by experts on all the important issues with III–V electronic materials and properties.

12

THE Si–SiO$_2$ INTERFACE AND OTHER MOSFET INTERFACES

12.1 INTRODUCTION

In the microelectronics industry the most important interface, the Si–SiO$_2$ interface, is the heart of the MOSFET. As discussed in Chapter 11 the importance of the Si–SiO$_2$ interface derives from the ability to electronically passivate the Si surface via the growth of SiO$_2$. When the Si surface is electronically passivated using high temperature thermal oxidation of Si in pure O$_2$, the Si surface states are reduced from around 10^{15} eV^{-1} cm^{-2} to below 10^{10} eV^{-1} cm^{-2} or about 1 in 10^5 Si bonds per centimeter squared at the surface are rendered electronically innocuous. Since the mid-1960s considerable effort has been expended in researching the details about the electronic passivation of Si and other semiconductors, yielding numerous new studies and techniques that were focused on the problem. A few examples were discussed in Chapters 5, 6, 7, and 11. In the following paragraphs some more of the key findings about the Si–SiO$_2$ interface are presented that have not already been discussed elsewhere in this book. In addition the other MOSFET interfaces are discussed. Specifically, the metal–semiconductor interfaces that appear as the source and drain contacts, and semiconductor–semiconductor interfaces that are sometimes used in more complex MOSFET-based devices are discussed. It is revealed that for metal–semiconductor junctions, interface states can dominate the barrier for highly resistive Schottky contacts while semiconductor–semiconductor interfaces can be prepared without any significant contribution from interfaces states.

Surfaces, Interfaces, and Thin Films for Microelectronics. By Eugene A. Irene
Copyright © 2008 John Wiley & Sons, Inc.

This chapter commences with a brief review of spectroscopic studies of the Si–SiO$_2$ interface using mainly XPS, which probes the chemical nature of the interface in an effort to elucidate the chemical pathways that lead to electronic passivation. This is followed by a discussion of several powerful but less common techniques that further probe other aspects of the Si–SiO$_2$ interface. For example, by optically (not physically) removing the SiO$_2$ from the Si surface, the Si–SiO$_2$ interface can be probed with high sensitivity, and this is accomplished by using a novel spectroscopic immersion ellipsometry (SIE) technique. Electron tunneling across the Si–SiO$_2$ interface can further reveal the nature of the interface such as roughness and can even be used to accurately determine the thickness of SiO$_2$ films that are less than a few nanometers thick, which is virtually impossible to do by any other technique. Lastly, compressive stress in SiO$_2$ near the Si–SiO$_2$ interface was shown to arise from both the thermal expansion and lattice mismatch between Si and SiO$_2$ at the interface. At the least this stress causes a density gradient in the SiO$_2$ that decreases outward from the Si surface into the oxide. The remaining sections of this chapter introduce the basic ideas about ideal metal–semiconductor interfaces and semiconductor–semiconductor interfaces, and this is followed by a discussion of real metal–semiconductor and semiconductor–semiconductor interfaces with several examples from MOSFET technologies.

12.2 NATURE OF THE Si–SiO$_2$ INTERFACE

Some 20 years after the beginning of the microelectronics revolution, a substantive review of the results of the chemical analysis of the Si–SiO$_2$ interface was published (1). Much of the work discussed in this report is that of the authors of the review, who have made many contributions to this area, but there are also many references to the thinking about the Si–SiO$_2$ interface that existed up to the time of the report. Since 1986 there have been a large number of additional papers and reviews, but these studies have added only little to the understanding beyond that contained in the 1986 report (1). Therefore, the 1986 report, from here on called the Grunthaner report, will serve as the main reference to the following paragraphs, keeping in mind that some details in the earlier report have been updated in the more current literature, and many more studies in this area have been undertaken since 1986.

First, the Grunthaner report (1) outlines the methodology that they and others have used to acquire and analyze XPS spectra from thin films and interfaces. In this context it must be remembered that the escape depth for the photoelectrons is around 2 nm, so to examine interfaces the films must be very thin to start with or the films must be thinned. Thinning in itself presents a formidable challenge because thinning techniques such as ion milling or sputtering (discussed in Chapters 3 and 10) and chemical etching can create artifacts. Ion sputtering can alter the stoichiometry via preferential sputtering of one kind of atom over another and chemical exposure can lead to the deposition of impurities from the etchant solutions as well as etching nonuniformities. This problem was addressed through the use of a nonaqueous chemical etchant, and the substrate to be etched was spun under an etchant stream. For the SiO$_2$

12.2 NATURE OF THE Si–SiO$_2$ INTERFACE

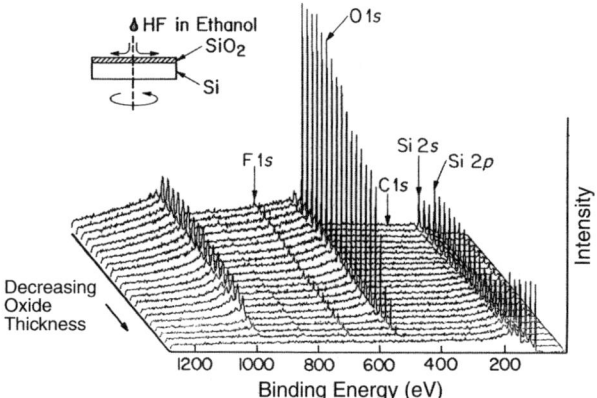

FIGURE 12.1 XPS spectra taken from successive chemical etching of SiO$_2$ on Si using the spin method. The inset shows the experimental setup for the etching. [Adapted from Grunthaner and Grunthaner (1), Figure 10.]

film etching, ethanol was used as the solvent and HF as the etchant, and the solution was used in small drops (about 20 µL) with the sample being spun at 3600 to 10,000 rpm in a dry N$_2$ ambient, as illustrated in the inset in Fig. 12.1. XPS characterization of etched samples is performed at various stages of etching, and Fig. 12.1 shows typical XPS results for etching a 20-nm SiO$_2$ film on Si in terms of Si 2s and 2p and O 1s core levels and the F 1s from the etchant. As the Si substrate is neared, there is a decrease in the O 1s intensity as expected. A more interesting result is seen for the evolution of the Si 2p spectra with the extent of etching, as shown in Fig. 12.2 where the Si in SiO$_2$ that shows a chemical shift near 104 eV decreases as the oxide is

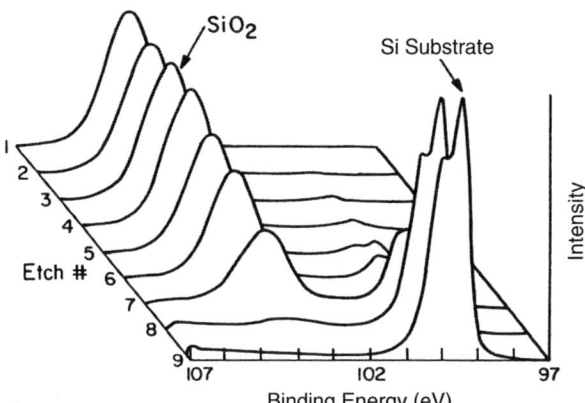

FIGURE 12.2 XPS Si 2p spectra taken from successive chemical etching of SiO$_2$ on Si using the spin method as was done in Fig. 12.1. [Adapted from Grunthaner and Grunthaner (1), Figure 11.]

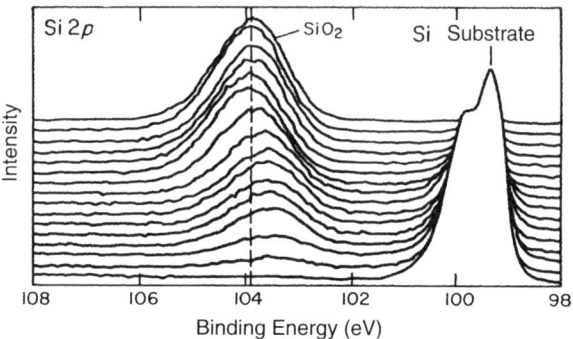

FIGURE 12.3 XPS Si $2p$ spectra taken from successive chemical etching of SiO_2 on Si using the spin method with a view to show binding energy changes. [Adapted from Grunthaner and Grunthaner (1), Figure 12.]

etched while the Si from the substrate near 99 eV increases. Figure 12.3 shows another view of similar spectra to those in Fig. 12.2 but with a more readable energy scale. The doublet appearing near 99 eV arises from photoemitted electrons from the Si substrate while the broad peak at 104 eV is from SiO_2. As the SiO_2 is etched, there is a shift of the 104-eV peak toward lower binding energy below about 5 nm SiO_2 thickness while the position of the 99-eV peak is constant. This interesting observation is the heart of the assertion made in the Grunthaner report (1) that the interface is chemically different than the bulk SiO_2. However, this conclusion can be reached only after eliminating shifts that can arise from charging effects (an electron beam charges the sample and then alters the energy of the photoemitted electrons) and initial (due to charge exchange among atoms at the surface) and final state effects (due to the photoemission process there is a reorganization of the remaining electrons that causes binding energy shifts).

Experimental precautions were taken to avoid sample surface charging effects. A consideration of the literature led to the conclusion that of the 0.45 shift observed only about 0.05 eV arises from final state effects. While initial state effects are more difficult to quantify, it was concluded that bond angle changes can give rise to the shifts, and this notion was incorporated into an interface model. In further support of the notion that the $Si-SiO_2$ interface region is structurally distinct, angle-resolved photoemission was also performed. For these measurements the effective mean free path (previously referred to in Chapter 5 as the electron escape depth) λ_{eff} is a function of the mean free path at $90°$ take-off angle θ, λ and the angle θ as

$$\lambda_{eff} = \lambda \sin \theta \tag{12.1}$$

Figure 12.4 displays XPS Si $2p$ spectra for two take-off angles for a 3.8-nm SiO_2 film on Si. Considering a mean free path of around 3 nm, at $30°$ take-off angle, the effective mean free path is half that at $90°$ and thus the interface is sampled less than the

12.2 NATURE OF THE Si–SiO$_2$ INTERFACE

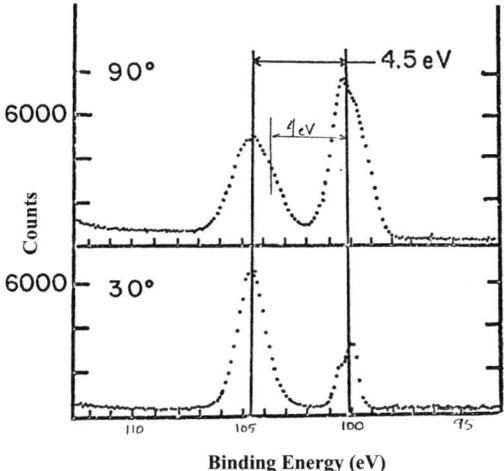

FIGURE 12.4 XPS Si 2p spectra taken on a 3.8-nm SiO$_2$ film on Si at two take-off angles 30° and 90°. [Adapted from Grunthaner and Grunthaner (1), Figure 17.]

bulk oxide. Figure 12.4 shows that when more of the interface is sampled a new feature shifted about 4 eV from the 100-eV peak is seen that indicates a chemically distinct moiety near the Si–SiO$_2$ interface.

All the evidence cited in the Grunthaner report (1) from a variety of spectroscopies (AES, RBS, SE, tunneling, diffraction, IR, contact angle, etch rate, etc.) in addition to XPS indicate that the Si–SiO$_2$ interfacial region is chemically different from bulk SiO$_2$ and furthermore the main region of this chemical difference extends to about 1 to 2 monolayers from the Si surface or about 0.2 to 0.4 nm. The following question remains: What is the chemical nature of the Si–SiO$_2$ interface region?

One way to approach this problem using XPS that yields the chemical environment is to consider the possible valence states for Si at the Si–SiO$_2$ interface. Figure 12.5 shows the five possible states that are differentiated by the bonding to oxygen, and the charges are not taken literally to be that amount of charge on the Si atoms. The suboxide configurations are indicated as the three states in between 0 and +4. Figure 12.6a shows the Si spectrum for a thin thermal SiO$_2$ film on Si,

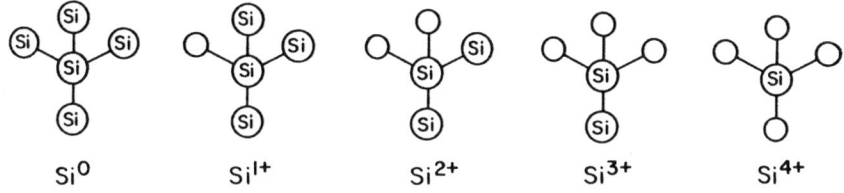

FIGURE 12.5 Five possible charge states for Si at the Si–SiO$_2$ interface. [Adapted from Grunthaner and Grunthaner (1), Figure 34.]

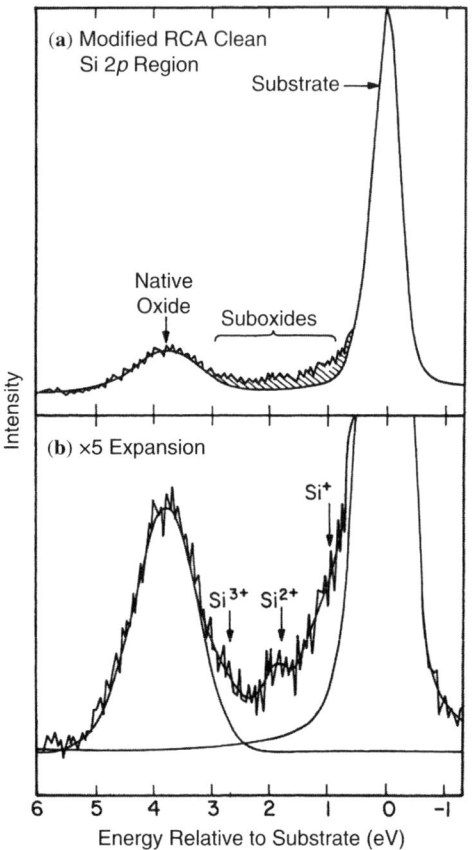

FIGURE 12.6 XPS spectra of Si 2p region. (a) Cleaned and thermally oxidized Si and (b) magnification of the suboxide region. [Adapted from Grunthaner and Grunthaner (1), Figure 35.]

and Fig. 12.6b shows the same spectrum with the region between the oxide and substrate peaks expanded. Small peaks or shoulders are seen in the expanded version that can be attributed to the suboxides +1, +2, and +3, as indicated in Fig. 12.6. From least-squares minimization fits quantitative information was also obtained. A measure of the distribution of the suboxides in the oxide can be obtained from the ratios of the suboxide intensity to the substrate intensity as a function of thickness. Figure 12.7a shows spectra as a function of SiO_2 film thickness. It is seen that the largest intensity suboxide is Si^+, and this state remains as the oxide is thinned. The normalized intensities versus oxide thickness are shown in Fig. 12.8 with the oxide thicknesses reported as the ratio of the thickness to the average electron escape depth, $\langle \lambda_{oxide} \rangle$. This representation indicates that the escape depth can vary with oxide thickness. Also the species near the interface are characterized by the larger changes in the intensities nearer the interface. Figure 12.8 shows that the

12.2 NATURE OF THE Si–SiO$_2$ INTERFACE

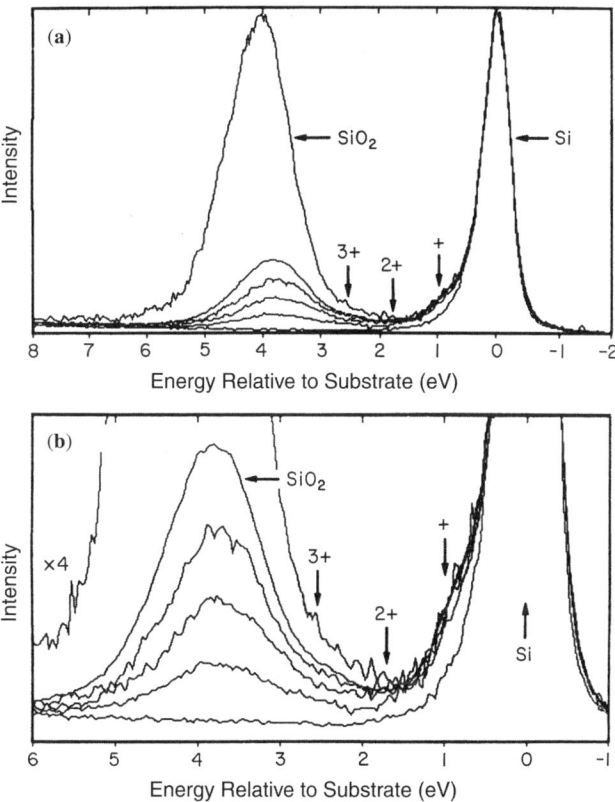

FIGURE 12.7 XPS spectra of SiO$_2$ films of various thickness on Si(111) in the Si 2p region with (**b**) a 4× expansion of (**a**). [Adapted from Grunthaner and Grunthaner (1), Figure 36.]

Si$^+$ is independent of oxide thickness down to about $0.2\langle\lambda_{oxide}\rangle$ or about 0.6 nm, and thus this species is within this thickness from the interface. Small amounts of Si^{2+} are found nearly independent of film thickness, and Si^{3+} shows a smaller change with thickness near the interface than Si$^+$. This indicates that the Si^{3+} valence state extends further into the oxide than the Si$^+$ state out to around 3 nm. It is also observed that the states depend on the Si orientation, as shown in Fig. 12.9, in particular the Si$^+$ and Si^{2+} states. This variation with Si substrate orientation is expected if the various oxidation states are near the Si–SiO$_2$ interface. More insight is obtained by considering the bonding for various states at the Si–SiO$_2$ interface.

Figure 12.10a shows the difference in bonding across the Si–SiO$_2$ interface for the Si(111) and Si(100) interfaces. It is seen that for Si(100) there are two Si–O bonds corresponding to the Si^{2+} state and two Si–Si bonds while for the Si(111) there are one Si–O and three Si–Si bonds corresponding to the Si$^+$ state. The preponderances of these states for the two orientations are concordant with the calculations summarized in Fig. 12.9. From TEM observations of steps on the Si surfaces a model for the

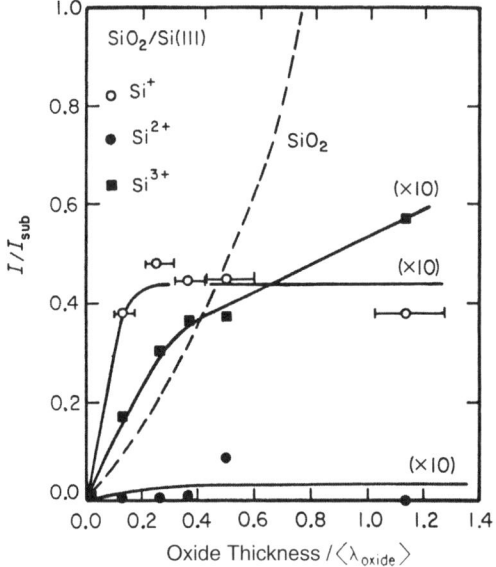

FIGURE 12.8 Normalized intensities versus the ratio of oxide thickness to the average electron escape depth, $\langle \lambda_{oxide} \rangle$. [Adapted from Grunthaner and Grunthaner (1), Figure 37.]

Si(100) surface indicates the preponderance of the Si^{2+} state, but around steps other states can occur as well, albeit in lower concentrations.

In summary XPS along with other techniques indicate that there is a chemical interface of several oxide layers that contains suboxide species Si^{+}, Si^{2+}, and Si^{3+} states. The Grunthaner report (1) also examines the processing dependence of these states and therefore the chemical interface.

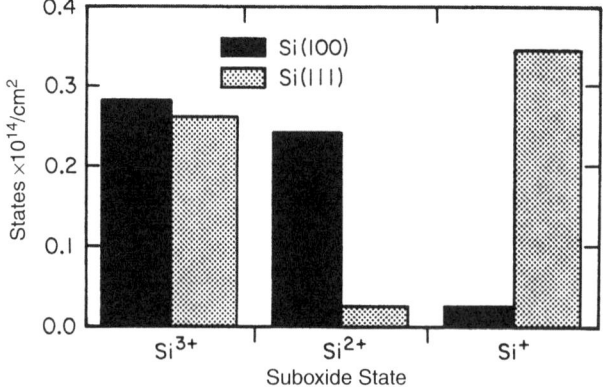

FIGURE 12.9 Calculation of the suboxide areal densities for SiO_2 on Si(111) and Si(100) surfaces. [Adapted from Grunthaner and Grunthaner (1), Figure 38.]

12.3 OTHER TECHNIQUES FOR INTERFACE STUDIES   445

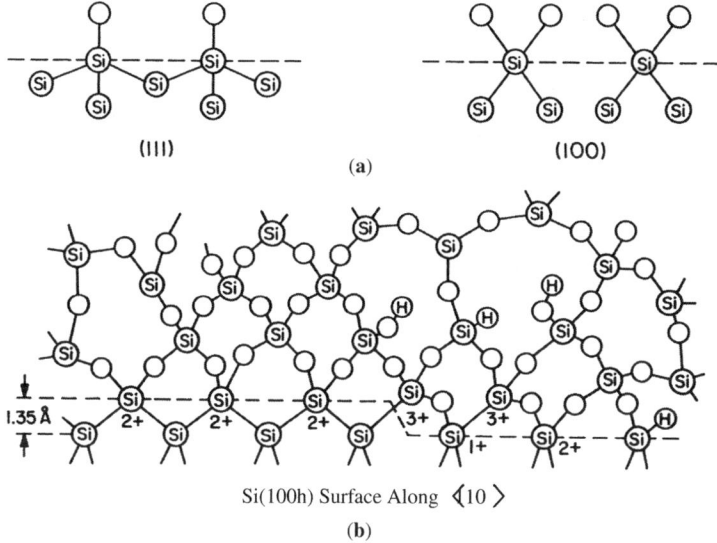

FIGURE 12.10 (a) Bonding of Si and O across the Si–SiO$_2$ interface for the Si(111) and Si(100) surfaces and (b) an interface model for bonding of the Si(100) surface. [Adapted from Grunthaner and Grunthaner (1), Figure 39.]

12.3 OTHER TECHNIQUES FOR INTERFACE STUDIES

12.3.1 Spectroscopic Immersion Ellipsometry (SIE)

Cited in the Grunthaner report (1) are several references to the use of ellipsometry and various spectroscopies for the elucidation of the nature of the Si–SiO$_2$ interface. However, conventional ellipsometry performed on this interface in air ambient is not particularly sensitive to the interface because the interface provides only a small optical contrast from the overlying SiO$_2$. However, in two studies (2,3), an ellipsometry technique was demonstrated that dramatically improves the optical sensitivity to the interfacial region between a dielectric film and substrate. Essentially, a film with substrate is immersed in a liquid that refractive index matches to the film, thereby optically removing the film from the measurement (the film is not physically removed). In addition, the use of spectroscopy and multiple angles of incidence provide sufficient specification of the interface parameters, which along with the enhanced sensitivity to the interface, enables the optical measurement of the interfacial properties. Theoretical and experimental verification was provided along with application to the Si–SiO$_2$ interface.

As discussed many times throughout this text, there are many different ways to study the interface region between a film and substrate. One way discussed in Chapter 9 is to use SE in air or vacuum ambients, as illustrated in Fig. 12.11a. The disadvantage of this method is that an accurate characterization of the ultra-thin interfacial transition layer is complicated by the optical responses of both the relatively

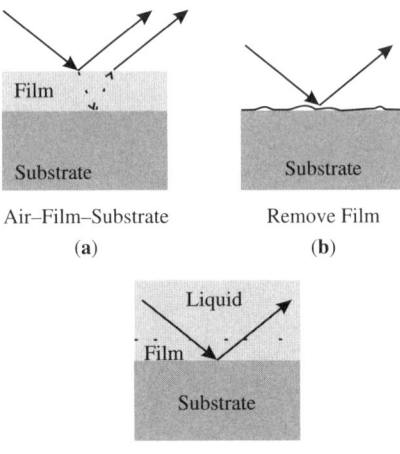

FIGURE 12.11 Methods to analyze an interface: (**a**) optically with overlayer film in place, (**b**) remove overlayer, and (**c**) immersion, where overlayer is optically removed using index matching liquid.

thick overlayer and the air or vacuum–film interface. Another method is to remove the overlayer physically or chemically and then probe the interface, as shown in Fig. 12.11b. This method was discussed above where chemical etching was used. However, this method could alter the interface region, and it is destructive. To improve sensitivity and preserve the original interface, a novel immersion ellipsometry technique mentioned above was developed in which variable angle of incidence SE measurements are performed with the sample submerged in a transparent liquid ambient that has optical properties that match those of the SiO_2 overlayer, as shown in Fig. 12.11c. Thus, in this spectroscopic immersion ellipsometry (SIE) technique the overlayer film is "optically" (not physically) eliminated, and the probing light beam becomes highly sensitive to the interface properties.

The use of the SIE technique hinges on finding suitable immersion liquids that not only match the index of films but also are nonreactive and transparent over a sufficiently wide wavelength range to permit measurement. Generally, it is difficult to achieve a perfect refractive index match for the liquid ambient, n_0, and the SiO_2 overlayer, n_{ov}, over a broad spectral range. Therefore, small deviations can and have been accounted for in the analysis. Carbon tetrachloride (CCl_4) is a suitable immersion liquid for matching index to SiO_2 films. The refractive index of the liquid ambient, n_0, was calculated using a Cauchy dispersion formula, taking into account temperature:

$$n_0 = n_\infty + \frac{a}{\lambda^2} + \frac{\partial n}{\partial T} \Delta T \qquad (12.2)$$

12.3 OTHER TECHNIQUES FOR INTERFACE STUDIES

where λ is the wavelength of probing beam (in angstroms), $n_\infty = 1.4427$, $a = 5.15 \times 10^5$ Å^{-2} at $T = 24.8°C$, and $\partial n/\partial T = 0.00055$ at $T = 20°C$ for pure CCl$_4$. It is known that the refractive index of thin-film SiO$_2$ overlayers, n_{ov}, depends on thickness, L_{ov}, oxidation temperature, T_{ox}, and preparation conditions such as preoxidation cleaning and oxidation ambient. The spectral dependence of the nonannealed thin-film SiO$_2$ refractive index, $n_{ov}^0(L_{ov}, T_{ox}, \lambda)$, for a substrate with particular orientation, for example, the Si(100) orientation, was calculated from a single term Sellmeier approximation:

$$n_{ov}^0(L_{ov}, T_{ov}, N, \lambda) = 1 + \frac{A(L_{ov}, T_{ov}, N)\lambda^2}{\lambda^2 - \lambda_0^2(L_{ov}, T_{ov}, N)} \tag{12.3}$$

where $A(L_{ov}, T_{ox}, N)$ and $\lambda_0(L_{ov}, T_{ox}, N)$ are dispersion parameters that are also dependent on the overlayer thickness and preparation conditions and with values $A = 1.15$ and $\lambda_0 = 92.3$ nm for 25 to 35-nm-thick SiO$_2$ films thermally grown at 800°C on Si(100).

Using a mixture of different liquids, it is possible to cover a range of refractive indexes of the ambient to that for SiO$_2$. Figure 12.12 shows the spectral dependencies of the refractive index of pure carbon tetrachloride, CCl$_4$, and mixture of CCl$_4$ with benzene, C$_6$H$_6$. Both CCl$_4$ and C$_6$H$_6$ are nonpolar organic liquids that do not interact with SiO$_2$. Figure 12.12 also shows the spectral dependencies of the refractive index of bulk and thin SiO$_2$ films calculated from Equation 12.3. With the use of pure CCl$_4$, no residue or damage was found after long exposures. To analyze the sensitivity of the immersion technique, the interface in the Si/SiO$_2$ system was assumed to include interface roughness and an interface suboxide transition zone, as shown in

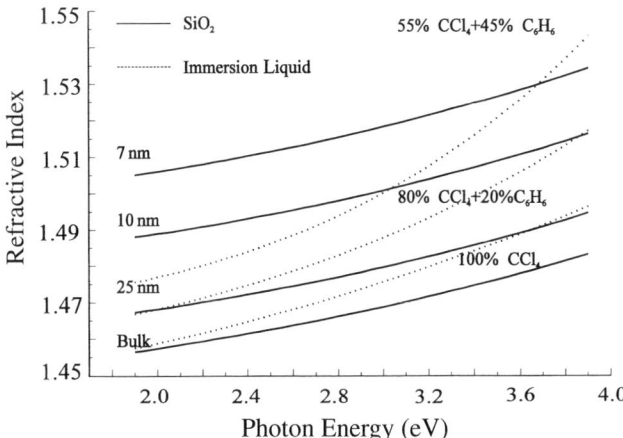

FIGURE 12.12 Refractive index versus photon energy of CCl$_4$, C$_6$H$_6$, and mixtures (dotted lines) along with bulk and thin film SiO$_2$ values (solid lines). [Adapted from Yakovlev and Irene (2), Figure 2.]

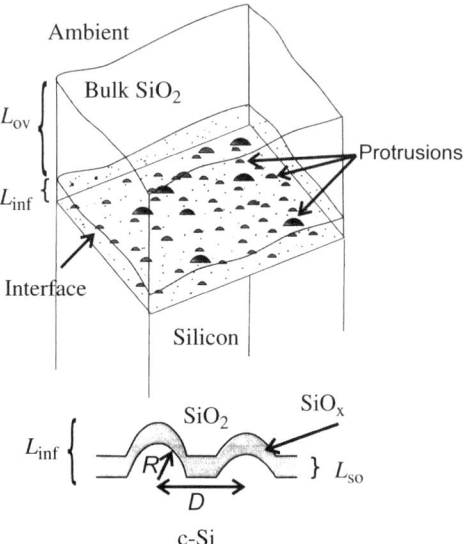

FIGURE 12.13 Interface model for the Si–SiO$_2$ interface. This interface layer (L_{inf}) made up of Si roughness (R) and suboxide (L_{so}) on an Si substrate. The bulk SiO$_2$ film has a thickness L_{ov}. [Adapted from Yakovlev and Irene (2), Figure 1.]

Figure 12.13. The interface microroughness with an effective height of 0.2 nm and composition of 50% c-Si and 50% suboxide SiO$_x$, with $x = 0.4$, and the Bruggeman effective medium approximation (BEMA) was used to calculate a dielectric function of the mixture. The interface suboxide transition zone was composed of 0.6 nm-thick SiO$_{0.4}$. Figure 12.14 shows simulation results for the air–20 nm SiO$_2$–Si system with and without the 0.8-nm interface in terms of $\Psi(E)$ and $\Delta(E)$, where E is the energy of probing light in the energy range 2.5 to 4.0 eV. It is seen that, for the air ambient, there is very low sensitivity in both Δ and Ψ to the presence of an interface layer different from SiO$_2$. Figure 12.15 shows the analogous simulation of the CCl$_4$–20 nm SiO$_2$–Si system. The interface sensitivity of Δ is dramatically improved by a factor near 10 times, using the CCl$_4$ ambient while for Ψ the enhancement is much smaller. The enhancement in sensitivity is shown more clearly in Fig. 12.16, which plots the relative interface sensitivity, which is given as

$$\delta\Delta(E) = \Delta_0(E) - \Delta_{inf}(E) \qquad (12.4)$$

for air and CCl$_4$ ambients. $\Delta_0(E)$ and $\Delta_{inf}(E)$ were calculated without and with an interface layer, respectively, and a similar formula is used for Ψ.

In the study of the Si–SiO$_2$ interface, the working model for the interface between crystalline Si substrate and amorphous SiO$_2$ film was shown as Fig. 12.13. The transition region has a structure with two major components: the "physical" interface and the "chemical" interface. The physical interface can be represented by

12.3 OTHER TECHNIQUES FOR INTERFACE STUDIES

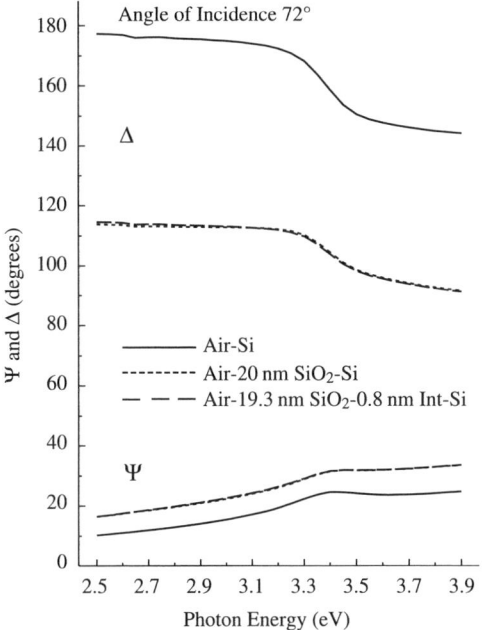

FIGURE 12.14 Calculated ellipsometric parameters vs. probing photon energy for a system with and without a distinct interface layer in ambient air. [Adapted from Yakovlev and Irene (2), Figure 3.]

microroughness or protrusions of Si into the oxide. The chemical interface, as discussed in the previous section, consists of a suboxide, SiO_x with $0 < x < 2$. For the case of nitridation of Si the interface layer could be a nitride or even an oxynitride. The crystalline silicon protrusions were described as hemispheres with an average radius R, and which form a hexagonal network with an average distance D between centers. The protrusions and the region between them are covered by a layer of suboxide assumed to be SiO (i.e., $x = 1$) with an average thickness of L_{SiO}. An effective interface thickness is given as

$$L_{inf} = R + L_{SiO} \tag{12.5}$$

The BEMA was used to calculate the effective dielectric function of the interface. The evolution of the Si–SiO$_2$ interface as a function of high temperature annealing (750 to 1000°C) was investigated by SIE. Figure 12.17 shows raw Δ data with the significant changes, and Fig. 12.18 shows the data in terms of an effective relative interface parameter defined as

$$\delta\Delta_{inf}(T_{an}, t_{an}) = \delta\Delta^{exp}(T_{an}, t_{an}) - \Delta_0^{exp} - \delta\Delta_{ov}^{cal}(T_{an}, t_{an}) \tag{12.6}$$

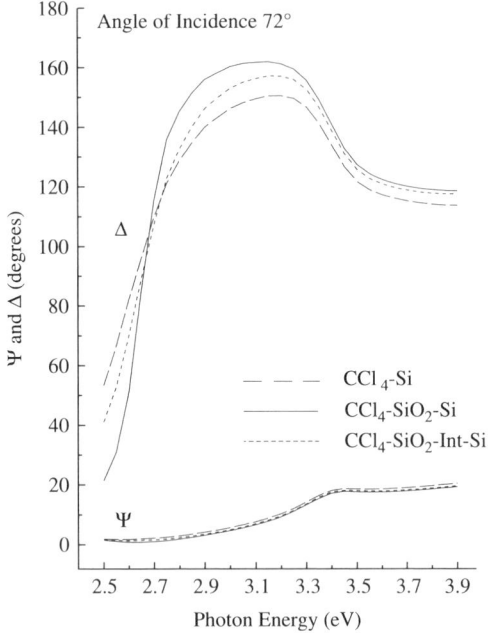

FIGURE 12.15 Calculated ellipsometric parameters vs. probing photon energy for a system with and without a distinct interface layer (Int) in CCl. [Adapted from Yakovlev and Irene (2), Figure 4.]

where $\Delta^{\text{exp}}(T_{\text{an}}, t_{\text{an}})$ is the experimental ellipsometric angle Δ at an annealing temperature and time, Δ_0 is the ellipsometric angle for an unannealed sample, and the term $\delta\Delta_{\text{ov}}^{\text{cal}}(T_{\text{an}}, t_{\text{an}})$ is the overlayer relaxation correction that corrects for the very small change in SiO_2 with annealing. This term is always less than a few percent of the other changes and thus could be omitted. Distinct changes are observed for anneal temperatures above 900°C, which corresponds to strain relaxation temperatures as

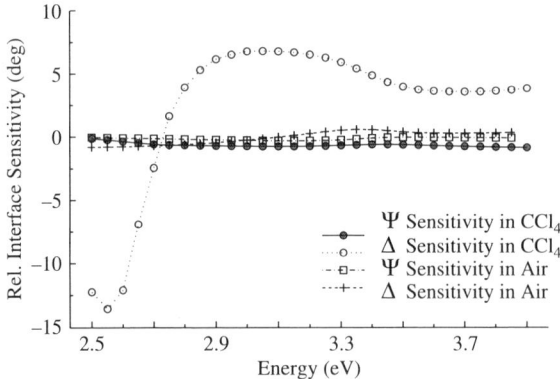

FIGURE 12.16 Relative interface sensitivity in ambient air and CCl_4.

12.3 OTHER TECHNIQUES FOR INTERFACE STUDIES

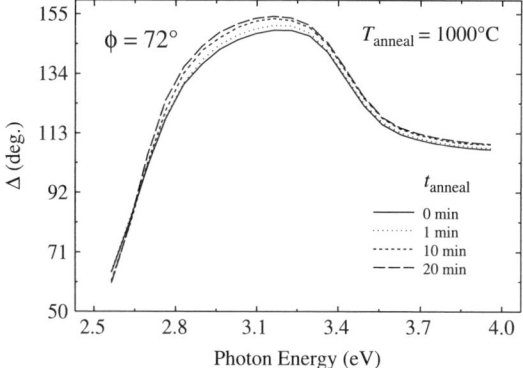

FIGURE 12.17 Dependence of Δ on incident photon energy for various interface annealing times. The measurements were performed on a 3-nm SiO_2 film on Si immersed in CCl_4 after the film was annealed at the specified time and cooling from the 1000°C anneal temperature. [Adapted from Yakovlev et al. (3), Figure 1.]

discussed above. Figure 12.19 shows modeled data in terms of the interface thickness defined above as L_{inf}, which displays the temperature–time-dependent shrinkage of the interface with annealing. For short annealing times, a rapid change in the interface is observed that correlates with the disappearance of protrusions, followed by a slower

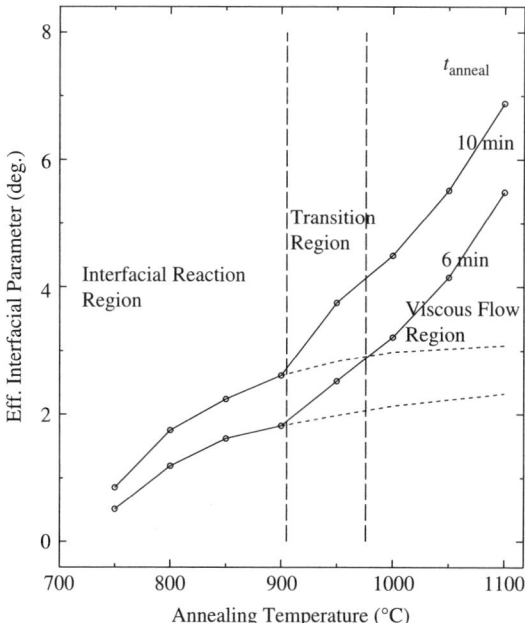

FIGURE 12.18 Experimental dependence of the effective interface parameter D_{inf} on annealing temperature for two annealing times. [Adapted from Yakovlev et al. (3), Figure 3.]

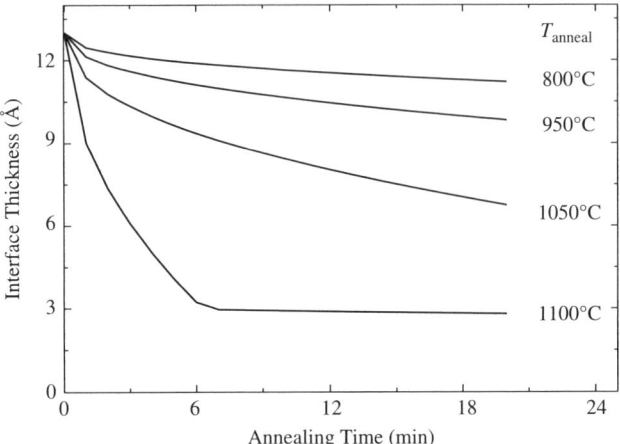

FIGURE 12.19 Dependence of the effective interface thickness on annealing time at a number of annealing temperatures. [Adapted from Yakovlev et al. (3), Figure 6.]

change that correlates with the disappearance of the suboxide. At high annealing temperatures viscous relaxation dominates, while at low annealing temperatures the suboxide reduction is apparent. With the use of the above optical model, it was reported that the thickness of the SiO layer at the interface, L_{SiO}, for all <100>, <110>, and <111> Si substrate orientations increased slightly, and the average radius of the crystalline silicon protrusions, R, decreased with the thickening of the SiO_2 overlayer. This yields an overall decrease in the interface layer (L_{inf}) as is seen in Fig. 12.19. These results are consistent with the well-accepted linear-parabolic (LP) Si oxidation model discussed in Chapter 10, which yields an accurate representation of the growth of SiO_2 on Si over a wide range of thickness, temperature, and oxidant partial pressures. Also, as discussed in Chapter 4, these results are concordant with Si surface smoothing from oxidation.

The instrumentation for SIE includes an SE system and a special immersion cell. Figure 12.20 shows one possible setup where a vertical ellipsometer bench was outfitted with the usual components for a rotating analyzer spectroscopic ellipsometer (see Chapter 9). A fused silica immersion cell, as shown in Fig. 12.20, was designed for the variable angle of incidence and spectroscopic measurements. The main feature of the cell is that the two optically flat and annealed fused silica plates, serving as the entrance and exit windows, are connected rigidly to the polarizer and analyzer arms, but not rigidly to the cell. The angular orientation of the windows was adjusted at the straight-through position of the ellipsometer ($\phi = 90°$) to avoid any deviation of the incident light beam. The windows must be orthogonal to the beam for the immersion measurements, otherwise the media with different indexes of refraction on each side of the windows will alter the direction of the beam, hence the angle of incidence by an amount proportional to the amount of deviation from orthogonal. Two fitted metallic tubes in each arm permit some lateral movement of the window position without a change of the window tilt. The cell is rigidly attached to a stage and connected to

12.3 OTHER TECHNIQUES FOR INTERFACE STUDIES

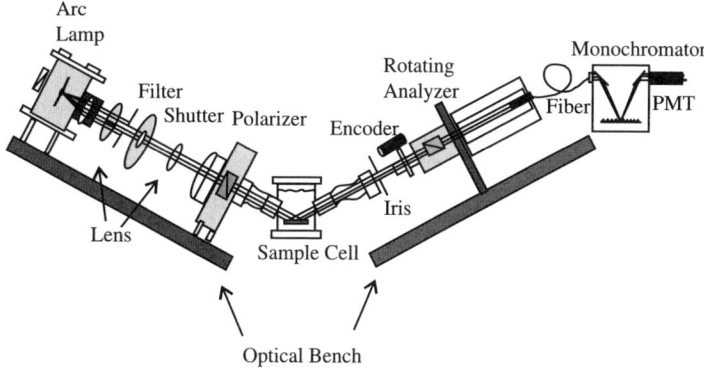

FIGURE 12.20 Rotating analyzer spectroscopic ellipsometer with immersion ellipsometry sample cell.

the tubes using flexible tubing that is inert to the immersion liquids. A change of ϕ in the range of 67° to 90° was possible while maintaining the window alignment precision during the immersion measurements. Table 12.1 indicates some liquids that can be used to match various SiO_2 film thicknesses. The refractive index for thermally grown SiO_2 varies with thickness with the higher index near to the $Si-SiO_2$ interface (see discussion below of tunneling current measurements used with SE to obtain the refractive indexes for ultra-thin SiO_2 films).

12.3.2 Tunneling Currents for Measurement of Refractive Index, Film Thickness, and Roughness

12.3.2.1 Tunneling Current Oscillation Measurement As mentioned above, the thickness required for gate oxide films for advanced MOSFET devices has gradually been decreased by designers to well below 5 nm or the ultra-thin-film regime. However, despite definite device advantages the ultra-thin dielectric films display significant quantum mechanical tunneling current, sometimes referred to as leakage current that is detrimental to both device operation and reliability. One means to avoid the large currents associated with ultra-thin films is to employ high K dielectrics

TABLE 12.1 Refractive Index Values for Immersion Liquids Used for SIE of SiO_2 Films

SiO_2 Thickness (nm)	Refractive Index SiO_2	Liquid	Refractive Index Liquid	dn/dT
20	1.465–1.475	Carbon tetrachloride	1.460	0.00055
10	1.485–1.495	Glycerine	1.475	
		Toluene	1.497	0.00056
7	1.515–1.525	Benzene	1.501	0.00063

that enable the use of thicker films, as discussed in Chapter 11. In addition to the detrimental tunneling conduction, there are some useful aspects to the tunneling current that will be discussed in some detail below. In fact, it will be shown that a measurement of the tunneling current can lead to accurate film thicknesses in this ultra-thin-film tunneling regime, and the obtained thicknesses can in turn be used to obtain the ultra-thin-film refractive index n. Film thickness and refractive index are difficult to measure accurately in the ultra-thin-film thickness regime. First, we consider some aspects of quantum mechanical tunneling that can lead to the useful measurements and then the specific applications.

As discussed in Chapter 6 quantum mechanical tunneling refers to the phenomenon of a particle traveling through a potential energy barrier that has energy greater than that of the particle. The wavelike property of electrons allows the penetration of such a barrier with a finite probability density of the particle as a decaying evanescent wave into an energy-forbidden region, as shown in Fig. 6.5. If the barrier is thin enough, the probability density will not completely decay, and there can be significant penetration through the barrier.

For an electron incident upon a potential barrier, its position in space is definable. This localized wave function with a definable position is called a wave packet. The electron wave packet incident on the insulator in a MOS device can tunnel through the triangular portion of the barrier when the applied bias is greater than the semiconductor–oxide work function potential. Figure 12.21 shows an energy level diagram for an Al/SiO$_2$/Si MOS capacitor with a negative bias voltage applied at the Al electrode. V_{ox} is the voltage drop across the SiO$_2$ film and Φ_M is the energy barrier at the Al–SiO$_2$ interface. For $|V_{ox}| < \Phi_M$, the electrons from the gate tunnel through the oxide film into empty states in the Si substrate, and the tunneling distance is equal to the SiO$_2$ film thickness (L_{ox}). For $|V_{ox}| > \Phi_M$, the electrons tunnel into the conduction band of the SiO$_2$ film with a tunneling distance that decreases with an increasing oxide electric field. In this regime, the Fowler–Nordheim (FN) theory predicts the current density (J_0) to be

$$J_0 = AF^2 e^{-C/F} \tag{12.7}$$

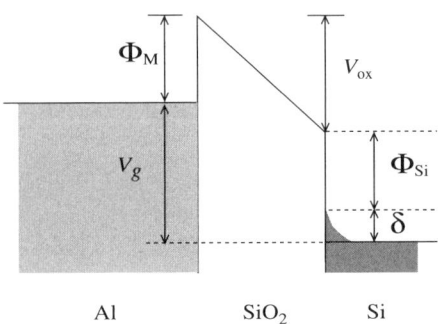

FIGURE 12.21 Electronic energy level diagram for an Al/SiO$_2$/Si MOS capacitor with the negatively biased metal gate.

12.3 OTHER TECHNIQUES FOR INTERFACE STUDIES

where F is the electric field in the oxide. A and C are constants determined from the band structure of the device. The Fowler–Nordheim equation (Equation 12.7) was derived using the Wentzel–Kramers–Brillouin (WKB) approximation that neglects electron wave interference effects. With an applied gate bias greater than the Al–SiO$_2$ work function (above about 4 V) on the MOS structure, charge will be injected into the conduction band of the insulator (SiO$_2$). Electrons can tunnel through the triangular potential barrier and propagate ballistically through the conduction band of the insulator toward the semiconductor substrate that is held in accumulation. The idealized electron potential energy barrier in the MOS system is a triangular barrier when a strong electric field is applied across the device. The abrupt change in potential at the Si–SiO$_2$ interface enables the coherent reflection of electron waves from the silicon surface. Because of electron reflections from the anode interface, the MOS device acts as an interference cavity for the electron waves (analogous to an optical etalon). The constructive and destructive interference of the electron waves leads to an oscillatory behavior superimposed on the background FN tunneling current. If the path through the insulator layer is long (thick film), the electrons will undergo scattering and the coherence may be lost. For thin films (<5 nm) where there are few inelastic phase-destroying collisions, the wave packets will interfere and the resulting interference pattern results in oscillations superimposed on the FN tunneling currents. The oscillations are sensitive to small changes in the electron potential near the collecting interface and the position and amplitudes of the oscillations contain information about the structure and composition of the Si–SiO$_2$ interface. Much of the experimental research on the appearance and use of the FN tunneling current oscillations was due to Maserjian and colleagues [see e.g., the review by Maserjian (4) The references therein].

The exact solution for an electron tunneling through a trapezoidal barrier yields the current density J as

$$J = BJ_0 \tag{12.8}$$

where J_0 is given above (Equation 12.7) and B is called the modulation factor and is given as

$$B = \left[\text{Ai}(-ax)^2 + \left(\frac{a}{\mathbf{k}}\right)^2 \text{Ai}'(-ax)^2 \right]^{-1} \tag{12.9}$$

where Ai and Ai' are, respectively, the Airy function and its derivative; a is the normalizing coefficient for the Airy function and is defined as $a = (2\,m_{cb}qF/h_2)^{1/3}$, where m_{cb} is the average effective mass of an electron in the conduction band of SiO$_2$ and q is the magnitude of the electronic charge; $\mathbf{k}$ is the electron wave vector in the Si at the interface, and x is distance traveled by the electron in the conduction band of SiO$_2$. (A solution to the quantum mechanical tunneling oscillations problem can be found in Gundlach (5) and Alferieff and Duke (6). Typically, B, the oscillatory component of the current, is extracted from the measured tunneling I–V curve and is usually much smaller than the dc component of the tunneling current. Therefore, small errors in the analysis can lead to significant errors in the estimation of the oscillations. However, a sensitive and robust method was developed (7) for extracting B from the FN tunneling current. This method does not require *a priori* knowledge of film thickness and oxide field as did earlier analyses. Hence the new method is not only more accurate than the traditional method but also can be used to obtain the film thickness in the ultra-thin-film regime that is troublesome using other techniques.

The Zafar et al. method (7) of analysis can be demonstrated starting with a measured I–V curve shown in the inset of Fig. 12.22 and the FN Equation 12.7, where the variable F_{cal} was numerically calculated in place of the oxide field as a function of measured current density. The small current oscillations can be seen in the I–V data in Fig. 12.22. Since the WKB FN equation neglects the interference effect of electron waves, F_{cal} has a small oscillatory component B_F, where B_F is related to the current modulation factor B as follows:

$$\ln B = 2\ln\left(1 + \frac{B_F}{F}\right) + \frac{CB_F}{F^2} \qquad (12.10)$$

where F is the nonoscillatory component of F_{cal} and is the actual oxide field. F_{cal} is shown in Fig. 12.22 and can be written as

$$F_{\text{cal}} = B_F + F \qquad (12.11)$$

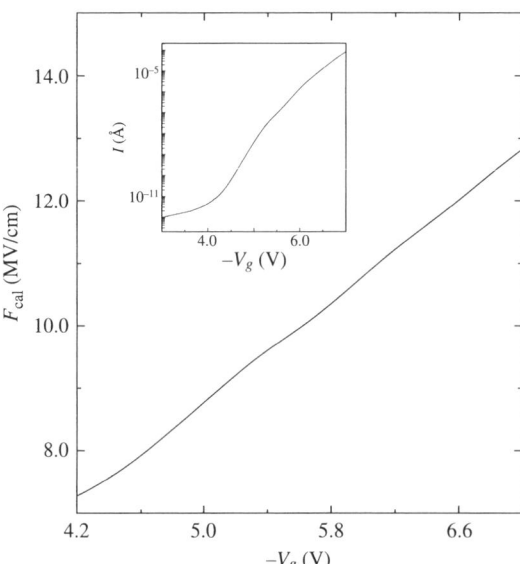

FIGURE 12.22 F_{cal} vs. applied gate voltage (V_g) where F_{cal} is obtained from the I–V data in the inset. [Adapted from Zafar et al. (7), Figure 3a.]

12.3 OTHER TECHNIQUES FOR INTERFACE STUDIES

Therefore, Equation 12.10 can be written as

$$\ln B = 2\ln\left(1 + \frac{F_{cal} - F}{F}\right) + \frac{C(F_{cal} - F)}{F^2} \qquad (12.12)$$

Equation 12.12 is then used to calculate $\ln B$, which starts with a calculation of F. Since F is the straight line passing through the nodes of the oscillations in F_{cal} (i.e., $F = F_{cal}$ at the nodes of the oscillations in F_{cal}), an equation for F can be obtained from the values of F_{cal} and V_g at the nodes. The positions of the nodes on the V_g axis are determined by numerically differentiating F_{cal} with respect to V_g, and the positions of the maxima and minima of dF_{cal}/dV_g are the nodes of the oscillations in F_{cal}. Using the values of F_{cal} corresponding to the nodes, the equation for F is obtained by fitting the data points to a straight line. Once F is obtained, $\ln B$ is calculated using Equation 12.12. Figure 12.23 shows the oscillatory component of the tunneling current as obtained from this analysis where 2 is the amplitude of an oscillation.

12.3.2.2 Application to SiO₂: Refractive Index, Film Thickness, and Roughness As discussed above, the FN tunneling current passing through thin SiO₂ films has a small oscillatory component. The phase and the amplitude of the current oscillations depend on various properties of the SiO₂, including barrier heights, oxide thickness, and roughness at the interfaces, and, therefore, a determination of the FN oscillations can lead to ultra-thin-film parameters. The focus of

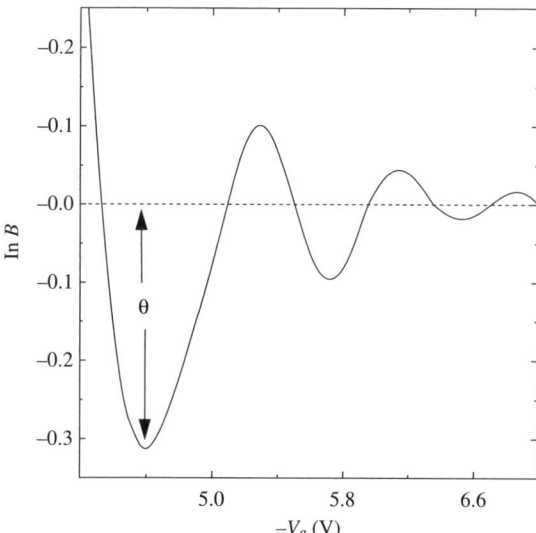

FIGURE 12.23 Oscillatory component ($\ln B$) of the tunneling current with V_g. θ is the amplitude of the oscillation. [Adapted from Zafar et al. (7), Figure 3b.]

this section is on the dependence of the oscillatory component on oxide thickness, SiO_2–Si interface roughness and film refractive index and is based primarily on two publications (7, 8). The SiO_2 film thickness study (8) shows that the oscillation amplitude decreases exponentially as the distance traveled by the electron in the conduction band of SiO_2 (the film thickness) increases. This decrease is attributed to the scattering effect, and from the results (discussed below) the mean free path of electrons in SiO_2 is estimated to be about 0.6 nm. The roughness study (7) was performed on thermally oxidized purposely roughened Si surfaces, and the roughness was independently characterized by AFM. In this study (7) the oscillations in the FN tunneling currents showed no dependence on the interface roughness, indicating that the roughness at the Si surface has little affect on the thickness uniformity of the thermally grown thin oxide films, and that oxide films grow uniformly over the roughened Si surfaces for at least the first 5 nm. This finding differs from the predictions of decreasing oscillation amplitude with increases in roughness at the SiO_2–Si interface. However, a later study that was performed using higher frequency roughness did show decreases in oscillation amplitude and this study will also be discussed below.

All the tunneling measurements were performed on metal–oxide–Si (MOS) capacitors. For roughness studies the Si surfaces were purposely roughened using a solution of HNO_3, HF, and CH_3COOH in volume ratios of 30 : 20 : 40 at room temperature. The roughness at the Si wafer surface could not be precisely controlled but could be varied by varying the chemical etching time, which was from 0 to 60 s. After etching, the wafers were rinsed in deionized water and dried in flowing N_2. These wafers were thermally oxidized at around 800°C in clean dry oxygen to produce about 5-nm-thick SiO_2 films. After oxidation, 200-nm-thick Al gates with an area of about 8×10^{-4} cm^2 were evaporated through a shadow mask. As a final step, the wafers were subjected to a postmetallization anneal at 400°C for 20 min in forming gas consisting of 10% H_2 in Ar. The thicknesses of the SiO_2 films ranged between 4.0 and 5.5 nm as first approximated using high frequency capacitance–voltage (C–V) measurements and then accurately measured using the tunneling current oscillations as discussed below.

Current versus voltage (I–V) measurements on the $Al/SiO_2/Si$ capacitors were obtained using a programmable step voltage supply with the current measurement sensitivity of about 5×10^{-13} A. A gate voltage, V_g, was applied to the metal contact and ranged from -3.0 to -7.5 V in steps of 0.01 V; current measurement was made at each voltage step. For the assessment of roughness at the SiO_2–Si interface, AFM measurements were made on the roughened Si surfaces before and after oxidation.

Figure 12.24 shows typical I–V measurements for oxide films with thicknesses varying between 4.2 to 5.2 nm (42 to 52 Å). A careful observation of this data reveals the small undulations that are larger for the thinner films. Note that the original studies discussed used angstrom units for thickness rather than nanometers, and thus the units are now switched to conform to the original results that are shown. For $|V_g| < 4.0$ V, a displacement current of about 10^{-12} A is measured and is subtracted from the measured current to obtain an accurate estimate of the tunneling current. For 42 and 45 Å films, the current increases sharply at around $V_g = -4.2$ V, marking the

12.3 OTHER TECHNIQUES FOR INTERFACE STUDIES 459

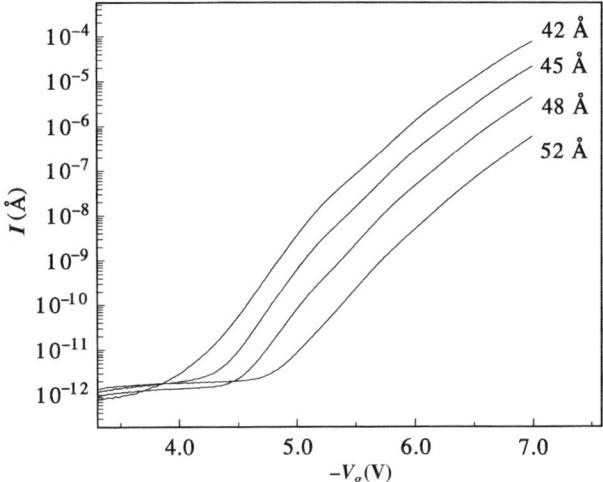

FIGURE 12.24 Current (I) vs. applied gate voltage V_g for various thickness SiO$_2$ film on Si. [Adapted from Zafar et al. (7), Figure 4.]

onset of the FN tunneling, and the FN onset is approximately equal to the barrier height voltage at the SiO$_2$–Si interface. For thicker oxides, the sharp increase in the current occurs at higher voltages. This is because the tunneling current at $|V_g| \approx 4.2$ V is smaller than the displacement current, and as a result the displacement current masks the onset of the FN tunneling. For $|V_g| > 4.2$ V it can be seen in Fig. 12.24 that the tunneling current has a small but noticeable oscillatory component that is the focus of the study.

Figure 12.25 displays the oscillatory component as ln B versus the oxide field F for five SiO$_2$ thicknesses. Since the tunneling current decreases with an increase in oxide film thickness, low field oscillations could not be measured for films much thicker than those shown in Fig. 12.25. For example, the first minimum could not be measured for 45- and 48-Å-thick oxide films, whereas both the first minimum and maximum could not be measured for the 52-Å-thick film. From Fig. 12.25 we see first that the oscillations shift toward lower fields as the oxide thickness increases in agreement with the predictions of the FN theory. Comparing oscillations for 44- and 42-Å-thick oxide films, it is seen that the 2-Å increase in thickness shifts the first minimum by 0.5 MV/cm, which is easily measured and is therefore sensitive to thickness and suggests a method for measuring the thickness provided that the oscillatory current analysis is independent of film thickness as is the new method described above. Also the oscillation amplitude decreases with increasing field for all films and this damping effect is larger for the thinner oxides and is discussed further below.

Figure 12.26 shows the dependence of the oscillation amplitude (θ) on L_{cb} for SiO$_2$ films of varying thicknesses (L_{ox}). L_{cb} is defined as the distance traveled by the electron in the conduction band of the SiO$_2$ and can be written as $L_{cb} = L_{ox} - \Phi_M/eF$, where e is the magnitude of an electronic charge and F is the oxide field.

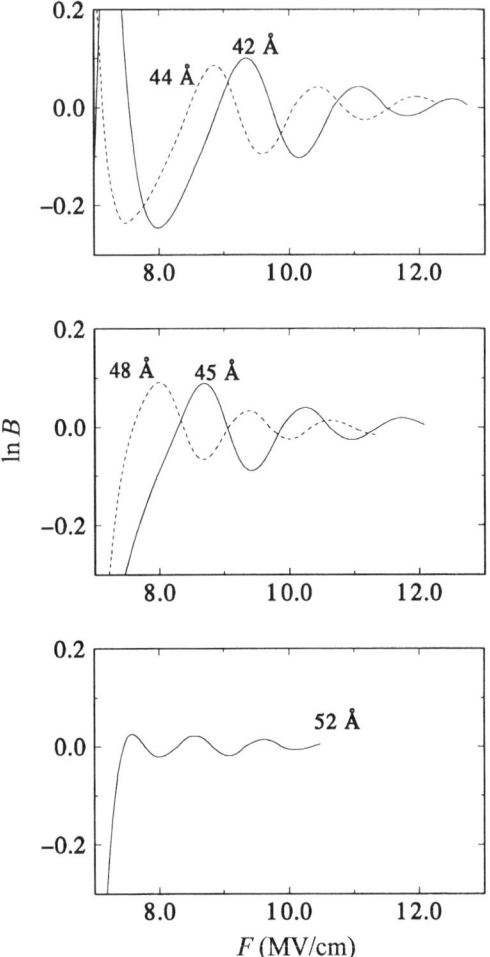

FIGURE 12.25 Ocillatory component of the current ln B vs. the oxide electric field F for various thickness SiO_2 film on Si. [Adapted from Zafar et al. (7), Figure 5.]

From these results the oscillation amplitude was observed to decrease exponentially with L_{cb} for all oxide films and can be attributed to the scattering of electrons in the conduction band of SiO_2 that destroys the phase coherence of the incident and reflected electron waves and θ can be written as

$$\theta = \theta_0 e^{-(L_{cb}/\lambda)} \tag{12.13}$$

where θ_0 is the oscillation amplitude without scattering and λ is the mean free path of the electron in the conduction band of SiO_2. λ and θ_0 can be estimated from the best-fit straight line in Fig. 12.26, and λ was found to be about 6 Å, in agreement with previously reported values, and can be interpreted as a bulk property of SiO_2.

12.3 OTHER TECHNIQUES FOR INTERFACE STUDIES

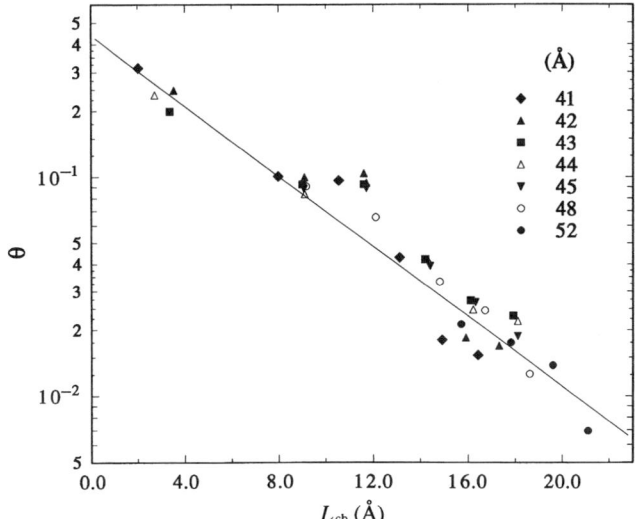

FIGURE 12.26 Amplitude of the quantum oscillations θ as a function of the distance the electrons travel in the conduction band of the SiO_2, L_{cb}. [Adapted from Zafar et al. (7), Figure 6.]

An Si surface roughness study was performed to verify the predictions of the Sune et al. model (9), which assumes that roughness at the SiO_2–Si interface would induce a distribution in oxide thickness as shown in Fig. 12.27a. This SiO_2 thickness distribution would cause the electron waves in different parts of the SiO_2 film to travel different distances through the film and therefore be out of phase upon exiting the film, and thus the resulting oscillations would be damped. From this reasoning, samples with rougher interfaces are predicted to have smaller oscillations compared to those with smoother interfaces. Figure 12.28a to 12.28c shows the AFM images of the Si surfaces that were purposely roughened by acid etching for the FN tunneling study. The number and height of the peaks increased as the acid etch time increased.

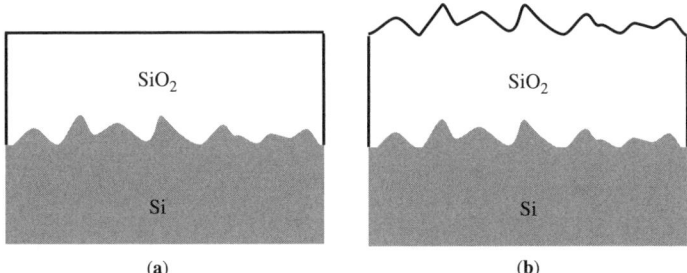

FIGURE 12.27 (a) Sune et al. model (see text) where oxide surface is smooth but Si interface is rough and (b) shows a conformal oxide on rough Si. [Adapted from Zafar et al. (7), Figure 7.]

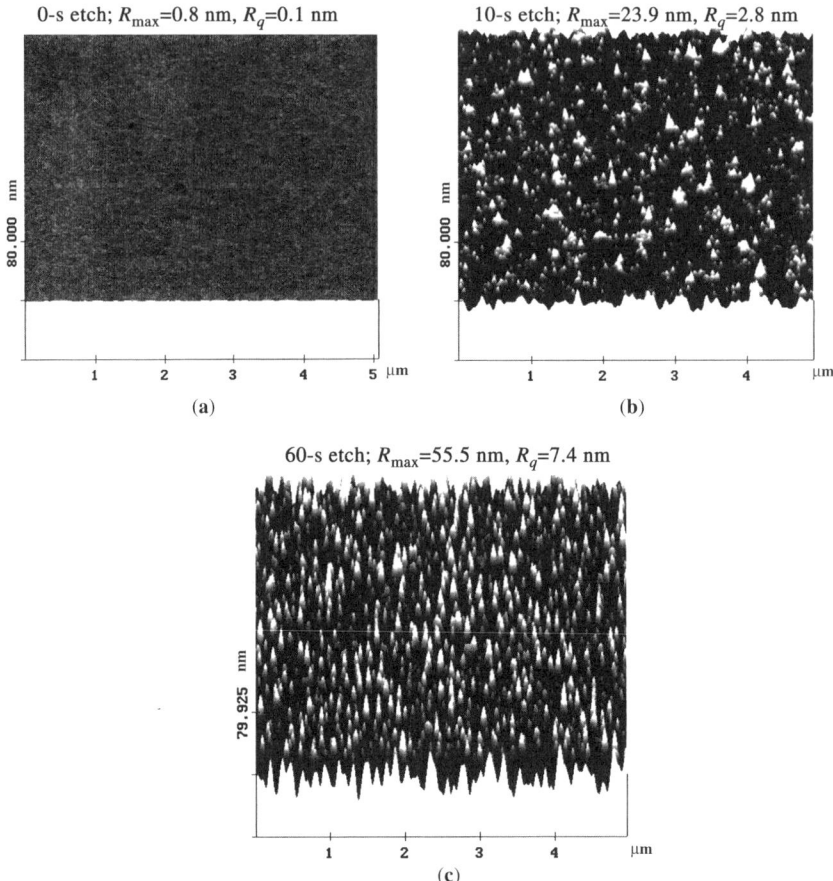

FIGURE 12.28 AFM images of Si surface purposely roughened using chemical etching and with R_q and R_{max} given as roughness parameters. [Adapted from Zafar et al. (7), Figure 8.]

The surface corresponding to no acid etch has almost no measurable features as compared to those corresponding to 10-s and 60-s etch times. The AFM results were analyzed by calculating two roughness parameters, R_q and R_{max}, which are also reported in Fig. 12.28. Recall from Chapter 4 that R_q is the root mean square (rms) value for roughness and is defined as the standard deviation of the vertical height of the points on the surface, and R_{max} is the difference between the highest and lowest points in the scan range. It is seen that both R_q and R_{max} increase as the acid etch time increases, indicating an increase in Si surface roughness as measured by these parameters. To determine whether the oxidation process to produce 50 Å SiO_2 altered the roughness, the samples were measured by AFM after removal of the SiO_2 by HF exposure, and no change was observed in the AFM images. Figure 12.29 compares the oscillatory component of the tunneling current in the oxide films with the same effective thickness, but of varying interface

12.3 OTHER TECHNIQUES FOR INTERFACE STUDIES

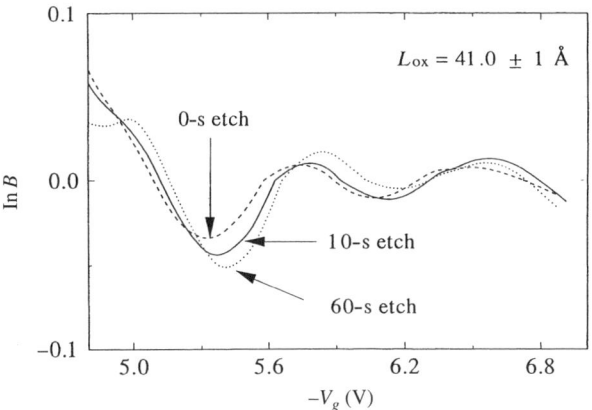

FIGURE 12.29 Comparison of the oscillatory component of the tunnel current oscillations (ln B) for various interfacial roughness levels from Fig. 12.28. [Adapted from Zafar et al. (7), Figure 9.]

roughness. The oscillations show almost no dependence on the SiO_2–Si interface roughness used in this study. This observation suggests that initial Si surface roughness does not induce a distribution in film thickness of thermally grown thin (<45 Å) SiO_2 films, and that thin thermal oxide films grow uniformly over the roughened Si surface, as shown in Fig. 12.27b. This conclusion about thickness uniformity is also compatible with thermal oxidation studies for SiO_2, where it is reported that Si oxidizes via the inward diffusion of oxidant through the growing oxide, resulting in a uniformly thick conformal SiO_2 film on Si (see Chapter 10).

However, in a later related study (10) of the effect of interfacial roughness on FN tunneling, a decrease in the amplitude of Fowler–Nordheim current oscillations (FNCOs) was reported for MOS devices made using Si surfaces that had a high spatial complexity roughness, namely a high fractal value (see Chapter 4 for the discussion of fractals). A model to explain the observations was proposed based on nonuniform Si reactivity, which yields SiO_2 thickness fluctuations. The previous study (7) discussed above that used lower spatial complexity roughness have shown no measurable changes in FNCOs resulting from the oxidation of purposely roughened Si surfaces. The spatial complexity of the purposely roughened Si surfaces was compared using the fractal dimension (D_F) as obtained from AFM measurements.

The samples used for this study were prepared similarly to the samples used for FN tunneling studies (7, 8) discussed above. After the I–V measurements the MOS capacitors were then prepared for AFM roughness measurements by dipping in 48% HF for 20 s for the removal of the Al gate electrode and the SiO_2 film so as to expose the bare Si surface. It has been shown previously that etching in 48% HF up to 1 min does not produce any AFM detectable changes to the Si surface. The results of two sample sets, A and B, which were prepared similarly but independently, each with a control, are compared and discussed below. The FN tunneling current data was analyzed using the Zafar et al. (7) procedure discussed and cited above.

Figure 12.30a and 12.30b show the measured I–V data for set A and set B samples, respectively. AFM images and AFM data analysis of the samples in set A and set B are displayed in Fig. 12.31 and 12.32, respectively. The I–V results for the rough set A sample shown in Fig. 12.31a clearly shows a larger current throughout the voltage range compared with the control. This can possibly be explained by the facts that the rough sample has a larger surface area and that there can be electric field intensification at surface asperities (see Chapter 4). For the samples from set B in Fig. 12.31b the situation is more complicated. In this case the magnitude of the electric field enhancement is thought to depend on both the shape and distribution of the surface roughness features as was pointed out in Chapter 4 (see Chapter 4, Fig. 4.17 and 4.18 and the associated discussion of asperity effects). As discussed in Chapter 4 a high spatial frequency (high D_F), with many closely spaced asperities, can cancel the electric field enhancement.

Figure 12.33 shows the extracted oscillatory components $\ln B$ for sample set A and set B. The oxide film thickness of the controls for the two sample sets were

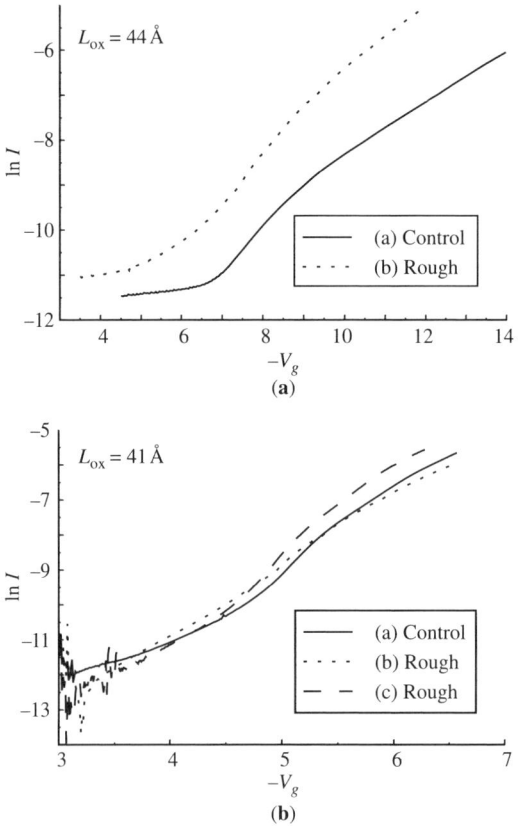

FIGURE 12.30 Current ($\ln I$) vs. applied gate voltage (V_g) for (**a**) sample set A and (**b**) samples set B. [Adapted from Lai and Irene (10), Figures 2 and 3.]

12.3 OTHER TECHNIQUES FOR INTERFACE STUDIES

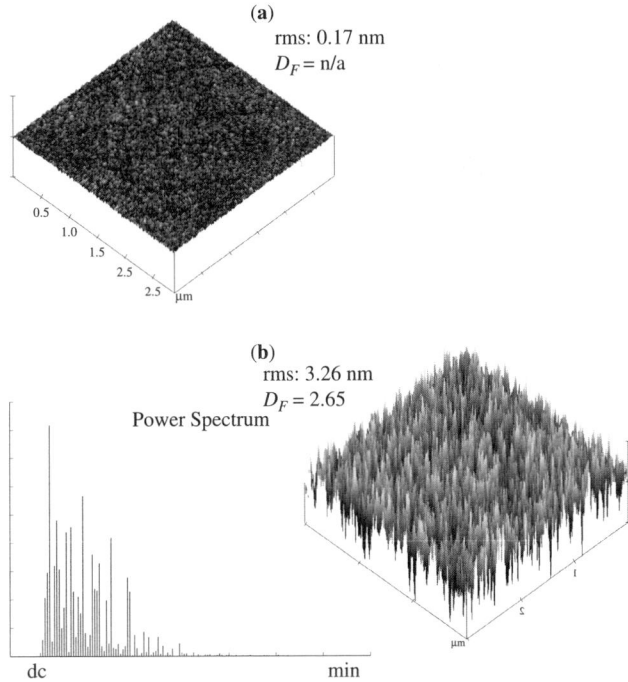

FIGURE 12.31 AFM results for set A samples with 10-nm height range: (**a**) control and (**b**) roughened sample with power spectrum. [Adapted from Lai and Irene (10), Figure 4.]

determined using the FN tunneling current method developed by Zafar et al. (7) and discussed above. Figure 12.33 shows that the magnitude of the oscillations for the roughest sample from each set is smaller than those for the control, thus indicating that in contrast to an earlier study (7) discussed above, interface roughness has an effect on the magnitude of the oscillatory component of the FN tunneling current. It should be noted that the oscillations for the "Rough 2" sample in set B are about the same as the FNCOs for the set B control. The AFM images and the power spectra of AFM profiles for the rough samples (Fig. 12.31 and 12.32) show that surfaces with high spatial complexity with fractal dimensions of 2.65 and 2.71, respectively, and the respective power spectra, with many high frequency components, are related to the decrease in the oscillation amplitude, whereas the sample with relatively low surface complexity with fractal dimension of 2.52 and a power spectrum indicating less high frequency features causes no change to the oscillatory component magnitude as compared to the control. Table 12.2 summarizes the results and shows that the sample having the highest D_F with the most high frequency roughness features has the largest reduction in the magnitude of quantum oscillations of $\sim 66\%$ as compared to its corresponding control. Also, the extent of FNCOs reduction decreases as D_F decreases.

As discussed above, Sune et al. (9) had predicted that Si surface roughness would result in damping of the oscillations from the FN tunneling current for thin SiO_2

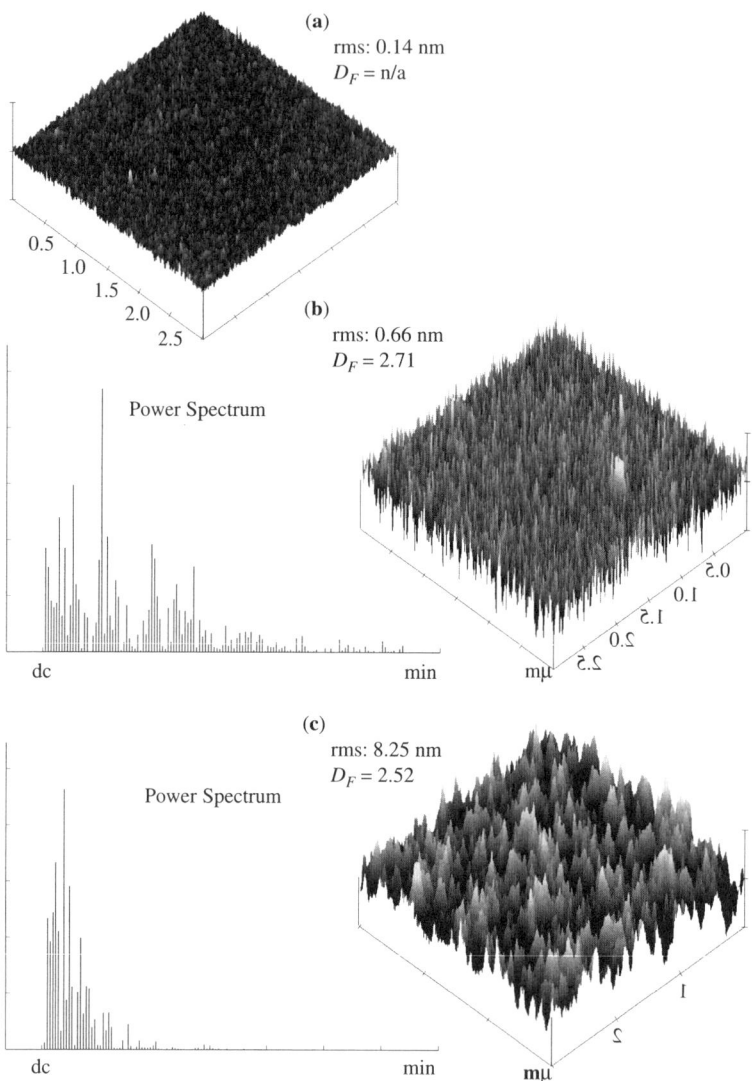

FIGURE 12.32 AFM results for set B samples with 3-nm height range for (a) and (b) and 30-nm range for (c): (a) control and (b) and (c) roughened samples with power spectrum. [Adapted from Lai and Irene (10), Figure 5.]

films, due to the production of a distribution of oxide thickness that results from the oxidation of a rough surface as, shown in Fig. 12.27a. This thickness distribution would in turn cause electron wave destructive interference. However, it is not clear in the Sune et al. (9) model how the thickness distribution shown in Fig. 12.27a would result from Si oxidation, namely, a flat top SiO_2 surface with a rough Si–SiO_2 interface. Thermal oxidation of Si occurs via the inward migration of

12.3 OTHER TECHNIQUES FOR INTERFACE STUDIES

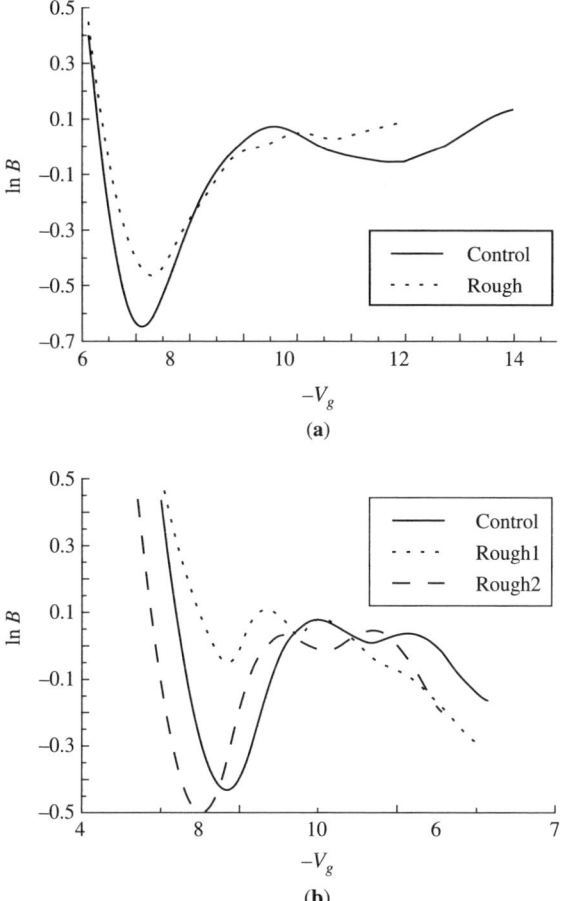

FIGURE 12.33 Oscillatory component of the FN tunneling current (ln B) vs. applied gate voltage (V_g) for (**a**) sample set A and (**b**) sample set B. [Adapted from Lai and Irene (10), Figures 6 and 7.]

TABLE 12.2 Comparison of Magnitude of Oscillations and Fractal Dimensions for Set A and B Roughened Samples

Roughened Samples	Magnitude of Oscillations Compared to Corresponding Controls	Fractal Dimension
Set A (rough)	~18% reduced	2.65
Set B (rough 1)	~66% reduced	2.71
Set B (rough 2)	~0% reduced	2.52

Adapted from Lai and Irene (10), Table 1.

oxidant through the growing oxide, where it reacts at the Si surface (see Chapter 10). Hence, one would expect conformal SiO_2, as shown in Fig. 12.27b, to result in no oxide thickness distribution. In fact the Zafar et al. (7) study has shown that with large surface roughness features there is no dependence on the oscillation magnitude on surface roughness, which means that the oxide film is grown uniformly over the roughened Si surface. The result is supported in the above discussed study (7) for the sample having the AFM image shown in Fig. 12.32c, which exhibits high rms roughness of 8.25 nm, but relatively low fractal dimension of 2.52, and mainly low frequency features in the power spectrum, which shows similar oscillations for the rough sample and the control.

To explain the FNCOs decrease in magnitude for the samples with the highest D_F or spatial frequency, it was proposed that in the initial Si oxidation regime, where the kinetics are dominated by the interface reaction between Si and oxidant (for SiO_2 film thickness less than 20 nm as discussed in Chapter 10), the oxidation is not uniform for high spatial frequency roughness. This nonuniformity of reactivity is anticipated from the Kelvin equation (see a discussion of the Kelvin equation in Chapter 3, e.g., Equations 3.39, 3.40, and 3.41) as follows:

$$\Delta G = \frac{2V_m \gamma}{r} \qquad (12.14)$$

where ΔG is the change of surface energy, γ is the surface energy, V_m is the molar volume, and r is the radius of curvature of a surface feature, and it teaches that the local chemical potential, here reactivity, is greater for small sharp features where r is small. That this actually occurs for the oxidation of roughened Si surfaces was experimentally verified. With the use of Fig. 12.34, it was proposed that the sharp feature at the top of Fig. 12.34 (circle in black) adjacent to another roughness feature yields a thicker oxide adjacent to a thinner oxide (within the circled region), which can cause destructive interference of the electron waves. For this model, it is also required that the sharp features are not isolated so that the interference occurs within the lateral coherence length for the electron wave packet. Thus, the proposed model hinges on the comparison of the spatial frequency of the roughness with the lateral coherence length of the electron wave. Maserjian (unpublished personal communication) actually estimated the lateral coherence for electrons to be about 2 nm for a 5-nm-thick SiO_2 film. This means that roughness on a lateral spatial scale of less than 2 nm would cause wave packet interference effects.

In the Lai et al. (10) study it was shown that there is a relationship between interface roughness and the FN tunneling current oscillations. A decrease in the FNCO amplitude is observed when the Si surface roughness features are sufficiently small to cause oxidation rate variations at the interface, which in turn create local nonuniformities in the oxide thickness. However, when the roughness features are large, there is no effect on FNCOs because the oxide film thickness is uniform and the phase coherence of the propagating electron waves is maintained.

The Zafar et al. (8) oxide film thickness independent method discussed above for extracting the small FNCOs from the measured I–V data was shown to be

12.3 OTHER TECHNIQUES FOR INTERFACE STUDIES

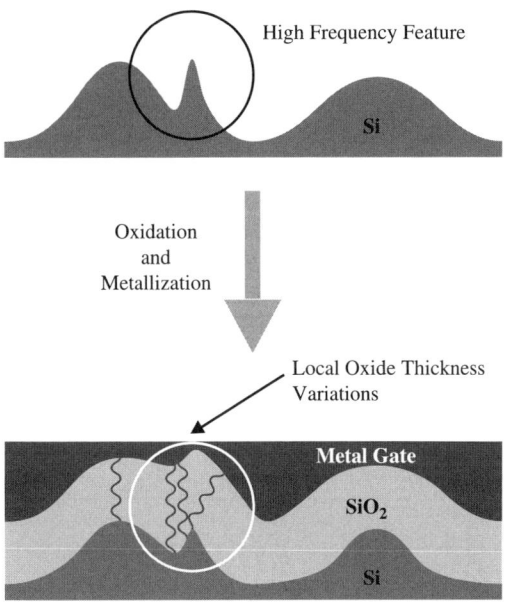

FIGURE 12.34 High frequency model leading to local thickness fluctuations. [Adapted from Lai and Irene (10), Figure 10.]

useful for determining the oxide thickness for ultra-thin SiO_2 films on Si, and this aspect of the FN technique is now discussed further. As discussed above, the oscillatory component of the tunneling current is extracted from the I–V data. Using Equation 12.7 the variable F_{cal}, which is used in place of the oxide field, is calculated as a function of the measured current density. The parameters used in the calculations to follow are $\Phi_M = 3.15$ eV and $m_{bg} = 0.42$ eV for Al gate contacts, and $\Phi_M = 2.93$ eV and $m_{bg} = 0.50$ for N^+ polycrystalline Si gate contacts. As was discussed above and shown for specific samples in Fig. 12.35, the current density J for three samples in Fig. 12.35a show small oscillatory components that are extracted and shown in Fig. 12.35b. As was discussed above F_{cal} as given by Equation 12.11 consists of an oscillatory (B_F) and a nonoscillatory (F) component with F being the actual applied oxide field and B_F is the oscillatory component of the tunneling current displayed as $\ln B$ and given by Equation 12.12. To calculate $\ln B$, F is first obtained. Since $F = F_{cal}$ at the nodes of oscillations in F_{cal}, an equation for F can be obtained by fitting the values of F_{cal} at the nodes to a straight line. The nodes are located from a numerical differentiation of F_{cal}. Once the nodes are located, the equation for F is obtained. Since $B_F = F_{cal} - F$, B_F is readily obtained. Substituting the values of F and B_F in Equation 12.10, the oscillations in the tunneling current are obtained.

Figure 12.35b shows the extracted oscillatory component as a function of the oxide field for samples with different film thicknesses. Curves *a* and *c* correspond to thinnest and thickest oxide films, respectively. The oscillations are observed to

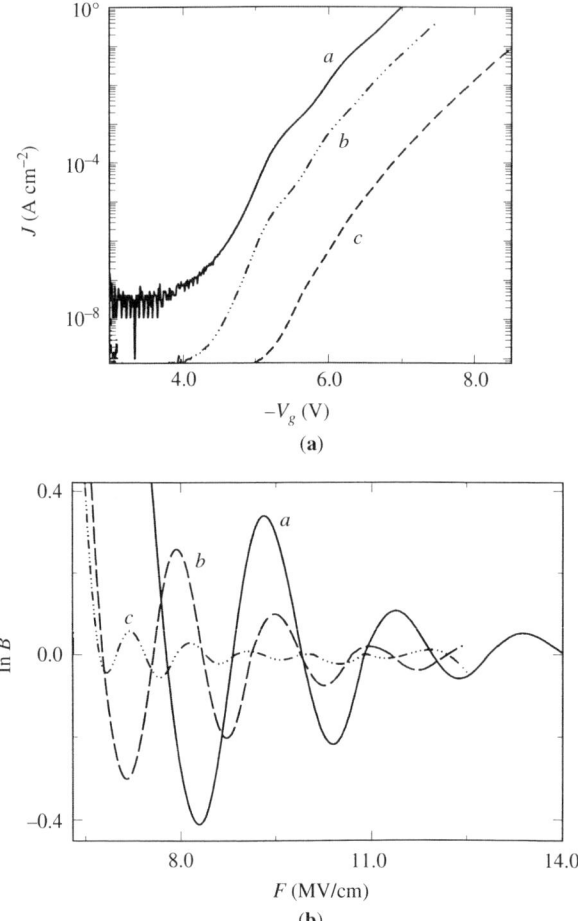

FIGURE 12.35 (a) Experimental current density (J) vs. gate voltage for three ultra-thin SiO$_2$ capacitors (a,b,c) and (b) extracted oscillatory component B vs. electric field F using the Zafar et al. method. [Adapted from Zafar et al. (8), Figures 1 and 2.]

shift toward lower fields with increasing oxide thicknesses, and the oscillation amplitude decreases with increasing oxide field that is attributed to an increase in electron scattering. The fields (F_n) at which various extrema (with subscript n) occur are obtained directly from Fig. 12.35b and are used for calculating the film thickness L_{ox} and electron effective mass in the SiO$_2$ conduction band m_{cb}. Substituting the measured F_n in the condition for the extrema,

$$K_n = \left(\frac{8\pi^2 q m_{cb} F_n}{h^2}\right)^{1/3} \left(L_{ox} - \frac{\Phi_M}{qF_n}\right) \tag{12.15}$$

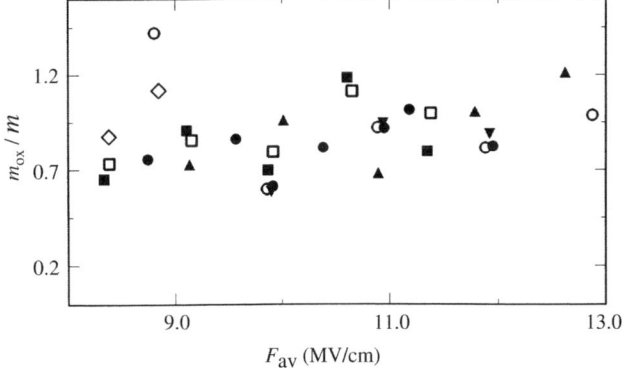

FIGURE 12.36 Ratio of effective mass m_{ox} in the SiO$_2$ conduction to the free electron mass m as a function of the average field between to adjacent extrema F_{av}. [Adapted from Zafar et al. (8), Figure 3.]

where for minima F_n is the field corresponding to the nth minimum and K_n is the nth zero of the derivative of the Airy function and for maxima F_n is the field for the nth maxima with K_n the nth maxima of the Airy function. Assuming that m_{cb} is constant between any two consecutive extrema, m_{cb} is obtained and shown in Fig. 12.36 where the ratio m_{cb}/m is plotted as a function of F_{av}, which is the average electric field between two consecutive extrema. The value for $m_{cb} = 0.86$ m is consistent with a previously published value and shows no dependence on the oxide field or the growth process for SiO$_2$ films. This value of m_{cb} along with the values of Φ_M for different gate materials are used in Equation 12.15 to calculate F_n as a function of oxide thicknesses for various extrema. Figure 12.37 displays the calculated values

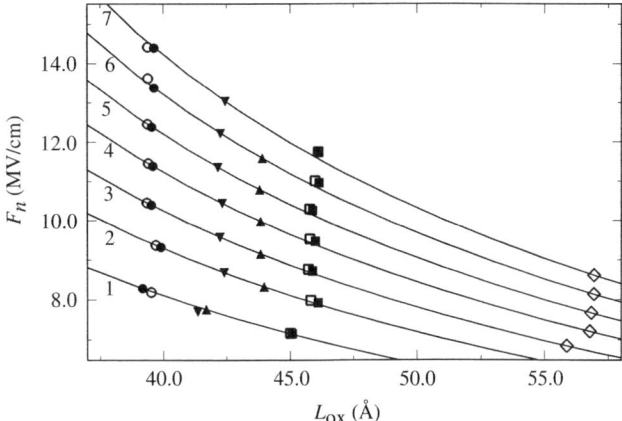

FIGURE 12.37 F_n from Equation 12.15 vs. oxide thickness L_{ox} for the extrema labeled 1 through 5 at left. The different symbols represent results from different samples. [Adapted from Zafar et al. (8), Figure 4.]

for F_n with $\Phi_M = 3.15$ eV for Al indicated by lines with the extrema number at the left side of the curves. The measured F_n from Fig. 12.37 are denoted by different symbols for different samples. For each sample the measured values of F_n are matched with calculated F_n, and a vertical line is best fit through the data points with the x intercept to obtain the oxide thickness L_{ox}. In the above estimation for L_{ox} the first measured extremum was not used because of the errors associated with the small FN tunneling current. The random error in L_{ox} was found to be less than 0.5%. Also it is seen that a thickness change of 0.1 nm produces significant shifts (>0.3 MV/cm) in F_n, and this shift increases as the oxide thickness decreases and based on this thickness changes of 0.1 nm are readily measured.

Following the work above where accurate ultra-thin SiO_2 film thicknesses were obtained, another study (11) was reported in which FNCOs were used first to accurately determine the thicknesses of ultra-thin SiO_2 films. Then using the FN-determined thicknesses as input, the real part of the refractive index n for thin SiO_2 films was determined in the range of 4 to 6 nm and with the use of precision single wavelength ellipsometry. An average value for this refractive index was found to be 1.894 ± 0.110. This value was shown to yield SiO_2 thicknesses to an accuracy of ± 0.1 nm. Typically, ellipsometry is employed as the technique used to measure film thicknesses and refractive indexes. However, for films less than 30 nm ellipsometry is unreliable for simultaneously extracting refractive index and thickness from the measured data.

In the Hebert et al. study (11) the FN current versus voltage measurements were obtained on two separately prepared sample sets (I and II) of N^+-poly/SiO_2/Si capacitors. Both set I and set II capacitors consisted of SiO_2 films that were grown by thermal oxidation at 900°C on P-type (100) Si wafers processed at two different facilities. The areas of these capacitors were 20 μm^2 for set I and 10 μm^2 for set II. The oxide thicknesses for these capacitors varied from 3.9 to 6.0 nm. Both the SiO_2 thicknesses as obtained from the FNCO measurements and ellipsomtetric measurements were made on the two sample sets. For the set I samples, ellipsometric measurements were taken on monitor wafers that did not contain capacitors, but had been processed identically to each wafer containing the capacitors. For set II, the optical and electrical measurements were taken on the same wafer and the optical measurements were made in an area adjacent to the capacitors. The ellipsometric measurements were made using a manual high precision single-wavelength ellipsometer in the polarizer–compensator–sample–analyzer (PCSA) configuration with a 632.8-nm laser light source (see Chapter 9). The ellipsometer was carefully calibrated, and reference samples of thermal SiO_2 on Si were used to check the angle of incidence and refractive index throughout the study. The ellipsometric measurables Δ and Ψ and the angle of incidence ϕ can be measured to about $0.02°, 0.01°$, and $0.01°$, respectively. Two-zone ellipsometric measurements were made in order to cancel systematic errors in the optical components. An iterative data reduction routine was used to calculate film thickness and refractive index from the ellipsometric measurables Δ and Ψ using a single-film optical model. All calculations used the Si refractive index value at 632.8 nm of $3.865 - i0.018$. At least three ellipsometric measurements were taken on each sample to obtain an estimate of the random error in Δ and Ψ.

12.3 OTHER TECHNIQUES FOR INTERFACE STUDIES

A single-film optical model was used with the assumption that k for SiO_2 is equal to zero. Since there are reports that indicate the presence of an interface layer between Si and SiO_2 consisting of atomically mixed Si and O of stoichiometry SiO_x, which would yield $k > 0$ for $0 < x < 2$, the measurements were also interpreted using $k = 0.018$ (the value for Si). This upper value was chosen since it is unlikely that k would be larger than the absorption coefficient for bare Si. Table 12.3 summarizes the results for both set I and set II samples. Column 2 shows the SiO_2 film thicknesses obtained from the FNCOs. From the average Δ and Ψ values for each sample and with the thicknesses obtained from the current oscillation measurements, calculated refractive indexes with standard deviations are in column 3. As a check, column 4 displays the back-calculated oxide thicknesses with standard deviations using each calculated refractive index from the minimum and maximum Δ and Ψ values. Agreement to within ± 0.1 nm is typically seen. Column 5 shows the calculated average refractive index values using the average Δ and Ψ values and ± 0.1-nm error in current oscillation thickness.

It is seen from this study that both sets of ultra-thin samples have refractive index values greater than that of bulk SiO_2 (1.465) in accord with previous studies (1) that indicate that the Si–SiO_2 interface is structurally, physically, and optically different than bulk SiO_2 and that it has an intrinsic film stress (discussed in the following section) that yields higher refractive index values near the interface. The average refractive index for set I is 1.816 ± 0.088 and for set II is 1.972 ± 0.065. The difference can be attributed to the random error in refractive index.

Also in this study SiO_2 films were prepared by thermal oxidation with thicknesses ranging from 4 to 150 nm at processing temperatures of 700 to 1000°C. From

TABLE 12.3 Refractive Indexes Calculated from the FN L_{ox} and the Ellipsometric Δ and Ψ values[a]

	L_{ox} (Å) (FN)	Av. n (FN L_{ox}, Δ, Ψ)	Av. L_{ox} (Å) (from n)	Av. n (± 0.1 nm)	L_{ox} (Å) ($n = 1.894$; $k = 0$)	L_{ox} (Å) ($n = 1.894$, $k = 0.018$)
Set I	39.3	1.737 ± 0.008	39.3 ± 0.00	1.757 ± 0.105	38.2	38.3
	40.4	1.855 ± 0.170	40.5 ± 0.01	1.835 ± 0.132	40.0	40.0
	44.0	1.809 ± 0.042	44.0 ± 0.00	1.833 ± 0.131	43.5	43.6
	45.0	1.700 ± 0.010	45.0 ± 0.00	1.707 ± 0.067	43.3	43.3
	46.0	1.853 ± 0.011	46.0 ± 0.01	1.859 ± 0.116	45.8	45.9
	48.0	1.943 ± 0.048	48.2 ± 0.02	1.913 ± 0.104	48.3	48.4
Set II	40.0	1.976 ± 0.000	41.1 ± 0.01	1.977 ± 0.001	41.3	41.3
	40.0	1.976 ± 0.000	40.9 ± 0.00	1.976 ± 0.000	41.0	41.0
	46.0	2.084 ± 0.006	46.0 ± 0.00	2.096 ± 0.165	45.8	45.9
	46.0	1.978 ± 0.001	46.6 ± 0.00	1.912 ± 0.093	46.7	46.8
	60.0	1.925 ± 0.084	60.5 ± 0.05	1.929 ± 0.089	60.6	60.6
	60.0	1.891 ± 0.136	61.3 ± 0.16	1.948 ± 0.061	61.1	61.6

[a] Back calculated thicknesses from the calculated refractive indexes are also shown.
Adapted from Hebert et al. (11), Table 1.

refractive index measurements it was observed that as the oxide thickness increases, the value of $n = 1.465$, which is the bulk SiO_2 refractive index, is reached at around 30 nm. Also it was found that by using the bulk SiO_2 value of $n = 1.465$ for the 5 nm SiO_2 films with the average Δ and Ψ values, an average error in thickness 20% larger than the thicknesses from the current oscillations is obtained. For SiO_2 films in the 4 to 6-nm thickness range the refractive index was found to be 1.894 ± 0.110 for 632.8-nm light, and when this value for n was combined with the SiO_2 bulk film value of 1.465 for film thicknesses greater than about 20 nm and with the index for bare Si (3.865) at 632.8-nm light, a best-fit formula was obtained:

$$n = 1.465 + 2.400 \exp(-3.913 \times 10^{-2} L_{ox}) \qquad (12.16)$$

In a more recent study (12) a more accurate formula for n for thin SiO_2 films was obtained using spectroscopic ellipsometry measurements and starting from the Hebert et al. (11) formula above for n based on FN tunneling current oscillations measurements. To obtain the new refractive index parameters, an iterative procedure was used starting with the Hebert et al. (11) FNCO-derived thickness interpolation formula 12.16 for n along with new SE data. The new SE measurements were obtained at various incidence angles (near the principle angle) on SiO_2 films ranging in thickness from 1.8 to 8 nm. The iterative procedure was then used to optimize the SiO_2 refractive index values obtained from the voluminous SE data. The new n values are useful for extracting more accurate ultra-thin SiO_2 film thicknesses from ellipsometry measurements.

For this study two independent sample sets of ultra-thin SiO_2 films on Si substrates were prepared. For sample set 1 single-crystal (100)-oriented boron-doped P-type Si wafers with resistivity of about 1 to 2 Ω-cm were used. The wafers were cleaned using a modified RCA cleaning procedure, which also included a 48% HF dip to remove native oxide and which was followed by a deionized water rinse. The cleaned wafers were dried with N_2 gas and then loaded onto a fused silica boat and placed in a double-wall fused silica horizontal furnace tube for the thermal oxidation at 800°C in pure oxygen ambient. After the oxide has grown to the desired thicknesses, the samples are cooled down at N_2 ambient and then taken out of the furnace for SE measurements. For sample set 2 thermal SiO_2 films on Si wafers in the thickness range of 2 to 8 nm were prepared in a cleanroom Si processing facility. These samples were also prepared on low resistivity (< 10 Ω-cm) Si(100) single-crystal substrates using thermal oxidation at 800°C in dry O_2.

An SE spectrum was obtained from the measurements at various incident photon energies. A close examination of Δ, Ψ space for the ultra-thin-film regime revealed that Δ, Ψ trajectories for films of different index converge toward zero film thickness (see Fig. 9.16). Hence, to separate thickness from index (the product is measured; see Equation 9.7) a measurement precision for Δ and Ψ needs to be less than the best precision that ellipsometers are capable of achieving (approximately 0.01°). Thus, if n can be supplied to the analysis from an independent measurement, then L_f can

12.3 OTHER TECHNIQUES FOR INTERFACE STUDIES

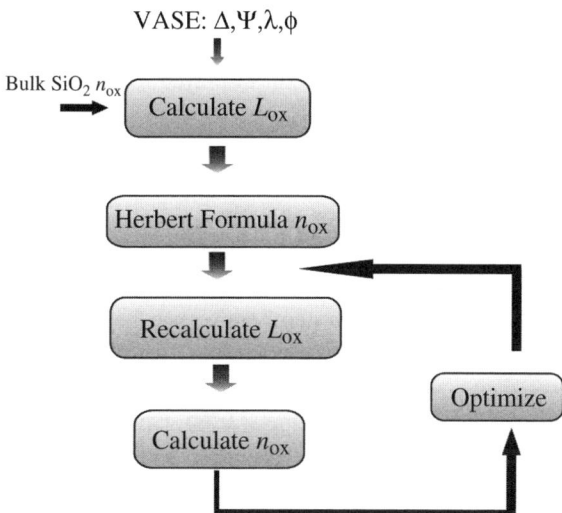

FIGURE 12.38 VASE analysis algorithm. [Adapted from Wang and Irene (12), Figure 1.]

be optimized from the voluminous SE data using an iterative procedure shown in Fig. 12.38. For the SE measurements performed, Δ and Ψ were measured over the 1.5 to 5.5-eV energy range and the variable angle SE (VASE) over the ϕ_0 range of 70.00° to 75.00° in 1° increments. SE spectra were modeled using the air–SiO$_2$–Si three-phase optical system described above. VASE was used to improve the precision of extracting parameters. Likewise SE, as opposed to single-wavelength ellipsometry, also improves the parameter extraction and reduces the correlation among unknown parameters. Correlation between refractive index arises because in ellipsometry the optical path through a material is measured, and the optical path is essentially the product of thickness and complex index, thus,

$$\frac{\partial \Psi}{\partial L} = \frac{\partial \Psi}{\partial n} \quad \text{and} \quad \frac{\partial \Delta}{\partial L} = \frac{\partial \Delta}{\partial n} \qquad (12.17)$$

The sensitivity of ellipsometry is optimized near the principle angle of incidence, which is about $\phi_0 = 76.0°$ for bare Si. Thus, the range of angle of incidence was chosen to be 70.00° to 75.00°, and we expect greater sensitivity near the larger ϕ_0 values. For the SE results presented below, two ellipsometers were used. One commercial variable angle SE instrument with about 0.01° precision in Δ and Ψ and in angle of incidence ϕ_0. The other is a custom-built ellipsometer with about the same high precision with careful alignment. Also, both instruments gave the same results within the errors above when compared on the same samples.

In the previously discussed Hebert et al. (11) study, SiO$_2$ samples in the 4- to 6-nm range were measured both by single-wavelength ellipsometry to obtain Δ and Ψ and by FN tunneling to determine the oxide thickness. To improve upon the Hebert et al.

(11) results, VASE data was obtained from 8 ultra-thin SiO_2 films, and then these data were used along with Hebert et al. (11) n parameter results. The iterative procedure used can be understood using the algorithm displayed in Fig. 12.38 as follows:

1. Perform VASE on a series of ultra-thin SiO_2 films with thickness from 1.4 to about 8 nm, at incidence angles of 70.00°, 71.00°, 72.00°, 73.00°, 74.00°, and 75.00°.
2. Using the VASE measured Δ and Ψ, calculate the film thickness using the refractive index for bulk SiO_2 films.
3. With the initial thicknesses from step 2 above, use the Hebert et al. (11) formula (Equation 12.16) to obtain corrected refractive index parameters for each sample.
4. Using the new refractive index parameter from step 3, recalculate the film thicknesses as in step 2.
5. Repeat steps 3 and 4 until a nonvarying result for the n parameter is obtained.

Figure 12.39 shows the ellipsometric data in terms of Δ both calculated (solid lines) and experimental, versus wavelength at various ϕ_0's for the 3-nm sample. Sharper changes in Δ are observed for the higher angles of incidence, which confirms that higher sensitivity is obtained as the principal angle is approached. However, insignificant differences were found for the refractive index parameters in the ϕ_0 range used in this study. Figure 12.40 shows the best-fit results obtained in the present study in the 1.4- to 8-nm SiO_2 thickness range (solid line with data points) along with the Hebert et al. (11) results (dashed line) for comparison.

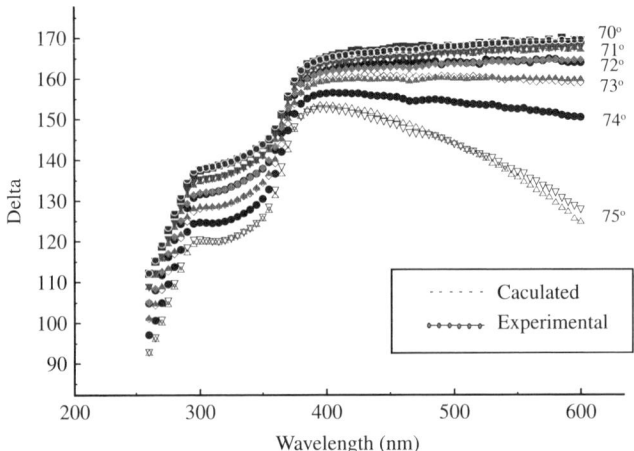

FIGURE 12.39 Delta vs. wavelength for 3-nm sample. [Adapted from Wang and Irene (12), Figure 2.]

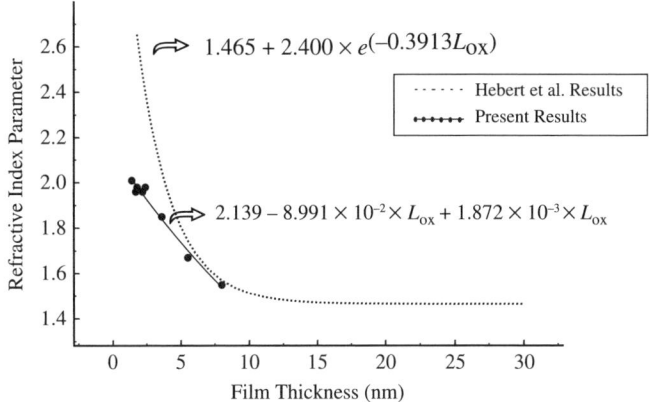

FIGURE 12.40 Best-fit refractive index results along with previous Hebert et al. results. [Adapted from Wang and Irene (12), Figure 3.]

The best-fit index parameter formula from Fig. 12.40 for 1.4- to 8-nm thick SiO_2 films was as follows:

$$n = 2.139 - 8.991 \times 10^{-2} L_{ox} + 1.872 \times 10^{-3} L_{ox}^2 \qquad (12.18)$$

It is seen that the index parameter increases with decreasing film thickness, which is in qualitative accord with previous results. However, the present results yield a systematically smaller refractive index parameter than for the Hebert et al. (11) study. This indicates that there is a more gradual change in the index parameter near the interface, namely the film is more oxidelike, followed by an abrupt change in the oxide within a few tenths of a nanometer of the interface. Essentially, a sharp optical interface is indicated. The two results are quite similar from about 4 nm and upward in thickness. In one earlier investigation using single-wavelength ($\lambda =$ 546.1 nm) ellipsometry alone and performed on thermally oxidized Si, it was estimated that the interface is 0.6 nm thick with a refractive index of 2.8 (real part), and in another study SE measurements were performed on thermally oxidized Si with SiO_2 thicknesses of 120 to 160 nm, an interface transition region were found at about 0.7 nm that consists of atomically mixed Si and O with average stoichiometry of $SiO_{0.4}$. The present study also suggests a sharp transition layer where the index varies from Si (near 4.0) to about 2.0 in less than 1 nm followed by a more gradual change.

To check the constancy of the n values from Equation 12.18, the n values obtained from the above described algorithm were first used in Equation 12.18 to obtain L_{ox} values. Then the L_{ox} values obtained from the analysis were reinserted in Equation 12.18 to recalculate n values, which were then compared to the original n values. Typically, less than 3% deviation was found for the re-determined n values, which will yield less than 1% change in L_{ox} for films around 2 nm and less than 2% for

TABLE 12.4 Comparison of Present and Previous Results with Conditions: Film with $\lambda = 632.8$ nm, $\phi_0 = 70.0°$, on Si

SiO$_2$ Thickness (nm)	Index Parameter, n from Eq. 12.18	Index Parameter, n from Eq. 12.16	SiO$_2$ Thickness from Eq. 12.16 (nm)	SiO$_2$ Thickness from Bulk Index (nm)
1.00	2.05	3.09	1.69	1.20
2.00	1.97	2.56	2.14	2.40
3.00	1.89	2.21	3.07	3.60
4.00	1.81	1.97	3.95	4.80
5.00	1.74	1.80	4.93	5.84
6.00	1.67	1.69	5.95	6.85
7.00	1.60	1.62	6.93	7.73
8.00	1.54	1.57	7.84	8.51

films around 8 nm thick. Table 12.4 compares the n and L_{ox} values obtained using the new results of the present study, the Hebert et al. (11) results, and the thick-film SiO$_2$ index of refraction (1.465). To construct Table 12.4, a set of Δ and Ψ values were calculated from the L_{ox} and n values obtained from Equation 12.18 (the first two columns). Then the Δ, Ψ values were analyzed using the n values from the Hebert et al. (11) Equation 12.16 (the third column) to yield the thickness values in the fourth column. The last column displays the data analyzed using the bulk SiO$_2$ refractive index. First, for films less than 3 nm the Hebert et al. (11) results are nonphysical because the exponential formula yields too large of an n parameter to be analyzed from the Δ and Ψ values, which were derived from Equation 12.18. The bulk SiO$_2$ film index values yield about a 20% error in L_{ox}. For films greater than 3 nm the new results and the Hebert et al. (11) results are similar with only a few percent difference. The bulk index can only be safely used for films greater than 10 nm thick.

12.3.3 Interfacial and Film Stress

When a solid film grows or is deposited upon a solid surface, it is likely that an interfacial stress will develop. In this section several studies are discussed that deal with the interfacial stress that occurs at the Si–SiO$_2$ interface. Before a discussion of the stress and its implications on the properties of the Si–SiO$_2$ interface, a brief general discussion of film stress is undertaken.

Stress is the buildup of forces that are not accommodated by the material, and in the elastic limit the stress is given by Hooke's law:

$$\sigma = E\varepsilon \quad (12.19)$$

where σ is the stress in units of force per area, ε is the strain or deformation, and E is the elastic modulus and often referred to as Young's modulus, and E is a material

12.3 OTHER TECHNIQUES FOR INTERFACE STUDIES

property. The residual or total stress (σ_T) that one observes, for example, as a result of film growth is the stress that derives from the algebraic sum of intrinsic stress (σ_i) and thermal stress (σ_{th}) components and is given as

$$\sigma_T = \sigma_i + \sigma_{th} \tag{12.20}$$

The thermal expansion stress is a result of the difference in thermal expansion coefficients, α, between SiO_2 and Si, $\Delta\alpha$, and the change in temperature ΔT. σ_{th} is proportional to the product of $\Delta\alpha$ and ΔT as

$$\sigma_{th} = \frac{E_f \Delta\alpha \Delta T}{1 - v_f} \tag{12.21}$$

where E_f is Young's modulus for the film and v_f is the film Poisson ratio. The intrinsic stress can have various origins. One origin is due to the change in the molar volume when, for example, Si is converted to SiO_2 via oxidation at the Si surface as was discussed in Chapter 10. Each atom of Si at the surface is converted to a molecule of SiO_2 that has a 2.2× or 120% larger volume than the Si atom from whence it came. In this case σ_i is given as

$$\sigma_i \propto \Delta \overline{V} \tag{12.22}$$

where $\Delta \overline{V}$ is the molar volume change that occurs upon conversion of Si to SiO_2. Another origin is due to the lattice parameter difference for epitaxial growth across an interface and given as

$$\sigma_i \propto \Delta a_0 \tag{12.23}$$

where Δa_0 is the difference in lattice parameters for film and substrate. Stress can be compressive (negative in sign) or tensile (positive in sign). The quantification of stress is largely a function of the specific geometry, and film stress can be measured by first determining the strain imparted to the substrate as a result of the film stress. Stoney's formula can often be used to quantify stress that obtains for a relatively thin uniform film on a relatively thick substrate (at least 10× thicker than the film) and the radius of curvature R of the substrate is measured. For this case Stoney's formula is

$$\sigma_T = \frac{E_s L_s^2}{6(1 - v_s) L_f R} \tag{12.24}$$

where the subscripts f and s correspond to the film and substrate, respectively, and the L's are thicknesses.

Early studies of SiO_2 film stress were performed at room temperature and on films grown on Si using oxidation temperatures greater than 1000°C, which were appropriate to the technology at that time. These studies concordantly reported that the measured residual room temperature compressive film stress could be explained both in magnitude and sign (tensile or compressive) based on the thermal expansion stress (σ_{th}), which develops upon cooling from the oxidation temperature to room temperature (ΔT), and as a result of the difference in thermal expansion coefficients, $\Delta \alpha$, between SiO$_2$ and Si. σ_{th} is proportional to the product of $\Delta \alpha$ and ΔT as given by Equation 12.21. Since at the oxidation temperature $\Delta T = 0$, the thermal component of the stress, σ_{th}, would be zero during oxidation, and thus the thermal component of stress could not be implicated in any stress-driven oxidation models. However, there were some early studies that indicated that an intrinsic compressive stress (σ_i) exists for oxidation temperatures below about 1000°C and that σ_i increased in magnitude with decreasing oxidation temperatures. Figure 12.41 shows a measurement of the total film stress σ_T and calculated values for the thermal stress σ_{th} from one study (13), and Fig. 12.42 shows the resultant intrinsic stress σ_i in this study obtained using Equation 12.20 and the results in Fig. 12.41. For these experiments SiO$_2$ was grown on Si at various oxidation temperatures, but the stress is actually measured at room temperature after the sample is rapidly cooled. Later *in situ* stress measurements will be discussed where the stress is measured at the oxidation temperature (where $\sigma_{th} = 0$). The reported stress was measured using a double-laser beam deflection technique to determine the radius of curvature of a film-covered Si substrate. This technique will be discussed further below when *in situ* stress measurements are

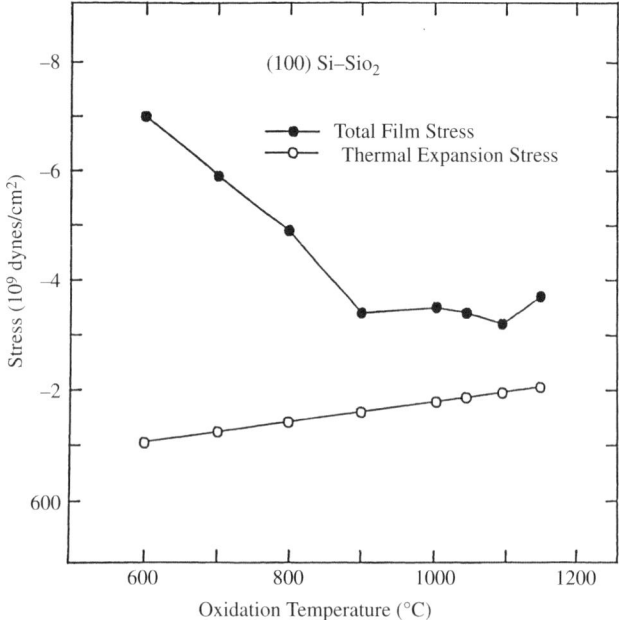

FIGURE 12.41 Total measured SiO$_2$ film stress (measured at room temperature) and calculated thermal stress for SiO$_2$ films on Si(100) grown at various oxidation temperatures. [Adapted from Kobeda and Irene (13), Figure 3.]

12.3 OTHER TECHNIQUES FOR INTERFACE STUDIES

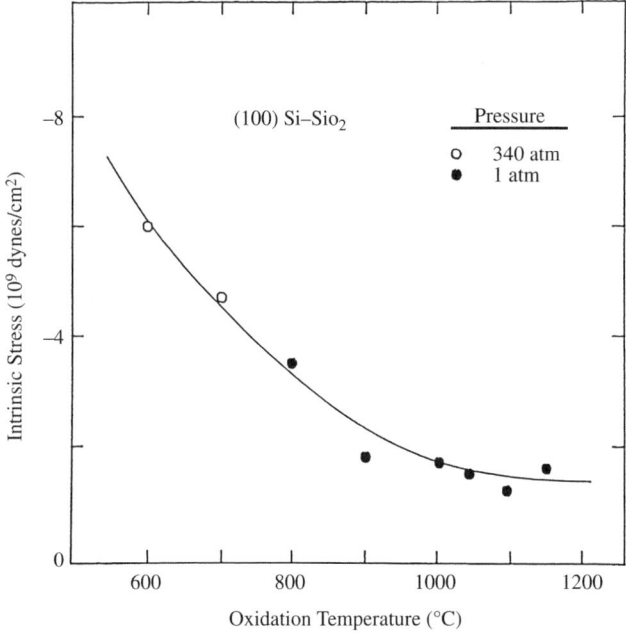

FIGURE 12.42 Intrinsic SiO_2 film stress obtained using the stress values in Fig. 11.41 and Equation 12.20. [Adapted from Kobeda and Irene (13), Figure 4.]

discussed. Based on this kind of data, the existence and temperature variation of σ_i was confirmed and a model was proposed (14) that explained both the appearance of the intrinsic stress and also the simultaneous appearance of an increased SiO_2 film density that had the same Si oxidation temperature dependence as σ_i. The model, called the *viscous flow model*, is explained with the use of Fig. 12.43. The

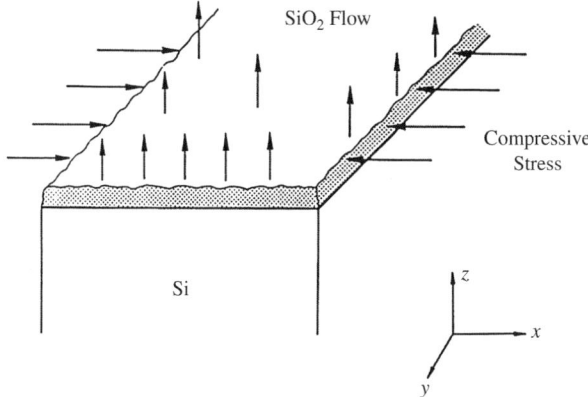

FIGURE 12.43 Viscous flow representation for SiO_2 films on Si surface. [Adapted from Irene et al. (14), Figure 1.]

molar volume requirement (120% increase in molar volume when an atom of Si converts into a molecule of SiO_2 as mentioned above) can be met for SiO_2 formation on the Si surface by the expansion of the as-formed SiO_2 film into the free direction, which in the case of an SiO_2 film forming on a flat smooth Si surface is the direction normal to the Si surface (z in Fig. 12.43). If all of the oxide produced by the chemical reaction at the interface flows into this free direction, then the isotropic expansion created by the conversion of Si into SiO_2 will occur without stress. The viscous flow model assumes that this free direction is "found" by the mechanism of viscous flow at the high oxidation temperatures above 1000°C where the oxide viscosity is sufficiently low. The oxide is constrained by adhesion in the plane of the Si surface and thus can only readily flow into the normal direction. The constraint in the lateral direction and flow in the normal direction can be analogized as the flow of toothpaste from a tube as the tube is compressed normal to the direction of flow. This results in a biaxial compressive stress in the SiO_2 film and a stress gradient in this film normal to the Si surface. However, at lower oxidation temperatures, the higher oxide viscosity precludes easy flow within the time frame for oxidation, and an intrinsic stress, σ_i, develops. Since the oxide viscosity increases as the temperature decreases, it then follows that the intrinsic stress that develops should also increase with decreasing oxidation temperature, as is observed (Fig. 12.42). Along with the observation of the intrinsic stress and its temperature dependence is the parallel observation of an increase of the SiO_2 film density with decreasing oxidation temperature. Using the viscous flow model, the densification of SiO_2 can be understood as the accommodation of the SiO_2 film growth system to the accumulation of stress, namely the system attains as small a volume as possible so as to minimize the compressive stress. Although the SiO_2 network is quite open, only a small density increase is permitted before large repulsive forces are encountered. Between the oxidation temperatures of 1100 and 700°C about 3% increase in density is observed. Later it was reported (15) that n is larger near the Si–SiO_2 interface as is the film density. The density was obtained directly from measurements of the film volume and mass. Figure 12.44a shows the measured film stress for various Si substrate orientations as a function of oxidation temperature, 12.44b shows the SiO_2 film refractive index from SE measurements, and 12.44c shows the density obtained from n measurements, all plotted as a function of oxidation temperature. The same density change as a function of oxidation temperature was found using infrared (IR) spectroscopy techniques (18). The IR spectra, as shown in Fig. 12.45 for SiO_2, prepared by thermal oxidation at three oxidation temperatures shows a shift toward lower wavenumber, for the 1075 cm^{-1} band (note that wavenumber is 1/wavelength λ and therefore the spatial analog of frequency v). This band is associated with the Si–O–Si bond angle, Θ, which is the angle between adjacent SiO_4 tetrahedra and is a measure of the Si–Si distance that relates directly to the SiO_2 density. The lower the wavenumber the smaller is Θ, and hence the smaller is the Si–Si distance and the higher is the film density. From the IR, a 2 to 3% increase in density is obtained in substantial agreement with earlier measurements.

It is also interesting to note the Si substrate orientation dependence of the interfacial stress displayed in Fig. 12.44a. In Chapters 1 and 10 the Si orientation

12.3 OTHER TECHNIQUES FOR INTERFACE STUDIES

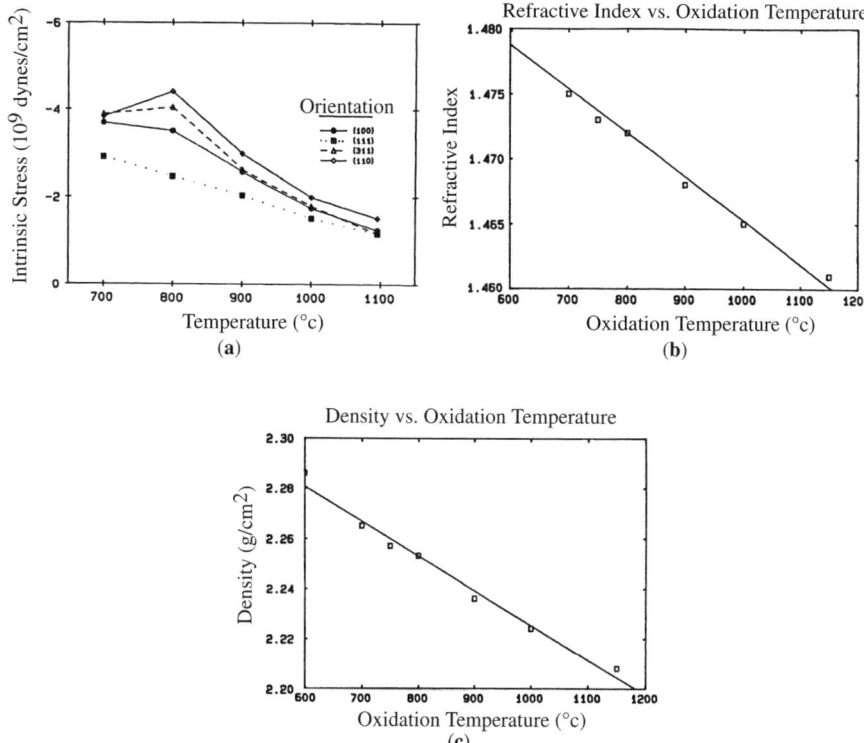

FIGURE 12.44 (a) Intrinsic film stress for several Si orientations as a function of oxidation temperature; (b) and (c) SiO$_2$ refractive index and density as a function of temperature respectively. [Adapted from Kobeda and Irene (16), Figure 1 and Irene (17), Figures 1 and 2.]

dependence of Si oxidation rate was discussed, and it was mentioned that in the initial regime the oxidation rate follows Si atom density (see Fig. 1.6) for the various orientations, but after a certain thickness the order changes. It is thought that this order change to where the Si(111) oxidizes fastest after 20 nm or so is due to the reduced stress at the interface that gives rise to enhanced transport of oxidant to the Si surface for the oxidation reaction.

In situ stress measurements during the oxidation of Si can be used to verify the values for both the intrinsic stress that occurs as a result of the film formation and commensurate molar volume change across the growth interface and thermal stress that develops upon cooling from the oxidation temperature to room temperature. There are many techniques that have been used to measure film stress and *in situ* film stress. In one study of the oxidation of Si (19) a laser beam reflection technique was used. Rather than using a single optical beam to measure the deflection of a film-covered substrate that can be subject to many errors, a parallel laser beam reflection apparatus was used that can measure wafer curvature resulting from the thermal oxidation of Si. Figure 12.46a shows the essential elements of this technique.

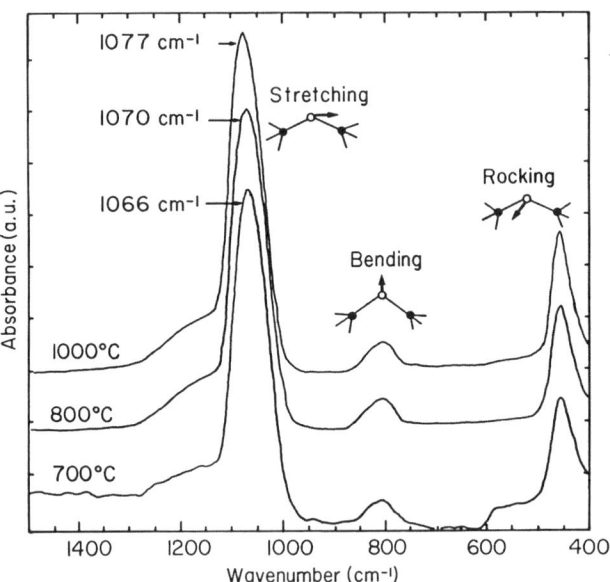

FIGURE 12.45 IR absorption for three SiO_2 films grown at different temperatures on Si. The molecular features that give rise to specific adsorptions are also shown. [Adapted from Lucovsky et al. (18), Figure 2.]

A plate-type beam splitter (BS_1) splits light from an HeNe laser (632.8 nm) into two separate beams made parallel by aligning the mirrors (M_1 and M_2) and the reflecting prism. The parallel beams are incident upon the sample, which is mounted in a nearly free-standing way on a translation stage. Upon reflection from the sample the beams are directed to another mirror (M_3) several meters away with a second beam splitter (BS_2) to increase the reflected path length, thereby providing leverage to any beam deviations caused by substrate curvature. The radius of curvature of the sample is calculated from the following equation:

$$R = \frac{2Lx}{\delta} \quad (12.25)$$

where L is the reflected path length, x the initial beam separation, and δ the deviation from x. Any initial sample curvature is subtracted by remeasuring the bare substrate after removing the film. The apparatus is calibrated with the use of precision ground mirrors of known radius of curvature typically $R = 15$ to 50 m. This range is a result of 100 to 1000-nm-thick SiO_2 films grown on 200-μm-thick Si slices and yielded errors less than 5% when compared with the calibration standards. In addition, the parallelism of the beams was checked periodically by mounting an optical flat in place of the sample on the translation stage. For substrate thicknesses ≤ 200 μm, stresses on the order of 10^9 dynes/cm^2 can easily be measured.

12.3 OTHER TECHNIQUES FOR INTERFACE STUDIES

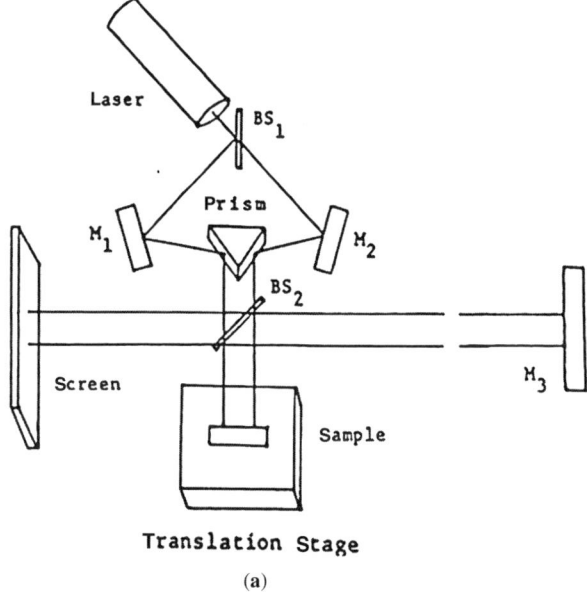

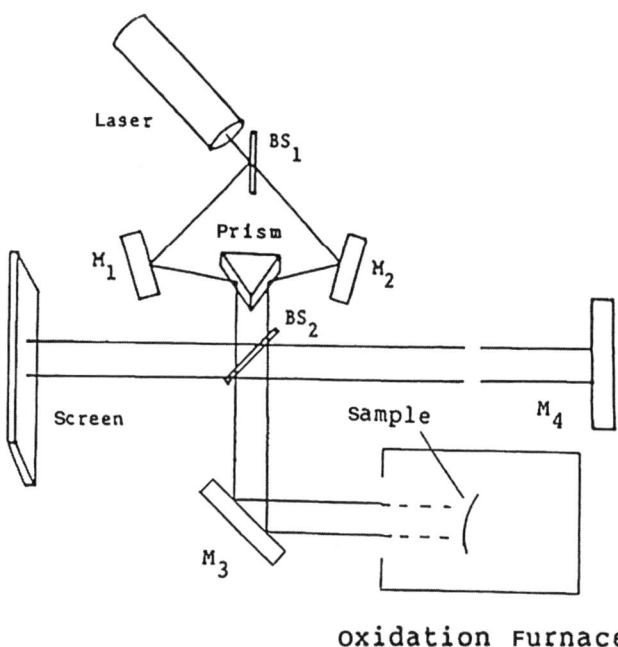

FIGURE 12.46 Reflected dual-laser beam technique to measure Si substrate strain as a result of a thin film (**a**) *Ex situ* apparatus and (**b**) used for *in situ* analyses while Si is being oxidized. [Adapted from Kobeda and Irene (19), Figure 1.]

Results show a higher compressive intrinsic film stress near the Si–SiO$_2$ interface and a lower stress in the bulk of the SiO$_2$ film. Thermal stress is also measured by monitoring the SiO$_2$ film-covered Si substrate curvature changes during temperature excursions. The results of this study are in general agreement with previous *ex situ* stress measurements and may indicate that film stress, being highest at the interface, influences the interface reaction in Si oxidation. The results shown below indicate initially high intrinsic film stress that undergoes relaxation as oxidation proceeds.

All film stress measurements were carried out using commercially available P-type, 2 Ω-cm, Si(100) wafers, which were 5 cm in diameter and 75 μm thick but which were cleaved to form 3 × 4 cm samples. Typical Si wafers are millimeters thick and so thin wafers were used so as to maximize bending from the film stress. Prior to cleaving, the original wafers received a preoxidation at 1000°C to prevent any contamination from Si dust as a result of scribing. After removal of this protective oxide in concentrated HF, the samples were cleaned using a modified RCA method prior to final experimental oxidations.

Stress measurements at the oxidation temperature were performed using a modification of the laser reflection technique mentioned above and shown in Fig. 12.46a. The modified apparatus is depicted in Fig. 12.46b and includes an oxidation furnace. The essential elements of the apparatus include a plate-type beam splitter (BS$_1$) that splits light from an HeNe laser (6328 Å) into two orthogonal beams that are made parallel by aligning the mirrors (M$_1$ and M$_2$) and the reflecting prism. The beams are transmitted through a second beam splitter (BS$_2$) and reflected either from a sample on a translation stage for *ex situ* measurements in Fig. 12.46a or for *in situ* measurements into an oxidation furnace using the mirror M$_3$ in Fig. 12.46b. For *ex situ* and *in situ* measurements, respectively, after reflecting from the sample, the beams return to either the beam splitter BS$_2$ or to mirror M$_3$ and are reflected to a larger flat mirror M$_3$ or M$_4$ by the beam splitter BS$_2$ to increase the reflection path length and hence the measurement sensitivity. The beams are then returned to a measurement screen to determine the separation distance, which can be used to calculate the wafer curvature. If oxidation results in warpage of the Si wafer, then a deviation in the initial separation of the beams is observed. Calibration procedures were described above.

The oxidation furnace, which was placed in the path of the parallel light beams in the stress measurement apparatus, consisted of a resistively heated fused silica tube. A constant flow of gaseous N$_2$ taken from a liquid N$_2$ source was maintained inside the furnace except during oxidation where ultra-high purity O$_2$ (less than 1 ppm H$_2$O and 0.5 ppm hydrocarbons) was used. All furnace components were constructed of fused silica. All temperatures are reported with a variation of less than 3°C, which is attributed to the small size of the hot zone. The water content inside the furnace was monitored constantly in the effluent gas using a capacitance hygrometer. The furnace end cap had an optically flat strain-annealed Pyrex window to allow entrance and exit of the laser beams without deviation or significant absorption loss.

For intrinsic stress measurements at the oxidation temperature, samples were preoxidized at 1100°C to about 4000 Å to provide a thick backside oxide, and the front side oxide was removed using an HF-based etchant. Since under the oxidation

12.3 OTHER TECHNIQUES FOR INTERFACE STUDIES

conditions used in the study, the rate of oxidation for a thick film on the back side of the wafer is negligible compared to the bare front side and the film is in a relatively stress-free state when grown at high temperatures, any curvature changes that occur during oxidation could be directly attributed to film growth on the front surface. These samples with grown oxide on one side were also cleaned as usual except that the HF dip was omitted.

The furnace was kept at room temperature in flowing N_2 during alignment to minimize oxide growth resulting from back diffusion of O_2 and H_2O from the laboratory ambient into the furnace while the end cap was off. After alignment of the sample, the furnace was ramped to the oxidation temperature and allowed to stabilize to measure the initial curvature point. Then the ambient was switched to O_2 and the curvature was monitored with time. Since the film thickness could not be measured directly during oxidation with this system, it was estimated from previous experiments. Thickness estimations were verified on the *in situ* measured samples using ellipsometry after the oxidation was complete, yielding average thickness errors less than 10% that translates to about 10% error in stress.

The thermal expansion stress was obtained *in situ* by measuring the wafer curvature during thermal treatment in the presence of an N_2 ambient. With no oxidation taking place, the change in curvature with temperature is directly related to the thermal expansion stress component. The *in situ* experiments allow for the measurement of intrinsic stress separate from the thermal stress. The curvature changes in the samples that occur during oxidation were monitored at the oxidation temperature where $\Delta T = 0$ and hence $\sigma_{th} = 0$, as seen in Equation 12.21 for σ_{th}. The thermal stress develops as the sample cools down from the oxidation temperature, resulting in compression in the oxide since the coefficient of thermal expansion α_{Si} for Si is greater than α_{SiO_2} for SiO_2. When total stress is measured at room temperature, the intrinsic stress is obtained by subtracting the thermal stress from the total film stress using Equation 12.20. The value for σ_f is obtained using Equation 12.24.

Figure 12.47 shows a comparison of *in situ* measurements made during oxidation to measurements obtained *ex situ* sample by sample in an earlier study (20) at room temperature. The *in situ* data is the average of three runs with an average error of about 0.6×10^9 dynes/cm^2. With about the same error for the previous data, the *in situ* and *ex situ* experiments yield the same compressive σ_i results. It is noticeable that all the data show a greater change in intrinsic stress with thickness in the early stages of oxidation. The systematic difference in the magnitude of the *in situ* and *ex situ* results was attributed to the fact that the *in situ* oxidations contained about 15 ppm H_2O, while the *ex situ* experiments were done in a conventional double-walled furnace having less than 5 ppm H_2O. H_2O is known to cause stress relaxation effects in SiO_2. In addition the thick backside SiO_2 film may contain adsorbed H_2O, which is released during oxidation and may affect the total curvature. Figure 12.48 shows the measured thermal stress as a function of oxidation temperature, where the slope yields the product $E \Delta\alpha/(1-\nu)$. A least-squares fit to the data in Fig. 12.48 resulted in a slope whose value was less than 10% different from the literature values of E and ν. The slight curvature in the data is possibly the result of the presence of about 25 ppm of H_2O inside the furnace during the experiment, which may result

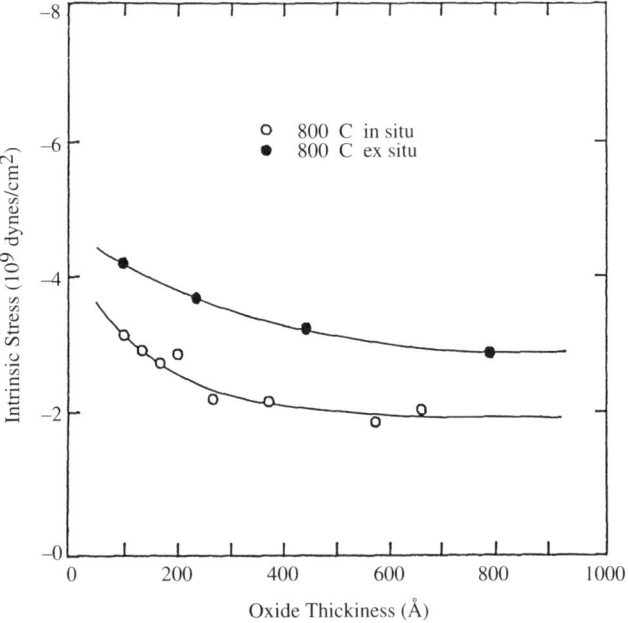

FIGURE 12.47 Comparison of *in situ* and *ex situ* intrinsic stress measurements for SiO_2 films on Si substrates. [Adapted from Kobeda and Irene (19), Figure 2.]

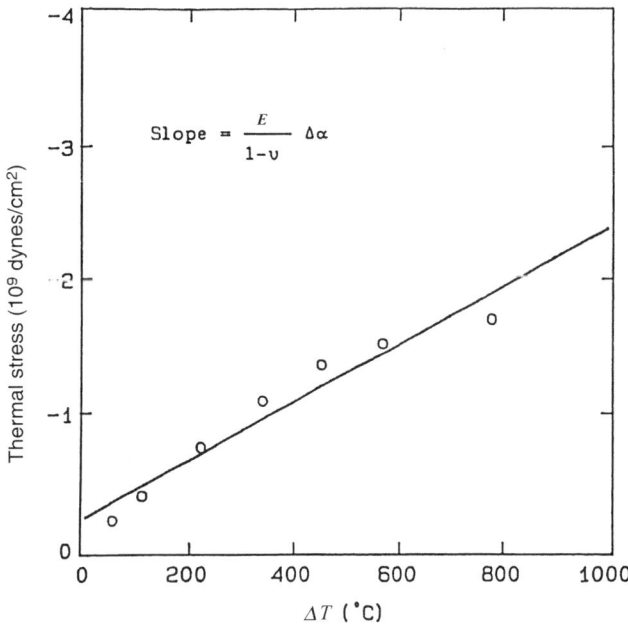

FIGURE 12.48 Thermal stress versus ΔT. [Adapted from Kobeda and Irene (19), Figure 3.]

12.3 OTHER TECHNIQUES FOR INTERFACE STUDIES

in a small amount of oxide growth at the higher temperatures. This growth will effectively reduce the curvature change.

It is seen from this study that σ_i is larger nearer the Si–SiO$_2$ interface where SiO$_2$ is produced via the interface reaction between Si and oxidant. The origin of σ_i has been attributed to the partial confinement of the molar volume change accompanying the transformation according to the viscous flow mechanism discussed above. Based on the fact that the maximum σ_i found from extrapolation to zero SiO$_2$ thickness is substantially less that the theoretical value of more than 10^{11} dynes/cm^2, an initially fast stress relaxation was suggested for the SiO$_2$ as it is formed, which is followed by a much slower relaxation that is characteristic of the bulk SiO$_2$ viscosity. Other studies have suggested the existence of a time-dependent viscosity where as the oxide grows the viscosity increases from an initial low value for the as-forming oxide, which is not yet relaxed into the high viscosity form characteristic of the nearly relaxed network structure. Further relaxation occurs for the already formed SiO$_2$ as this oxide continuously exposed to the oxidation temperature and pushed out from the interface by newly forming SiO$_2$, hence yielding the profile in Fig. 12.47. Corroboration of the stress profile in Fig. 12.47 was reported (21,22) from IR spectroscopy measurements of SiO$_2$, as discussed above using Fig. 12.45. Essentially, it was reported that near the Si–SiO$_2$ interface the angle between adjacent tetrahedra became gradually smaller, indicating a greater packing of SiO$_4$ tetrahedra near the Si–SiO$_2$ interface and with about the same thickness scale as in Fig. 11.47 for the 800°C oxidation. Also as seen in Fig. 12.47 the extent of the higher stress regime for σ_i is several hundred angstroms. This is similar to the extent of the initially rapid oxidation regime, L_i, which is observed for the thermal oxidation of Si and which was found not to conform to the kinetics predicted from the linear parabolic model for Si oxidation as discussed in Chapter 10. The qualitative conformity of the rapid oxidation regime with the high stress regime supports the recently proposed stress-enhanced oxidation model that was proposed as an attempt to explain the effects of σ_i on the interface reaction during thermal oxidation. This model assumes proportionality between the linear oxidation rate constant, k_1, as obtained from the linear parabolic model for oxidation, and stress divided by viscosity as

$$k_1 \propto \frac{\sigma_i}{\eta} \tag{12.26}$$

where η is the oxide viscosity. This model is derived from the notion that the relaxation of the forming SiO$_2$ in the direction normal to the Si surface, the free direction, which is given by the ratio σ_i/η in the Maxwell model for a solid, also influences the oxidation rate at the interface. This form is reasoned because the stress relaxation process parallels the growth of oxide in the normal direction to the Si surface, and relaxation away from the interface facilitates the formation of new oxide layers. The necessary volume for the new layers of oxide is created by the flow of the already formed SiO$_2$ away from the interface. The higher the stress and the lower the SiO$_2$ viscosity, the faster is the flow of formed oxide from the interfacial region, and hence the more volume is available to the newly formed oxide. The above-cited IR studies also show that the bond angle profiles match the stress profiles at all temperatures studied, yielding sharper profiles for higher oxidation temperatures where relaxation takes place more readily. Thus, the extent and the evolution with temperature of the stress profiles are corroborated by and consistent with several recent studies.

12.4 OTHER MICROELECTRONICS INTERFACES

12.4.1 Introduction

With the pervasive MOSFET theme used throughout this text, the various fundamental surface and interface issues were introduced and discussed in Part I with more direct MOSFET applications in Part II. Significant parts of Chapters 5, 7, 11, and 12 deal specifically with the dielectric–semiconductor interface that is crucial to the MOSFET. The justification for this coverage lies with the fact that the MOSFET devices depicted in Figs. 4.16 and 7.5 with associated discussions show that this device operates based on the ability to alter carrier type and concentrations at the dielectric (SiO_2)–semiconductor (Si) interface. Furthermore and most relevant to a text on surfaces and interfaces is the fact that the MOSFET device operation is intimately entwined with the surface and interface properties (structure, thermodynamic properties, roughness, interface electronic states, interface charges, etc.). The interface in MOSFETs that most vividly illustrates the effect of surface and interface condition on resultant properties is the dielectric–semiconductor interface. Consequently, this interface has received almost exclusive attention throughout the text. However, as seen in the MOSFET figures cited above, there are other interfaces that comprise the MOSFET device, namely the metal–semiconductor interface and the metal–dielectric interface, both of which enable electrical contact to the device. Also in more complex MOSFET configurations and in bipolar and other transistors the semiconductor–semiconductor interface plays an essential role in device fabrication and operation.

In the following sections some of the interface and surface issues associated with the metal–semiconductor and semiconductor–semiconductor interfaces will be presented since these interfaces are the next most important interfaces in MOSFET technology. Other related technologies and interface issues will also be mentioned as appropriate. The intent is to further elucidate the surface and interface topics already covered and not to dwell on the associated and often complex technologies that would not be afforded adequate coverage in the brief discussions herein.

12.4.2 Electronic Characteristics of Junctions

To begin to understand the issues associated with metal–semiconductor and semiconductor–semiconductor interfaces, it is useful to commence with the parallel band picture as shown in Fig. 12.49 for two metals that are first apart (Fig. 12.49a) and then form a junction (Fig. 12.49b). In Fig. 12.49a each metal displays a characteristic Fermi level E_F that is the approximate position of the highest energy electrons

12.4 OTHER MICROELECTRONICS INTERFACES

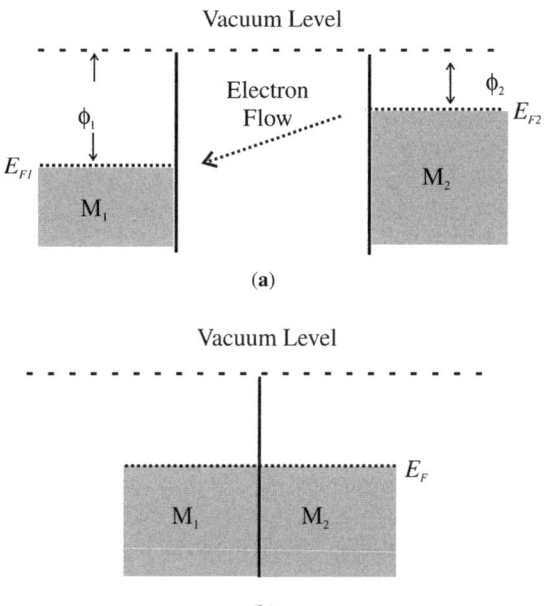

FIGURE 12.49 (a) Two separated different metals with Fermi levels (E_F) and work functions (ϕ) indicated; (b) the same metals as in (a) but after joining and at equilibrium.

(neglecting temperature and the effective mass ratio's) in that material. On the left is one metal M_1 and on the right another metal M_2 and each with a different Fermi level E_F. As was introduced in Chapter 5, the minimum energy required to emit an electron from E_F to an infinite distance from the metal (essentially to emit an electron from the metal) is called the work function ϕ_M. The zero of energy is set at infinite distance and is oftentimes labeled as the vacuum level. Thus, the electrons bound by a material are at negative energies with respect to the vacuum level. When these two depicted metals are joined, electrons from the metal with the higher E_F, metal M_2, flow to M_1 that has a lower E_F as indicated by the arrow. Electrons endeavor to occupy the lowest allowed vacant states. Thus, metals M_2 and M_1 that were neutral before joining now become charged. M_1, which gains electrons, becomes negative while M_2, which loses electrons, becomes positive. Likewise E_{F1} for M_1 rises while E_{F2} for M_2 drops. When all the electrons flow from higher to lower energy, equilibrium will result where the E_F's will level at $E_{F1} = E_{F2} = E_F$ as is shown in Fig. 12.49b. The difference in potential generated between M_1 and M_2 as a result of the equilibration of the Fermi levels is called the contact potential ϕ_{1-2} and is given as

$$\phi_{1-2} = \phi_1 - \phi_2 \qquad (12.27)$$

Because the contact potential results in no difference in Fermi levels, it is difficult to measure experimentally as a potential difference in an external circuit. Also, the

492   THE Si–SiO$_2$ INTERFACE AND OTHER MOSFET INTERFACES

charge on the metals cannot create an electric field in the metals. In Chapter 5 the Kelvin method was discussed where the potential difference in Equation 12.27 above can be measured and from which the work function could be obtained using a known reference electrode. The simple idea of the equalization of the Fermi levels pervades the formation of all ideal or nearly ideal junctions.

12.4.3 Ideal Metal–Semiconductor Junctions

While the formation of other ideal junctions follows the treatment above for a metal–metal junction, an important difference between metal–metal and metal–semiconductor junctions (and metal–insulator, insulator–semiconductor, and semiconductor–semiconductor junctions as well) is that the semiconductor (and insulator) unlike a metal can support an internal electric field as a result of the presence of excess charge as was discussed in Chapter 7. The region in the semiconductor where the electric field is generated is called the space charge region, and the evolution of the space charge region can be understood using Fig. 12.50. Figure 12.50a as for the metal case above shows a separated parallel band picture of a metal on the left and a N-type semiconductor on the right. The semiconductor has a work function ϕ_S defined as the distance in energy from E_F to the vacuum

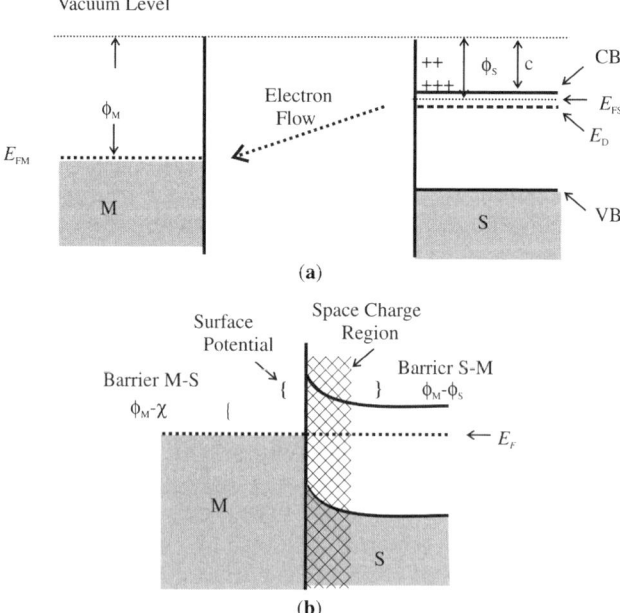

FIGURE 12.50 Separated metal (M) and N-type semiconductor (S) with Fermi levels (E_F), work functions (ϕ), electron affinity (χ), band edges (E_{vb} and E_{cb}), and donor level (E_D) with $E_{FM} < E_{FS}$; (b) M and S after contact and equilibration with band bending and the development of a space charge region (cross-hatched) and a surface potential.

12.4 OTHER MICROELECTRONICS INTERFACES

level as for a metal. In this case for an N-type semiconductor, E_F is between the donor level and the conduction band (labeled as E_{FS}). The distance between the conduction band edge and the vacuum level is called the electron affinity χ. With the given E_F's (E_{FM} is the original Fermi level for the metal) the arrow in Fig. 12.50a indicates the direction for electron flow from the CB of the semiconductor to the metal in order to reach equilibrium. The $+$ charges in the semiconductor indicate the establishment of the space charge layer in the semiconductor. As electrons flow from the semiconductor to the metal, the E_F's will tend to equilibrate as discussed above for the metal–metal junction. Thus, as before, the E_F for the semiconductor drops and the metal E_F rises until equilibrium is achieved, and the result at equilibrium is shown in Fig. 12.50b. The electrons that flow to the metal come from the ionized donors in the semiconductor that are left behind primarily in the region of the semiconductor near the junction. This cross-hatched region in Fig. 12.50b becomes depleted of the majority carrier electrons in the N-type semiconductor, and consequently this region assumes a positive charge. This residual positive charge in the region depleted of majority carriers, the depletion region, is so indicated on the resulting energy band diagram as an upward hill toward the junction. This indicates that it becomes more difficult for more electrons to leave the semiconductor, and its magnitude is equivalent to the semiconductor surface potential ψ_s reached at equilibrium. Also, this surface potential provides a barrier to electron flow from the semiconductor to the metal. In the opposite direction a somewhat larger barrier for electron flow is seen. The electron current flow is proportional to e^{-E_B} where E_B is the barrier energy to electron flow to an allowed state. Therefore, we can write for current flow where current is the change in charge per time:

$$I = Ae^{-E_B/kT} \qquad (12.28)$$

It should be noted that the barriers arise from the position of the initial Fermi levels and the electron band structure and that the application of an external potential (called a bias) can alter the barriers and enhance or retard current flow. The preferential current flow in one direction is called rectification.

Figure 12.50 depicts the case of a metal–N-type semiconductor junction where $\phi_M > \phi_S$. There are three other cases to consider: $\phi_M < \phi_S$ with an N-type semiconductor and $\phi_M > \phi_S$ and $\phi_M < \phi_S$ for P-type semiconductors.

Figure 12.51 displays the case for $\phi_M < \phi_S$ with an N-type semiconductor. It is seen that electrons achieve equilibrium by flowing from M to S thereby leaving a space charge region in the semiconductor that is electron rich and thus negatively charged. The electrons in this region can easily migrate elsewhere to reduce the space charge and thus the bands in the semiconductor bend downward indicating a "down hill" migration toward the metal. In the other direction, namely M to S, there is a barrier whose height depends on the specific values for the ϕ's but is nevertheless uphill. So in this case electron current can flow more readily in one direction.

Figures 12.52 and 12.53 display the cases of $\phi_M > \phi_S$ and $\phi_M < \phi_S$, respectively, for P-type semiconductors. Figure 12.52 for $\phi_M > \phi_S$ shows the electron energy bands in the space charge region bending upward indicating a barrier for electron

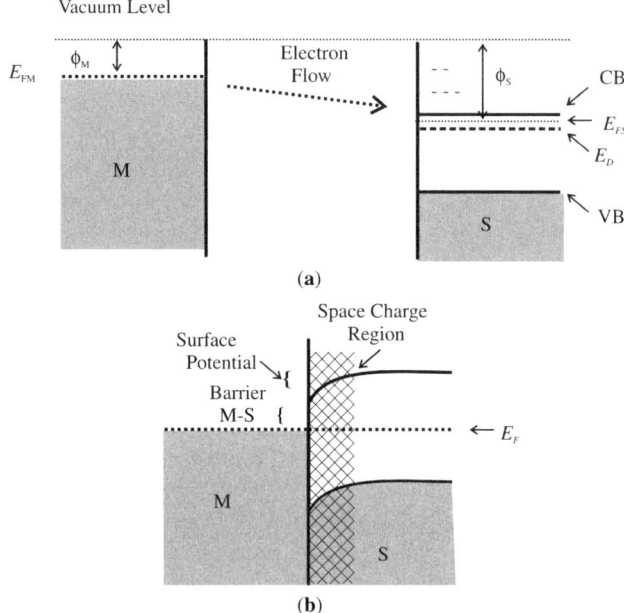

FIGURE 12.51 (a) Separated metal (M) and N-type semiconductor (S) with Fermi levels (E_F), work functions (ϕ), band edges (E_{vb} and E_{cb}), and donor level (E_D) with $E_{FM} > E_{FS}$; (b) M and S after contact and equilibration with band bending and the development of a space charge region (cross-hatched) and a surface potential.

flow from S to M. However, the majority carriers in the P-type semiconductor are holes. Thus, the upward bent bands for electrons are actually down hill for holes, and therefore holes can flow readily from S to M, but electrons from the metal are impeded by the barrier from M to S. In Fig. 12.53 for $\phi_M < \phi_S$, the electron energy bands in the space charge region of the semiconductor are bending downward, and hence this is actually a barrier for the majority carrier holes in the semiconductor. Electrons from the metal can easily pass into the semiconductor.

Those junctions where majority carriers can pass easily are called *ohmic* contacts where Ohm's law is obeyed. Those junctions where majority carriers must overcome exponential barriers are called *Schottky* contacts. For contact to MOSFET devices it is usually desirable to have ohmic contacts. Schottky contacts are used in devices where a gate dielectric that can electronically passivate the interface electronic states is not available such as III to V semiconductors, and the gate metal is directly deposited upon the semiconductor in the gate region. The device appears similar to the MOSFETs depicted in Fig. 4.16 and 7.5 but without the gate dielectric layer between the gate metal and the semiconductor substrate. The Schottky barrier acts like a dielectric and enables an operating electric field to be developed at the channel; otherwise the device operates similar to a MOSFET. This kind of Schottky barrier device is a metal–semiconductor field–effect transistor with the acronym MESFET.

12.4 OTHER MICROELECTRONICS INTERFACES

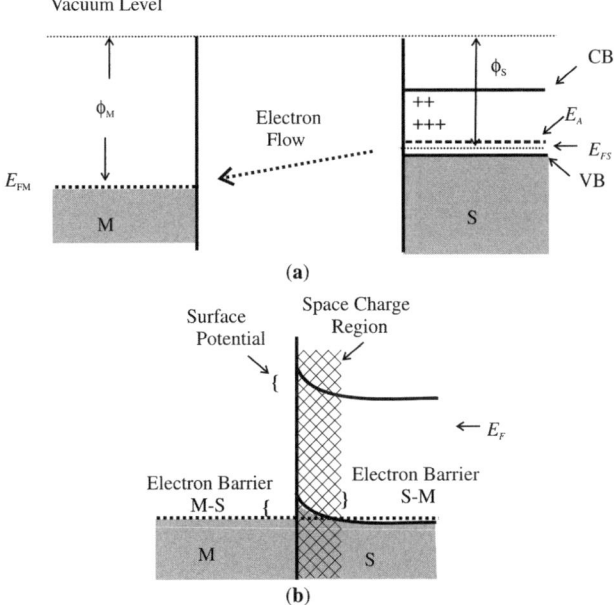

FIGURE 12.52 (a) Separated metal (M) and P-type semiconductor (S) with Fermi levels (E_F), work functions (ϕ), band edges (E_{vb} and E_{cb}), and acceptor level (E_A) with $E_{FM} < E_{FS}$; (b) M and S after contact and equilibration with band bending and the development of a space charge region (cross-hatched) and a surface potential.

12.4.4 Ideal Semiconductor–Semiconductor PN Junctions

A PN junction is made by contacting a P-type semiconductor with an N-type semiconductor. As before, with metals electrons will flow so as to achieve the lowest energy configuration consistent with the appropriate physics. Thus, if we join an N- and P-doped semiconductor, say Si, then the donor level will be higher than the acceptor levels, as shown in Fig. 12.54a. Electrons flowing from N-type to P-type would reduce the majority carriers on the N side, but because the electrons would fall into holes on the P-type side, the majority carriers on the P side are also reduced. Thus, the region in between the P and N semiconductors, the junction region, is devoid or nearly devoid of carriers. This region is called the depletion width or region, as shown in Fig. 12.54b. Electrons from the N-type side have a large barrier to overcome to traverse the barrier, as do holes from the P-type side. Thus, it would be difficult for current flow in either direction. The PN junction appears in many devices. However, to render this junction useful, typically an external potential or bias is applied. Figure 12.55a shows the same PN junction as in Fig. 12.54b but with a forward bias, that is, a + potential on the P-type and a − potential on the N-type. The effect of the bias is to push the majority carriers together, reduce the depletion width, and reduce the barriers. The opposite occurs with reverse bias, as shown in

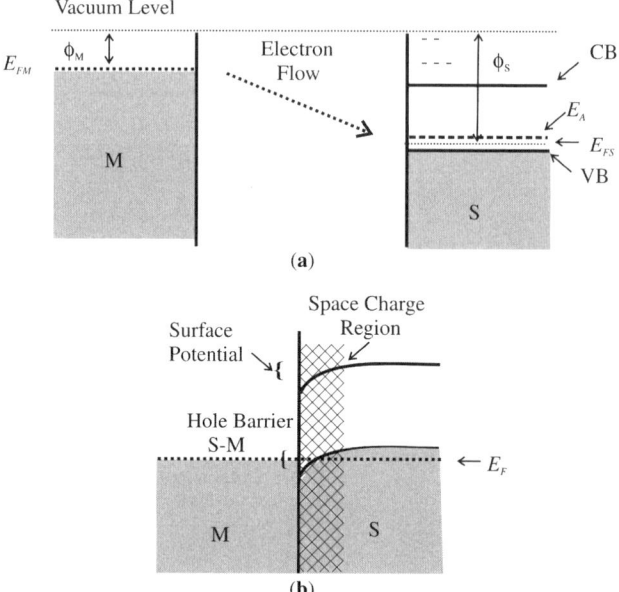

FIGURE 12.53 (a) Separated metal (M) and P-type semiconductor (S) with Fermi levels (E_F), work functions (ϕ), band edges (E_{vb} and E_{cb}), and acceptor level (E_A) with $E_{FM} > E_{FS}$; (b) M and S after contact and equilibration with band bending and the development of a space charge region (cross-hatched) and a surface potential.

Fig. 12.55b. The bias gives rise to rectifier devices and when combined with another PN junction results in a transistor.

12.4.5 Nonideal Junctions

The discussions above for the formation of ideal junctions yielded a procedure that enables a determination of junction barriers and other characteristics. This straightforward approach is based on the commonly accepted electron energy band theory. However, while the discussions are both logical and based on scientific fact, it is found that more often than not the experimental barrier heights are different than those found using the theory outlined above. In fact, in many instances experimental barriers are found to be independent of the metal used and sometimes dependent on the exact experimental conditions for the junction formation. The observed behavior for real junctions begs the following question: Why is there a difference between the straightforward theory presented above and reality?

One answer to this question lies with the fact that electron band theory applicable to bulk materials was used rather than surface-modified electron band theory, as was done in Chapter 5 where the changing surface potential gave rise to intrinsic surface states. Furthermore it was implicitly assumed that perfect surfaces were used to form perfect interfaces. The interfaces were assumed to be sharp and included only the two

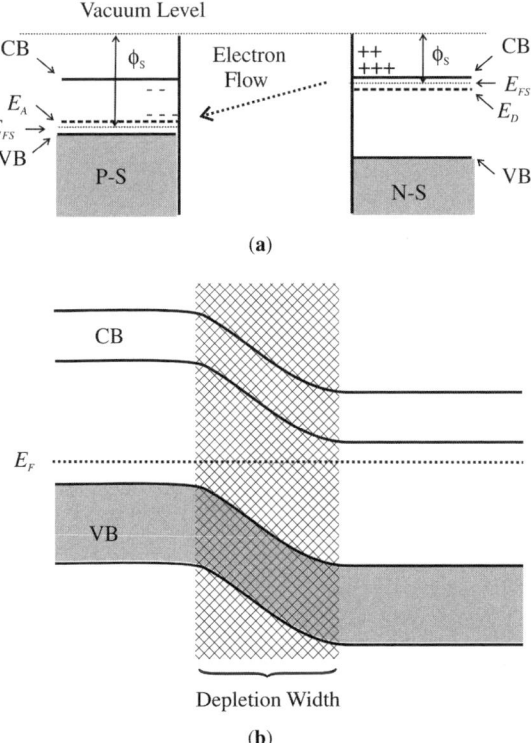

FIGURE 12.54 (a) Separated P- and N-type semiconductor with Fermi levels (E_F), work functions (ϕ), band edges (E_{vb} and E_{cb}), and doping levels (E_A and E_D); (b) P-S and N-S after contact and equilibration with the development of a space charge region (cross-hatched) that is depleted of carriers.

materials in question, that is, no impurities or other substances of any kind. Thus there were no extrinsic electronic interfaces states. The only electronic states that were considered were those that arose from the bulk potentials for pure materials. From Chapter 5 it was clear that surface electronic states must also be considered and that these surface states arise from the termination of the respective materials (intrinsic states) and from defects and impurities (extrinsic states). In addition many examples in Chapters 6 and 11 indicated that when materials form a junction intermixing and/or interaction is expected that yields compounds and mixtures that are different from either of the original junction constituents. Therefore, in general, interface electronic states are anticipated to occur for the formation of junctions between dissimilar materials and even similar and identical materials. In this latter case for the formation of an Si PN junction, impurities or crystallographic defects can yield interface states, although if the junction is formed carefully using well-known epitaxial processes and proper cleanliness, this situation is the least likely to be dominated by interface electronic states. As alluded to in Chapter 5 and discussed more fully in

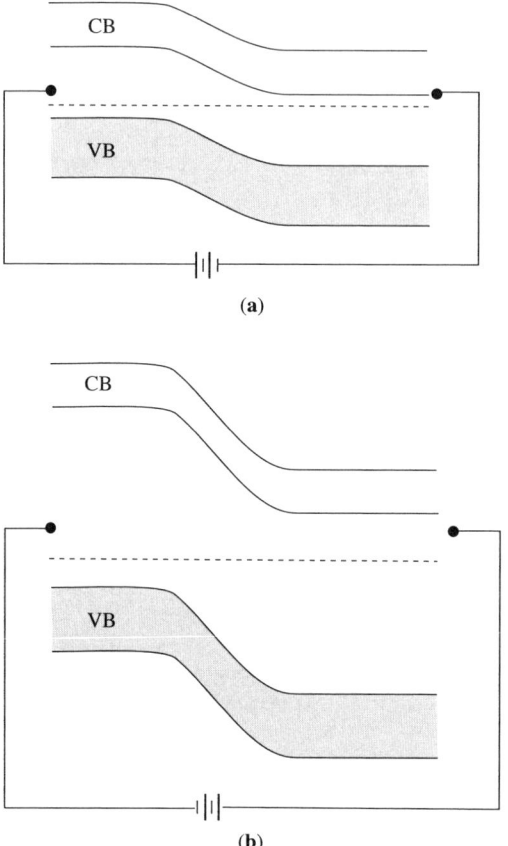

FIGURE 12.55 (a) Joined P- and N-type semiconductors with forward bias; (b) joined P- and N-type semiconductors with reverse bias.

Chapter 7 (Section 7.4), the appearance of a large number of interface states from whatever origin can pin the Fermi level of the system, whereupon the potential at the interface and hence the barrier is determined by the interface states and to a much lesser extent by the materials. The nature and magnitude of the interface states is dependent on the materials and on processing conditions. However, at this time a number of theories addressing these issues have had limited success, and there is no single theory that can predict barrier heights for all the important electronic materials.

Despite the fact that in many if not most of the important electronic material cases the barrier heights for the relevant junctions cannot be predicted from simple theory, great strides have been made in formulating recipes and procedures for forming reproducible and stable junctions based mostly on empirical observations.

In a recent poignant application to GaAs technology, Lodha et al. (23) reported the unpinning of the GaAs Fermi level with the use of a 3.5-nm low temperature deposited GaAs capping layer deposited upon N-type GaAs(100) substrates, and the

12.4 OTHER MICROELECTRONICS INTERFACES

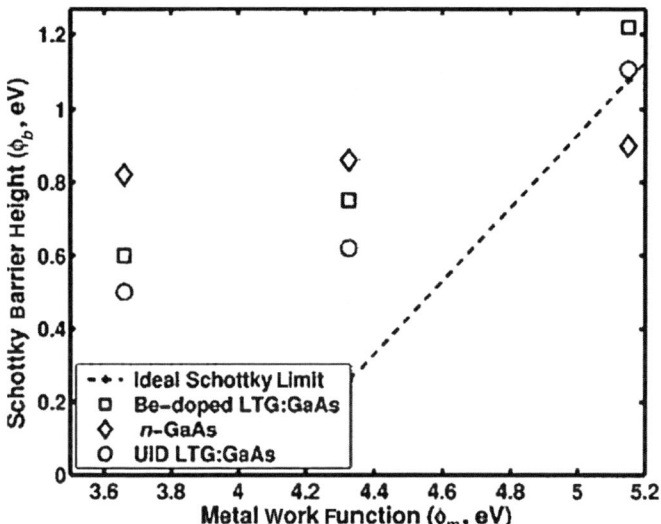

FIGURE 12.56 Barrier heights for N-type GaAs (◇) and N-type GaAs with Be-doped LTG GsAs (□) layer and unintentionally doped (UID) LTG layer (○). Also shown is the ideal Schtokky junction formation (dashed line). Three different metals were used: Mg ($\phi_M = 3.66$ eV), Ti ($\phi_M = 4.33$ eV) and Ni ($\phi_M = 5.15$ eV). [Adapted from Lodha et al. (23), Figure 3.]

capping layer was then used to form the metal to GaAs junction. As mentioned above, this Schottky junction is used for MESFET device applications. Typically, metal contacts to GaAs have barrier heights of about 0.8 eV regardless of the metal used. As discussed above, this indicates that the Fermi level of the GaAs is pinned and does not measurably vary with contact with metals with different ϕ_M. It was found that a GaAs capping layer deposited using molecular beam epitaxy (MBE, see Chapter 10) at low temperatures of about 250°C oxidized more slowly in air. For III to V semiconductors the oxidation causes the buildup of extrinsic surface states that can pin the Fermi level. It was suggested that the slow oxidation of this low temperature grown (LTG) GaAs was due to the As excess.

Figure 12.56 shows results of the measured barrier heights ϕ_b for three different samples and using three metals to form contacts for each sample. First, the control sample was simply the starting N-type GaAs(100) substrate. The results in Fig. 12.56 for this sample shows almost no variation of ϕ_b with the metal used for the junction and hence that the GaAs Fermi level is pinned. Indeed a value of nearly 10^{14} interface electronic states/eV·cm² was found for this sample. With the LTG capping layer followed by junction formation with the three metals, the interface state levels dropped by slightly more than an order of magnitude, and as seen in the data in Fig. 12.56 a distinct variation of ϕ_b with the metal is evident, indicating that the Fermi level is at least somewhat unpinned.

The above-cited publication (23) displays a procedure to prepare Schottky contacts with controllable barriers that depend on the metal as was indicated in the simple theory above. On the other hand, ohmic contacts are also desirable to source and drain regions of GaAs-based MESFETs. These contacts have been prepared by depositing the desired metal and then annealing the sample to form a highly conducting alloy. For example, in one study (24) Au–Ge–Ni and Au–Ge–Pt films were deposited onto N-type GaAs and heat treated. Ohmic contacts were found. Many other examples exist in the literature. The problem has been that while low resistivity contacts could be formed the contacts were rough and not suitable for optical applications. To remedy this problem, procedures have been developed to deposit ohmic contacts that do not require annealing. In one study (25) Ti/Au alloys were deposited upon both N- and P-type GaAs using MBE and produced low resistivity ohmic contacts that were smooth and specular. It should be mentioned that for these ohmic contacts interface electronic states are not relevant since there is no focus on controlling barriers as was for the Schottky barrier contacts discussed above.

The formation and application of semiconductor–semiconductor junctions has received considerable attention in virtually all aspects of modern device science and engineering. The materials that are used for these junctions include Si and Si–Ge alloys within Si technology and III–V and II–VI materials as well and both crystalline and amorphous morphologies. The materials issues and device issues are vast. Since the main theme for applications is the MOSFET and the main material is Si or Si based, the focus here for semiconductor–semiconductor junctions is Si-based and MOSFET-related applications. For MOSFET-based applications of semiconductor–semiconductor junctions, it is usually desired to form an epitaxial film upon a Si substrate. For the case of Si–Si junctions the purpose is usually to achieve different doping type in the film and substrate. The materials issues are minimal in achieving high quality epitaxy since the materials are the same for this homoepitaxy. The usual issue with epitaxy is cleanliness, which means the maintenance of ultra-high vacuum and proper substrate preparation. Heteroepitaxy is rarely desired in Si technology, but one relatively modern and major technology evolves around the preparation of Si–Ge alloy epitaxial films upon an Si(100) substrate. This technology affords significant MOSFET speed enhancements for both P- and N-type Si-based MOSFETs. It should also be noted that Si–Ge epitaxial films on Si have application for bipolar transistors that can be integrated with N- and P-channel MOSFETs (CMOS) to form so-called BiCMOS circuitry, which are used in many consumer applications such as cell phones and GPS systems and compete favorably with III–V materials that have previously been used for these applications.

In a recent relevant application for P-channel MOSFETs that rely on hole mobility, the hole mobility has been dramatically improved using epitaxial Ge–Si alloys (26). In Chapter 7 CMOS devices were mentioned as being comprised of an N-channel and a P-channel MOSFET connected in an inverter configuration, and because they draw little power to operate are the heart of modern computer memories. These devices are inherently asymmetric due to the disparate mobility of holes ($470 \text{ cm}^2/\text{V} \cdot \text{s}$) in the P-channel MOSFET and electrons in the N-channel MOSFET ($1400 \text{ cm}^2/\text{V} \cdot \text{s}$). With the hole mobility in Ge being nearly $5\times$ higher than in Si suggests the use of a Ge layer to serve as the channel in P-channel MOSFETs. Unfortunately, Ge cannot be grown epitaxially on Si due to the more than 4% lattice mismatch. However, the lattice mismatch can be modulated using an $Si_{1-x}-Ge_x$ alloy where x varies from 0 to 1, which grades the lattice mismatch from 0 to about 4%. Without repeating all the details, the channel was fabricated using a high Si alloy near the Si surface with a larger Ge content alloy away from the Si–alloy interface. In this way the hole mobility in CMOS was improved to about $1000\,cm^2/V\cdot s$ at room temperature. This clever processing used an ultra clean vacuum CVD apparatus in which composition and doping control was high. Similar heterostructures have been fabricated for many other materials systems, in particular III–V materials due to the many potential applications. The message here is that unlike the case for real metal–semiconductor junctions, semiconductor–semiconductor junction science and technology is highly developed and predictable and not usually limited by interface electronic states as are the other junctions.

REFERENCES

1. Grunthaner FJ, Grunthaner PJ. Mater Sci Repts 1986;1:65.
2. Yakovlev VA, Irene EA. J Electrochem Soc 1992;139:1450.
3. Yakovlev VA, Liu Q, Irene EA. J Vac Sci Technol A 1992;10:427.
4. Maserjian J. In:Helms CR, Deal BE, editors. The physics and chemistry of SiO_2 and the Si–SiO_2 interface. New York: Plenum; 1988;505.
5. Gundlach KH. Solid-State Electron 1966;9:949.
6. Alferieff ME, Duke CB. Chem Phys 1967;46:938.
7. Zafar S, Liu Q, Irene EA. J Vac Sci Technol A 1995;13(1):47.
8. Zafar S, Conrad KC, Liu Q, Irene EA, Hames G, Kuehn R, Wortman JJ. Appl Phys Lett 1995;67:1031.
9. Sune J, Placencia I, Farras E, Barniol N, Aymerich X. Phys Status Solidi 1988;109:496.
10. Lai L, Irene EA. J Appl Phys 2000;87:1159.
11. Herbert KJ, Zafar S, Irene EA, Kuehn R, McCarthy TE, Dermirlioglu EK. Appl Phys Lett 1996;68:266.
12. Wang Y, Irene EA. J Vac Sci Technol B 2000;18:279.
13. Kobeda E, Irene EA. J Vac Sci Technol B 1986;4:720.
14. Irene EA, Tierney E, Angillelo J. J Electrochem Soc 1982;129:2594.
15. Irene EA, Dong D, Zeto RJ. J Electrochem Soc 1980;127:396.
16. Kobeda E, Irene EA. J Vac Sci Technol B 1987;5:15.
17. Irene EA. Phil Mag 1987;55:131.
18. Lucovsky G, Mantini MJ, Srivastava JK, Irene EA. J Vac Sci Technol B 1987;5:530.
19. Kobeda E, Irene EA. J Vac Sci Technol B 1989;7:163.
20. Kobeda E, Irene EA. J Vac Sci Technol B 1988;6:574.
21. Fitch JT, Kobeda E, Irene EA. In:Helms CR, Deal BE, editors. The physics and chemistry of SiO_2 and the Si-SiO_2 interface. New York: Plenum; 1988; p. 139.
22. Fitch JT, Kobeda E, Lucovsky G, Irene EA. J Vac Sci Technol B 1989;7:153.

23. Lodha S, Janes DB, Chen N. Appl Phys Lett 2000;80:4452.
24. Witmer M, Finsted T, Nicolet MA. J Vac Sci Technol 1977;14:935.
25. Patkar MP, Chin TP, Woodall JM, Lundstrom MS, Melloch MR. Appl Phys Lett 1995;66:1412.
26. Ismail K, Chu JO, Meyerson BS. Appl Phys Lett 1994;64:3124.

SUGGESTED READING

H. R. Huff and D. C. Gilmer, (eds.), 2005. *High Dielectric Constant Materials*, Springer. The beginning of this new book covers the reasons for the use of the $Si-SiO_2$ interface so pervasively in microelectronics. Later chapters discuss the search (as yet incomplete search) for a substitute for SiO_2. The chapters are written by experts and there are extensive references.

INDEX

Absorption constant, versus incident photon energy, 355
Actual force, stress and film thickness, 17
Adsorption, 229–255
 collision outcomes, 229
 definition, 229
Adsorption isotherms, 238–243
AES. *See* Auger electron spectroscopy (AES)
AFM (atomic force microscopy), 317, 318
 See Atomic force microscopy (AFM)
Al/SiO$_2$/Si MOs capacitor, electron energy diagram, 454
ALD (atomic layer deposition), 345–349
ALE (atomic layer epitaxy), 345–349
Ambient exposure, to create a surface, 8–9
Anneling time, interface thickness, 452
Astigmatism, in atomic resolution, 49
Atom density calculation, Si, 13–14
Atomic force microscopy (AFM), 159–160, 176–181, 317, 318
 force interactions, 178
 light reflection sensing method, 180
 operation, 177–181
 schematic diagram, 177
 tunneling current, 179
 versus STM, 176
Atomic layer deposition (ALD), 345–349
Atomic layer epitaxy (ALE), 345–349
Atomic versus energy separation diagram, 135
Attractive dispersion, 230
Auger electron spectroscopy (AES), 181–184
 sputter profiles, 183
 versus photoionization, 182
 versus x-ray fluorescence, 182
Auger emissions, 181

Band bending, 138–139
Bare Si, MSRI spectra, 196
Barium strontium titanate (BST)
 film sputter, 194
 interface with Si, 405–419
 $K_{HK}(K_{int})$ dependence, 412
 MSRI spectra, 408
 optical properties, 405–406
 Si MSRI results, 407
 Si real-time SE results, 406, 407

Surfaces Interfaces and Thin Films for Microelectronics. By Eugene A. Irene
Copyright © 2008 John Wiley & Sons, Inc.

Barrier heights, 236
  N-type GaAs, 499
BEMA (Bruggeman effective medium approximation), 272
  real-time data fit, ECR plasma oxidation results, 371
BET formula, 243
BET isotherm, 242
Binding potential
  electronic states, 7
  graded, 7
  surface atom versus bulk atom, 7
Boltzmann factor, 248
Bonding at Si(111) surface diagram, 77
Bound electron problem, 164
Box-counting method for D_F, 105
Bragg's law, 22
Brewster's angle, 277–278
Bruggeman effective medium approximation (BEMA), 272, 313
BST (barium strontium titanate)
  interface with Si, 405–419
  $K_{HK}(K_{int})$ dependence, 412
  MSRI spectra, 408
  optical properties, 405–406
  Si MSRI results, 407
  Si real-time Se results, 406, 407
BST deposited on base Si, 195
BST sputter on Si, 195
BST/SiO_2/Si structure, EFTEM images, 416
Bubble radius, 70–71
Bulk properties versus surface properties, 4

c direction, diffraction from surfaces, 32
Capillarity
  curved surfaces, 69–73
  definitions, 67–69
  principles of, 67–73
Capillary rise, 71
Capillary tube
  diagram, 70
  immersed in liquid, 72
Cauchy dispersion formula, 446
CCD (charge-coupled devices), 280
Charge distribution, Jellium model, 136
Charge-coupled devices (CCD), 280
Charged surfaces, 199–228

Chemical reactivity, roughness effects, 115–120
Chemical vapor deposition (CVP) system, 329–333
  chemical reactions, 332
  components of, 330
  generalized system, 329
  low pressure, 332
  plasma assisted, 332
  precursor materials, 332–333
    halides, 332
    hydrides, 332
    metal organic compounds, 332
  radio frequency (RF) heated low pressure, 330
Chemisorption, 233–236
  Morse potential, 233
Chromatic aberration, in atomic resolution, 48
Clausius–Massotti equation, 271
Cleavage, to create surface, 8
  in ambient, 8
  in reactive environment, 8
Cleaved Ge and GaAs, photoemission results, 358
Cleaved Si(111), photoemission results, 357
Collisions with a surface, resulting processes, 230
Compensator, definition, 279
Complex refractive index, 262
Complimentary MOSFET (CMOS) geometry, 210
Compressive stress, 479
Contact angle
  observations, 84
  roughness effects, 112–113
Coulomb's law, 200
Cross sectional transmission electron microscopy (XTEM), 51–52
  sample preparation, 52
Crystalline material
  diffraction techniques, 21
  surface structure study, 21
c-Si as function of temperature, dielectric function, 366
Cubic material γ plot, 76
Curved surfaces, 69–73
Curved surfaces, capillarity, 69–73

CVD (plasma chemical vapor deposition) system, 312
CVP (chemical vapor deposition) system, 329–333
  chemical reactions, 332
  components of, 330
  generalized system, 329
  precursor materials, 332–333
    halides, 332
    hydrides, 332
    metal organic compounds, 332
  radio frequency (RF) heated low pressure, 330

Dangling bonds surface condition, 137
DC (diamond cubic) structure, 12
Desorption, temperature programmed, 248–255
Detectors, ellipsometry, 280
Diamond cubic (DC) structure, 12
  Si, 13
Dielectric function
  of Si, 288, 289
  of c-Si as function of temperature, 366
  of Si(100) as function of oxidation, 367
Dielectric response function, 264
Dielectric optical properties, 270
Diffraction from surfaces, 31–45
  c direction, 32
  electromagnetic radiation selection, 34
  electron microscopy, 45–63
  low energy electron diffraction, 34–38
  nonnormal incidence, 35
  normal incidence, 35
  RESP and Ewald construction, 31–33
Diffraction geometry, multiple planes, 34
Diffraction techniques, 27–30
  Laue method, 29
  powder method, 28–29
  rotating crystal method, 28
Dilation symmetry, fractal objects, 99–100
Direct current sputtering, 339
Direct reaction, chemical vapor deposition (CVP) system, 332
Direct recoil spectroscopy, 190
Disproportionation, chemical vapor deposition (CVP) system, 332

E-beam evaporation, 337
ECR (electron cyclotron resonance), 312
  microwave plasma system, schematic, 365
  plasma system, schematic, 365
ECR plasma grown SiO_2 film, XPS spectra, 369
ECR plasma oxidation of InP, 385–387
  thickness versus time, 390
  XPS spectra, 388, 391
ECR plasma oxidation of Si
  optical models, 368
  versus GaAs, 397
ECR plasma oxidation results
  BEMA fits to real-time data, 371
ECR plasma oxide grown on InP, etch results, 87
ECR plasma SiO_2
  on N-type Ge, 380
  on N-type Si, 379
ECR plasma systems, 342
Effective medium approximations (EMA), 271–274
EFTEM images, $BST/SiO_2/Si$ structure, 416
Electric field lines, 202, 203
Electric field on MOS structure, 221
Electromagnetic radiation (emr), 26
Electron cyclotron resonance (ECR). See ECR
Electron escape depth, versus electron energy, 36
Electron mean free path, versus electron energy, 36
Electron microscopy, 45–63
Electron microscopy. See Transmission electron microscopy (TEM); Scanning electron microscopy (SEM)
Electronic field lines, smooth versus rough surfaces, 111
Electronic passivation, 84, 361–363
Electronic properties of materials, roughness effects, 108–111
Electronic states, interface, 351–361
Electronic structure of surfaces, 4
Electrostatics, Poisson equation and, 200–205
Ellipsometer, polarizer, compensator, sample, analyzer (PCSA) configuration, 274
Ellipsometer, rotating analyzer spectroscopic, 453

Ellipsometers, automated, 265
Ellipsometry
  alignment, 286–287
  calibration, 286–287
  charge-coupled devices, 280
  compensators and retarders, 279
  definition, 258–263
  detectors, 280
  measurements, 263–264, 287–292
    accuracy, 264–266
    Si surface, 288
    surface with overlayer, 289–292
  null, 280–282
  optical models, 266–274
  phase modulation (PME), 283
  photomultiplier tubes (PMTs), 280
  rotating analyzer (RAE), 283, 285
    components, 286
  rotating compensator (RCE), 283
  rotating element, 282–284
  rotating polarizer (RPE), 283
  semiconductor photodiodes, 280
  Si oxidation experiments, 16, 17
  single-wave (SWE), 284–285
  small surface changes and, 265
  spectroscopic (SE), 284–286
Ellipsometry techniques, manual and automated, 274–287
EMA (effective medium approximations), 271–274
Energy bands, 135–136
  band bending, 138–139
  extrinsic surface states, 137–138
  incident photons and, 151
  work functions, 140
Energy
  surface, 5–6
  versus atomic separation diagram, 135
Epitaxy, 322
  bond stability, 323
  layer-by-layer growth, 323
  molecular beam (MBE), 323
  types of, 322–323
Equilibria, three phase, 82
Equilibrium, vapor–solid, 236–238
Evaporation
  definition, 334
  physical vapor deposition (PVP) system 334–337

Ewald construction, 26–27, 31–33
  c direction, 32
  RESP, 31–33
  RESP with lattice rods and Ewald sphere, 33
  two-dimensional RESP, 27
  See also Diffraction techniques; Diffraction from surfaces.
Extrinsic surface states, 137–138

Fermi level work functions, in metals, 172
Fermi planning, MOSFETs and, 209–213
Field emission microscope
  drawing, 144
  observations, 145, 146
Field emission surface electronic state, 139–146
Film and interfacial stress, 478–490
Film deposition, 329–349
  chemical vapor deposition (CVP) system, 329–333
Film formation, 321–349
  dissimilar materials, 323
  epitaxy, 322
  gas-phase reaction, 324
  mechanisms, 324
  nucleation, 321
  reaction at surface, 324
  reaction with surface, 324
Film growth, 324–329
  Henry law dissolution of O_2, 326
  process, 325
Finite solids, Kronig–Penney (KP) model, 124–131
Fixed oxide charges, 219, 225
  versus interface-trapped charge, 227
Fournier transform (FT), 95
Fractal description, roughness, 99–106
Fractal dimension, 99
  extraction from experimental data, 104–106
Fractal profiles, Weierstrass–Mandelbrot model, 101–102
Fractals, 99–103
  Koch curve, 101
  self-similarity, 100
Franck–van der Merwe growth, 308
Fresnel reflection, 261

INDEX 507

FT (Fournier transform), 95
FTIR spectra, c-Ge, 374

Ga and As $3d$ XPS spectra, 398
Ga and As XPS spectra, 396
GaAs
  major orientations, 395
  optical model, 395
GaAs binary and ternary reference oxides, 399
GaAs oxidation, effect of temperature, 394
GaAs passivation, thermal and ECR plasma oxidation and SiO_2 deposition, 392–403
GaAs(110), photoemission results, 359
Gas-phase reaction, film formation, 324
Ge oxidation results, 373
Ge passivation, thermal and ECR plasma oxidation and SiO_2 deposition, 372–382
Ge(111), photoemission results, 359
Geometric structure of surfaces, 3–4
Glan-Taylor prism, 277, 278
Glan-Thompson prism, 277

Halides, chemical vapor deposition (CVP) system, 332
Henry's law, dissolution of O_2, 326
Heterogeneous nucleation, 308–312
  Franck–van der Merwe growth, 308
  Stranski–Krastnov nucleation, 309–310, 311
  Vollmer–Weber theory, 309
Hf, time-of-flight MSRI spectra, versus bare Si, 428
HfO_2
  C-V measurements, 427–429
  index of refraction, 427
  interface quality comparison, 430
  optical properties, 426–427
  as potential high K dielectrics, 426–429
High K dielectrics, potential films, 419–434
High static dielectric constant gate oxides, 403–434
Homogeneous nucleation, 299–307
  mixing, 304–305
  Vollmer–Weber theory, 305–307

Hooke's law, 478
Hydrides, in chemical vapor deposition (CVP) system, 332
Hydrolysis, chemical vapor deposition (CVP) system, 332

Ideal metal, semiconductor junctions, 492–494
Ideal semiconductor-semiconductor PN junctions, 495–496
Idealized surfaces, 5, 6, 9
Immersion liquids for SiO_2 films, refractive indices, 453
Impurities, surface condition, 137
Incident electron beam, effects of, 46
Incident light/reflected light ratio, 355
Incident photo energy versus absorption constant, 355
Infinite solids, Kronig–Penney (KP) model, 124–131
Inhomogeneous film on substrate, optical model, 268, 269
InP passivation
  etch results, 387
  SIMS data, 383–384
  thermal and ECR plasma oxidation and SiO_2 deposition, 382–392
InP thermal oxidation model, 386
InP(100) in O_2, thermal oxidation, 382
InP, pseudodielectric functions, 386
Integrated circuits, MOSFETs, 352
Integrated ion sputter deposition system, 192
Interaction energy versus separation distance, 231
Interfaces, 79–83
  analysis methods, 446
  flat, 80
Interface charge
  MOSFET measurements, 213–228
  oxide charges, 219–224
  versus fixed oxide charge, 227
Interface electronic states, 351–361
Interface layer, intermixed, effects, 404
Interface reaction, BST on Si in presence of O_2, 416
Interface study techniques, 445–490
  spectroscopic immersion ellipsometry (SIE), 445–453

Interface trapped charges, 219
  density samples, 228
Interfacial and film stress, 478–490
Intermixed interface layer, effects of, 404
Inverse photoelectron spectroscopy (IPES), 151–155
Ion scattering, 184–197
  low energy ions, 184
Ion scattering spectroscopy (ISS), 184–185, 189–190
  solid surface, 186
IPES (inverse photoelectron spectroscopy), 151–155
Isotherms, adsorption, 238–243
Isotropic media, Jones matrix, 276–277
ISS. See Ion scattering spectroscopy.

Jellium model, charge distribution, 136
Joined P- and N-type semiconductors, 498
Jones matrix, 275
  isotropic media, 276
Junctions
  electronic characteristics, 490–492
  ideal metal, 492–494
  ideal semiconductor-semiconductor, 495–496
  metal-N-type semiconductor, 493
  nonideal, 496–501

k space (wave vector representation), 30–31
Kelvin equation, 80, 81
Kelvin probe measurement, surface electronic states, 146–148
$K_{HK}(K_{int})$ dependence, 411
  BST thin films, 412
  calculation parameters, 411
  STO thin films, 412
Kink site, in idealized surface model, 10
Knudsen cell, molecular beam epitaxy (MBE), 338
Koch curve, fractals, 101
KP (Kronig–Penney) model, 124–134
  finite solids, 131–134
  infinite solids, 124–131
$K_{STO}(K_{int})$ dependence, 413
Kurtosis, roughness feature, 93–94

Langmuir isotherm, 239, 245
Langmuir–Blodgett method, organic films, 344–345
Langmuir–Hinshelwood (LH) mechanism, 245
Laser ablation, 337
Laue diffraction method, 29
  "white" radiation, 29
Layer-by-layer growth, epitaxy, 323
Ledges, in idealized surface model, 10
  one direction, 11
  two directions, 11
LEED. See Low energy electron diffraction.
Lennard-Jones potential curve, 233
Light, transverse wave nature, 258
Light reflection sensing method, atomic force microscopy (AFM), 180
Linear polarizer, 277–279
  mechanism of operation, 277
Linear-parabolic (LP) model, 15
London forces, 230
Lorentz–Lorenz equation, 271
Low energy electron diffraction (LEED), 11, 34–38
  fundamental lattices, 43
  geometry, 37, 40
  indexing patterns, 40–45
  nomenclature, 40–45
  pattern and reconstruction, 38–40
  reconstruction, 41
  Si patterns, 44
Low index planes, 10
LP (linear parabolic) model, 15
LPCVP (low pressure chemical vapor deposition) system, 332

Mass spectroscopy of recoiled ions, 190–192
  applications, 192–197
Maxwell–Boltzmann distribution of velocities, 334
Maxwell–Garnet equation, 271–272
MBE (molecular beam epitaxy), 323
Metal organic compounds, chemical vapor deposition (CVP) system, 332
Metal wires, calculations, 78
Metal–metal junction, 490–492
Metal–N-type semiconductor junction, 493

INDEX 509

Metal–oxide–semiconductor field-effect transistors (MOSFETs), 15
Metal–semiconductor junction, 492–494
Metal–vacuum interface, 143
MgO_2 films
   cross-sectional TEM on Si, 431
   C-V results, 433
   electronic results, 433
   energy-filtered cross-sectional TEM on Si, 432
   potential high K dielectrics, 429–434
   refractive indices, 432
Microwave electron cyclotron resonance (ECR) plasma system, schematic, 365
Miller index planes, 11
Mobile oxide charges, 219
Molecular beam epitaxy (MBE), 323
   Knudsen cell, 338
   physical vapor deposition (PVP) system, 337–338
Monatomic solid, structure, 7
Morse potential, 233
MOS capacitor, 212, 213–215, 217
   PECVD SiO_2 films, 376–377
MOS structure, electric field, 221
MOSFET device
   description, 108–111
   N-channel, 109
MOSFETs (metal–oxide–semiconductor field-effect transistors), 15
   Fermi planning, 209–213
   integrated circuits, 352
   interface charge, 213–228
   pinned Fermi level, 361
   Si-based, 84
   size reduction, 403
MSRI spectra
   Si, 196
   Zr on Si, 196
Multiple films on substrate, optical model, 268

N- and P-type joined semiconductor, 498
N- and P-type separated semiconductor, 497
N-channel MOSCFET, 210
N-channel MOSFET device, 109
Nearest-neighbor interactions in surface energy, simple, 5–6

Nitridation, chemical vapor deposition (CVP) system, 332
Noncrystalline material, surface structure study, 21
Nonideal junctions, 496–501
N-type and separated metal semiconductor, 494
N-type GaAs, barrier heights, 499
Nucleation, 298–321
   heterogeneous, 308–312
   homogenous, 299–307
   mixing, 304–305
   Vollmer–Weber theory, 305–307
Nucleation studies, 312–321
   real-time ellipsometry (RTE), 312
Nuclei formation reaction, 298, 299
Nucleus, cap-shaped, 310
Null ellipsometry, 280–282

Optical model
   films and surfaces, 266–274
   inhomogeneous film on substrate, 268, 269
   multiple films on substrate, 268
   parameters from regression analysis, 315
   rough film on substrate, 268
   uniform single film on substrate, 268
Optical properties of materials
   dielectric, 270
   roughness effects, 106–108
Organic films
   Languir–Blodgett method, 344–345
   self-assembly method, 343–345
   spin-casting method, 345, 346
Oxidation
   chemical vapor deposition (CVP) system, 332
   single-crystal Si, 15–18
Oxide charges, 219–224
Oxide trapped charges, 219
Oxides, grown on c-Ge, 375
Oxidized polycrystalline Si films, transmission electron microscopy (TEM), 54

P- and N-type joined semiconductor, 498
P- and N-type separated semiconductor, 497

PACVP (plasma assisted chemical vapor deposition) system, 332
Passivation, electronic, 361–363
Passivation studies, semiconductor, 363–403
  Si thermal and plasma oxidation, 363–372
P-channel MOSFET, 209, 210
PCSA (polarizer, compensator, sample, analyzer) ellipsometer configuration, 272
Phase modulation ellipsometry (PME), 283
Photodiodes, semiconductor, 280
Photoemission results
  cleaved Ge and GaAs, 358
  cleaved Si(111), 357
  GaAs(110), 359
  Ge(111), 359
  S(111), 360
Photoemission spectrometer, 150
Photoionization, versus AES, 182
Photomultiplier tubes (PMTs), 280
Physical vapor deposition (PVP) system, 329, 333–349
  evaporation, 334–337
  generalized method, 333
  molecular beam epitaxy, 337–338
  plasma deposition, 340–343
  process, 333
  sputtering, 338–340
Physiosorption, 230–233
Plasma chemical vapor deposition (CVD) system, 312
Plasma deposition, physical vapor deposition (PVP) system, 340–343
PME (phase modulation ellipsometry), 283
PMT (photomultiplier tube), 280
PN junctions, ideal semiconductor-semiconductor, 495–496
Poisson equation, 199–200
  electrostatics and, 200–205
  simple solutions, 205–209
Polarized light
  definition, 258
  incident linear, 260
  linear, 258–259
  monochromatic wave, 259

Polarizer, compensator, sample, analyzer (PCSA) ellipsometer configuration, 274
Polycrystalline Si film, transmission electron microscopy (TEM), 53
Porous Si partly oxidized, transmission electron microscopy (TEM), 55
Powder diffraction technique, 28–29
Power spectral density (PSD), 95
  plots of various surfaces, 97, 98
PSD (power spectral density), 95
Pt/HfO$_2$/Si sample, 429
Pt/ZrO$_2$/Si sample, 424
P-type and separated metal semiconductor and PVP (physical vapor deposition) system, 329, 333–349
  evaporation, 334–337
  generalized method, 333
  molecular beam epitaxy, 337–338
  plasma deposition, 340–343
  process, 333
  sputtering, 338–340
Pyrolysis, chemical vapor deposition (CVP) system, 332

Radio frequency (RF) heated low pressure system, 329–333
RAE (rotating analyzer ellipsometry), 283–285
  components, 286
RBS. *See* Rutherford backscattering spectroscopy (RBS)
RCE (rotating compensator ellipsometry), 283
Reactive gas on metal surfaces, 233–235
Real-time ellipsometry (RTE), 312, 319
Reciprocal lattice (REL), 22
Reciprocal space (RESP) 22–26
  c direction, 32
  definition, 22–23
  Ewald construction, 31–33
  lattice rods and Ewald sphere, 33
  lattice rods, 33
  points in, 26
  representations, 23
  See also Diffraction from surfaces.
Reconstructed surface, 7

INDEX 511

Reflected dual-laser technique, Si substrate strain measurement, 485
Reflected light/incident light ratio, 355
Reflection
  bare substrate, 261
  film covered substrate, 261
  smooth versus rough surface, 107
Refractive index, complex, 262
Regression analysis
  flowchart, 273
  parameters in optical models, 315
REL (reciprocal lattice), 22
Residual or total stress, 479
RESP. *See* Reciprocal space.
Retarder, definition, 279
RF magnetron sputtering, 339
RF plasma methods, 340–341
Risers, in idealized surface model, 10
Rochon prism, 277
Rotating analyzer ellipsometry (RAE), 283–285
  components, 286
Rotating analyzer spectroscopic ellipsometer, 453
Rotating compensator ellipsometry (RCE), 283
Rotating crystal diffraction method, 28
Rotating element ellipsometry, 282–284
Rotating polarizer ellipsometry (RPE), 283
Rough film on substrate, optical model, 268
Rough surface
  computer simulated, 104
  side view, 18
  TLK model, 19
Roughness, 89–121
  fractal description, 99–106
  height parameters, 90–92
  intuitiveness, 89
  kurtosis, 93
  period, 92–93
  periodic surface profile, 91, 92
  shape of features, 93–95
  skewness, 93
  statistical descriptors, 95–99
    Fournier transform, 95
  surface, 18–19
  wavelength, 92–93
Roughness effects, 106–121
  chemical reactivity, 115–120
  contact angle, 112–113
  electronic field lines, smooth versus rough surfaces, 111
  electronic properties, 108–111
  optical properties of materials, 106–108
  surface thermodynamic properties, 113–115
Roughness levels, tunnel current oscillation comparison, 463
Roughness parameters, 90–106
Roughness profile, 93, 94
RPE (rotating polarizer ellipsometry), 283
RTE (real-time ellipsometry), 312, 319
Rutherford backscattering spectroscopy (RBS), 185

S(111) photoemission results, 360
S/V ratio, explanation, 6
Scanning electron microscopy (SEM), 58–63
  doped polycrystalline Si, 61, 62
  major components, 60
  results, 63
Scanning probe microscopy (SPMs), 19, 89–90, 159–181
  applications, 172–176
  electron tunneling, 161–168
  information from, 173
  modes of operation, 170–171
  operation, 168–171
  schematic, 169
  versus PES and IPES, 175
Schrödinger equation, 161
SE (spectroscopic ellipsometry), 284–286
Secondary ion mass spectrometry (SIMS), 185
Self-assembly method, organic films, 343–345
Self-similarity, surface profile, 96
SEM (scanning electron microscopy), 58–63
Semiconductor junctions, ideal metal, 492–494

Semiconductor passivation studies, 363–403
  GaAs passivation, 392–403
  Ge passivation, 372–382
  InP passivation, 382–392
  Si thermal and plasma oxidation, 363–372
Semiconductors, electron and hole mobilities, 362
Separated metal and N-type semiconductor, 494
Separated metal and P-type semiconductor, 495, 496
Separated P- and N-type semiconductor, 497
Separation distance versus interaction energy, 231
Si diamond cubic (DC) structure, 13
Si LEED patterns, 44, 45
Si oxidation experiments, ellipsometry, 16
Si oxidation system, double-walled, 328
Si single-crystal orientations
  oxidation, 15–18
  pure O_2, 16, 17
Si- SiO_2 system, measurement of charges, 224–228
Si substrate strain measurement, reflected dual-laser technique, 485
Si surface bonds, 15–16
Si surface
  ellipsometry measurement, 288
  roughened, AFM images, 462
Si thermal and plasma oxidation, 363–372
Si(100) as function of oxidation, dielectric function, 367
Si(100)/SiO_2 BST structure, TEM cross-sectional images, 416
Si, intrinsic film stress, various orientations, 483
SIE (spectroscopic immersion ellipsometry), 445–453
  immersion liquids, 446–447
Silicon. *See* Si.
SIMS. *See* Secondary ion mass spectrometry.
SIMS depth profiles, InP thermal oxidation, 384
Single film on substrate model, 267–268
Single-wave ellipsometry (SWE), 284–285

Singular surfaces, atoms in, 12
SiO_2
  refractive index, 457
  roughness, 457
SiO_2 capacitors, ultra-thin, current density versus gate voltage, 470
SiO_2 film
  IR absorption, 484
  stress measurements, *in situ* versus *ex situ*, 488
  thickness, 457
  ultra-thin, 403–404
  XPS spectra for ECR plasma grown, 369
SiO_2–Si interface application, 83–87
Si-rich SiO_2 film, transmission electron microscopy (TEM), 56
Si–SiO_2 interface
  BEMA calculations, 449–450
  bonding differences, 443–444
  Grunthaner report, 438–439
  interface model, 448–449
  mean free path, 440
  MOSFET, 437–438
  nature of, 438–444
  possible Si charge states, 441
  spin method etching, 439, 440
  Sune et al. model, 461
  XPS Si $2p$ spectra, 439, 442
  XPS spectra, 439
    SiO_2 films, 443
Si–water interaction, 83
Skewness, roughness features, 93
Snell's law of refraction, 279
Solid surface
  definition, 3
Solids
  surface energy, calculations, 73–79
  three-dimensional version, 6, 7
Spectroscopic ellipsometer (SE) system, 192, 284–286
Spectroscopic immersion ellipsometry (SIE), 445–453
  immersion liquids, 446–447
Spherical aberration, in atomic resolution, 48
Spin casting method, organic films, 345, 346
SPM (scanning probe microscopy), 19, 89–90

INDEX

Scanning probe microscopy. *See* SPM.
Sputtering
  physical vapor deposition (PVP) system, 338–340
  schematic of process, 339
Stepped surface structure diagram, 75
Steps, in idealized surface model, 10
STM. *See* Scanning tunnel microscopy (STM)
Stranski–Krastnov nucleation, 309–310, 311
Stress
  compressive versus tensile, 479
  definition, 478
  Hooke's law, 478
  residual or total, 479
  thermal expansion, 479SiO_2 film stress, early studies, 479–481
Structure of surface, 3
Stylus profiler, schematic, 160
Suboxide areal densities, SiO_2 on Si(111), 444
Substrate and surface film interface, 4–5
Substrate model, single film 267–268
Sune et al. model, Si–SiO_2 interface, 461
Surface
  diffraction from. *See* Diffraction from surfaces.
  definition, 3, 5–7, 65
  preparing, 8–9
  simple models, 9–18
Surface area creation, liquid and solid, 66
Surface atom site, 10
Surface bonding, 3
Surface conditions, new surface states, 137
Surface coverage versus pressure, 239
Surface diffraction, electromagnetic radiation selection, 34
Surface electronic states, 123–156
  Kronig–Penney (KP) model, 124–134
  measurement, 139–156
    field emission, 139–146
    Kelvin probe, 146–148
    photoemission, 148–156
    thermionic, 139–146

Surface energy, 5, 66
  cost of, 5
  description, 5
  solids, calculations, 73–79
  thermodynamic approach, 65–67
Surface film and substrate interface, 4–5
Surface-mediated reactions, 244
Surface profile
  complex, 93, 94
  self-similarity, 96
Surface properties, 3
Surface reaction mechanisms, 243–247
Surface roughness, 18–19
  See also Roughness.
Surface-sensitive measuring techniques, 4
Surface stages, energy band formation, 135–136
Surface state charge, versus surface potential, 354
Surface stress, 66
Surface tension, 66
Surface terminations, low energy films adsorbed on metal surfaces, 86
Surface topology/morphology, 159–181
Surface with overlayer, ellipsometry measurement, 289–292
Surface/volume ratio, explanation, 6
SWE (single-wave ellipsometry), 284–285

TDS (thermal desorption spectroscopy), 248–255
Teflon–water interaction, 83
TEM (transmission electron microscopy), 46–58
TEM cross-sectional images, Si(100)/SiO_2 BST structure, 416
Temperature-programmed desorption, 248–255
Tensile stress, 479
Terrace-ledge-kink (TLK) model, 9–18
  roughness, 19
Terraces, in idealized surface model, 10
Thermal desorption spectroscopy (TDS), 248–255
Thermal expansion stress, 479
Thermal oxidation stress, 16–17
Thermal SiO_2 substrate values, conductance method, 419

Thermionic surface electronic state, 139–146
Thermodynamic properties, roughness effects, 113–115
Thermodynamics
  of liquids, 66
  of surfaces and interfaces, 65–88
Thickness fluctuation, high frequency model, 469
Three phase equilibria, 82
Time-of-flight ion scattering and recall spectrometry (ToF-IASRS), 185–188
  definition, 185
  system, 192
Time-of-flight MSRI spectra, bare Si versus Hf, 428
TLK (terrace-ledge-kink) model, 9–18
  roughness, 19
ToF-IASRS. *See* Time-of-flight ion scattering and recall spectrometry
Transmission electron microscopy (TEM), 46–58
  aberrations in atomic resolution, 48–49
  chemical etching, 51
  columns, 46–47
  cross-sectional (XTEM), 51–52
    sample preparation, 52
  oxidized polycrystalline Si films, 54
  polycrystalline Si film, 53
  porous Si partly oxidized, 55
  results, 52–58
  sample preparation, 49–52
  sample size, 49
  sample thickness, 49
  Si-rich SiO_2 film, 56
  techniques, 47–48
  thinning methods, 49–51
  XTEM of ZrO_2, 58
  YBCO, 57
Tunneling, tip and surface, 167

Ultra-thin SiO_2 films, 403–404
Ultraviolet photoemission spectroscopy (UPS), 151–155
  spectra of Ni surface, 153
Uniform single film on substrate, optical model, 268

UPS (ultraviolet photoemission spectroscopy), 151–155
  spectra of Ni surface, 153

Vacancy, surface condition, 137
Vacuum–metal interface, 143
van der Waals forces, 230
Vapor–solid equilibrium at surfaces, 236–238
Variable angle SE (VASE) analysis algorithm, 475
VASE (variable angle SE) analysis algorithm, 475
Vicinal plane, 11
Vicinal surface of Si, atom density calculation, 13–14
Vollmer–Weber theory, 305–307, 309

Water–Si interaction, 83
Water–Teflon interaction, 83
Wave vector representation (**k** space), 30–31
Weierstrass–Mandelbrot model, fractal profiles, 101–102
"White" radiation, Laue diffraction method, 29
Wallston prism, 277
Work functions, 142
  energy bands, 140
  metal–vacuum interface, 143

XPS (X-ray photoelectron spectroscopy), 156
XPS spectra, 154, 155
  ECR plasma grown SiO_2 film, 369
  Ge $3d$, 374
X-ray diffraction, 21
X-ray fluorescence, versus AES, 182
X-ray photoelectron spectroscopy (XPS), 156
  spectra, 154, 155
XTEM (cross-sectional transmission electron microscopy), 51–52
  sample preparation, 52
XTEM of ZrO_2, transmission electron microscopy (TEM), 58

YBCO, transmission electron microscopy (TEM), 57

INDEX   515

Young equation, 310
Young-Laplace equation, 80

Zr on Si, MSRI spectra, 196, 422
ZrO_2

index of refraction, 421
interface quality comparison, 425
as potential high K dielectrics, 419–426
ZrO_2/Si interface, HREM cross-sectional image, 423